DISSOCIATIVE RECOMBINATION OF MOLECULAR IONS

Dissociative recombination (DR) of molecular ions with electrons is a complex, poorly understood molecular process. Its critical role as a neutralizing agent in the Earth's upper atmosphere is now well established and its occurrence in many natural and laboratory produced plasmas has been a strong motivation for studying the event. For the first time, theoretical concepts, experimental methodology, and applications are united in one book, revealing the governing principles behind the gas-phase reaction. The book takes the reader through the intellectual challenges posed, describing in detail dissociation mechanisms, dynamics, diatomic and polyatomic ions, and related processes, including dissociative excitation, ionpair formation and photodissociation. With the final chapter dedicated to applications in astrophysics, atmospheric science, plasma physics, and fusion research, this is a focused, definitive guide to a fundamental molecular process. The book will appeal to academics within physics, physical chemistry, and related sciences.

MATS LARSSON is a Professor and Experimentalist in the Physics Department at Stockholm University. He obtained his Ph.D. in physics in the Research Institute of Physics and Stockholm University. His research interests include primary chemical reactions, interstellar chemistry, and molecular spectroscopy. He was a member of the Physics and Mathematics Committee of the Swedish Natural Science Research Council from 1989 to 1995. He was made chair of the Research Committee of the Swedish National Space Board in 2001, and was also chair of the Evaluation Committee for Atomic and Molecular Physics, Fusion Research and Plasma Physics of the Swedish Research Council from 2001 to 2003.

ANN E. OREL is a Professor and Chair in the Department of Applied Sciences at the University of California, Davis. She obtained her Ph.D. in chemistry at the University of California, Berkeley and was made a fellow of the American Physical Society in 2000. Her research interests include theoretical atomic and molecular physics and computational science.

DISSOCIATIVE RECOMBINATION OF MOLECULAR IONS

MATS LARSSON
Stockholm University

and

ANN E. OREL
University of California, Davis

CAMBRIDGE UNIVERSITY PRESS
Cambridge, New York, Melbourne, Madrid, Cape Town, Singapore, São Paulo

Cambridge University Press
The Edinburgh Building, Cambridge CB2 8RU, UK

Published in the United States of America by Cambridge University Press, New York

www.cambridge.org

© M. Larsson and A. E. Orel 2008

This publication is in copyright. Subject to statutory exception
and to the provisions of relevant collective licensing agreements,
no reproduction of any part may take place without
the written permission of Cambridge University Press.

First published 2008

Printed in the United Kingdom at the University Press, Cambridge

A catalog record for this publication is available from the British Library

ISBN 978-0-521-82819-2 hardback

Cambridge University Press has no responsibility for the persistence or accuracy of URLs for external or third-party internet websites referred to in this publication, and does not guarantee that any content on such websites is, or will remain, accurate or appropriate.

We would like to dedicate this book to Sheldon Datz, who was responsible for introducing us to this interesting area of physics.

Contents

Preface		page ix
1	Introduction	1
	1.1 History 1900–1950	1
	1.2 History 1950–1970	5
	1.3 History 1970–1990	7
	1.4 History 1990–present	10
2	Experimental methods	11
	2.1 Merged beams	11
	2.2 Ion storage rings	30
	2.3 Stationary afterglow technique	51
	2.4 Flowing afterglow technique	59
	2.5 Shock-tube technique	68
3	Theoretical methods	70
	3.1 Introduction	70
	3.2 What is a resonance?	75
	3.3 Formal resonance theory	78
	3.4 Resonance parameters and structure	83
	3.5 Nonadiabatic couplings	89
	3.6 Calculation of dynamics	93
4	The H_2^+ molecule	104
5	Diatomic hydride ions	119
	5.1 HeH^+	119
	5.2 NeH^+, ArH^+, KrH^+, and XeH^+	132
	5.3 CH^+	133
	5.4 NH^+ and OH^+	139
	5.5 LiH^+	140

6	Diatomic ions	143
	6.1 Rare-gas dimer ions: He_2^+, Ne_2^+, Ar_2^+, Kr_2^+, Xe_2^+	143
	6.2 The atmospheric ions: O_2^+, N_2^+, and NO^+	154
	6.3 Other diatomic ions	180
7	The H_3^+ molecule	184
	7.1 History of H_3^+	184
	7.2 The dissociative recombination of H_3^+	186
8	Polyatomic ions	227
	8.1 Dissociation dynamics in recombination of XH_2^+ ions (X = C, N, O, S, P)	227
	8.2 Astrophysical molecular ions	244
	8.3 Cluster ions	267
	8.4 Hydrocarbon ions	277
	8.5 Other polyatomic ions	283
	8.6 Electron capture dissociation	283
9	Related processes	287
	9.1 Dissociative excitation and ionization of molecular ions	288
	9.2 Ion-pair production	294
	9.3 Electron impact detachment of negative ions	296
	9.4 Electron–molecule scattering; dissociative attachment	300
	9.5 Photodissociation and photoionization	308
10	Applications	315
	10.1 Molecular astrophysics	315
	10.2 Atmospheric physics and chemistry	319
	10.3 Plasma physics and fusion research	320
	References	321
	Index	377

Preface

This research monograph provides a single-volume description of the dissociative recombination of molecular ions with electrons. Since this is one of the most complex gas-phase processes, its study is a challenge to theorists and experimentalists alike. The theory, experiment, and applications of dissociative recombination are scattered in the scientific literature as original research articles, conference proceedings, and review articles. This book brings this information together in a single work for the first time.

The book is intended for researchers and Ph.D. students in the fields of atomic and molecular physics, chemical physics and physical chemistry, molecular astrophysics, atmospheric physics, and other areas of science where electrons and molecular ions are important.

This book was written during a period when each of us had several other commitments which slowed down the writing. One of us (AEO) was department chair at UC Davis essentially during the entire writing process, and ML chaired committees for the Swedish Space Board and the Swedish Research Council.

We are grateful for the hospitality of the Institute for Atomic and Molecular Physics (ITAMP) at the Harvard-Smithsonian Center for Astrophysics and Harvard University Physics Department (Kate Kirby, Hussein Sadeghpour), the Cluster Research Laboratory, Toyota Technological Institute, Tokyo (Tamotsu Kondow), and the University of Chicago (Takeshi Oka), all of which provided excellent working conditions for us when we needed to get away from our home institutions to focus on writing.

Several people have assisted us in reading part of the book and making valuable suggestions: Alex Dalgarno, Shirzad Kalhori, Holger Kreckel, Åsa Larson, Valery Ngassam, Takeshi Oka, Jeanna Royal, Albert Viggiano and Vitali Zhaunerchyk. We offer them our sincerest thanks for their help.

Finally we would like to thank Rainer Johnson, Brian Mitchell, Ioan Schneider, Andreas Wolf, Chris Greene, and the members of our research groups for access to material prior to publication.

1
Introduction

This book is focused on a single molecular process, dissociative recombination, and it may at first seem surprising that this topic can fill a whole book. As we shall see, it is not surprising when the complexity and applicability of the process are taken into account, and when the formidable challenges the process has provided to both experimenters and theorists are considered. This book brings together all the information we have on dissociative recombination in a single source, something which so far has been missing from the scientific literature.

A free electron which has a positive kinetic energy recombines with a positive atomic or molecular ion if its energy can be removed, so that it can enter a bound state. In the absence of a third body that can absorb the excess energy, the energy can be carried away by a photon. This is the only option for an atomic ion, and the process is inefficient. A molecular ion can make use of its internal structure and transfer the electron to a bound state while breaking one or several chemical bonds. This is a very efficient process that has taken its name from the fact that the capture of the electron is stabilized by dissociation. It is a primary chemical process, but is rarely described in chemical textbooks. It is the most complex of gas-phase reactions leading to the production of neutral atoms and molecules.

1.1 History 1900–1950

The process by which molecular ions recombine with electrons was for a long time strongly linked to discussions about the Earth's ionosphere. As we shall see later, it was not until after World War II that the research area initiated by the classic first recombination experiment of Thomson and Rutherford (1896) began to make contributions to the understanding of dissociative recombination. Initially, this research area was named "electrical discharges in gases"; it gradually transformed into gaseous electronics and plasma physics (see e.g. Loeb (1939) for the research in "gas discharge" physics before World War II).

The possibility of a conducting layer in the Earth's atmosphere was first put forward in 1902 by Lodge (1902). He realized that solar radiation causes ionization, and that the presence of free electrons converts the atmosphere into a feeble conductor. His short note in *Nature* was inspired by an attempt by Joly (1902) to explain by invoking the aether how in December 1901 Marconi was able send radio waves (or wireless telegraphy, as it was also called) from Cornwall in England to St John's in Newfoundland (Canada) around the curved Earth (see Marconi (1910)). Lodge starts his note by saying that "I can assure Prof. Joly that his explanation will not do" (Lodge 1902, p. 222). The contribution by Lodge is less well known than the two independent suggestions of a conducting layer by Kennelly (1902) and by Heaviside (1902). The conducting layer soon became known as the Heaviside layer, the Kennelly–Heaviside layer, or the Heaviside–Kennelly layer. In the 1920s, Appleton and Barnett (1925a,b) provided the first experimental evidence of the Heaviside layer, and they also discovered a second layer at a higher altitude. Appleton was awarded the Nobel prize for physics in 1947 for this achievement. The second layer was for some time known as the Appleton layer, and in the citation for the Nobel prize it was stated that it was awarded "for his investigations of the upper atmosphere, especially for the discovery of the so-called Appleton layer." In the late 1920s the ionized layers began to be called the ionosphere, and Appleton used the labels E and F for the Heaviside and Appleton layers, respectively. Curiously, whereas Appleton named his Nobel lecture "The ionosphere" (Appleton 1949), this term was not used in the presentation of his Nobel prize by Hulthén (1949), a member of the Nobel Committee; instead "the Heaviside layer" and "the Appleton layer" were used. Using modern terminology, the ionosphere is a weakly ionized plasma embedded in the thermosphere, the hot, tenuous region above 80 km. The D region is about 75–95 km above the ground, the E region is 95–150 km above the ground, and the F1 and F2 layers are 150–200 km and 250–360 km above the ground, respectively.

The auroral green line at 5577 Å was first studied by Ångström (1869), but it was many decades before it was established that the origin of the green line is atomic oxygen (McLennan & Shrum 1925). Kaplan (1931) proposed that the auroral green line arises if ionized molecular oxygen recombines with electrons, so that atomic oxygen in the ^1S state is formed. A transition from the ^1S state to a lower state would then give rise to radiation at 5577 Å; this is the first mention of dissociative recombination. It is known today that dissociative recombination is not the dominant source of the $O(^1S \rightarrow ^1D)$ emission at 5577 Å in the aurora; energy transfer from excited molecular nitrogen to atomic oxygen is the currently favored principal source (Rees 1984). Nevertheless, as suggested by Kaplan dissociative recombination is involved in the faint glow from the Earth's atmosphere known as the airglow. At 200 km, dissociative recombination of O_2^+ gives rise to the

$O(^1S \to {}^1D)$ transition at $\lambda = 5577$ Å. Because of its weakness, the study of the airglow spectrum was conducted much later than the auroral spectral studies. The auroral green line and the green airglow arise from the same transition in atomic oxygen, even if the precursors are different.

During the 1930s, much progress was made in understanding the ionosphere. Appleton (1949) developed techniques for exploring the ionosphere by means of reflection of radio waves, and these techniques were also adopted by other researchers. It was realized that electrons are primarily created by ionization by solar radiation in the ultraviolet, and that the ionization reaches a maximum about noon, falls off as sunset approaches, and continues to decrease during the night. Appleton (1937) summarized the knowledge of the ionosphere as of 1937 in his Bakerian lecture.

But if the source of electrons is photoionization, what could the sink of electrons be? And what process could account for the removal of electrons with a rate coefficient of the order of 10^{-8} cm^3 s^{-1}? Attachment of electrons to neutral atoms and molecules was put forward as one possibility, but in the long run this was difficult to reconcile with the results from the radio wave reflection measurements. Appleton (1937) favored recombination between electrons and positive ions, but did not try to sort out what the process could be at a detailed atomic and molecular level.

By the end of the 1930s atomic and molecular spectroscopy had a long history of making important contributions to space physics, but then atomic and molecular collision physics began to make an impact. Massey (1937) realized that a theoretical description of the upper atmosphere, with its resemblance to a gas in a low-pressure discharge source, necessarily would have to include individual collision processes that occur in such systems. The theory of layer formation was worked out by Chapman (1931), who showed that if ionizing radiation enters the atmosphere from above, the decrease in density of atmospheric constituents as a function of height, and hence the decrease in ionization density, is balanced by the decreasing radiation density as a function of decreasing height, and leads to the formation of a layer with a sharp ionization density maximum. Massey (1937) identified processes that could lead to the removal of electrons, and dismissed dissociative recombination of O_2^+, which was the molecular ion explicitly considered, for two reasons: it would require electrons within a narrow energy range in order to allow O_2^+ to capture an electron into an unstable O_2 state far above the dissociation limit, and it would have to rely on the weak interaction between electronic and nuclear motion. Furthermore, radiative recombination could be dismissed as well, since recombination by emission of a photon proceeds with a rate coefficient of the order of 10^{-12} cm^3 s^{-1}. This lack of a suitable electron−ion recombination process did not pose any problems to Massey at the time, since his conclusion of a 100:1 ratio, λ, between negative ions and electrons was in agreement with Chapman's theory of the variation of the Earth's magnetic field (Chapman 1931). It was also the prevailing view in the community working with electrical discharge that electron–ion recombination was an improbable process, and that the so-called volume recombination between ions (positive and negative) was the normal neutralizing process (Loeb 1939). Furthermore, the pioneering experiment by Kenty (1928), the first to provide a measurement of a rate coefficient that could be unambiguously ascribed to electron–ion recombination, had given 2×10^{-10} cm^3 s^{-1} as an upper limit to the recombination rate coefficient of Ar$^+$. Thus, Massey (1937) concluded that electrons are removed in the ionosphere by attachment followed by recombination of negative and positive ions.

New observational data by Appleton and Weekes (1939) made it difficult to uphold the view of a large excess of negative ions with respect to electrons in the ionosphere, and Massey's "back-of-the-envelope" estimates of the cross section for electron attachment to atomic oxygen did not survive more detailed quantum mechanical calculations (Bates & Massey 1943a). In the expression for an apparent recombination coefficient for electrons, $\alpha = \alpha_e + \lambda \alpha_i$, where α_e is the electron–ion recombination coefficient and α_i is that of mutual neutralization of positive and negative ions, the maximum value of λ was estimated to be 0.5 (Bates & Massey 1943b). This was possible to reconcile with the negative ion theory since the decrease in λ from 100 to 0.5 was compensated by an increase of α_i (Bates & Massey 1943a) as compared with Massey's (1937) original estimate. After a more careful analysis, however, λ was found to be only about 10^{-3} (Bates & Massey 1946), a value too small to be compensated by an increase of α_i. Forced to dismiss electron attachment to neutral atoms and molecules as a process for electron removal in the ionosphere, Bates and Massey (1946, p. 285) stressed that "[While] there are, as has been insisted, grave difficulties in the present theory of the ionized layers..." They did not consider dissociative recombination as being a fast process, but acknowledged that many difficulties would disappear if it were. In their second paper on this subject (Bates & Massey 1947), they discussed the process

$$O_2^+ + e^- \rightarrow O' + O'' \tag{1.1}$$

(O' and O'', using the notation in the original paper, refer to unspecified electronic states of the oxygen atom products) and concluded that it may, after all, proceed rapidly. The arguments, however, were indirect, and they admitted that there was not enough evidence to reach a firm conclusion.

The discussions about dissociative recombination had hitherto been driven entirely by problems related to the upper atmosphere (Kaplan 1931, Massey 1937, Bates & Massey 1943b, 1946, 1947). The breakthrough in establishing that dissociative recombination is a rapid process, however, came as a result of laboratory work at MIT that benefited from microwave equipment donated to MIT by the US army as thanks for the MIT contribution to the US war effort. Biondi and Brown (1949a) applied microwave techniques in order to study simultaneous (ambipolar) diffusion of electrons and positive ions in a pure helium afterglow plasma, making use of the superior accuracy of this new technique in measuring the electron density without perturbing the plasma. The analysis of the diffusion coefficient was complicated by the presence of electron–positive ion recombination, and they found that their data were best fitted if the recombination coefficient was 1.7×10^{-8} cm^3 s^{-1}. Bates (1950a) pointed out that such a large rate coefficient would most likely derive from He$_2^+$, which may be formed in the discharge.

Six months after their article on ambipolar diffusion, Biondi and Brown (1949b) published an article in which they reported recombination rate coefficients for a number of monatomic and diatomic gases. They found rate coefficients in the 10^{-8}–10^{-6} cm^3 s^{-1} range, and concluded that these coefficients were much larger than those predicted by the theory of radiative recombination. Microwave experiments at nearby Harvard University, which also included optical detection of the afterglow, gave similar results (Holt *et al.* 1950, Johnson, McClure, & Holt 1950). The results from these new experiments inspired Bates (1950b) to write his seminal article in which he explained how dissociative recombination of molecular ions with electrons can be very fast.

It is interesting to note that Bates's (1950a) explanation that He$_2^+$ could give a rate coefficient of the order of 10^{-8} cm^3 s^{-1} is incorrect, but was based on an experiment (Biondi & Brown 1949a) which is now known also to be incorrect! In fact, He$_2^+$ is one of the few molecular ions which recombines very slowly (Carata, Orel, & Suzor-Weiner 1999), and the rate coefficient given by Biondi and Brown (1949a) was not properly corrected for diffusion loss.

1.2 History 1950–1970

Bates (1950b) avoided the problem of the weak coupling between electronic and nuclear motions by invoking a two-step mechanism in which the electron first is captured by the molecular ion so that an unstable neutral molecule is formed, followed by rapid dissociation of the neutral molecule into its atomic constituents. The incoming electron interacts not with the heavy nuclei but with the electron cloud, and the fast dissociation prevents the electron from being transferred back to the continuum by autoionization. Bates derived an approximate formula for the rate coefficient, and by means of order of magnitude estimates, he arrived at a tentative rate coefficient of 10^{-7} cm^3 s^{-1}. But he also acknowledged the tremendous

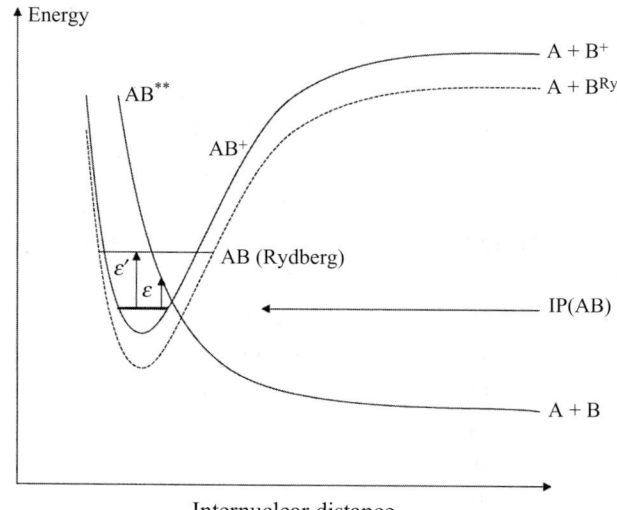

Figure 1.1 The dissociative recombination mechanism as proposed by Bates (1950b). The molecular ion AB$^+$ populating its lowest vibrational level collides with an electron with kinetic energy ε. The electron's kinetic energy is converted to electronic excitation energy, and the electron is captured by the ion so that a doubly excited neutral molecule AB** is formed. The molecule is unstable with respect to autoionization as long as the total energy of AB** is larger than the ionization energy, but because of the rapid movement of atoms A and B along the strongly repulsive curve, autoionization becomes prohibited, and the electron capture is stabilized. The cross section, σ, is proportional to $1/\varepsilon$.

difficulties of an accurate computation, and for the rest of the 1950s the development of the field of dissociative recombination was driven almost entirely by experiments (Bates & Dalgarno 1962).

Figure 1.1 shows a schematic representation of the potential energy curves involved in the dissociative recombination process. It follows from the figure that atoms A and B will receive kinetic energy in excess of thermal energy. If either A or B is produced in an excited state, spectral line broadening would result. Thus, it was early realized that optical observations of the afterglow would be one way of establishing the mechanism proposed by Bates (1950b). The other obvious route was to ascertain that the afterglow contained molecular ions, something which required the addition of a mass spectrometer to the afterglow apparatus. In practice, these experimental additions to the afterglow technique took some time before being fully implemented and giving conclusive results. A helium afterglow was used to search for broadening of the $\lambda = 5876$ Å ($3\ ^3D \to 2\ ^3P$) line, but because of the very slow recombination of He$_2^+$ with thermal electrons (unknown at the time), the results were inconclusive (Biondi 1956). Although a mass spectrometer was added to an afterglow apparatus at an early stage

(Phelps & Brown 1952), most afterglow studies were performed without mass identification.

In the 1960s, Connor and Biondi (1965) and Frommhold and Biondi (1969) carried out a series of afterglow studies of Ne_2^+ that finally established that emission lines from the afterglow are composed of a broad part arising from dissociating, energetic Ne atoms, and a narrower part arising from thermal Ne atoms that have transferred most of their kinetic energy to residual atoms before radiating.

An important addition to Bates's (1950b) mechanism was made in 1968 by Bardsley (1968b), although the idea had already been outlined in the abstract book from the *Fifth International Conference on the Physics of Electronic and Atomic Collisions* in Leningrad, USSR, in 1967 independently by Bardsley (1967) and Chen and Mittleman (1967). They proposed that the electron can also surrender its kinetic energy by exciting a vibrational mode in the molecule while being captured into a Rydberg state which is member of a Rydberg series that converges to the ion core. In a second step, the Rydberg state is predissociated by an electronically doubly-excited, repulsive state. Bardsley (1967, 1968a,b) labeled this mechanism as the indirect one, to distinguish it from the direct mechanism of Bates (1950b). It is interesting to note how Chen and Mittleman (1967) assessed the complexity of the process: "Thus we are faced with an extremely complex problem the results of which depend critically upon the details of molecular states which are prohibitively difficult to obtain for even the simplest molecule."

By the end of the 1960s, only the rare gas dimers, the most important atmospheric ions, such as N_2^+, O_2^+, and NO^+, and a few other systems had been investigated experimentally. No *ab initio* calculations had been performed. Two comprehensive reviews had been published (Danilov & Ivanov-Kholodny 1965, Bardsley & Biondi 1970), along with a shorter review by Biondi (1964), and a theoretical review of significant importance to dissociative recombination (O'Malley 1971).

1.3 History 1970–1990

One ion is conspicuously missing from the list given at the end of the previous section – H_2^+. In its capacity as the simplest molecule structure, it is ideal for a comparison of experiment and theory. The problem of studying dissociative recombination of H_2^+ in a decaying plasma is that it is very rapidly converted to H_3^+ by the reaction $H_2^+ + H_2 \rightarrow H_3^+ + H$. Thus, the techniques developed up to 1970, which were all based on the reaction rate coefficient being obtained by measuring the decay in concentration of charged particles when an ionizing agent was removed, were unsuitable for the study of H_2^+. In the early and mid 1970s, several new techniques were developed which aimed at measurement of cross section rather than rate coefficient.

Peart and Dolder (1971) developed a technique in which a beam of electrons crossed a beam of ions at a $10°$ angle, and used it to measure the cross section for dissociative recombination of D_2^+ (Peart & Dolder 1973a) and H_2^+ (Peart & Dolder 1974a). Dunn and coworkers (Phaneuf, Crandall, & Dunn 1975, Vogler & Dunn 1975) used crossed beams to study the atomic products following dissociative recombination of D_2^+. Finally, McGowan, Caudona, and Keyser (1976) developed a merged-beam technique, which they used to study the dissociative recombination of a range of molecular ions, including H_2^+.

Walls and Dunn (1974) combined a static ion trap with an electron beam to measure the cross section for dissociative recombination of O_2^+ and NO^+. This is the first application of a technique which makes use of stored ions. Mathur, Kahn, and Hasted developed a different type of trap technique (1978).

There was also a development of afterglow techniques. A shock-tube technique was developed by Fox and Hobson (1966) and Cunningham and Hobson (1969, 1972a), and allowed studies of recombination at elevated electron *and* ion temperatures. In the context of elevated temperatures, the studies of recombination in flames should also be mentioned (see e.g. Butler and Hayhurst (1996)). Sauer and Mulac (1971, 1972, 1974) performed a few recombination rate coefficient measurements by observing the time-dependent emission from recombination end products using the pulse radiolysis technique. (In pulse radiolysis, an important technique in radiation chemistry, the ionizing source is pulses of electrons, usually from Van de Graaf accelerators of energy 2−4 MeV; a single-volume description is given by Matheson and Dorfman (1969).) Building on the flowing afterglow technique developed by Ferguson and coworkers (Fehsenfeld *et al.* 1965, Goldan *et al.* 1965) to study ion−neutral reactions, Smith and Adams (1983) developed the technique to also allow studies of positive ion−negative ion reactions. In a flowing afterglow, a flow tube with a large Roots-type pump is used to force a carrier gas to flow towards the pump. Ionization occurs upstream of the flow by chemi-ionization (ion−neutral reactions) or some other type of ionization. In order to study positive ion−negative ion reactions, Smith and Adams supplied their flow tube with a Langmuir probe, which allowed the measurement of the charge density as a function of position along the tube. With some modification, the flowing afterglow/Langmuir probe (FALP) technique could be used to measure dissociative recombination rate coefficients (Alge, Adams, & Smith 1983). It is less well known that Mahdavi, Hasted, and Nakshbandi (1971) performed a flowing afterglow study of recombination more than a decade before the FALP technique appeared. Apparently only one experiment was performed by Hasted's group.

During the period leading up to 1970, the focus had been on measuring rate coefficients and establishing the mechanism. The afterglow techniques now started also to address the question of product state distributions. To begin with the atomic states into which a diatomic ion recombines were identified (Zipf 1970). Later, the breakup of polyatomic molecular ions in dissociative recombination was studied by optical techniques (Vallée et al. 1986), as reviewed in detail by Adams (1992). The beam techniques also addressed these problems for diatomic (Phaneuf, Crandall, & Dunn 1975, Vogler & Dunn 1975) and polyatomic (Mitchell et al. 1983) systems, but in contrast to the afterglow techniques, these were the only efforts with beam techniques until the advent of the ion storage rings.

Holt et al. (1950) had used optical methods to monitor the decaying plasma in an afterglow. A modern version of this method was employed by Amano (1988, 1990), who used an infrared laser to monitor the ion concentration in an afterglow. With a narrow band laser, Amano could for the first time measure the disappearance of ions in specific quantum states.

The rapid experimental development of the field of recombination during the 1970s and 1980s is well described in a number of review articles (Biondi 1973, Dolder & Peart 1976, 1986, Berry & Leach 1979, Eletskii & Smirnov 1982, Mitchell & McGowan 1983, Compton & Bardsley 1984, McGowan & Mitchell 1984, Mitchell 1986, 1990a,b, Johnsen 1987, Adams & Smith 1988a,b).

There was also an impressive development of theoretical methods during this period, partly inspired by the data for the simplest molecular ion. Nielsen and Berry (1971) performed the first *ab initio* calculations on H_2^+, but included only the direct mechanism. Bottcher developed a projection operator formalism (Bottcher 1974) and applied it to H_2^+ (Bottcher 1976). Although both the direct and indirect mechanisms were included, they were treated separately. Lee (1977) and Giusti (1980) showed how multichannel quantum-defect theory (MQDT) can be used to treat dissociative recombination, including a unified treatment of the direct and indirect processes. Giusti-Suzor, Bardsley, and Derkits (1983) showed that configuration interaction theory and MQDT are two alternative ways of treating dissociative recombination, and applied both methods to H_2^+, thus making possible a comparison of experiment (Auerbach et al. 1977) and theory. Although a strict comparison was impaired by the undetermined H_2^+ vibrational distribution in the experiment, theory was nevertheless able to support the existence of the narrow window resonances observed in the experiment, and could show how they arise from interferences between direct and indirect dissociative recombination. Such resonances also emerged from O'Malley's (1981) treatment of the direct mechanism, in which he allowed for the Rydberg states involved in the indirect process to play a role. A more detailed comparison would take several more years to realize, including efforts to produce H_2^+ primarily in its lowest vibrational level, however, the essential physics is captured in the papers of Giusti (1980) and Giusti-Suzor, Bardsley, and Derkits (1983). This work provided the framework for understanding dissociative recombination. The theoretical development has been reviewed by Giusti-Suzor (1986) and Guberman (1986).

Although H_2^+ is the ideal system for comparison of experiment and theory, other molecular systems occupied more important roles in atmospheric physics and astrophysics. The need for a consistent approach to the construction of potential curves of relevance to dissociative recombination was quickly realized (Guberman 1983a).

1.4 History 1990–present

The development of heavy-ion storage rings for atomic and molecular physics (see e.g. Larsson (1995a) for a review) led to a boost of the study of dissociative recombination. This development will not be described here, but rather in the subsequent chapters of this book. The first papers on the subject based on work using ion storage rings were published in a single issue of *Physical Review Letters* (Tanabe *et al.* 1993, Forck *et al.* 1993b, Larsson *et al.* 1993a). The development of the techniques established prior to 1990 will also be described.

The crossing of potential curves, as shown in Fig. 1.1, exercised a strong influence on the scientific community involved in dissociative recombination research (too strong, Bates (1994) would argue). In the early 1990s, an increasing amount of experimental data made it difficult to uphold the view that a crossing of the ion potential curve by a neutral state potential curve is required to drive dissociative recombination. Estimates based on semiclassical treatment (Bates 1992b, 1993a) and *ab initio* calculations (Guberman 1994, Sarpal, Tennyson, & Morgan 1994) made it clear that dissociative recombination can be quite effective even in the absence of a curve crossing.

Conferences with dissociative recombination as the central theme have been organized from 1988 onwards in an ad hoc fashion, and the proceedings from these conferences give a very good coverage of the development of the field (Mitchell & Guberman 1989, Rowe, Mitchell, & Canosa 1993, Zajfman *et al.* 1996, Larsson, Mitchell, & Schneider 2000, Guberman 2001, 2003a, Wolf, Lammich, & Schmelcher 2005). Several review articles also describe the development (Adams 1992, 1993, Glosik 1992, Bates 1994, Flannery 1994, Mitchell 1995, Larsson 1995a,b, 1997, 2000a, 2001, Larsson & Thomas 2001, Johnsen & Mitchell 1998, Adams, Babcock, & McLain 2003, Guberman 2003b, Petrie & Bohme 2003, Florescu-Mitchell & Mitchell 2006, Adams, Poterya, & Babcock 2006). The review by Florescu-Mitchell and Mitchell (2006) contains a complete listing of experimental results until the end of 2005. A database which builds on the compilation by Florescu-Mitchell and Mitchell (2006) is available at URL: http://mol.physto.se/DRdatabase.

2
Experimental methods

Several experimental methods were mentioned in Chapter 1. In this chapter most of these methods will be described in more detail. They represent the experimental methods that have produced the vast majority of the dissociative recombination data.

2.1 Merged beams

2.1.1 Kinematics and resolution

It follows from Chapter 1 that the cross section for dissociative recombination is inversely proportional to the kinetic energy of the incident electron, and that resonances in the cross section may occur because of interferences between the indirect and the direct process. For these reasons, the ideal method for detailed studies of the cross section is the merging of beams of electrons and beams of molecular ions. The application of merged beams to the study of atomic and molecular collision processes has been comprehensively reviewed by Phaneuf et al. (1999).

A perfect experiment by means of merged beams would consist of a monoenergetic beam of ions interacting, at an angle of $0°$, with a monoenergetic beam of electrons over a precisely known distance, and with no other atoms or molecules present. The beams overlap perfectly, and the velocity and currents of the beams are precisely known. The neutral reaction products are detected with 100% efficiency. This ideal situation can never be realized, of course. A beam of charged particles has an energy spread, it is difficult to keep the angle of intersection at exactly $0°$, and no matter how good the vacuum, there will be residual gas molecules present in the interaction region.

In many experiments in, for example, particle physics, it is desirable that a large fraction of the particle energy is converted into interaction energy. In merged-beam experiments, only a small fraction of the available laboratory energy is converted into interaction energy.

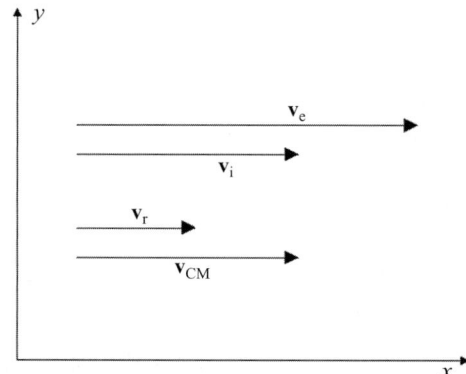

Figure 2.1 Vector diagram showing the velocities of the electrons (v_e), ions (v_i), center-of-mass (v_{CM}), and the relative velocity (v_r). It is based on the fact that the ion is much heavier than the electron, hence $v_{CM} = |\mathbf{v}_{CM}| \approx v_i$.

In relativistic kinematics, it is appropriate to describe the interaction in the center-of-momentum frame. In the nonrelativistic limit, the interaction energy is described in the center-of-mass frame of reference. When a beam of electrons with speed v_e ($= |\mathbf{v}_e|$ in Fig. 2.1) in the laboratory frame and a beam of ions with speed v_i in the same frame intersect at an angle θ, the relative speed of the two beams is given by

$$v_r = (v_e^2 + v_i^2 - 2v_e v_i \cos\theta)^{1/2}. \tag{2.1}$$

The electrons' kinetic energy is $E_e = (m_e v_e^2)/2$ and that of the ions is $E_i = (m_i v_i^2)/2$, where m_e and m_i are the electron and the ion masses, respectively. The total kinetic energy in the center-of-mass frame (i.e., the interaction energy) is given by

$$E_{CM} = \frac{1}{2}\mu v_r^2 = \mu \left[\frac{E_e}{m_e} + \frac{E_i}{m_i} - 2\left(\frac{E_e E_i}{m_e m_i}\right)^{1/2} \cos\theta \right], \tag{2.2}$$

where $\mu = m_e m_i/(m_e + m_i)$ is the reduced mass. When the two beams are merged, θ is equal to zero and $\cos\theta = 1$. Equation (2.2) is then expressed as

$$E_{CM} = \mu \left[\left(\frac{E_e}{m_e}\right)^{1/2} - \left(\frac{E_i}{m_i}\right)^{1/2} \right]^2. \tag{2.3}$$

Since for any molecular ion $m_i \gg m_e$, in practice $\mu \approx m_e$ and $v_{CM} \approx v_i$. Thus, conceptually, we can regard the center-of-mass frame as identical with a reference frame in which the ions are at rest. The speed of the electrons in the ions' rest frame becomes identical with the relative speed. In practice, Eq. (2.3) should be replaced by its relativistic form (Schennach et al. 1994) when the laboratory electron energy

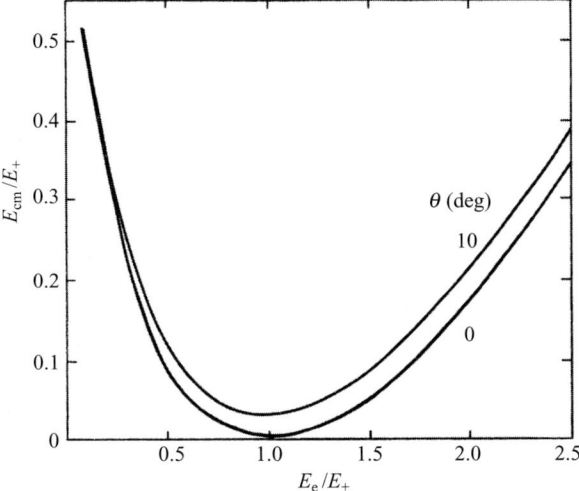

Figure 2.2 The center-of-mass energy as a function of electron energy, with both scales normalized to E_+ for $\theta = 0°$ and $10°$. The speed of the electron beam is larger than that of the ion beam when $E_e/E_+ > 1$. Only for the merged-beam case ($\theta = 0°$) is zero center-of-mass energy accessible. (Reproduced with permission from Auerbach et al., "Merged electron–ion beam experiments I. Methods and measurements of (e–H_2^+) and (e–H_3^+) dissociative recombination cross section," J. Phys. B **10**, pp. 3797–3820 (1977), IOP Publishing Limited.)

is in the keV range. Equation (2.2) shows that $E_{CM} = 0$ when $(E_e/m_e) = (E_i/m_i)$, which means that collisions at zero energy are possible under the assumed ideal conditions, i.e., merged monoenergetic beams and $\theta \equiv 0$. If μ is replaced by m_e in Eq. (2.3) and we define $E_+ = (m_e/m_i)E_i$, Eq. (2.3) takes the form

$$E_{CM} = (E_+ - E_e)^{1/2}. \qquad (2.4)$$

In an experiment with merged beams, the ion energy E_+ is usually kept fixed, whereas the electron energy E_e is changed in order to induce collisions at different center-of-mass energies. Figure 2.2 shows E_{CM} normalized to E_+ as a function of (E_e/E_+). The electrons can move faster or slower than the ions, a situation which is symmetric with respect to velocities but not with respect to energies, as shown in Fig. 2.2. Two major advantages with merged beams are illustrated in Fig. 2.2; first, it is possible to study, in principle, collisions at zero center-of-mass energy; second, very-low-energy collisions can be obtained by means of changes in the laboratory electron energy E_e that are much larger than E_{CM}.

So far we have considered an ideal case. Different approaches have been used to analyze the consequences of departure from ideal conditions. One approach is to use Eq. (2.2) as the starting point, make use of the fact that $\sin\theta \approx \theta$ for merged beams, and take the partial derivatives with respect to E_e, E_i, and θ to obtain an expression for ΔE_{CM}:

$$\Delta E_{CM}/E_+ = |1 - (E_+/E_e)^{1/2}|\Delta E_e/E_+ + |1 - (E_e/E_+)^{1/2}|\Delta E_i/E_i \\ + 2(E_e/E_+)^{1/2}\theta\Delta\theta, \quad (2.5)$$

where ΔE_e and ΔE_i are the energy spreads in the electron and ion beams, respectively, and $\Delta\theta$ is the angular spread in the electron beam.

Auerbach et al. (1977) used this expression in order to analyze the resolution in their merged electron–ion beam experiment (MEIBE) at the University of Western Ontario in Canada. In this single-pass merged-beam apparatus, the ions were accelerated to 440 keV, which gave an equivalent electron energy of 120 eV at $E_{CM} = 0$ for an H_2^+ beam. Auerbach et al. (1977) estimated ΔE_e and ΔE_i to be 0.1 eV and 200 eV, respectively, and calculated ΔE_{CM} for various combinations of $\theta\Delta\theta$. Owing to the deamplification effect of the first and second terms on the right hand side of Eq. (2.5), both of which become zero when $E_e = E_+$, the resolution is dominated by the $\theta\Delta\theta$ term for realistic values of θ and $\Delta\theta$. Using the appearance of the cross section for dissociative recombination of H_2^+ around $E_{CM} = 0$, Auerbach et al. (1977) estimated the resolution to be 20 meV near zero center-of-mass energy, arising from $\theta = 0.5°$ and $\Delta\theta = 0.5°$ (expressed in radians in Eq. (2.5)). Later, the resolution was improved to 5 meV at $E_{CM} = 0.01$ eV (Mul and McGowan 1979a), presumably by careful alignment of the beams. The high resolution was confirmed when 5 meV wide resonances were observed in dissociative recombination of H_2^+ in its lowest vibrational level (Van der Donk et al. 1991). Even narrower resonances have been observed in MEIBE (Rogelstad et al. 1997), but these are probably due to some laser-selective effects.

The analysis based on Eq. (2.5) implies that the arrow illustrating the electrons in Fig. 2.1 makes a slight angle (θ) with respect to the arrow illustrating the ions, that the $\mathbf{v}_e$ arrow is composed of a distribution of arrows spanning an angle $\Delta\theta$, and that there is an uncertainty in the length of the $\mathbf{v}_e$ and $\mathbf{v}_i$ arrows. If we assume that the average velocity of the electrons is parallel to the x-axis in Fig. 2.1, the only contribution to a velocity spread in the y-direction comes from the angular distribution $\Delta\theta$. This is a simplified description of an electron beam. A more realistic approach was used by Dittner et al. (1986) in their analysis of dielectronic recombination measurements of the Na-like ions P^{4+}, S^{5+}, and Cl^{6+}. In order to follow their procedure, we need to look closer at how an electron beam is formed, and how the rate of electron–ion interaction is measured.

Figure 2.3 shows a schematic electron gun heated to a temperature which results in emission of electrons. The electrons are emitted from the surface of a thermionic cathode, which is heated to a temperature of about 1000 K, and form a cloud in front of the cathode, with a temperature equal to the cathode temperature,

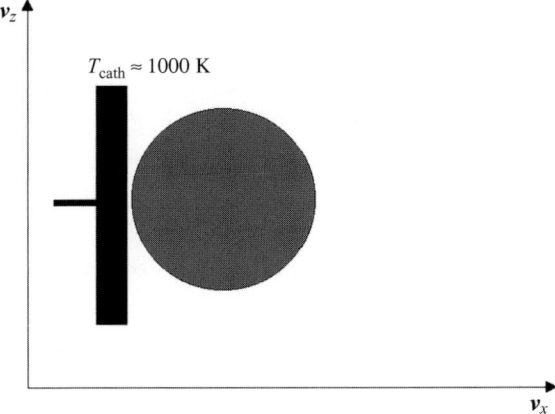

Figure 2.3 Electrons are emitted from the surface of a thermionic cathode heated to about 1000 K. An electron cloud is formed in front of the cathode, with a spherical, Maxwellian-shaped velocity distribution ($\sigma_{vx} = \sigma_{vy} = \sigma_{vz}$, see text), and an electron temperature equal to the cathode temperature. If the cathode is on a negative potential, electrons can be extracted from the cloud by nearby anodes. The filled circle illustrates the isotropic electron-velocity distribution.

i.e., $T_e = T_{\text{cath}}$. The electron velocity distribution $f(\mathbf{v}_e)$ can be assumed to be of Maxwellian shape:

$$f(\mathbf{v}_e) = \exp\{-[(v_x^2/\sigma_{vx}^2) + (v_y^2/\sigma_{vy}^2) + (v_z^2/\sigma_{vz}^2)]\}/\sigma_{vx}\sigma_{vy}\sigma_{vz}\pi\sqrt{\pi}, \qquad (2.6)$$

where σ_{vx}, σ_{vy}, and σ_{vz} are the root-mean-square speeds in the x-, y-, and z-directions, respectively, and $f(\mathbf{v})$ is normalized to unity: $\int f(\mathbf{v}) d^3v = 1$. For a spherical distribution, for which $\sigma_{vx} = \sigma_{vy} = \sigma_{vz} = \sigma_v$, Eq. (2.6) takes the form

$$f(\mathbf{v}_e) = \frac{\exp(-v^2/\sigma_v^2)}{\sigma_v^3 \pi \sqrt{\pi}} = \left(\frac{m_e c^2}{2kT}\right)^{3/2} \frac{\exp(-m_e v^2/2kT)}{c^3 \pi \sqrt{\pi}}. \qquad (2.7)$$

where $kT = m_e \sigma_v^2/2$ is the electron temperature per degree of freedom.

If the cathode is on a negative potential and anodes are positioned close to the cathode in the x direction, electrons will be extracted from the cloud and form an electron beam. In order to prevent the electron beam from blowing up because of the space-charge effects, a longitudinal magnetic field is used to guide the electrons. When the electrons have reached their terminal velocity, the velocity distribution in the longitudinal (x) direction has been reduced significantly because of the kinematic compression effect. To understand how this effect arises, consider two electrons in the cloud in Fig. 2.3, one with speed $v_{x1}(0) = 0$ and kinetic energy $E_1(0) = 0$

along the x-axis (we neglect for the moment motions along the y- and the z-axis), and the other with speed $v_{x2}(0) > 0$ and kinetic energy $E_2(0) = m_e v_{x2}(0)^2/2$. Now, the cathode potential is changed to $-U$ V, so that the two electrons are accelerated to the new energies $E_1 = eU = m_e v_{x1}^2/2$ and $E_2 = m_e v_{x2}(0)^2/2 + eU = m_e v_{x2}^2/2$. The difference in energy after acceleration is $E_2 - E_1 = m_e v_{x2}(0)^2/2$, which gives $v_{x2}^2 - v_{x1}^2 = v_{x2}(0)^2$ and $\Delta v = v_{x2} - v_{x1} = v_{x2}(0)^2/2v_x$, where $v_x = (v_{x2} + v_{x1})/2$. Thus, $\Delta v = v_{x2}(0)(m_e v_{x2}(0)^2/8eU)$. If we assume that $m_e v_{x2}(0)^2/2$ is approximately equal to the thermal energy $E_{\text{thermal}} = kT_{\text{cath}} = kT_e$ of the electron cloud before acceleration, then $\Delta v = (E_{\text{thermal}}/4eU) v_{x2}(0)$. This gives a compression of the longitudinal velocity distribution by a factor of 10^{-4} if $E_{\text{thermal}} = 0.1$ eV and $eU = 250$ eV. The treatment is simplified but includes the essential physics. Since the electrons are accelerated only in one direction, the velocity distributions in the y- and z-directions are unaffected. In practice the longitudinal velocity spread is higher than theoretically predicted. Each electron has a potential energy from the electrostatic field of the other electrons. If this potential after acceleration is larger than the spread in kinetic energy, the latter increases until the two are in equilibrium.

Let $\sigma_{v\perp} = \sigma_{vy} = \sigma_{vz}$ and $\sigma_{v\|} = \sigma_{vx}$, where, as we have seen, $\sigma_{v\|} \ll \sigma_{v\perp}$; the anisotropic, "flattened," electron velocity distribution can now be expressed as

$$f(\mathbf{v}_e) = \frac{m_e}{2\pi k T_\perp} \exp\left(-\frac{m_e v_\perp^2}{2kT_\perp}\right) \sqrt{\frac{m_e}{2\pi k T_\|}} \exp\left(-\frac{m_e v_\|^2}{2kT_\|}\right), \qquad (2.8)$$

where $v_\perp^2 = v_y^2 + v_z^2$ and $v_\| = v_x$, and $2kT_\perp = m_e \sigma_{v\perp}^2$, $2kT_\| = m_e \sigma_{v\|}^2$. An ion interacting with electrons having the distribution given in Eq. (2.8) encounters the situation shown in Fig. 2.4. As earlier, we assume that the ion is in rest in the center-of-mass frame, and that the interaction energy is given by $E_{\text{CM}} = m_e v_r^2/2$. The velocity v_d in Fig. 2.4 is the detuning velocity, which is determined by the acceleration voltage of the electron gun cathode; it is equal to the difference between the electron and ion speeds in the longitudinal direction, i.e. $v_d = |v_{e\|} - v_{i\|}|$ (in the laboratory frame) $= |v_{e\|}|$ (in the center-of-mass frame). The detuning energy $E_d = m_e v_d^2/2$ is approximately equal to E_{CM} for large v_d. When E_d is small, v_d can be much smaller than v_r, which is then dominated by the transverse electron velocity spread $v_\perp$. Since it is E_d that is controlled in a merged-beam experiment, cross sections are often displayed with respect to E_d. A word of caution, the terminology in the literature differs. For example, in the paper by Schennach *et al.* (1994), the detuning energy as defined here is referred to as the relative energy.

The electron velocity distribution given in Eq. (2.8) is expressed in the electron reference frame. Figure 2.4 shows the situation from the ion's point of view, i.e., from the center-of-mass frame, which we, for simplicity and to a very good

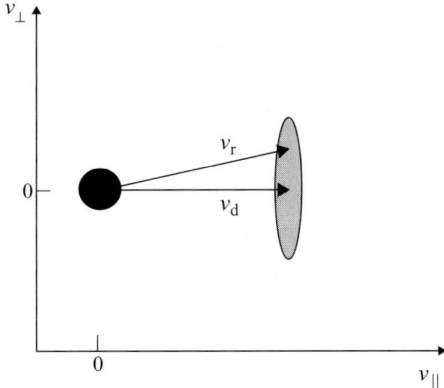

Figure 2.4 An ion at rest in the center-of-mass frame interacts with electrons characterized by an oblate, flattened velocity distribution given mathematically in Eq. (2.8) in the electron's rest frame and in Eq. (2.9) in the center-of-mass frame (the ion's rest frame). When $v_d \to 0$, $v_r \approx v_\perp$.

approximation, assume is the same as the ion's rest frame. It is of course possible to use Eq. (2.8) to develop the formalism using the rest frame of the electron; however, we will continue to use the center-of-mass frame as reference, which means that Eq. (2.8) takes the form

$$f(\mathbf{v}_e, v_d) = \frac{m_e}{2\pi k T_\perp} \exp\left(-\frac{m_e v_\perp^2}{2k T_\perp}\right) \sqrt{\frac{m_e}{2\pi k T_\parallel}} \exp\left(-\frac{m_e(v_\parallel - v_d)^2}{2k T_\parallel}\right). \quad (2.9)$$

In order to express the rate of interaction between electrons and ions, we need to consider the spatial electron density $n_e(\mathbf{r})$, the spatial ion density $n_i(\mathbf{r})$, the volume of interaction V, the relative velocity between ion and electron $\mathbf{v}_{rel}$ and its absolute value $v_{rel} = |\mathbf{v}_{rel}|$, and the velocity-dependent cross section $\sigma(v_{rel})$; the rate R (in units of s^{-1}) is then given by

$$R = \int_V \int_{v_{rel}} n_e(\mathbf{r}) n_i(\mathbf{r}) \sigma(v_{rel}) v_{rel} f(v_{rel}) d^3 v_{rel} d^3 r. \quad (2.10)$$

Dittner *et al.* (1986) made the assumption that their ion beam was very well defined, so that only the electrons contributed to the distribution function, i.e., $f(\mathbf{v}_{rel}) = f(\mathbf{v}_e, v_d)$ as given in Eq. (2.9). For an electron beam which is uniform in space, and with a diameter larger than that of the ion beam, Eq. (2.10) simplifies considerably. Integration over n_i gives the number of ions present in the interaction volume, and n_e is a number which can be taken outside of the integral. The number of ions can be expressed as $I_i l / eq v_i$, where I_i is the ion current, l is the length of the interaction region, q is the charge state (for small molecules usually 1 or 2), and e is the

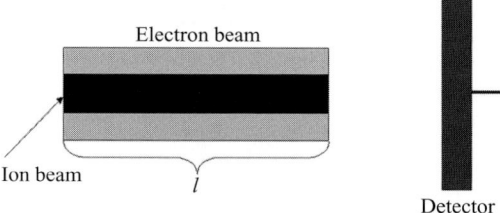

Figure 2.5 A monoenergetic ion beam passes through an electron beam with uniform charge distribution and with an electron-velocity distribution given by Eq. (2.9). Neutral particles are formed in the interaction region of length l and hit the detector, which, assuming 100% detection efficiency, would register R events per unit time, where R is given by Eq. (2.12).

electron charge. The rate coefficient $\alpha(v_d)$ (usually expressed in cubic centimeters per second) is defined as

$$\alpha(v_d) = \langle v_r \sigma \rangle = \int_0^\infty \sigma(v_r) v_r f(\mathbf{v}_e, v_d) d^3 v_r, \qquad (2.11)$$

where $f(\mathbf{v}_e, v_d)$ is given in Eq. (2.9), $v_r^2 = v_\parallel^2 + v_\perp^2$, and $d^3 v_r = 2\pi v_\perp dv_\perp v_\parallel$. We can now write Eq. (2.10) in a much simpler form:

$$R = \frac{\alpha(v_d) n_e I_i l}{eq v_i}. \qquad (2.12)$$

Figure 2.5 illustrates the experimental situation that leads to the expression given above (Eq. (2.12)). If all neutral particles arising from recombination in the interaction region hit the detector and are detected with 100% efficiency, the count rate R is given by Eq. (2.12). Dittner et al. derived Eq. (2.12) and used it to analyze their dielectronic recombination data, and this approach has been adopted in subsequent work by other researchers (Andersen & Bolko 1990, Schennach et al. 1994). Owing to the compression technique used by Dittner et al. (1986), their electron beam was characterized by $kT_\perp \approx 1$ eV. Andersen and Bolko (1990) used an electron beam with a transverse temperature essentially determined by the cathode temperature and obtained $kT_\perp = 0.15$ eV and $kT_\parallel = 2$ meV. As we shall see in Section 2.2 on ion storage rings, much lower temperatures can be obtained by means of special beam handling techniques.

It is immediately clear from Fig. 2.4 that the resolving power of the merged-beam experiment will depend on the size of the electron cloud, i.e., on the velocity spread in the transverse and longitudinal directions. The effect of the finite electron velocity distribution can be illustrated in different ways. Let us assume that the interaction of an electron and an atomic ion A^+ leads to the formation of A^0 by electron capture and subsequent stabilization by photon emission, a process known as dielectronic recombination. The process is resonant, and we assume that the

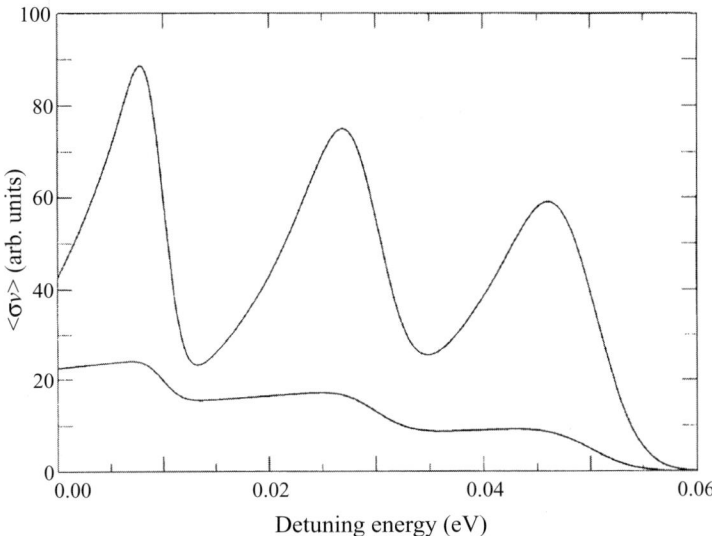

Figure 2.6 Two different fictive measurements of δ-function resonances at 10, 30 and 50 meV. The detuning energy, E_d, is given on the x-axis and the rate coefficient as defined in Eq. (2.11), $\alpha(v_d)$, on the y-axis. The upper curve was obtained with $kT_\perp = 10$ meV and $kT_\parallel = 0.1$ meV and the lower one with $kT_\perp = 100$ meV and $kT_\parallel = 0.1$ meV. (Reproduced with permission from Danared 1995.)

cross section can be approximated by a delta function of the interaction energy E_{CM} with a resonance at E_0:

$$\sigma(E_{\text{CM}}) = \sigma_0 \delta(E_{\text{CM}} - E_0). \tag{2.13}$$

If this cross section is inserted in Eq. (2.11), the integral can be evaluated analytically (Andersen 1993), which gives the rate coefficient $\alpha(v_d)$ expressed in the transverse and longitudinal temperatures as

$$\alpha(v_d) = \frac{\sigma_0 v_0}{2\lambda k T_\perp} \exp\left[\frac{-m_e}{2k T_\perp}\left(v_0^2 - \frac{v_d^2}{\lambda^2}\right)\right]$$
$$\times \left[\text{erf}\left(\sqrt{\frac{m_e}{2k T_\parallel}} \frac{v_d + \lambda^2 v_0}{\lambda}\right) - \text{erf}\left(\sqrt{\frac{m_e}{2k T_\parallel}} \frac{v_d - \lambda^2 v_0}{\lambda}\right)\right], \tag{2.14}$$

where erf(x) is the error funtion, $v_0 = (2E_0/m_e)^{1/2}$, $\lambda = (1 - T_\parallel/T_\perp)^{1/2}$, and $T_\parallel \ll T_\perp$. Figure 2.6 shows $\alpha(v_d)$ as a function of E_d for $kT_\parallel = 0.1$ meV and $kT_\perp = 10$ meV and 100 meV, and for three δ-function resonances located at 10, 30, and 50 meV. The effect of reducing the transverse electron temperature by a factor of 10 is clearly shown, as is the possibility of using the resonances to determine the electron temperature (Andersen & Bolko 1990).

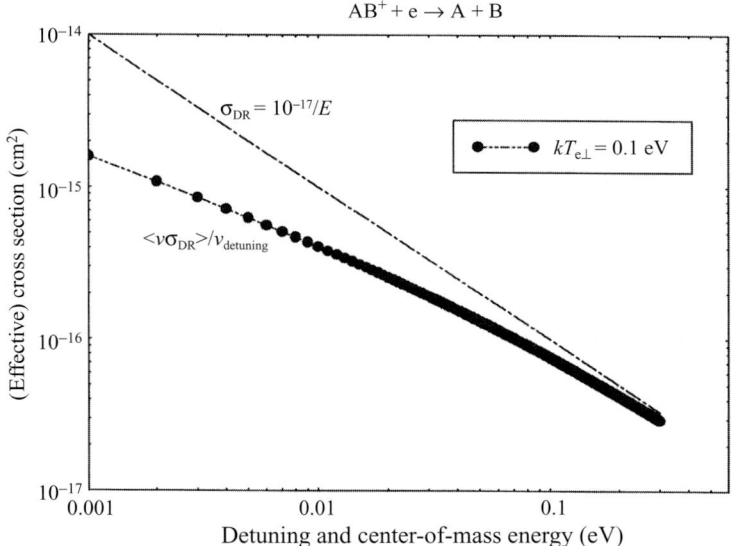

Figure 2.7 Effect of transverse electron temperature. The lower curve represents the measured effective cross section, $\langle v_r\sigma\rangle/v_d$, for a $kT_\perp = 100$ meV electron beam ($kT_\| \equiv 0$ for simplicity) and a cross section which is $10^{-17}/E_{CM}$. The x-axis is the detuning energy for the effective cross section, $\langle v_r\sigma\rangle/v_d$, and the center-of-mass energy for the true cross section, σ. (Reprinted from *Int. J. Mass Spectrom. Ion Proc.* **149/150**, M. Larsson, "Dissociative recombination in ion storage rings", pp. 403–414, Copyright (1995), with permission from Elsevier.)

Dissociative recombination of molecular ions does not generate δ-function cross sections, but rather cross sections of the shape $\sigma(E_{CM}) = 1/E_{CM}$ with or without resonances superimposed. Figure 2.7 shows the effect of the transverse electron temperature in such a case.

It remains to be discussed how the 5 meV resolution obtained by Mul and McGowan (1979a) can be reconciled with the reasoning leading up to Eq. (2.14). In order to do this, we need to take a closer look at the apparatus used in Western Ontario (Auerbach *et al.* 1977), the only single-pass merged-beam machine that has been used to study dissociative recombination. Figure 2.8 shows a schematic illustration of the merged electron and ion beams apparatus developed and used at the University of Western Ontario in Canada (Auerbach *et al.* 1977).

It is obvious that Fig. 2.5 is a simplification; the beams are perfectly merged, but the illustration says nothing about how the beams are merged and de-merged. Auerbach *et al.* (1977) solved this challenging problem by means of trochoidal electron analyzers. The concept derives from Stamatovic and Schultz (1968, 1970), who realized that electrons moving along a z-axis in an axial magnetic field could be monochromatized by passing through a region where they are also exposed to a

2.1 Merged beams

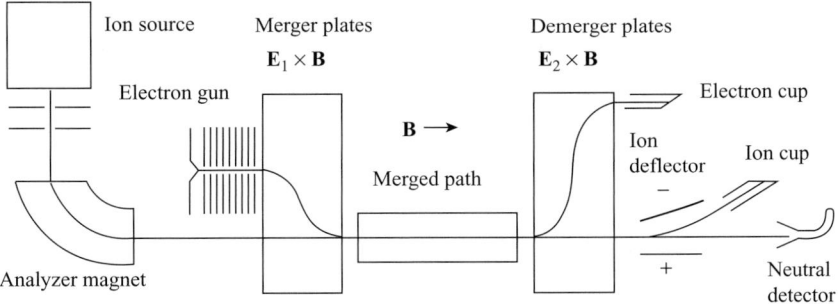

Figure 2.8 Schematic illustration of MEIBE, the single-pass merged-beam apparatus that was used at the University of Western Ontario in Canada to study electron–molecular ion dissociative recombination. The merged path is 7.62 cm long. The ion beam is produced in a 450 keV Van de Graaf accelerator (High Voltage Engineering Corporation Model AN400). Typical electron currents I_e are 15 to 30 μA, whereas the ion current is kept below 5 nA. (Reproduced with permission from R. A. Phaneuf, C. C. Havener, G. H. Dunn, and A. Müller, "Merged-beams experiments in atomic and molecular physics," *Rep. Prog. Phys.* **62**, pp. 1143–1180 (1999), IOP Publishing Limited.)

perpendicular electric field. In the $\mathbf{E} \times \mathbf{B}$ field the electrons exhibit cycloidal motion, which is a combination of a steady motion along the z-axis, a cyclotron motion around the magnetic field, and a drift perpendicular to $\mathbf{E}$ and $\mathbf{B}$. Electrons with different velocities spend different times in the $\mathbf{E} \times \mathbf{B}$ region and thus dispersion occurs. The drift causes the electrons to exit the crossed field region displaced from where they entered. Stamatovic and Schultz (1968, 1970) obtained an electron energy resolution (FWHM) of 0.02 eV. The trochoidal analyzer shown in Fig. 2.8 not only provides a solution to the merging and de-merging problem, but also gives an excellent energy resolution. Stamatovic and Schultz (1970) showed that the energy resolution is proportional to the sum of the entrance and exit apertures, which means that one can obtain a high-energy resolution by using small apertures. Auerbach *et al.* (1977) pointed out that no attempt was made to obtain a highly monochromatic electron beam; in a later paper (Mul & McGowan 1979a), the resolution had been improved from 20 meV to 5 meV at low center-of-mass energy and, although it was not explicitly stated, such improvements were most likely to have been achieved by a better collimation of the beam. Not only does this make the electron beam more monochromatic, it also reduces the $\Delta\theta$ factor in Eq. (2.5).

2.1.2 Cross section measurement

In the previous section, Eqs. (2.10)–(2.12) described how an expression for the rate of interaction, R, can be developed for the simplified situation shown in Fig. 2.5.

Let us now return to the situation shown in Fig. 2.1, i.e., we assume that there is no velocity spread in either the electron or the ion beam. However, we make no assumption concerning the overlap of the merged beams, nor do we assume that the beams are uniform in space.

The rate of interaction dR in a volume element dV is defined through the relation

$$\frac{dR}{dV} = n_e n_i v_r \sigma, \tag{2.15}$$

where the quantities on the right hand side are defined in the previous section. The particle densities can be expressed in terms of particle flux, j_e and j_i, by the relations

$$n_e = \frac{j_e}{v_e e} \tag{2.16}$$

and

$$n_i = \frac{j_i}{v_i e}. \tag{2.17}$$

The electron and ion currents are obtained by integrating the particle flux over the cross sectional areas of the beams:

$$I_e = \int_{S_e} j_e dS_e, \tag{2.18}$$

$$I_i = \int_{S_i} j_i dS_i, \tag{2.19}$$

where S_e and S_i are the cross sectional areas of the electron and the ion beams, respectively. Integrating Eq. (2.15) over the entire interaction volume and making use of Eqs. (2.16)–(2.19) gives

$$R = \int_V \frac{dR}{dV} dV = \frac{v_r}{v_e v_i} \sigma \frac{I_e I_i}{e^2} \frac{\iiint j_e(x,y,z) j_i(x,y,z) dx dy dz}{\int_{S_e} j_e dS_e \int_{S_i} j_i dS_i}. \tag{2.20}$$

The factor

$$F = \left(\frac{\iiint j_e(x,y,z) j_i(x,y,z) dx dy dz}{\int_{S_e} j_e dS_e \int_{S_i} j_i dS_i} \right)^{-1}, \tag{2.21}$$

is called the form factor. Since the beams are merged, the form factor is sometimes defined after integration over the merged path (along the x-axis in Fig. 2.1):

$$F^* = Fl = \left(\frac{\iint j_e(y,z) j_i(y,z) dy dz}{\int_{S_e} j_e dS_e \int_{S_i} j_i dS_i} \right)^{-1}, \qquad (2.22)$$

where l is the length of the merged region. The cross section can be expressed as

$$\sigma = \frac{R v_e v_i e^2 F^*}{v_r I_e I_i l}, \qquad (2.23)$$

which is the form given by Auerbach *et al.* (1977) and Phaneuf *et al.* (1999).

It remains to be seen how Eq. (2.23) is related to Eq. (2.12). The latter equation was derived based on the assumption that the electron beam is uniform and completely overlaps the ion beam. This means that j_e can be taken outside the integration sign and F^* is reduced to

$$F^* = \frac{\int_{S_e} j_e dS_e \int_{S_i} j_i dS_i}{\iint j_e(y,z) j_i(y,z) dy dz} = \frac{\iint j_e(y,z) dy dz \iint j_i(y,z) dy dz}{\iint j_e(y,z) j_i(y,z) dy dz}$$

$$= \frac{j_e \iint dy dz \iint j_i(y,z) dy dz}{j_e \iint j_i(y,z) dy dz}. \qquad (2.24)$$

Thus, the form factor F^* is reduced to the area of the electron beam, A_e. Finally, by making use of

$$n_e = \frac{I_e}{v_e A_e e}, \qquad (2.25)$$

we obtain

$$R = \frac{v_r \sigma n_e I_i l}{e v_i}. \qquad (2.26)$$

This expression is identical to Eq. (2.12) when the beams are assumed to be monoenergetic, in which case $\alpha(v_d) \equiv v_r \sigma$, and the ion charge state is $q = 1$.

It is obvious from Eq. (2.23) that it is of critical importance to determine the form factor in order to measure absolute cross sections, and from Eq. (2.26) it follows that it is advantageous if the ion beam can be merged inside a uniform electron beam, since the determination of the form factor then becomes trivial. This is also true in the reverse situation, i.e., a large uniform ion beam and a smaller, totally overlapped electron beam, but this is technically more difficult to achieve.

The form factor can be determined by measuring the horizontal and vertical beam profiles along the axis of the merged beams. In the single-pass apparatus shown in Fig. 2.8 this was done by means of a scanning system, which measures the beam profiles at three different locations in the section where the electron and ion beams are merged. The procedure has been described in detail by Keyser *et al.* (1979). There is a factor of 2 error in this paper, which means that cross sections measured between 1977 and 1985 using MEIBE are a factor of 2 too large (Mitchell 1990a).

A prerequisite for determining an absolute cross section is that the electron and ion currents are known. It follows from Fig. 2.8 that these currents are measured by means of Faraday cups. A single-pass merged-beam experiment has the advantage that the ion current can be measured with a Faraday cup, which is a simple, accurate, and robust technique, whereas an experiment in which the ion beam circulates in a closed orbit requires more advanced techniques. This will be described in more detail in Section 2.2.

The neutral particles produced in dissociative recombination events must be counted in order to determine an absolute cross section, as follows from Eq. (2.23). It is obviously an advantage if all neutrals produced in the interaction region hit the detector that the detector has 100% detection efficiency and that background processes that can obscure the recombination signal are negligible. This ideal situation can be achieved in some cases in ion storage rings, but not in the single-pass apparatus of Auerbach *et al.* (1977). The first point poses no problem; if the detector is sufficiently large, it will collect all neutrals produced in the interaction region. The neutral detector in Fig. 2.8 is a surface barrier detector, which is shown schematically in Figure 2.9. It works in the following way. The neutral particles impinging on the detector create electron–hole pairs in the solid semiconductor material (silicon), and these electron–hole pairs are collected by an electric field. A thin layer of gold is evaporated onto the silicon crystal, so that a semiconductor junction is created. A depletion zone is formed at the interface in which there are no mobile charge carriers. Ionizing radiation entering the depletion zone creates electron–hole pairs that are removed from the zone by an electric field, and a current signal proportional to the ionizing radiation is formed. The depletion zone can be extended by applying a reverse-bias voltage to the junction, usually of the order of 50 V. If the depletion zone is sufficiently thick, the particles are stopped and electron–hole pairs are created, with the number of pairs proportional to the particle energy. If electrodes are placed on either side of the junction, a voltage difference proportional to the particle energy will occur. After amplification, the signal is transferred to a multichannel analyzer. A much more detailed description is given by Leo (1987). Figure 2.10 shows a pulse-height spectrum taken from a study by Mitchell and Hus (1985) of the dissociative recombination of CO^+ using MEIBE. In this experiment, a 400 keV beam of CO^+ was merged with a beam of 8 eV electrons. Dissociative recombination leads to the formation of C atoms and O atoms, and they impinge on the detector virtually simultaneously. The combined effect of a hit by these

2.1 Merged beams

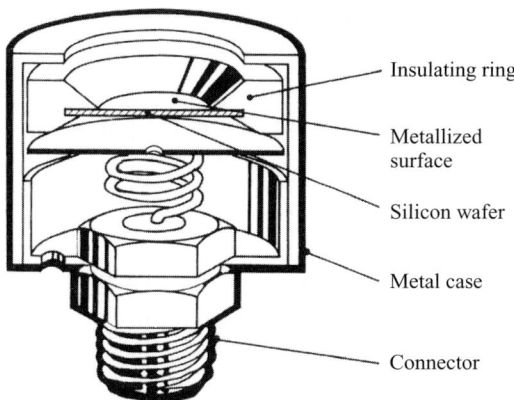

Figure 2.9 Schematic diagram of a surface barrier detector. The semiconductor material is silicon, which has the advantage that it can be operated at room temperature (other materials require cooling in order to operate). The silicon crystal is sensitive to radiation damage, which limits the long term use of the detector. (Reproduced from *Techniques for Nuclear and Particle Physics Experiments*, 1987, W. R. Leo, Springer-Verlag, Berlin, Fig. 10.12 (from EG&G – Ortec Catalog 1983–1984). With kind permission of Springer Science and Business Media.)

two atoms is a peak corresponding to the full beam energy, 400 keV. Although the interaction region is kept at ultrahigh vacuum, 5×10^{-10} Torr, the residual gas is a target which is denser than the electron beam. Collisions of CO^+ ions with residual gas molecules (probably mainly molecular hydrogen) lead to the formation of both charged and neutral products. The ion deflector in Fig. 2.7 prevents the ions from reaching the surface barrier detector, so only neutral products contribute to the pulse height spectrum in Fig. 2.10. The following reactions contribute in addition to dissociative recombination:

$$CO^+ + M \rightarrow C^+ + O + M, \tag{2.27}$$
$$CO^+ + M \rightarrow C + O^+ + M, \tag{2.28}$$
$$CO^+ + M \rightarrow C + O + M^+, \tag{2.29}$$

where M is a residual gas molecule. The O and C atoms that result from the first two of these reactions give an unresolved peak around 200 keV (composed of peaks at 12/28 and 16/28 of the full beam energy). Thus, the energy resolving capability of a surface barrier detector makes it possible to separate the background from processes (2.27) and (2.28) from the signal by a simple pulse-height analysis. This is not possible with the background from process (2.29). There are two ways to reduce this background: lower the pressure in the interaction region, i.e., reduce the density of M, or increase the ion beam energy. Whereas the first action is obvious and needs no further comments, the second action requires a more detailed explanation. Charge-transfer processes such as (2.29) are strongly energy-dependent, as Stearns

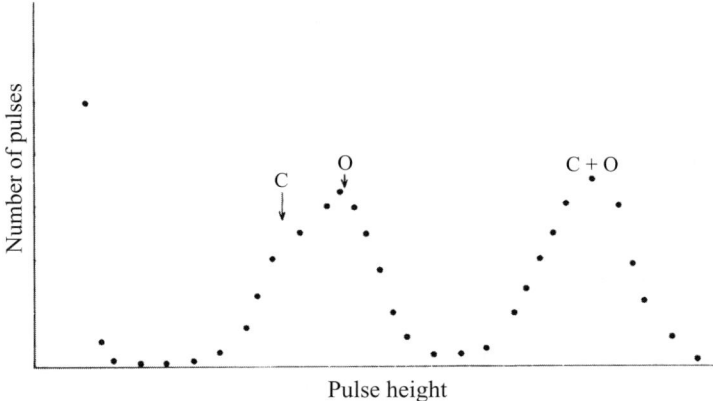

Figure 2.10 Pulse-height spectrum taken at MEIBE in a study of CO$^+$. The C and O peaks derive from collisions of CO$^+$ with residual gas molecules, whereas the peak at C+O is composed of events from dissociative recombination *and* collisions in the background gas. The latter process dominates and probably accounts for 85–90% of the peak. (Reproduced with permission from J. B. A Mitchell and H. Hus, "The dissociative recombination and excitation of CO$^+$," *J. Phys. B* **18**, pp. 547–555 (1985), IOP Publishing Limited.)

et al. (1971) found in an experiment in which they studied dissociation of fast HeH$^+$ ions passing through molecular hydrogen and other gases. The maximum energy available in MEIBE is 450 keV, but as we shall see in Section 2.2, the high energy available in storage rings leads to a strong reduction of the charge-transfer process. The signal-to-background in the CO$^+$ experiment (Mitchell & Hus 1985) is not given in the paper, but in an experiment with H$_2^+$ (Hus *et al.* 1988) it was reported as 18% at $E_{CM} = 0.01$ eV. Since 400 keV H$_2^+$ ions are faster than CO$^+$ ions at the same energy, it is likely that the signal-to-background ratio is lower for CO$^+$.

In experiments where four out of five events arise from background processes, and only one out of five is a recombination event, signal recovery is important. In merged-beam experiments this is usually done by modulating the two beams while recording the count rate. The modulation must be sufficiently rapid (typically 1 kHz) so that fluctuations in the density of the background gas are damped out. Phaneuf *et al.* (1999) gave an example (Figure 2.3 in their paper) of how the signal is recovered by modulating both beams and using gated counters. Auerbach *et al.* (1977) modulated only the electron beam and took the difference between the number of counts with $(S + B)$ and without (B) the electron beam. The true neutral signal is then $S_{\text{true}} = (S + B) - B$, with an uncertainty $\delta S_{\text{true}} = (S + 2B)^{1/2}$, as is characteristic of Poisson statistics.

The detection efficiency of the surface barrier detector in MEIBE is reported to be 100% (Auerbach *et al.* 1977), though this was inferred rather than measured.

This is probably a valid assumption considering the high ion beam energy. For a low-energy beam (<10 keV), for which surface barrier detectors are not used, the determination of the efficiency of the neutral-particle detector is a difficult task (Dunn et al. 1984). The assumption of 100% detection efficiency has been questioned for molecules close to the upper mass limit (32 amu) of MEIBE (Le Padellec, Sheehan, & Mitchell 1998), but measurements by Sheehan, Lennard, and Mitchell (2000) have shown that the detection efficiency is indeed 100% for carbon and oxygen in the energy range 0.2–1.0 MeV.

In a review, McGowan and Mitchell (1984) summarized advantages and disadvantages of the different techniques used to study dissociative recombination (all mentioned in Chapter 1). It is worth ending this section by reproducing this list of the advantages and disadvantages of the merged-beam technique.

> Advantages: (i) very wide energy range; (ii) high energy resolution; (iii) absolute cross section measured; (iv) product branching ratios can be measured; (v) long interaction path gives high signal count rate.
> Disadvantages: (i) up to now reactant-ion excitation states have not been well determined; (ii) in the present form, limited to molecular weights less than 32.

Point (i) of the advantages follows directly from Eq. (2.2); there are no technical problems in increasing the center-of-mass energy by just increasing the acceleration voltage of the electron gun. Points (ii) and (iii) have been discussed extensively, and point (iv) will be discussed in Section 2.2; point (v) is also obvious from Fig. 2.7, in particular if a comparison with the situation of inclined beams is made. Concerning the disadvantages, we will see in Section 2.2 how point (ii) is addressed by the development of ion storage rings. Point (i) is very important and deserves a subsection of its own.

2.1.3 Ion sources

In the previous sections we treated the molecular ion as a classical object with no internal structure. Following the procedures outlined in these sections, a merged-beam experiment will deliver a result which can be compared with a theoretical calculation, compared with other experimental results, or used in applications. The important point, however, is that a molecular ion is not a classical object but a quantum mechanical object with an internal energy structure. The molecular ion can contain rotational and vibrational excitation, and sometimes also electronic excitation. This causes problems in the situations discussed above, problems that are present for all experimental techniques, not just the one using merged beams. The better the internal state distribution is characterized in any given experiment, the more useful the result will be.

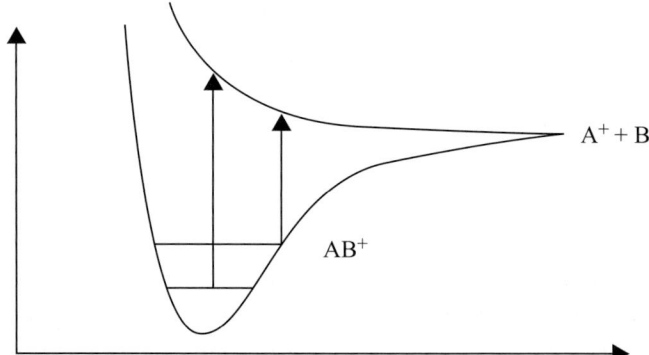

Figure 2.11 Generic potential curves for the molecular ion AB$^+$. The arrows show electronic excitation from the two lowest vibrational levels in the electronic ground state.

In the single-pass merged-beam experiment shown in Fig. 2.8, the time between ion creation and ion destruction is very short, typically in the microsecond range. Thus, the ions exposed to the electron beam will have the same internal state distribution as when they were formed in the ion source. In the original MEIBE (Auerbach *et al.* 1977), a radio-frequency discharge source was used to produce H_2^+ ions. It is known that such a source produces H_2^+ ions populating all bound vibrational levels, with a distribution probably similar to that measured by von Busch and Dunn (1972). Thus, the experiment is fairly well defined as far as the reactant-ion excitation states are concerned, but the result is not easy to compare with theoretical calculations. Auerbach *et al.* (1977) also studied H_3^+, but for this ion they had no idea about the internal state distribution.

Peart and Dolder (1974b) systematically explored five different ion sources in order to find the one which gave the optimal production of de-excited H_3^+ ions. They used the process of dissociative excitation to evaluate the performance of the ion sources:

$$H_3^+(v) + e^- \rightarrow H^+ + 2H + e^-. \tag{2.30}$$

This process has a threshold because the electron has to have a certain minimum energy in order to excite $H_3^+(v)$ to a repulsive, electronically excited state that leads to the formation of a proton. Since this procedure has also been used later for other molecular ions, Figure 2.11 shows the principle by means of generic potential curves for a hypothetical molecular ion AB$^+$. It illustrates how the onset of production of A$^+$ can be used as an indicator of the population of vibrational levels in the electronic ground state. It is a somewhat crude method which can be used to ensure that only the few lowest vibrational levels are populated, but not that the ion is in only its zeroth vibrational level. Peart and Dolder (1974b) found that a relatively high pressure of 0.1 Torr in the ion source caused collisional deactivation of vibrationally excited H_3^+, and used this source in a study of dissociative recombination of H_3^+ (Peart & Dolder 1974c).

The ion-pair formation process

$$H_3^+ + e^- \rightarrow H^- + H_2^+ \qquad (2.31)$$

was studied in a storage ring experiment (Kalhori *et al.* 2004), and a comparison was made with earlier results from inclined- and merged-beam experiments (Peart, Foster, & Dolder 1979, Yousif, Van der Donk, & Mitchell 1993). Peart, Foster, and Dolder (1979) used a low-pressure duoplasmatron ion source in their inclined-beam experiment, and clearly stated that they expected the pressure to be too low to cause collisional deactivation of the vibrational energy. Yousif, Van der Donk, and Mitchell (1993) used a high-pressure source which they anticipated would give H_3^+ ions in only their lowest vibrational level. However, a comparison with the storage ring data and theoretical calculations show that the H_3^+ ion beam used by Yousif, Van der Donk, and Mitchell was contaminated with vibrationally excited H_3^+ ions (see Section 7.2.4).

In the first merged-beam experiment on H_2^+, McGowan, Caudano, and Keyser (1976) used a mixture of hydrogen and helium in the ion source in order to take advantage of the exothermic reaction $H_2^+ (v) + He \rightarrow HeH^+ + H$ to remove vibrationally excited H_2^+. Since this reaction is exothermic only for $v > 2$, it was thought that a hydrogen/helium mixture in the ion source would result in a beam of H_2^+ ($v \leq 2$). The observation of resonances in the dissociative recombination cross section of H_2^+ was considered further evidence for successful removal of vibrationally excited H_2^+. In later experiments using MEIBE, the resonances could not be reproduced, and it has been speculated that the resonances observed in the first experiment (McGowan, Caudano, & Keyser 1976) were facilitated by a fortuitous choice of ion source conditions (Hus *et al.* 1988). Instead Sen, McGowan, and Mitchell (1987) and Hus *et al.* (1988) developed a radio-frequency trap source based on a design by Teloy and Gerlich (1974). The radio-frequency field confined the ions for milliseconds, which allowed a sufficient number of collisions to occur before the ions were extracted from the source. In a later experiment, the dissociative recombination cross section of H_2^+ ($v = 0$) was measured using MEIBE (Van der Donk *et al.* 1991) by a careful use of the trap source. The window resonances in the cross section were firmly established in the experiment, although their amplitudes remain difficult to reconcile with the resolution expected in MEIBE.

Other ion sources have been developed to be used in merged-beam experiments in ion storage rings, as discussed in the next section.

2.2 Ion storage rings

The merged- and inclined-beam techniques discussed in Section 2.1 are characterized by the fact that the ions are destroyed after having passed the interaction region. The destruction occurs when they are collected in a Faraday cup so that the current can be measured. In the 1980s the development of ion storage rings for atomic and molecular physics started, inspired by the storage ring at CERN for antiprotons, LEAR (Baird *et al.* 1990). The driving force behind this development was the desire to develop techniques applicable to the study of highly charged atomic ions, and the philosophy was taken from LEAR: if a particular ion species is difficult and/or expensive to produce, such as antiprotons or highly charged ions, a storage ring is a suitable device to achieve the enrichment and improvement of beam quality needed for experiments. This works well also for molecular ions (Larsson 2003).

An ion storage ring combines a source for ion production with a ring system in which the ions circulate in closed orbits for time scales much longer than the revolution time. This means that molecular ions passing an electron target without recombining return to the electron target after a short time ($\sim$1–10 µs) and can recombine during their second, third, etc. passage through the electron target. Since the electron beam in a storage ring intersects the ion beam at an angle $\theta = 0°$, the two beams are merged and the formulas developed in Section 2.1 apply. The ion storage rings address the two disadvantages of the merged-beam technique discussed in Section 2.1.2 in that they allow better control of the reactant-ion excitation states and push the mass limit upwards. But they also offer several other advantages, which will be discussed in Section 2.2.1. The ion storage rings built and commissioned in the late 1980s and early 1990s relied on magnetic fields for confining the ions, whereas the electrostatic storage rings of the late 1990s and early 2000s rely on electric fields. The difference is that magnetic confinement relies on the Lorentz force, which contains a velocity dependence, while electric confinement relies on the electric charge and the strength of the electric field. The velocity dependence of the Lorentz force causes problems for storage of heavy biomolecular ions.

The magnetic ion storage rings that have been used for the study of dissociative recombination are, in order of when they became operational: the Test Storage Ring (TSR) at the Max-Planck-Institute for Nuclear Physics in Heidelberg, Germany (Habs *et al.* 1989); the Test Accumulation Ring for the Numatron Accelerator Facility (TARN II) at the Institute for Nuclear Study of the University of Tokyo, Japan (Tanabe *et al.* 1991); the Aarhus Storage Ring Denmark (ASTRID), at the University of Aarhus, Denmark (Møller 1991); and the Cryogenic ion source Ring (CRYRING) at the Manne Siegbahn Laboratory, Stockholm University, Sweden (Abrahamsson *et al.* 1993). It is hardly a coincidence that most of the storage rings were built at institutes with a tradition in nuclear physics, and it is clear that inspiration and know-how was coming from CERN and other high-energy physics laboratories. It is an example of how the membership of small countries like Denmark and Sweden in an international high-energy physics laboratory (CERN) was of benefit to the accelerator physics and atomic and molecular physics communities in these countries. TARN II has now been decommissioned, and the future lifetime of CRYRING is unclear. ASTRID was designed for a dual purpose, as an ion storage ring and a synchrotron radiation source. Initially the ion and electron modes were used equally for beam time delivery to researchers, but now the electron mode dominates.

Since the vast majority of dissociative recombination studies have been performed at the magnetic storage rings listed above, this section will focus on magnetic rings and electrostatic storage rings will be described in less detail.

2.2.1 Magnetic ion storage rings

Figure 2.12 shows an artist's view of CRYRING at the Manne Siegbahn Laboratory. The other magnetic storage rings are very similar and differ in details only. Parameters for ASTRID, CRYRING, TARN II, and TSR are given in Table 2.1.

The TSR has an additional feature not available at the other storage rings, namely the possibility to extract part of the stored ion beam. The advantage is that the extracted beam can be used to probe the excitation states of the stored beam by means of a Coulomb explosion imaging (CEI) technique. The ions collide with a thin foil (<100 Å) which strips off all electrons, so that when the ions leave the foil they repel each other due to the Coulomb force; a few meters from the foil they have already separated by a few centimeters and are recorded by an imaging detector. Decoding the images allows the reconstruction of the ion's vibrational motion when it passed through the foil; in other words, the CEI allows a snapshot of the ions' internal state distribution at any given time.

At CRYRING, ions are produced in an ion source kept at a platform voltage of 40 kV, mass selected and passed through a radio-frequency quadrupole (RFQ) device. If the charge-to-mass ratio is larger than 0.25, the RFQ acts as an accelerator, and the ions leave the RFQ with an energy of 300 keV amu^{-1}. When the ratio is smaller than 0.25, which is the case for most molecular ions except the lightest, the RFQ acts as a focusing device and does not affect the ion energy. The next step is injection of ions into the ring. This is accomplished by multiturn injection, which involves the application of a time-dependent deformation of the main orbit, so that several orbits are stored side by side until the horizontal acceptance of the ring is filled. The procedure can be compared to winding a thread on a spool. The ions are now in the ring, circulating with an energy of either 40 keV amu^{-1} or 300 keV amu^{-1}. It is desirable to increase the energy before an experiment starts. The advantages of higher beam energy were discussed in Section 2.1.2.

Table 2.1. *Parameters for the ion storage rings that have been used to study dissociative recombination*

Storage ring	Circumference (in m)	Magnetic rigidity (T m)	Experiments started
ASTRID	40	2.0	1991
CRYRING	52	1.44	1992
TARN II[a]	78	6.1	1990
TSR[b]	55	1.5	1988

[a] Decommissioned.
[b] In the TSR it is possible to extract part of the beam. If the extracted ion beam is sent through a thin foil, the molecular ion Coulomb explodes, and the atomic ions can be measured with a position-sensitive detector (see Amitay *et al.* (1998, 1999)).

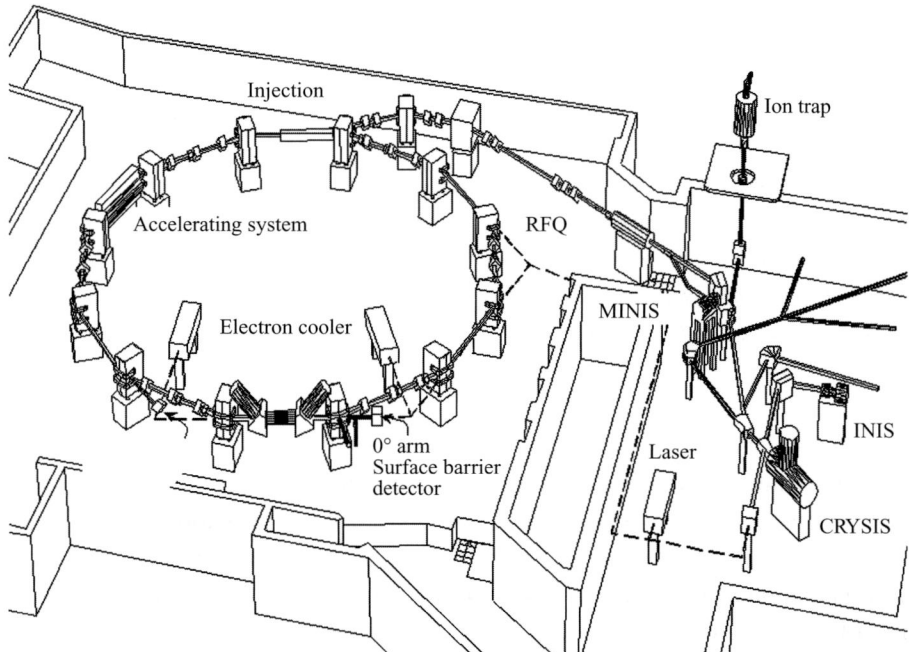

Figure 2.12 Schematic view of CRYRING. Molecular ions are created in the ion source MINIS, accelerated and mass selected. In some cases they are further accelerated by the Radio Frequency Quadrupole (RFQ), and injected into the ring. The interaction region, i.e. the region in the electron cooler where the ion and electron beams are merged, is 85 cm. (Reprinted from *Int. J. Mass Spectrom. Ion Proc.* **149/150**, M. Larsson, "Dissociative recombination in ion storage rings," pp. 403–414, Copyright (1995), with permission from Elsevier.)

The electron capture background is reduced and it is easier to produce a velocity-matched electron beam if the ions have a higher velocity. Additional advantages will be discussed below. Thus, the ions are accelerated in a radio frequency cavity, where an alternating electric field increases the velocity of the ions each time they pass the cavity. The ions are kept in closed orbits by dipole, quadrupole, and sextupole magnets, and in order to keep the ions in orbit during acceleration, the magnets must be operated synchronously with the increase of the ion energy. The maximum energy is determined by the magnetic rigidity, which is the product of the maximum magnetic field, B, and the radius of curvature, ρ, of the dipole magnets. In CRYRING, the radius of curvature is 1.2 m, and hence from the rigidity $B\rho$ given in Table 2.1, the maximum magnetic field is 1.2 T. The maximum ion beam energy, which is set by the magnetic rigidity, is given by $96(q^2/A)$ MeV, where q is the charge and A is the mass. The limitation with magnetic storage is now obvious; if A is increased the maximum energy is lowered, which means that the ions move more slowly. At some point the ions will move too slowly for it to be possible to keep them in orbits around the ring. It is difficult to estimate the value of A at which beam instabilities will occur; this is something which must be determined empirically. In comparison, electrostatic storage is velocity-independent and there is no upper mass limit (if we disregard the gravitational force).

Acceleration of the ions takes about 1 second, after which the ions are left freely coasting in the ring. At this point it is interesting to look at the major loss mechanism. Process (2.29), electron capture from residual gas molecules, is significantly reduced at the high energy at which the ions are stored (3.4 MeV in the case of CO^+), but processes (2.27) and (2.28) together with charge stripping

$$CO^+ + M \rightarrow C^+ + O^+ + M \qquad (2.32)$$

lead to losses. These processes do not cause any problems as far as measurement of the dissociative recombination cross section is concerned, but severely limit the beam lifetime if the pressure in the ring's vacuum system is too high. CRYRING is usually kept at a pressure below 10^{-11} Torr, and the other rings in Table 2.1 are similarly operated at ultrahigh vacuum, and this is essential in allowing storage times of up to 1 minute.

Beam cooling

The storage of the ions in the form of a freely coasting beam over a time scale that can be made very long, tens of seconds, has the advantage that rotationally and vibrationally excited levels and electronically excited metastable states present in the stored ion beam are removed by spontaneous emission of radiation, provided that the radiative lifetimes are shorter than the storage time. This is always the case for infrared active vibrational modes in polyatomic molecular ions. Modes usually labeled as infrared inactive must be checked on an individual basis, and homonuclear diatomic ions such as N_2^+ and O_2^+ do not cool vibrationally as a result of storage. The rotational levels in diatomic molecules with a dipole moment will be populated essentially according to a Maxwell–Boltzmann distribution characterized by a temperature equal to the temperature of the vacuum system, i.e., 300 K, as demonstrated for CH^+ by a TSR team (Hechtfischer et al. 1998). But as shown by Kreckel et al. (2002), rotational lifetimes in an equilateral triangular molecule like H_3^+, which lacks a dipole moment, can be much longer than 1 minute. Thus, even with the obvious advantage offered by the storage ring technique of removing internal excitations, much care must be exercised in order to characterize the internal state distribution of the molecular ions as precisely as possible.

All the magnetic rings in Table 2.1 use an electron beam that serves both as an electron target and as a device for translational beam cooling. In TSR an additional electron beam has been installed, making it possible for the first time to separate the two functions. Figure 2.12 shows how one straight section of CRYRING is occupied by the electron cooler. Translational cooling of the stored ion beam occurs when the ion beam is merged with the cool electron beam and the two beam velocities are equal (i.e., $v_d = 0$ and $E_d = 0$). The velocity spread in the ion beam causes some ions to move faster than the electron beam and some ions to move slower. The Coulomb interaction between ions and electrons leads to a frictional force that decelerates those ions moving faster than the electrons and accelerates those ions moving slower. An alternative way of expressing this is that heat is transferred from the translationally hot ion beam to the cold electron beam, which leaves the interaction region slightly warmer than when it entered. The ion beam leaves the interaction regions slightly colder, makes one turn around the ring, and is again merged with the cool electron beam. Since this can be repeated at a megahertz rate for a light molecular ion, the phase-space volume of the ion beam is reduced within a few seconds leading to a stored beam of small momentum spread, small divergence, and small cross-sectional area.

Electron cooling has been reviewed by Parkhomchuk and Skrinsky (1991), and was first observed for molecular ions during the commissioning of CRYRING in 1992. Beam cooling in the longitudinal direction is observed by recording the frequency spectrum of the stored ion beam. Each particle in the beam gives rise to signal in a pick-up (Schottky detector) positioned in one straight section of the storage ring. In an ideal case, each ion gives a δ-function signal during each passage. By Fourier transforming this time-domain spectrum into the frequency domain, a contribution is obtained at each harmonic of the revolution frequency. The longitudinal momentum spread causes the ions to have slightly different revolution frequencies, something which gives rise to a Schottky band. Figure 2.13 shows Schottky spectra of an uncooled and a cooled 42 MeV D^+ beam in CRYRING

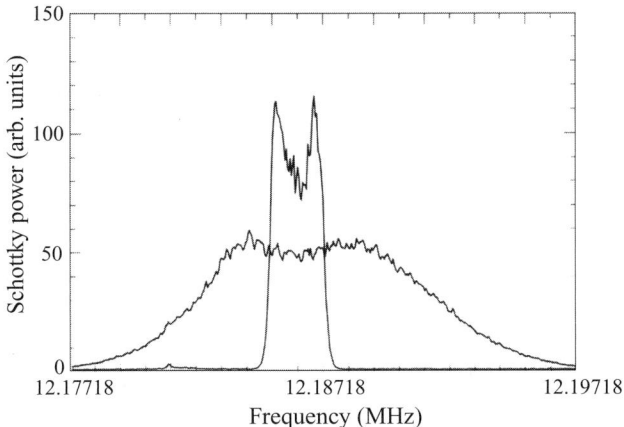

Figure 2.13 Schottky spectrum of uncooled and cooled 42 MeV D$^+$ beam in CRYRING. (Reproduced with permission from M. Larsson, "Atomic and molecular physics with ion storage rings," *Rep. Prog. Phys.* **58**, pp. 1267–1319 (1999), IOP Publishing Limited.)

obtained by detection of image charges induced in the electrodes of a Schottky detector.

Transverse cooling can be observed by using a detector as in Fig. 2.5 which records the position of dissociative recombination reaction products. The decrease of the ion beam diameter is immediately observed as a shrinking of the area hit by neutral particles. Beam cooling is not always essential, although always advantageous, in dissociative recombination experiments since the resolution in cross section measurements is primarily determined by the electron beam rather than the ion temperature. There are, however, examples where the experiments could not have been carried out without beam cooling. One such example is the determination of dissociative recombination cross section for H_2D^+ (Datz *et al.* 1995a) in CRYRING, another is the imaging of reaction products of H_2O^+ (Thomas *et al.* 2002), also in CRYRING. Datz *et al.* (1995a) studied dissociative recombination of H_2D^+ and were confronted with the problem of the presence of D_2^+ ions in the H_2D^+ beam. Since H_2D^+ was produced in the ion source by ion molecule reactions in a gas mixture of HD and H_2 gas, it was impossible to avoid also producing D_2^+. Because of the relative mass difference of 3.7×10^{-4} between the two types of ions, there was a slight difference in orbit around the rings and hence in the circulation time. The Schottky spectrum recorded for the electron cooled beam revealed two separate peaks, one arising from H_2D^+ ions and the other from D_2^+ ions. With the storage rings acting as a high-resolution mass spectrometer, Datz *et al.* (1995a) could optimize the ion source for minimum production of D_2^+ and measure the fraction of these ions in the H_2D^+ beam, something which made possible the first, and so far only, determination of the absolute cross section for dissociative recombination of H_2D^+.

The electron beam can also cause vibrational and rotational cooling. Superelastic collisions transfer vibrational energy from the ion to the electron:

$$H_2^+(v) + e^-(E) \to H_2^{**} \to H_2^+(v') + e^-(E + \Delta E)(v' < v, \Delta E > 0). \quad (2.33)$$

The initial step is the same as in dissociative recombination, but instead of stabilization of the electron capture by molecular dissociation, autoionization occurs and the molecular ion ends up in a lower vibrational level while the electron acquires kinetic energy.

Because radiative cooling is so effective for polar molecules, superelastic collisions occur experimentally for molecular ions lacking a dipole moment, i.e., molecules that do not cool radiatively. Effects of vibrational cooling were observed for D_2^+ (Larsson et al. 1994) and H_2^+ (Van der Zande et al. 1996), but the mechanism was not identified until later (Tanabe et al. 1999, Saito et al. 2000). Krohn et al. (2000) used the CEI technique to study superelastic collisions in H_2^+ in more detail and showed that they could account for the vibrational deexcitation of H_2^+. The work was extended to D_2^+ (Krohn et al. 2003). Studies of H_3^+ (Larsson et al. 2003) and D_2H^+ (Lammich et al. 2003a) showed that the rate of dissociative recombination was affected by how long the ions were stored, and that the rate was likely to depend on the rotational temperature. Evidence of a rotational temperature lower than the temperature of the vacuum system (300 K) was found for D_2H^+. Because in these experiments, electron beams with a temperature below 300 K were used, they lend support to the conjecture that rotational deexcitation is also induced by the electron beam.

To summarize, in an ion storage ring three types of cooling occur: radiative cooling which usually gives a final rotational and vibrational temperature of 300 K, i.e., the ions are in equilibrium with the blackbody radiation emitted from the vacuum pipes; electron cooling of the translational degrees of freedom, which leads to an ion beam of low remittance, i.e., low divergence, low momentum spread and small cross sectional area; and finally electron-induced vibrational and rotational cooling (superelastic collisions).

One may then ask the question: are there no heating mechanisms? Yes, there are. In the first place, an ion which has been cooled to a rotational temperature below 300 K, either by superelastic collisions or because it was cooled already in the ion source, heats up to 300 K provided that it couples to the blackbody background radiation. This has been demonstrated for D_2H^+ by Lammich et al. (2005). In this experiment the rate coefficient in the electron cooler (Eq. (2.11)) was used as an indicator of the rotational temperature. The ions were first cooled by superelastic collisions, then the electron beam was turned into a weak cooling mode, where it was set to $E_d = 10$ eV most of the time. With the rotational cooling source turned off most of the time, the rotational temperature started to increase owing to coupling to the background radiation. The experiment was repeated for H_3^+ and in this case also removal of the cooling source led to an increased rotational temperature. The blackbody radiation does not affect a symmetric molecule like H_3^+, so there must be an additional heating mechanism. The most likely explanation is collisions with residual gas molecules. Such collisions are known from experiments with atomic ions to cause a repopulation of atomic states (Mannervik *et al.* 1999). The effect is very small and shows up for H_3^+ only if the electron beam is tuned off from cooling. This means that in experiments where the cross section is determined as a function of detuning energy, the measurement scheme must be such that the electron cooler is set to $E_d \neq 0$ for time scales shorter than the heating time.

Resolution

The framework developed in Section 2.1.1 for merged beams is the same when applied to the multipass situation in a storage ring. In the first generation of ion storage ring experiments on dissociative recombination (see Larsson (1995a)) the resolution was essentially the same as that in the single-pass experiments of Andersen and Bolko (1990). A major improvement was made when Danared and coworkers (Danared 1993, Danared *et al.* 1994) realized that letting the electron beam expand in a decreasing guiding magnetic field could decrease the transverse electron temperature. The technique was first implemented at CRYRING (Danared *et al.* 1994), and soon after at ASTRID, TARN II, and the TSR. Consider the cathode in Fig. 2.3. If the area of the cathode surface is made 10 times smaller than in the original design, and the magnetic field is 10 times stronger, the electron beam can be expanded by letting it pass to a region of 10 times weaker magnetic field. The transition from the strong-field region to the weak-field region must take place over a distance which is long compared with wavelength of the cyclotron motion the individual electrons make around the magnetic field lines. In the original electron cooler design at CRYRING, a magnetic field of 0.03 T surrounded the cathode, and this field strength was maintained into the interaction region. The 4-cm diameter of the electron beam was maintained from the cathode to the interaction region. Danared *et al.* (1994) decreased the size of the cathode so that the electron beam emitted from the cathode had a diameter of $4/\sqrt{10} = 1.26$ cm while being guided by a 0.3 T magnetic field. The electron beam was then adiabatically transferred to a region where the magnetic field was 0.03 T and the diameter of the beam was increased to 4 cm. The passage of the electrons through a negative field gradient slows down the cyclotron motion. The ratio between the longitudinal magnetic field and the transverse energy spread is an adiabatic invariance, which means that the transverse electron temperature, $kT_\perp$, was reduced from 0.1 eV to 0.01 eV (Danared *et al.* 1994).

In a further development of the electron coolers, superconducting magnets were used to create even stronger magnetic fields, and hence an expansion factor of 100 became possible (Tanabe *et al.* 1995, Danared *et al.* 1998). The transverse electron temperature in CRYRING has been estimated to be $kT_\perp = 2 \pm 0.5$ meV, thus it is slightly higher than the theoretical value of 1.0 meV. The reason for this is that the space charge of the electron beam creates a radial electric field that gives a transverse drift motion to the electrons.

The adiabatic expansion technique has reached the limit set by technology; making the cathode even smaller and the magnetic field higher would be impractical. An alternative approach to making a cold electron beam is to use a laser-irradiated photocathode rather than a thermionic cathode (Habs *et al.* 1988). This avoids the heating of the cathode to 1200 K in order to release the electrons, and the electron cloud formed in front of the cathode is cold from the beginning. It may sound as if it is straightforward to implement this technique, however, extensive research and development were required before it was successfully used at the TSR. It is proper to call this electron beam a target, since it is not intended for phase-space cooling of the ion beam, but only to serve as the target in collision experiments. A conventional, thermionic cathode, electron cooler is used to provide phase-space cooling. The team at the TSR has achieved $kT_\perp = 0.5$ meV using the photocathode electron target (Orlov *et al.* 2005).

Cross section measurements

In Section 2.1.2, different expressions were derived to express the rate of interaction, R, in a merged beam. The first expression, Eq. (2.12), was derived under the assumption that the merged beam contains a small ion beam embedded in a uniform electron beam with a finite velocity distribution. A second expression, Eq. (2.23), was derived for monoenergetic beams, but with no assumptions made about how the two beams overlap. Finally we showed the connection between Eqs. (2.12) and (2.23) in the derivation of Eq. (2.26). In ion storage rings, the electron beam is approximately uniform and larger than the ion beam. Thus, Eq. (2.12) can be used as a starting point for deriving cross sections.

It is instructive to consider Fig. 2.5 in the discussion of how dissociative recombination absolute cross sections are measured in ion storage rings. In order of decreasing importance, the following points must be considered.

The number of ions in the interaction regions must be measured. This is more difficult than in a single-pass experiment since a Faraday cup cannot be used: inserting such a device in a storage ring would obviously destroy the circulating beam. Nondestructive beam current measurements in the ring require the use of current transformers (Unser 1981). These devices measure the magnetic field generated by the coasting ion beam. In storage ring experiments in the 1990s, dc current transformers were used, and their sensitivity made it difficult to measure beams smaller than a few hundred nanoamperes. Much progress has been made, and ac transformers are now capable of measuring beams of a few nanoamperes (Paal *et al.* 2003).

The length of the interaction region is not known exactly. Toroidal magnetic fields are used at both ends of the electron cooler in order to guide the electron beam into and out of the electron cooler. This is a technique for merging and demerging that lacks the advantages of the trochoidal electron analyzer and that creates a region where the beams are not parallel and an uncertainty in the length of the $\theta = 0°$ interaction region. In the region where the beams are not parallel, collisions at an energy higher than the nominal one will occur, and the reaction products from these collisions will hit the detector. The effect can be compensated for by modeling the toroidal regions using the measured longitudinal magnetic field in the toroidal region (Lampert *et al.* 1996, Al-Khalili *et al.* 2003).

The detuning energy, E_d, must be precisely known. The determination of E_d consists of two unrelated problems: determining the detuning velocity and hence the detuning energy, and *maintaining* the detuning energy at a fixed value during a measurement. The detuning energy is determined by the voltage applied to the electron cooler cathode corrected for a voltage drop across the cathode. The electron space charge causes the kinetic energy of the electrons in the beam to have a radial dependence. The cathode is operated in a space-charge limited mode and the longitudinal magnetic field prevents the electron beam from blowing up, which implies that the current density distribution is homogeneous. In the center of the electron beam, the kinetic energy is (Kilgus *et al.* 1992):

$$E_e = eU_c - \frac{I_e r_e m_e c^2}{ev_e}[1 + 2\ln(b/a)], \qquad (2.34)$$

where r_e is the classical electron radius, U_c is the applied cathode voltage minus a contact potential, b is the diameter of the beam tube, and a is the electron beam diameter. The electron beam can ionize residual gas molecules, in which case the ions trapped in the electron beam will compensate the electron space charge. This modifies Eq. (2.34) to

$$E_e = eU_c - \frac{I_e r_e m_e c^2 [1 - \xi(E_e)]}{ev_e}[1 + 2\ln(b/a)], \qquad (2.35)$$

where $\xi(E_e)$ is the correction due to trapped slow ions (Rosén *et al.* 1998a). In CRYRING, the electron energy in the beam center has been determined by measuring the ion velocity by means of the Schottky noise signal of the circulating beam and then making use of the circumference of the storage ring to obtain the ion velocity. This gives the electron energy at cooling conditions, i.e., when the two beams are velocity matched. The procedure allows an estimate of $\xi(E_e)$, which was found to be about 0.25 for an electron beam of 185 eV (Strömholm *et al.* 1997). One should note that ξ depends on E_e. The radial dependence of the electron energy is an additional reason to ascertain whether the beams are well aligned; if they are not ions traveling at an angle with respect to the electron beam experience a broader energy distribution.

The problem of maintaining the detuning energy is related to the frictional (drag) force discussed in the section on beam cooling. If the electron beam is kept at an energy which is higher than the ion beam energy, thus $E_d > 0$, the frictional force will accelerate the ion beam leading to a decrease of E_d while data are being recorded. Instead of a measurement at a well-defined value of E_d, a range of detuning energies is sampled. There are three methods of dealing for this problem. First, one can ramp the cathode voltage across $E_d = 0$ while counting the number of particles hitting the detector. The ramping makes it difficult for the ion beam to follow the rapidly changing speed of the electron beam, and closest to $E_d = 0$, where the drag force is largest, one can compensate for the effect during the data analysis by making use of the knowledge of the velocity dependence of the frictional force. A second method is to rapidly switch the electron cooler between two settings, one that gives the desired E_d and the other corresponding to cooling. The former method is more commonly used since it is less time consuming. Third, one can use two different electron beams, one for cooling and one as the target. This is clearly the best solution, since the electron cooler can be used to "lock" the coasting ion beam to a fixed energy, keeping it continuously cooled, while the measurements are done at the other electron beam, the target. This configuration has been installed at the TSR.

The detection efficiency can be lower than 100% for two different reasons. Neutral particles formed in the interaction region must pass a dipole magnet before they reach the detector. The high speed of the particles, up to 10% of the speed of light, in combination with the strong magnetic field in the dipole magnet, up to 1.2 T in CRYRING, leads to a motional electric field which can field ionize Rydberg atoms or molecules. The transverse electric field component is (Simonsson 1991):

$$E_\perp [\text{MV m}^{-1}] = 5A/3q \times E_i [\text{MeV amu}^{-1}]. \tag{2.36}$$

This is a more severe problem in studies of dielectronic recombination, where atoms in high Rydberg states are frequently formed. Atoms making an important contribution to the cross section are field ionized, which leads to a cut-off limit in the cross section.

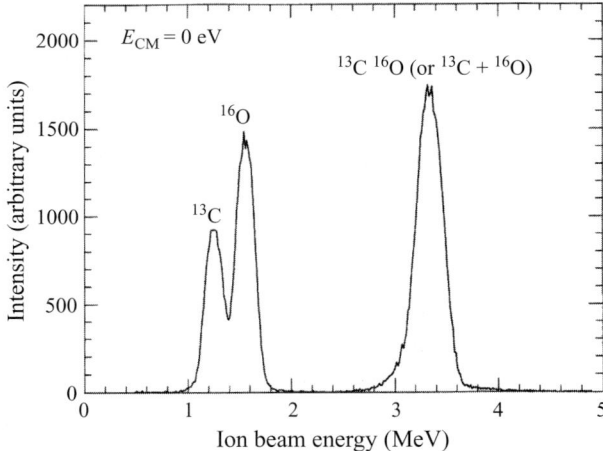

Figure 2.14 Pulse-height spectrum taken at CRYRING in a study of ^{13}CO$^+$. The C and O peaks derive from collisions of CO$^+$ with residual gas molecules, whereas the peak at C + O is totally dominated (>95%) by dissociative recombination events. It occurs at an energy corresponding to the full beam energy of 3.4 MeV. Also note the much improved resolution in the pulse height spectrum as a result of the high beam energy. (Reprinted figure with permission from S. Rosén, P. Peverall, M. Larsson, *et al.*, *Phys. Rev. A* **57**, pp. 4462–4471 (1998). Copyright (1998) by the American Physical Society.)

Another reason for the detection efficiency being lower than 100% occurs when a heavy molecule dissociates into an H atom and a heavy fragment, in which case the H atom takes most of the kinetic energy and can scatter so much in the perpendicular direction that it misses the detector. This can usually be modeled and adjustments made to compensate for it.

Finally there are situations in which the detection efficiency is purposely reduced. In measurements of product branching ratios, a grid with a transmission much less than 100% is placed in front of the detector. This technique is discussed in Section 2.2.1.

The electron beam current is easily measured because the electron beam is demerged from the interaction region and collected.

Figure 2.14 shows a pulse-height spectrum from a study of CO$^+$ in CRYRING. The C + O peak, which in Fig. 2.10 is totally dominated by collisions with residual gas molecules, is here dominated by dissociative recombination, and only a small correction for the electron capture process given in Eq. (2.29) is required. This correction can be made by tuning the electron beam to an energy at which the recombination cross section is negligible while recording the C + O peak.

Strictly speaking, the application of Eq. (2.12) leads to the determination of a rate coefficient, $\langle v_r \sigma \rangle$, rather than a cross section. When the detuning is larger than the transverse electron temperature, the cross section is obtained by dividing the rate coefficient by v_r. For detuning energies smaller than E_d, a deconvolution procedure is required. In the literature one finds different ways of displaying the results. Giving $\langle v_r \sigma \rangle$ on the y-axis is unambiguous, since it is the measured quantity, but can lead to problems if results from different storage rings with different electron temperatures are displayed on one and the same graph. Sometimes the effective cross section, $\langle v_r \sigma \rangle / v_d$, is displayed. Finally one can deconvolute σ and display it. Whatever quantity is used to display the results, it is clear that if the thermal rate coefficient is needed, which is often the case in applications, unfolding σ from $\langle v_r \sigma \rangle$ is required. The rate coefficient, $\alpha(T_e)$, for an isotropic Maxwellian electron-velocity distribution at an electron temperature T_e is obtained by integrating the energy-dependent cross section over the electron energy

$$\alpha(T_e) = \int_0^\infty \frac{8\pi m_e E_{CM}}{(2\pi m_e k T_e)^{3/2}} \sigma(E_{CM}) e^{-(E_{CM}/kT_e)} dE_{CM}. \tag{2.37}$$

This quantity will be referred to as the thermal rate coefficient or constant in order to avoid confusion with the rate coefficient measured in the electron target. Note also that E_{CM} rather than E_d is used. The detuning energy is replaced by the center-of-mass energy as a result of the deconvolution procedure. The advantage of knowing the cross section's energy dependence is that the thermal rate coefficient can be calculated for any desired electron temperature, at least in principle. In practice, the situation is somewhat more complicated. Consider Fig. 2.7, in which the result of a hypothetical measurement of a cross section $\sigma = 10^{-17}/E_{CM}$ is shown. If the transverse electron temperature is exactly 100 meV and the longitudinal temperature is 0, then the $10^{-17}/E_{CM}$ cross section follows directly as a result of the deconvolution procedure. In reality, the transverse temperature is not precisely known, the longitudinal temperature is not 0, and the true cross section may contain resonances at very low energy, which would not show up in the measurement. Let us first assume that the cross section is indeed without resonances, i.e., it follows a simple $1/E_{CM}$ dependence. The deconvolution procedure introduces an error in σ because of the uncertainty in the electron temperature. One way to deconvolute σ is to assume a model cross section, calculate $\langle v_r \sigma \rangle$ with the known parameter of the electron beam used in the experiment, and compare this with the experimental results. The model cross section is varied until optimal agreement with experiment is obtained. This approach was used for H_3^+ (Sundström *et al.* 1994b), and in the derivation of an expression for the cross section, $\sigma = (1.43 \pm 0.02) \times 10^{-15}/E^{1.15 \pm 0.02}$ cm^2

(where E is expressed in eV), the transverse electron temperature was assumed to be 0.125 eV, with the + sign corresponding to $kT_\perp = 0.15$ eV and the − sign to $kT_\perp = 0.10$ eV. The longitudinal electron temperature, which was of the order of 0.1 meV, did not influence the result and was in fact assumed to be zero in the deconvolution procedure.

Clearly the development of ultracold electron beams (Danared et al. 1994) has been important for confidence in the cross section recovery procedure from the measured rate coefficient (as defined in Eq. (2.11)). The cross section can be obtained directly by just dividing $\alpha(v_d)$ by v_d down to detuning energies approaching a few meV, and low-energy resonances can be identified in the same low-energy range. In a comparison of results for HD$^+$ from three different ion storage rings (Al-Khalili et al. 2003), it was shown that $\alpha(v_d)$ measured in CRYRING could be used to obtain the thermal rate coefficient for $T_e \geq 300$ K. The new electron target at the TSR (Orlov et al. 2005) will allow determination of the thermal rate coefficient for HD$^+$ to even lower temperatures without any deconvolution procedure. In some cases the rate coefficient is required for electron temperatures much lower than 300 K. McCall et al. (2004) used the ultracold electron beam in CRYRING in a study of H$_3^+$ and found that the recombination cross section below 0.1 meV was best described by $3.1 \times 10^{-15}/E^{0.76}$ cm^2 (note that the H$_3^+$ rotational distribution was different from the one in the Sundström et al. (1994b) experiment). The cross section did not follow a $1/E^n$ dependence above 0.1 meV, and only a few data points were measured below a detuning energy of 0.1 meV, which made it difficult to determine the exponent for E with confidence. A lower limit of 10^{-8} eV was used in the integration of Eq. (2.37). The extrapolation of the cross section to the lower integration limit was also made based on a $1/E$ cross section, which obviously leads to a substantial difference in the cross section at 10^{-8} eV. However, the thermal rate coefficient is virtually the same whether $1/E^{0.76}$ or $1/E$ is used except when T_e is smaller than 10 K. In the study of McCall et al. (2004), the rate coefficient at $T_e = 23$ K was required, and even at this low temperature the result was insensitive to the $1/E^n$ extrapolation below 0.1 meV. The transverse electron temperature in the experiment was $kT_\perp = 2.0 \pm 0.5$ meV, and the error in the transverse electron temperature was found to introduce a 6% uncertainty in the thermal rate coefficient at 30 K and 2% at 300 K.

Finally, we emphasize that the ion temperature is set by the conditions at which the cross section was measured in the storage ring, in many cases $T_{\text{vib}} = T_{\text{rot}} = 300$ K. Equation (2.37) can be used to calculate the thermal rate coefficient for electron temperatures much higher than 300 K, but if such a result is used in applications, one must bear in mind that Eq. (2.37) does not accommodate vibrational and rotational excitations that occur at elevated temperatures.

Ion sources

It was pointed out earlier that the ability to remove rovibronic excitations in the stored beam by radiative cooling is a distinct advantage with ion storage rings. Ions for which this does not work were mentioned, and superelastic collisions were discussed as a means of achieving rovibrational cooling. However, there are situations in which neither of these methods is suitable, and this has prompted the development of ion sources at storage rings. We will give three examples in this section.

A homonuclear molecular ion such as O_2^+ does not cool rotationally in a storage ring, and it is not known to what extent superelastic collisions are effective for cooling. Thus, in order to study dissociative recombination of O_2^+ occupying only its zeroth vibrational level, a high-pressure (> 0.1 Torr) hollow cathode ion source was developed and installed at CRYRING, which was capable of producing O_2^+ and collisionally quenching vibrationally excited levels before the ions were extracted from the source (Peverall *et al.* 2001). In a further study of O_2^+, dissociative recombination of vibrationally excited levels was studied. In order to control the vibrational excitations, an electron impact ion source was developed and characterized at SRI International by means of a charge-exchange experiment, before being shipped and installed at CRYRING (Petrignani *et al.* 2005c).

It was mentioned earlier that H_3^+ does not cool rotationally on the time scale of a storage ring experiment. Superelastic collisions can achieve cooling, but it takes a long time and leaves few ions in the ring for experiment. A better approach is to inject rotationally cold ions; they will not be heated by the background radiation, and when the beam is phase-space cooled by the electron beam, heating by residual gas collisions is counteracted by superelastic collisions. McCall *et al.* (2004) used a supersonic discharge source, which was characterized by laser spectroscopy before being installed at CRYRING. The rotational temperature was estimated to 20–60 K. The source does not deliver vibrationally relaxed ions, but this does not matter since vibrational relaxation is fast (about 1 s) on the storage ring timescale. A different approach was taken at the TSR, where a cryogenic 22-pole radio-frequency ion trap was constructed (Kreckel *et al.* 2005a). In the trap, the H_3^+ ions are cooled by collisions with helium buffer gas cooled to 10 K.

The results of the experiments using the ion sources discussed here will be described in Chapters 6 and 7.

Measurements of product state distributions and branching ratios

In many applications, information on the recombination products is equally as important as the cross section and rate coefficient. The ion storage ring technique is particularly suited for obtaining this information.

Consider Fig. 2.14, the pulse-height spectrum obtained for the dissociative recombination of CO$^+$. The peak at full beam energy of 3.3 MeV corresponds to the combined signal of a C atom carrying 1.48 MeV and an O atom carrying 1.82 MeV (the experiment was performed with ^{13}CO$^+$) impinging on the surface barrier detector within 1 ns. The time resolution of the detector is not high enough to allow a decomposition of the signal into two components. In other words, the signal does not disclose whether it is derived from a hit by C + O or by a CO molecule. In this particular case it is well known that the recombination products are a C atom and an O atom, but let us assume that the generic molecule ABC$^+$ recombines with an electron to form A + BC, AB + C, or A + B + C, or even the rearrangement products AC + B. The signal from the surface barrier detector gives no information about the breakup pattern.

A grid technique was developed at the single-pass merged-beam machine MEIBE in Western Ontario (Mitchell *et al.* 1983, Forand *et al.* 1985) based on a concept originating from Berkner *et al.* (1971). It was later implemented at CRYRING (Datz *et al.* 1995b) in an ingenious form specially suited for H$_3^+$. Most measurements at storage rings, however, use the grid technique in its original implementation (see Larsson & Thomas (2001) for a listing of all storage ring results until 2001, and Chapter 8). The grid technique at MEIBE suffered from signal-to-noise problems, which have been overcome at storage rings. Moreover, the increased resolution in the pulse-height spectrum at the high ion beam energies available in the storage ring is a further reason for the success.

If a grid with known transmission is inserted in front of the surface barrier detector when dissociative recombination of ABC$^+$ is studied, the grid will block some recombination products while others will go through one of the holes and hit the detector. If A is blocked and BC hits the detector, a pulse will occur at $(m_{BC}/m_{ABC}) \times$ (full beam energy). The probability for this to happen is $T(1-T)$. The peak appearing at this energy can also have a contribution from the decay channel A + B + C, where A is blocked and B and C go through the grid holes. The probability is then $T^2(1-T)$. Of course, the relative contribution to the $(m_{BC}/m_{ABC}) \times$ (full beam energy) will depend on the branching ratios for the two channels. This can be generalized:

$$\begin{aligned} ABC^+ + e^- &\rightarrow AB + C & (\alpha), \\ ABC^+ + e^- &\rightarrow A + BC & (\beta), \\ ABC^+ + e^- &\rightarrow A + B + C & (\gamma), \end{aligned} \quad (2.38)$$

where $\alpha + \beta + \gamma = 1$ (if we neglect the rearrangement channel AC + B). Let the number of counts in the full beam energy peak be labeled N(ABC), the number of counts in the peak $(m_{BC}/m_{ABC}) \times$ (full beam energy) be labeled N(BC), etc. Based on the reasoning given above, the following matrix equation can be constructed:

$$\begin{pmatrix} N(A+B+C) \\ N(B+C) \\ N(A+C) \\ N(A) \\ N(B) \\ N(C) \end{pmatrix} = \begin{pmatrix} T^2 & T^2 & T^3 \\ 0 & T(1-T) & T^2(1-T) \\ 0 & 0 & T^2 \\ 0 & T(1-T) & T(1-T)^2 \\ 0 & 0 & T(1-T)^2 \\ T(1-T) & 0 & T(1-T)^2 \end{pmatrix} \begin{pmatrix} \alpha \\ \beta \\ \gamma \end{pmatrix}. \quad (2.39)$$

A solution of this matrix equation gives directly the three branching ratios α, β, and γ. The virtue with this technique is that it allows all branching ratios to be determined in only a single measurement. Most measurements are performed at $E_d = 0$, where the cross section has its maximum and the best signal-to-noise conditions are at hand. The insertion of the grid will make some peaks coincide with peaks arising from collisions in the rest gas. Thus, a background subtraction is required for those peaks, which is usually not a problem at $E_d = 0$, but which becomes more difficult when $E_d \neq 0$ and the cross section is smaller.

The grid technique is powerful since it allows the determination of the full set of branching ratios, but it also has its shortcomings in that it does not provide any information about the dynamics of the dissociation process or information about the state of excitation of the recombination products. Let us consider the following process:

$$\begin{aligned} AB^+ + e^- &\rightarrow A^* + B + \text{KER (1)}, \\ AB^+ + e^- &\rightarrow A + B^* + \text{KER (2)}, \\ AB^+ + e^- &\rightarrow A + B + \text{KER (3)}, \end{aligned} \quad (2.40)$$

where * is used to denote an excited state and KER stands for kinetic energy release. The detector systems discussed so far cannot distinguish the three decay channels in reaction (2.40). Instead one makes use of the fact that the electronically excited products will carry less kinetic energy than the ground state products. Thus, they will separate less in space when flying out of the interaction region, and by recording the products with an imaging detector system, information about the state of excitation can be extracted. This is a general approach taken in chemical dynamics (Suits & Continetti 2001). Imaging of recombination products was started at the TSR (Zajfman et al. 1995) and CRYRING (van der Zande et al. 1996), although the first experiment at CRYRING relied on a rather simple, graded foil in order to obtain kinetic energy release information. Figure 2.15 shows the detector system now in use at CRYRING (Rosén et al. 1998b). It is similar to the system developed at the TSR (Zajfman et al. 1995, Amitay & Zajfman 1997). A fast neutral particle striking the first multichannel plate (MCP) gives rise to secondary electrons which are amplified further by the second and third MCPs. The electron cloud

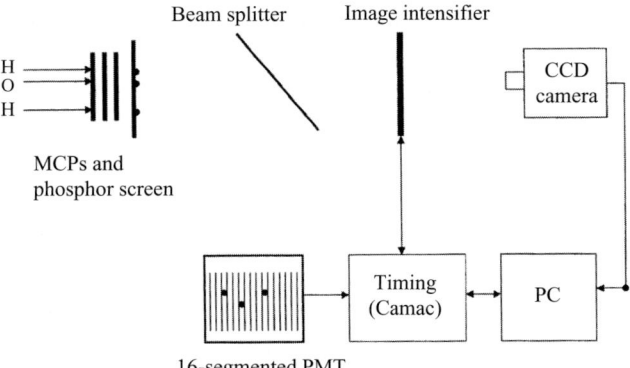

Figure 2.15 A simplified drawing of the imaging detector system at CRYRING. Dissociative recombination of H_2O^+ has generated the products H + H + O, and by the amplification of the multichannel plates (MCPs) a shower of electrons give rise to flashes on the phosphor screen. The illumination of three strips on the segmented photomultiplier (PMT) not only provides the timing information, but is also used to turn on the image intensifier, so that the position-of-arrival information is recorded by the charged-coupled-device (CCD) camera. Using the image intensifier as a shutter significantly reduces a background of random three-particle events. (Reproduced with permission from Larsson and Thomas 2001.)

emerging from the last MCP hits the phosphor screen, which thus serves as the anode, and gives rise to a flash. The position- and time-of-arrival of the electron cloud are encoded by the CCD camera and the segmented photomultiplier, respectively. The time resolution is about 0.5 ns, and the number of pixels in the CCD camera determines the position resolution – typically the position resolution is 100 μm. The detector at CRYRING has a 64 × 64 pixels camera and it is operated in a low-event probability, high repetition rate mode and data are acquired and stored on an event-by-event basis. Similar detector systems have been used at the other storage rings.

In addition, a detector has been developed that has made it possible to obtain the timing information using two CCD cameras that record the ratio of intensities of a pair of two-dimensional images (Strasser *et al.* 2000, Zajfman, Heber & Strasser 2002). Let us consider the case when two particles hit the phosphor screen with a certain time difference Δt. One of the CCD cameras measures the integrated intensity of the two flashes over the entire decay time of the phosphor screen, which can be on the order of 50 ns. The other CCD camera is gated to measure the integrated intensities over a time scale which is considerably shorter than 50 ns. The time difference Δt can then be retrieved from the four integrated intensities. In principle, the example can be generalized to any number of particles, which is the advantage with this camera system. So far, however, it has not been used much for the study of dissociative recombination, and the proof of the principle was demonstrated using the photodissociation of H_2^+.

A third type of detector system, which has been employed in a few experiments at CRYRING, uses delay lines as the anode (Dörner *et al.* 1996) and has the advantage of allowing a high count rate, on the order of 30 kHz (Österdahl *et al.* 2005).

Reactions (2.40) give rise to two hits on the imaging detector, which are separated by the distance D. This projected distance can be expressed in terms of the ion beam energy E_b, the kinetic energy release E_{KER}, the masses m_A and m_B, the distance L between the detector and the recombination event, and the angle θ between the molecular axis and the electron beam direction during dissociation:

$$D = \sqrt{\frac{E_{KER}}{E_b}} \frac{m_A + m_B}{\sqrt{m_A m_B}} L \sin\theta. \tag{2.41}$$

The electron cooler has a finite length, which in the case of CRYRING is 85 cm and the TSR 1.5 m. This introduces an uncertainty as to exactly where a recombination event takes place. The angle between the molecular axis and the electron beam direction will also differ from event to event, and this leads to a distribution of projected distances, $P(D)$. If the electron energy is so small that it can be neglected, the kinetic energy release is determined by the initial and final quantum energy states of the participating particles, i.e., the quantum state of the molecular ion and the quantum states of the two atomic products. For a given value of KER and isotropic fragmentation, the measured spectrum of transverse separations can be described by an analytic function (Zajfman *et al.* 1995):

$$P(D) = \frac{1}{D_{l+L} - D_L} \left\{ \arccos\left[\min\left(1, \frac{D}{D_{l+L}}\right)\right] \right.$$
$$\left. - \arccos\left[\min\left(1, \frac{D}{D_L}\right)\right] \right\}, \tag{2.42}$$

where D_{l+L} and D_L correspond to the maximum projected distance (i.e. $\theta = 90°$) between recombination fragments created at the entrance and exit of the interaction region. Expressions for distributions other than isotropic have been given by Amitay *et al.* (1996b). If fragmentation into several different product states occurs, the total spectrum is given by a summation of the individual distributions multiplied by the branching ratios. The fact that the kinetic energy release depends on the quantum state of the molecular ion makes the imaging technique important also for determining the internal state distribution of the molecular ion in the stored ion beam.

The analysis becomes more complicated when dissociative recombination of a polyatomic molecular ion is studied by means of an imaging system. First, the detector shown in Fig. 2.15 does not have chemical resolution, in other words, a hit on the detector gives the same flash whether it is a hit by atom A or atom B. In the diatomic case this is of no concern since only the distance is measured. For a triatomic ion ABC^+ it becomes of crucial importance to know where each atom hits. Three types of triatomic systems have been studied at storage rings, and will be discussed in later chapters. The first systems are H_3^+ and D_3^+ (Strasser et al. 2001, Strasser et al. 2002a), and here the chemical information is not an issue, the second systems are the H_3^+ isotopologs H_2D^+ and D_2H^+ (Strasser et al. 2004), for which less quantitative information was obtained as compared with H_3^+, and the third systems are of the type XH_2^+ (Datz et al. 2000b, Thomas et al. 2002, 2005a, Hellberg et al. 2005, Zhaunerchyk et al. 2005), where X is C, N, O, S, and P. In the XH_2^+ experiments, identification of the X atom is required, and the way to do that is to position a 5-mm-diameter foil in front of the center of the imaging detector. The thickness of the foil was chosen so that the X atom would go through, while the H atoms would not. For example, an ion beam of 250 keV amu^{-1} H_2O^+ results, after dissociative recombination leading to three-body breakup, in an O atom and two H atoms with energy 250 keV amu^{-1}. The range of H atoms in aluminum at this energy is 2.2 µm, whereas the range of O atoms is 3.2 µm. Thus, a foil of thickness 2.5 µm will absorb the H atoms while the O atoms will pass through. A hit behind the foil identifies an O atom.

More details of the imaging technique will be discussed in the chapters on individual molecules.

2.2.2 Electrostatic ion storage rings

The advantage with confining the ions by electric rather than magnetic fields is that in the former case, the force keeping the ions in closed orbits depends only on the electric field strength and the charge state of the ion. Thus, there is no dependence on the mass of the ion. This is in contrast to a magnetic ring, where the increase of ion size eventually will make it impossible to bend the ion beam in the dipole magnets, or the ion beam must be so slow that it is impossible to operate the ring under stable conditions. Whereas the electrostatic ring can be set for a given charge state but is independent of the mass, the magnetic ring must be fine tuned for any given mass. The arguments given here suggest that an electrostatic ring is the preferred choice to study dissociative recombination, but this is not correct. Electrostatic ion storage rings are operated at energies of typically tens of keV, which means that for the study of dissociative recombination some of the advantages with the magnetic rings are absent. The physics of electrostatic storage rings and other electrostatic trap devices has been reviewed by Andersen, Heber, and Zajfman (2004).

The first electrostatic ion storage ring to be constructed and commissioned was ELISA at the University of Aarhus (Møller 1997, Andersen et al. 2002); it has

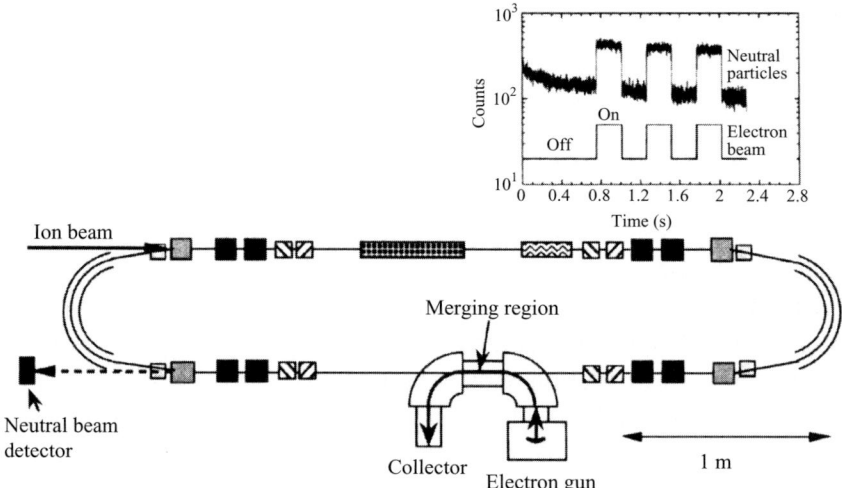

Figure 2.16 Schematic drawing of the KEK electrostatic storage ring. The ion beam energy is 20 keV, the electron beam current about 12 μA with a transverse temperature of 3 meV. The ring is kept at 5×10^{-11} Torr and the $1/e$ lifetime for a protonated peptide is just over 10 s. The inset shows how neutral particles are formed in the interaction region as a result of the electron beam being turned on. (Reprinted figure with permission from T. Tanabe, K. Noda, M. Saito, S. Lee, Y. Ito, & H. Takagi, *Phys. Rev. Lett.* **90**, 193201-1−4 (2003). Copyright (2003) by the American Physical Society.)

been used to study slow rearrangement processes (Jensen, Pedersen, & Andersen 2000), the photophysics of protein chromophores (Nielsen *et al.* 2001, Boyé *et al.* 2003), and the decay of cluster ions (Hansen *et al.* 2001, Tomita *et al.* 2001). ELISA has also been supplied with an electron target (Bluhme *et al.* 2004), but not in a configuration with merged beams; instead the electron target is mounted so that it crosses the stored ion beam at a 90° angle. This allows studies of electron scattering on anions. Such studies have also been performed in a linear ion trap configuration invented by Zajfman's group (Dahan *et al.* 1998) (see Chapter 9).

The first and so far only electrostatic storage ring with an electron beam merged with the ion beam is the one at KEK in Tsukuba, Japan (Tanabe *et al.* 2002, 2003), which is very similar to ELISA in design apart from the electron target. Whereas ELISA is 7.62 m in circumference, the KEK ring is 8.1 m. Thus, these rings are considerably smaller than the magnetic rings, and also less expensive to operate. Figure 2.16 shows the KEK electrostatic storage ring. It has been used for studies of both electrons interacting with positively charged peptide ions (Tanabe *et al.* 2003) and negatively charged DNA ions (Tanabe *et al.* 2004). Because of the low ion energy as compared with magnetic rings, 20 keV, it is difficult to work at velocity-matching conditions ($E_d = 0$) for biological molecules. For light ions such as H_2^+ and D_3^+, velocity matching occurs at electron energies of 5.45 eV and 1.82 eV, respectively (Tanabe et al. 2003). An electrostatic storage ring offers no particular advantages if one wishes to study dissociative recombination of H_2^+ and other light ions as compared with a magnetic storage ring. The advantage is for studies of heavy molecules, which are difficult or impossible to handle in a magnetic ring.

There are also electrostatic ion storage rings on the drawing board in Heidelberg (Zajfman et al. 2005) and Stockholm (Löfgren et al. 2006) and one is being commissioned at Tokyo Metropolitan University (Azuma, Tanuma, & Shiromaru 2004). One feature with these new rings is that they will be designed to operate at low temperatures, below 20 K, which will allow cold molecules to remain cold and not be heated up to room temperature.

2.3 Stationary afterglow technique

The afterglow techniques discussed in this and the following section build on an entirely different approach compared with merged beams. Instead of counting particles emerging from an interaction region, the decay of the electron density in an afterglow plasma is monitored. The stationary afterglow technique was the first method developed to study dissociative recombination (Biondi & Brown 1949a,b), and represented a big step forward with respect to earlier techniques of measuring recombination coefficients (Loeb (1955), note that volume recombination had been studied for more than half a century in 1950). The advantages with the stationary afterglow technique are that it is not limited in ion mass, the recombining ion is in a low vibrational level, and the excitation state of products can be obtained from the radiation of the decaying plasma. In this section we will describe in detail how the method works. A description of the early development of the technique has been given by Biondi (2003).

Figure 2.17 shows a simplified diagram of the stationary afterglow apparatus. To begin with we shall assume ideal conditions. Thus, a molecular gas of 100% purity is introduced into the microwave cavity by the gas handling system until the pressure is about 10–40 Torr. The microwave source generates a short pulse, 0.1–2 ms, of typically 50 W microwave power in the frequency range 3–10 GHz, which induces a breakdown in the gas and generates a plasma. The electron density after the short pulse is on the order of 10^9–10^{10} cm^{-3}. The electrons and ions come to thermal equilibrium with the neutral gas in a few tens of microseconds. The electrons recombine with the molecular ions and the electron and ion densities decrease as a function of time; eventually the gas is neutral again. In the ideal situation we consider

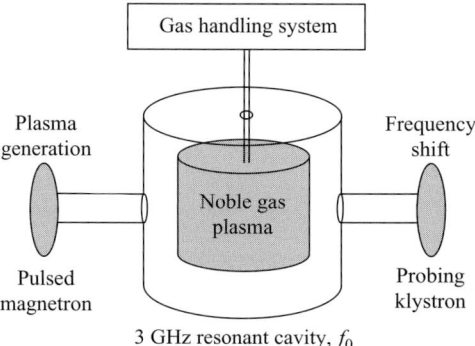

Figure 2.17 Simplified diagram of stationary afterglow apparatus. A microwave cavity is connected to a gas handling system and a source for generation of microwaves. A short pulse of microwaves generates a breakdown of the gas in the cavity and creates a plasma. The electron density is subsequently measured as a function of time after the ionizing pulse by means of a shift in the resonant wavelength of the microwave cavity. The microwaves are used both to generate the plasma and to measure the electron density. (Reproduced from *Dissociative Recombination of Molecular Ions with Electrons*, 2003, ed. S. L. Guberman, pp. 13–23, "Dissociative recombination of electrons and ions: The early experiments," M. A. Biondi, Figure 1. With kind permission of Springer Science and Business Media.)

here, the electron gas is characterized by an isotropic Maxwellian distribution, and if we assume that the gas and microwave cavity are kept at room temperature, the electron temperature is $T_e = 300$ K. Every single electron in the plasma is lost by dissociative recombination and the volume loss rate of electrons in the afterglow can be expressed as

$$\frac{dn_e}{dt} = -\alpha(T_e) n_e n_i, \tag{2.43}$$

where n_e is the electron density, n_i the ion density, and $\alpha(T_e)$ is the dissociative recombination rate coefficient defined in Eq. (2.37). In the present ideal case, we assume that $n_e = n_i$. The electron density in the afterglow plasma is measured by means of microwaves. The resonant wavelength of the microwave cavity before the discharge pulse is fired, λ_0, is shifted by the presence of electrons in the cavity, whereas the ions, with their much larger mass, give a negligible contribution to the shift. The relation between the electron density and the wavelength shift $\Delta\lambda$ is

$$\Delta\lambda = C\lambda^2 n_e, \tag{2.44}$$

where C is a geometric factor taking into account the distribution of the probing electric field in the cavity and the distribution of electron density; we assume initially that the electron density is uniform.

2.3 Stationary afterglow technique

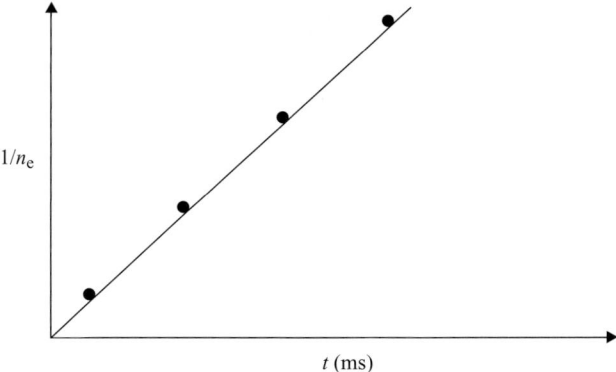

Figure 2.18 The results from a model stationary afterglow experiment. A microwave field is used to probe the wavelength shift caused by the electrons as a function of time after the plasma is created. The slope gives the rate coefficient $\alpha(T_e)$.

The sequence of the experiment is now clear. The discharge pulse is fired, and the electron density is probed by means of Eq. (2.44) as a function of time after the ionizing pulse has been terminated. The solution of Eq. (2.43) is

$$\frac{1}{n_e(t)} = \frac{1}{n_e(0)} + \alpha(T_e)t. \qquad (2.45)$$

Figure 2.18 shows how the data are usually displayed.

As presented here, the stationary afterglow technique is straightforward and gives the thermal rate coefficient as the slope of a linear plot. Yet, the first generation of experimental data coming from the technique developed by Biondi and Brown (1949b) was met by some scepticism from the gaseous electronics community (Loeb 1955). The fact that none of the results tabulated in the review by Bardsley and Biondi (1970) dates from the 1950s shows that it took time for the experimental details of the technique to be mastered. Much progress was made in the 1960s, and the results listed in Table 2.2 of Bardsley and Biondi (1970) for $T_e = 300$ K are reliable for Ne_2^+, Ar_2^+, N_2^+, O_2^+, and NO^+.

Several simplifications were made in the derivation of an expression for the rate coefficient. These were:

- The assumption that a molecular gas of 100% purity is present in the microwave cavity.
- The assumption that all electrons are lost by dissociative recombination, and that no electrons are produced in the afterglow plasma.
- The assumption that the electron density is uniform in the cavity.

2.3.1 Rate coefficient measurements near 300 K

It was early recognized that, in order to have some confidence in what ions are present in the afterglow plasma, a mass spectrometer connected to the microwave cavity is needed (Phelps & Brown 1952). But it was not until the 1960s that it was possible to install and use reliable mass spectrometers; Oskam (1969) has reviewed in detail the requirements of a mass spectrometer that is to be used for collision studies in an afterglow plasma. It is very difficult to avoid impurities in the gas handling system, and if these impurities lead to the formation of ions with a large recombination rate coefficient, and these ions are different from the one for which a measurement is intended, an incorrect result will be obtained. The early experiments using molecular gases such as nitrogen were flawed because it was not recognized that cluster ions such as N_3^+ and N_4^+ could form if the gas pressure was too high. In order to determine $\alpha(N_2^+)$, Kasner and Biondi (1965) used a mass spectrometer to monitor N_2^+ ions diffusing out to the walls and into the mass spectrometer. In order to avoid the formation of N_3^+ and N_4^+ ions, they used a low nitrogen pressure ($<10^{-2}$ Torr), and in order to minimize electron ambipolar diffusion, they added 20 Torr of neon as buffer gas. They found that the decay of N_2^+ ions under a wide range of experimental conditions followed the decay of the electron concentration.

Experiments with a noble gas such as neon are easier; it is straightforward to obtain very pure samples of neon and clustering is negligible. The Ne_2^+ ions are formed by the action of the microwave discharge, which initiates excitation and ionization that leads to molecular ion formation by collisions. It is somewhat more difficult to prepare samples of argon without impurities of nitrogen, and for krypton and xenon one must be careful to avoid some admixture of the other one. Helium is a special case, since He_2^+ recombines very slowly, and its rate coefficient is too small to be measured by an afterglow technique. The rate coefficient for Ne_2^+ was measured in 1949 by Biondi and Brown to be $(2.07 \pm 0.05) \times 10^{-7}$ cm^3 s^{-1}, a result that was in agreement with an independent stationary afterglow experiment which gave $\alpha = (2.2 \pm 0.2) \times 10^{-7}$ cm^3 s^{-1} (Oskam & Mittelstadt 1963). These early results are close to the present best value of $(1.75 \pm 0.2) \times 10^{-7}$ cm^3 s^{-1} (Johnsen 1987). Ne_2^+ is probably the ion most extensively studied by means of the stationary afterglow technique, and the results are consistent. It is a suitable system for comparison of results from different experimental techniques; however, there are as yet no merged-beam studies of Ne_2^+.

The somewhat lower value for Ne_2^+ quoted by Johnsen (1987) as compared with the first result (Biondi & Brown 1949b) is related to the second simplifying assumption, namely that all electrons are lost by dissociative recombination.

Equation (2.43) must be modified to contain additional source and sink terms:

$$\frac{dn_e(\mathbf{r}, t)}{dt} = \sum_i P_i - \sum_j L_j - \alpha(T_e)n_e^2(\mathbf{r}, t) + D_a \nabla^2 n_e(\mathbf{r}, t). \quad (2.46)$$

P_i and L_j describe the rates of processes leading to the production and loss of electrons, and the last term describes the ambipolar diffusion, i.e., the diffusion of electrons in the presence of their own space-charge field. The electron density is now also a function of the position, $\mathbf{r}$, in the cavity, and quasineutrality of the plasma implies $n_e \approx n_i$. The ambipolar diffusion coefficient, D_a, depends on the gas pressure. This suggests that by keeping the pressure high, D_a can be made very small; however, there is a trade-off. If the pressure is high, there is a risk that electrons are lost by three-body electron attachment:

$$A + e^- + M \rightarrow A^- + M. \quad (2.47)$$

Electron production after the discharge pulse has been turned off could be caused, for example, by Penning ionization. If a rare gas is used as buffer gas, the discharge pulse can excite these atoms to metastable states, followed by Penning ionization:

$$A + He^* \rightarrow A^+ + He + e^-. \quad (2.48)$$

In order to extract a reliable dissociative recombination rate coefficient from a measurement of the decay of the electron density, one must choose conditions so that additional processes taking place in the afterglow are minimized or at least can be modeled with some confidence. Ambipolar diffusion is essentially impossible to avoid, but if other source and sink terms can be made negligible, Eq. (2.46) reduces to

$$\frac{dn_e(\mathbf{r}, t)}{dt} = -\alpha(T_e)n_e^2(\mathbf{r}, t) + D_a \nabla^2 n_e(\mathbf{r}, t). \quad (2.49)$$

From the preceding paragraph, in which ambipolar diffusion is introduced, it becomes obvious that the assumption of a uniform charge distribution cannot be upheld. Instead the microwave field probes a weighted average of the electron density, $n_{\mu w}$, which is defined by

$$n_{\mu w} = \int_{vol} n_e(\mathbf{r}, t) E^2(\mathbf{r}) dV \bigg/ \int_{vol} E^2(\mathbf{r}) dV, \quad (2.50)$$

where $E(\mathbf{r})$ is the electric component of the probing microwave field amplitude in the cavity. In the early stationary afterglow works it was assumed that $n_e = n_{\mu w}$, but during the 1960s it was realized that this assumption leads to erroneous results and a procedure was developed to recover α by means of Eq. (2.49). If D_a is known and

α treated as a free parameter, Eq. (2.49) can be solved for a cylindrical geometry such as the one shown in Fig. 2.17, and the solution can be compared with measured data. The rate coefficient is then varied until optimal agreement with experimental data is obtained (Frommhold, Biondi, & Mehr 1968).

It took about 20 years to develop the stationary afterglow technique to the point where reliable rate coefficients could be measured at $T_e = 300$ K for the ionospheric ions and the rare gas ions. This illustrates the tremendous difficulties in measuring dissociative recombination rate coefficients. The next level of difficulty occurred when the temperature dependence of the rate coefficient was sought.

2.3.2 Rate coefficient measurements as function of electron temperature

In many applications the rate coefficient at an electron temperature that is different from room temperature is needed. Efforts to measure the rate coefficient as a function of electron temperature were initiated in the 1960s, when confidence had been gained in measurements near room temperature. Until this point in the discussion, the microwaves have served two purposes: to create a plasma in a short pulse, and to probe the electron density in the afterglow plasma. In order to measure the dependence of the rate coefficient on electron temperature, a third microwave mode is added: microwave heating. This means that two microwave fields are present in the afterglow, one high-Q microwave field for probing the electron density, and one low-Q field for heating the electrons. The massive molecular ions are essentially unaffected by the microwave field, which only interacts strongly with the electrons. Thus the ion temperature, T_i, remains approximately equal to the temperature of the neutral gas, i.e., $T_i = T_{gas} = 300$ K. This is similar to the situation in the merged-beam technique, where knowledge of the cross section allows the rate coefficient to be obtained over a wide range of temperatures, but the ion temperature is fixed.

In order to understand how the microwave heating technique works, we assume first that the gas is atomic, for example neon. The time varying microwave field induces an oscillatory motion in the electrons, which is randomized in elastic collisions with the neutral gas atoms. The elastic collisions lead to a small energy loss, which is balanced by the constantly present microwave field. It has been shown that the electron-velocity distribution under these conditions is nearly Maxwellian (Margenau 1946). Next we assume that the electron temperature is also uniform in the experimental cavity. In an experiment on Ne_2^+, Frommhold, Biondi, and Mehr (1968) found the rate coefficient of Ne_2^+ to vary as $T_e^{-0.43}$ over the range 300 K $\leq T_e \leq$ 11 000 K.

The situation changes when a molecular gas is present in the cavity. With its dense structure of energy levels, one can no longer assume that only elastic collisions occur. Rotational and vibrational excitations can occur as a result of collisions between electrons and the molecular gas:

$$A_2(v, J) + e^-(E) \to A_2(v', J') + e^-(E - \Delta E), \qquad (2.51)$$

where $v < v'$, $J < J'$, and $\Delta E > 0$. These collisions reduce the electron energy significantly below that in an atomic gas, which makes it very difficult to know at which electron temperature a measurement is performed. Penetrante and Bardsley (1986) pointed out that the effect is large even if an atomic buffer gas is used and the molecular component is small. The inelastic collisions also lead to deviation from a Maxwellian distribution unless the electron density is kept high at around 10^{10} cm^{-3}.

The assumption that the electron temperature is uniform in the cavity is a simplification. The heating microwave field is not uniformly distributed in the cavity, so that the electrons are heated most effectively where the electric component of the microwave field is largest. Johnsen (1987) discussed whether the heat conduction in the electron gas is sufficient and identified two limiting cases by considering the following. The number of collisions an electron has to make with the atomic gas in order to reach its equilibrium mean energy given by the local field is of the order of $n = M/m_e$, where M is the atomic mass. During the equilibration process, the electron diffuses a distance $\delta(2n)^{1/2}$, where δ is the electron mean-free path for momentum transfer. When the length $L = \delta(M/m_e)^{1/2}$ is small compared with the cavity radius, the temperature of the electron gas will be highly nonuniform. In the opposite case, the temperature distribution should be fairly uniform even if the heating microwave field is not. The arguments given above do not take into account a molecular component in the gas. Johnsen (1987) concluded that in many stationary afterglow experiments, the length was typically of the same order as the radius of the cavity, i.e., conditions between the limiting cases prevailed, and the assumption made in earlier work of a uniform temperature distribution was incorrect. This led Dulaney, Biondi, and Johnsen (1987) to remeasure NO$^+$, for which stationary afterglow results had been obtained earlier by Huang, Biondi, and Johnsen (1975) that were in disagreement with the ion trap results of Walls and Dunn (1974). Dulaney, Biondi, and Johnsen (1987) used a numerical model to take into account inelastic collisions and the nonuniformity of the heating microwave field, and succeeded in obtaining agreement with the ion trap experiment.

In some cases the reduction of the electron temperature owing to inelastic collisions (Eq. (2.50)) is so dramatic that it was first interpreted as lack of temperature dependence of the rate coefficient. Huang *et al.* (1978) studied recombination of the cluster ions H$_3$O$^+ \cdot$(H$_2$O)$_n$ and found no or very weak temperature dependence over the range 300–8000 K. Johnsen (1993b) has shown that the water cluster ions indeed have a temperature dependence of $T_e^{-0.5}$, and studies of NH$_4^+$(NH$_3$)$_n$

cluster ions, also previously believed to have no temperature dependence (Huang, Biondi, & Johnsen 1976), have shown that the rate coefficient has a $T_e^{-0.65}$ dependence (Skrzypkowski & Johnsen 1997). In these studies, radio-frequency heating at 14.6 MHz rather than microwave heating was applied, and ionization was done using a short flash of ultraviolet light. The electron density was also measured by a radio-frequency probe (Johnsen 1986). The radio-frequency heating method has the advantage that the electron−molecule collision rate is much higher than the radio-frequency instead of being much smaller, as is the case with microwave heating (Johnsen 1993a,b).

Zipf (1980a) performed pioneering work combining the stationary afterglow technique with laser techniques in order to obtain information on final atomic product states and the dependence of the rate coefficient on the vibrational excitation of the molecular ion. His experiment on N_2^+ (Zipf 1980a) revealed an additional problem (Johnsen 1987) with the microwave afterglow technique. Zipf (1980a) used laser-induced fluorescence to monitor the decay of N_2^+ in its $v = 0$, 1, and 2 levels. From the decay, Zipf inferred the rate coefficients for the vibrationally excited levels with respect to that for the zeroth level, and found them to be 13% and 26% larger, respectively. Later, it became clear that there is a strong coupling between the vibrational levels of the N_2^+ ions and the neutral nitrogen molecules by charge-exchange collisions, and that the microwave discharge pulse probably excited molecular nitrogen to a vibrational temperature of 1500 K, i.e., the same temperature as inferred by Zipf by laser-induced fluorescence for the ions. Instead of measuring the rate coefficients of isolated N_2^+ ions in vibrationally excited levels, he measured them for ions collisionally coupled to vibrationally excited molecular nitrogen (Johnsen 1987, 1989). In this context it should also be mentioned that Biondi, Zipf, and coworkers developed optical spectroscopy techniques in order to establish the dissociative recombination mechanism (Connor & Biondi 1965) and to identify recombination products (Zipf 1970, Gutcheck & Zipf 1973, Shiu, Biondi, & Sipler 1977, Shiu & Biondi 1977, 1978, Zipf 1979, 1980b). This type of work was later extended to the flowing afterglow technique, as described in Section 2.4.

An alternative to microwave heating is to change the temperature of the walls surrounding the gas. After a while, the gas will be in thermodynamic equilibrium with the wall, and one can assume that $T_{wall} = T_{gas} = T_i = T_e$. This approach was tried by Kasner (1967, 1968) and Kasner and Biondi (1968). It does not cover the same wide temperature range as microwave heating, but it should not be affected by the problems with microwave heating discussed above. The temperature dependence for Ne_2^+ was measured to be $T_e^{-(0.42 \pm 0.04)}$ in the range 295−503 K (Kasner 1968), which is consistent with the results of Frommhold, Biondi, and Mehr (1968). Also the results for O_2^+ (Kasner & Biondi 1968) seem to be consistent with other measurements (O_2^+ is discussed in Chapter 6), but the measurement of a temperature-independent rate coefficient for N_2^+ (Kasner 1967) in the range 205–480 K cannot be reconciled with any other measurement, and must be incorrect.

After several decades of efforts to measure dissociative recombination rate coefficients using the stationary afterglow technique by Biondi, Johnsen, and coworkers at the Physics Department of the University of Pittsburgh, it was eventually abandoned in favor of the flowing afterglow technique, which avoids some of the problems inherent with a stationary afterglow. However, the stationary afterglow technique has been resurrected in the laboratory of J. Glosík in Prague in the form of an advanced integrated stationary afterglow (AISA), which uses a Langmuir probe rather than a microwave field to measure the electron density (Glosík *et al.* 2000).

The stationary afterglow technique has produced many results on dissociative recombination of molecular ions, some reliable and some that must be treated with caution. The best source of information concerning the reliability of stationary afterglow data is the review by Johnsen (1987). In this book, the discussion of afterglow data, apart from the examples given in this section, will be given in the chapters about individual molecules.

2.4 Flowing afterglow technique

The flowing afterglow technique was initially developed for the study of ion–molecule reactions (Fehsenfeld *et al.* 1965), anticipated to be used to study dissociative recombination (Ferguson, Fehsenfeld, & Schmeltekopf 1965), and developed into a powerful tool in plasma chemistry (Smith & Adams 1979, Adams & Smith 1988a). It was implemented for the study of dissociative recombination in a largely forgotten work by Mahdavi, Hasted, and Nakshbandi (1971) and later by the Birmingham group: Smith *et al.* (1975), Alge, Smith, and Adams (1983), and Smith and Adams (1983). In the flowing afterglow technique, the ion formation phase is geometrically separated from the microwave ionizing discharge pulse, which prevents vibrational heating. Figure 2.19 shows a schematic diagram of a flowing afterglow apparatus.

The electron density is measured with a Langmuir probe. The probe is positioned in the afterglow plasma, where it is exposed to a flux of electrons and ions. The flux depends on the density and the temperature. The thermal velocity depends on the mass of the particle, which means that given that the ion and electron temperatures are the same, the electron flux is much higher than the ion flux. By applying different voltages to the probe, the electron density, electron temperature, and ion density can be measured, at least in principle. Johnsen *et al.* (1994) have shown that only

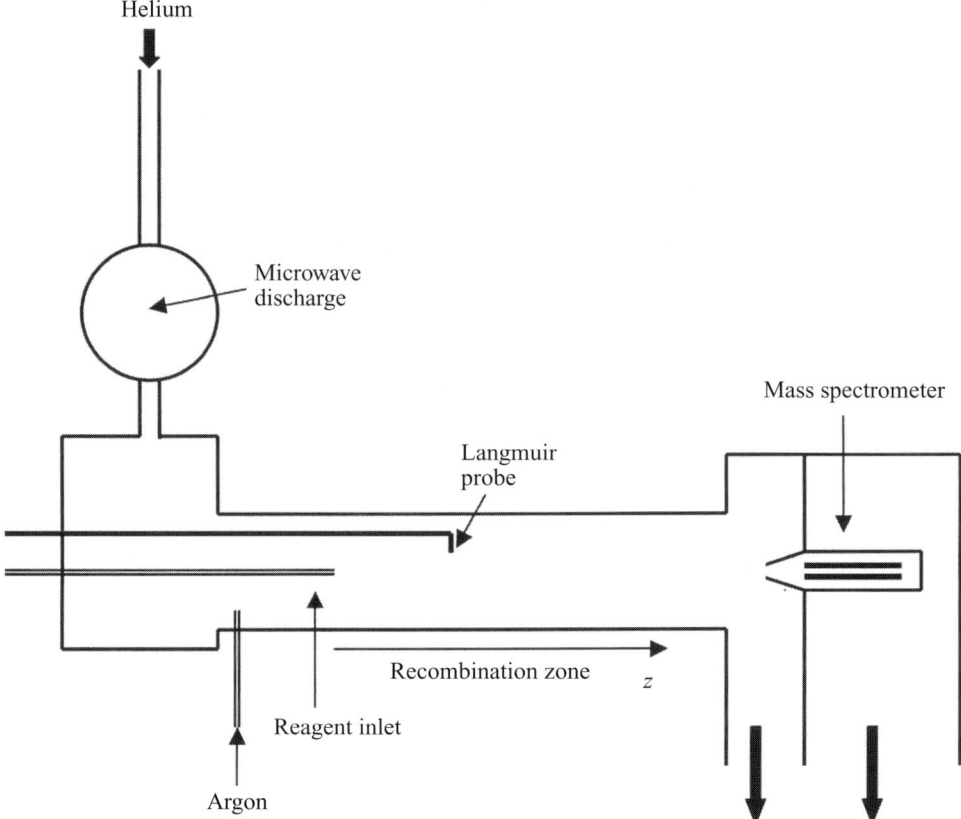

Figure 2.19 Schematic diagram of flowing afterglow apparatus (redrawn from Johnsen and Mitchell 1998). The flow tube is typically 1–2 m, with a diameter of 5–10 cm. Helium at a typical pressure of 0.5–1 Torr is the carrier gas, and in the microwave discharge, metastable and ionized helium is formed. Argon at a pressure of about 15% of the helium pressure is ionized by metastable helium by Penning ionization, i.e., $He^* + Ar \rightarrow He + Ar^+ + e^-$, or by the charge transfer, $Ar + He^+ \rightarrow Ar^+ + He$, or exchange reaction $He_2^+ + Ar \rightarrow Ar^+ + 2He$ (He_2^+ is formed in three-body reactions between He^+ and He), and acts as the ionizer by ion–molecule reactions: $Ar^+ + AB \rightarrow Ar + AB^+$. The recombination zone is shown with a coordinate axis (labelled z). The electron density in the recombination zone is measured by a movable Langmuir probe. The ion composition in the flowing afterglow plasma is determined by a mass spectrometer. The reagent inlet can be movable.

the electron density, not the ion density, can be measured with acceptable accuracy. Thus, in flowing afterglow experiments on dissociative recombination usually the electron density is measured as function of distance from the reagent inlet. It is possible to use several reactant gases fed into the flow at different positions if needed. The arrangement gives better control of the ion chemistry as compared

Table 2.2. *Flowing afterglow apparatuses used to study dissociative recombination*

Location	Principal investigator	Special feature	Reference
Birmingham, UK	Adams, N.G., Smith, D.	No longer used for dissociative recombination	Alge, Adams, & Smith 1983
Georgia, USA	Adams, N.G.	Variable temperature 80–600 K (VT-FALP)	Adams 1992
Prague, Czech Republic	Glosík, J.	High-pressure, maximum 20 Torr	Glosík & Plašil 2000
Pittsburgh, USA	Johnsen, R.	Optical spectrometer	Gougousi, Johnsen, & Golde 1995, Rosati, Johnsen, & Golde 2003
Rennes, France	Rowe, B.R., Mitchell, J.B.A.	Movable mass spectrometer (FALP-MS)	Mostefaoui et al. 1999
Fukuoka, Japan	Tsuji, M., Nishimura, Y.	Optical spectrometer	Tsuji et al. 1995

with the stationary afterglow technique. A further advantage is that the ions are transported to the mass spectrometer by the carrier gas, not by diffusion as in the stationary afterglow technique. Diffusion is slow at the fairly high pressures used to minimize electron diffusion in stationary afterglow, and the ions reaching the mass spectrometer may not give a correct representation of the ion composition in the middle of the afterglow plasma.

There are presently four flowing afterglow apparatuses in use for the study of dissociative recombination. The original flowing afterglow/Langmuir-probe (FALP) apparatus (Alge, Adams, & Smith 1983) is no longer used for recombination. Table 2.2 lists the flowing afterglow apparatuses that have been used to study dissociative recombination.

A new flowing afterglow technique has been developed at Rennes: flowing afterglow with photoions (FLAPI) (Novotný et al. 2005b,c).[1] It was developed in order to allow studies of polycyclic aromatic hydrocarbon (PAH) ions. In FLAPI, the helium plasma generated by microwave absorption is a source only of electrons, and not, as in FALP-MS, a source of both ions and electrons. Ionization of PAH molecules is instead accomplished by photons from a 157 nm fluorine excimer laser. This is a more efficient way of causing ionization and makes it possible to

---

[1] The acronym FIAPI is used in one paper by Novotný et al. (2005c). In the review by Florescu-Mitchell and Mitchell (2006), and in another paper by Novotný et al. (2005b), the acronym is FLAPI.

conduct experiments with smaller amounts of expensive PAHs. The number of excess electrons created by photoionization is small compared with those from the microwave-induced plasma. The electron density is measured by a Langmuir probe as a function of position in the flow tube, just as in the FALP technique.

2.4.1 Rate coefficient measurements

If no gas is introduced through the reagent inlet, the plasma will contain only Ar^+ ions, and possibly also He^+ and He_2^+ ions, neither of which recombines on the length scale of the recombination zone. Ambipolar diffusion of ions and electrons is the only loss process. In a generic experiment, the molecular gas AB is introduced through the reagent inlet, and ionized by the ion–molecule reaction $Ar^+ + AB \rightarrow Ar + AB^+$. Let us assume that AB^+ is the only molecular ion present in the flowing afterglow plasma and that it is lost mainly by dissociative recombination. The decay of the electron density then follows the equation

$$\frac{dn_e}{dt} = v_p \frac{dn_e}{dz} = D_a \nabla^2 n_e - \alpha(T_e) n_e n_i, \tag{2.52}$$

where z is the distance from the inlet of the reactant gas (see Fig. 2.19), v_p is the plasma flow velocity, and the first term on the right hand side describes the ambipolar diffusion. The electron diffusion can be minimized by keeping the helium buffer gas pressure high, and at a high electron density, say 10^{10} cm^{-3}, the plasma is dominated by recombination losses. The electrons are rapidly thermalized in collisions with the helium buffer gas. As in the stationary case, we assume quasineutrality, so that $n_e \approx n_i$, and Eq. (2.52) simplifies to

$$v_p \frac{dn_e}{dz} = -\alpha(T_e) n_e^2 \tag{2.53}$$

with the solution

$$\frac{1}{n_e(z)} - \frac{1}{n_e(0)} = \alpha(T_e) \frac{z}{v_p}, \tag{2.54}$$

where $n_e(0)$ is the electron density at $z = 0$. An ideal data set would look like the one shown in Fig. 2.20. In practice the inverse of the electron density will not be linear near $z = 0$. It takes some centimeters from the gas inlet for the ion–molecule reactions to consume the Ar^+ ions and the plasma starts to decay (Ar^+ recombines of course very slowly). Far downstream of the plasma, where the electron density has dropped to a small value, the plasma will become diffusion controlled, which leads to a departure from linearity also at large z. The analysis described here is the approach used by Alge, Adams, and Smith (1983). It can be difficult to identify the region where the plasma makes the transition from being recombination dominated

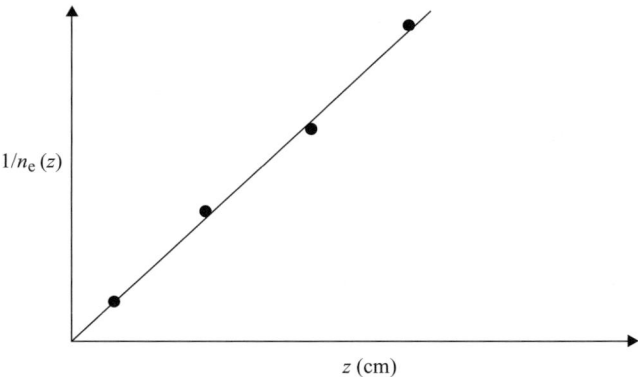

Figure 2.20 The results from an idealized flowing afterglow experiment. A Langmuir probe is used to measure the electron density as a function of distance from where the ions are created. The slope gives $\alpha(T_e)/v_p$. The plasma flow velocity, v_p, can be measured by modulating the plasma upstream and follow the disturbance along the flow tube with the Langmuir probe.

to being diffusion dominated. This has been discussed in detail by Gougousi, Golde, and Johnsen (1997).

The analysis given above does not include the possibility of the AB$^+$ ions undergoing ion–molecule reactions with an impurity gas I; in this case an additional term is needed:

$$v_p \frac{dn_i}{dz} = -\alpha(T_e) n_e n_i - k n_i n_I, \tag{2.55}$$

where k is the ion–molecule rate coefficient and n_I the density of the impurity gas. Integration of this equation gives

$$\ln \frac{n_i(z)}{n_i(0)} = -\frac{\alpha}{v_p} \int_0^z n_e dz - \frac{k n_I z}{v_p}, \tag{2.56}$$

where $n_i(0)$ is the ion density of AB$^+$ at $z = 0$. Ambipolar diffusion of ions to the walls in the recombination zone has been neglected in this expression, but would add a third term on the right hand side of Eq. (2.56). The data analysis must now rely on a measurement of $n_i(z)$ and $n_e(z)$. The Birmingham FALP (Alge, Adams, & Smith 1983) did include a mass spectrometer, but it was located at a fixed position downstream of the recombining plasma, hence $n_i(z)$ could not be measured. The FALP apparatus in Renne includes a movable mass spectrometer (Rowe et al. 1992). The ion density can be measured as a function of z if the mass spectrometer is movable, so that it can mass selectively sample the AB$^+$ ions along the recombination zone. The electron density as a function of z is measured by the Langmuir probe, and the integrated electron density is calculated. A plot of $\ln(n_i(z)/n_i(0))$ vs. $\int n_e dz$ gives α/v_p. The drawback is that the measurement is sensitive to the amount of impurity, which can be difficult to control. A more robust way of doing the measurement, which does not depend on the impurity, is to cause a variation in the electron density by changing the position of the microwave discharge along the vertical direction in Fig. 2.19, and by varying the microwave power. The measurements of n_e and n_i then take place at a fixed position z as a function of the position of the microwave discharge, and the term $kn_1 z/v_p$ is a constant. In practice, both methods have been used as a check for consistency.

If two ions are present in the afterglow, the relation $n_e = n_{i1} + n_{i2}$ holds, and coupled equations are obtained:

$$\frac{dn_{i1}}{dt} = -\alpha_1 n_e n_{i1} \quad \text{and} \quad \frac{dn_{i1}}{dt} = -\alpha_2 n_e n_{i2}. \tag{2.57}$$

The data analysis becomes more complicated. The rate coefficients α_1 and α_2, and the initial ion concentrations $n_{i1}(0)$ and $n_{i2}(0)$ can be varied as free parameters while Eq. (2.57) is integrated numerically until the best fit with the experimental data is obtained (Adams & Babcock 2005). This technique is suitable for the determination of the rate coefficient for different isomers present in the plasma. In such cases, a mass spectrometer is of little use because it could never distinguish between two isomers having identical mass. The decay of minority ions with a mass different from the dominant ion can be measured by a mass spectrometer, as has been demonstrated at the flowing afterglow apparatus in Rennes (Abouelaziz et al. (1993); see also Table 2.2). Apart from dissociative recombination, the minority ion can be lost by ion–molecule reactions with the reactant gas, and by ambipolar diffusion. The rate equation takes the form

$$\frac{dn_i}{dt} = \alpha n_e n_i - k n_i n_{RG} - \frac{D_a n_i}{\Lambda^2}, \tag{2.58}$$

where n_{RG} is the reactant gas density, k the ion–molecule rate coefficient, D_a the ambipolar diffusion coefficient, and Λ the characteristic diffusion length. This equation can be solved, which gives

$$v_p \ln \frac{n_i(z)}{n_i(z_0)} = -\alpha \int_{z_0}^{z} n_e dz - \left(k n_{RG} + \frac{D_a}{\Lambda^2}\right)(z - z_0). \tag{2.59}$$

The mass spectrometer is set to monitor n_i at a fixed position z while the electron density is varied. A plot of $\ln(n_i(z)/n_i(z_0))$ versus the integrated electron density gives a line with slope α/v_p (Lehfaoui et al. 1997).

The state of excitation of the AB^+ ions has not been discussed so far. It cannot be assumed that the charge transfer reaction $Ar^+ + AB \rightarrow Ar + AB^+$ produces AB^+ ions occupying only their lowest vibrational level; on the contrary, it is likely that vibrationally excited AB^+ is formed. The time the ions spend in the afterglow is too short to rely on relaxation by spontaneous emission. Instead vibrational de-excitation is assumed to occur in quenching reactions of type

$$AB^+(v') + AB \to AB^+(v) + AB^* \qquad v < v', \qquad (2.60)$$

where the vibrational excess energy in AB^+ is converted into excitation energy in the neutral molecule AB. Collisions with the helium buffer gas ensure rotational relaxation. In an experiment with a straightforward ion chemistry as for the generic example given here, it is fair to assume that $T_e = T_{rot} = T_{vib} = T_{gas} = 300$ K.

In some cases more complicated ion chemistry must be applied in order to form the desired ion. This is best described by an example. The astrophysically important ion HCO^+ cannot be made by a simple charge transfer reaction. Rowe *et al.* (1992) produced it by injecting a mixture of H_2 (90%) and CO (10%) through the reagent inlet. This led to the following reactions with Ar^+ and He^+:

$$Ar^+ + H_2 \to ArH^+ + H, \qquad (2.61a)$$
$$ArH^+ + H_2 \to H_3^+ + Ar, \qquad (2.61b)$$
$$Ar^+ + CO \to CO^+ + Ar, \qquad (2.61c)$$
$$CO^+ + H_2 \to HCO^+ + H, \qquad (2.61d)$$
$$He^+ + CO \to He + C^+ + O, \qquad (2.61e)$$
$$H_3^+ + CO \to HCO^+ + H_2. \qquad (2.61f)$$

There are no reasons to assume that Reactions (2.61d) and (2.61f) produce HCO^+ in its vibrational ground state. There are no HCO molecules that can vibrationally quench the ions, and the dominant molecular component, H_2, is ineffective in de-exciting HCO^+. The Renne group (Rowe *et al.* 1992) concluded that argon atoms can cause de-excitations, but that it requires about 100 collisions with argon for the deactivation to occur, and estimated that 7% of the ions populated the first bending vibrational level whereas 93% were in the ground vibrational level. Adams, Smith, and Elge (1984) also measured the rate coefficient in their FALP apparatus, but found a value a factor of 2 smaller than that of Rowe *et al.* (1992). The reason for this discrepancy remains unclear, although differences in vibrational population do not seem a plausible explanation. The HCO^+ ion is discussed in more detail in Chapter 8.

The considerable spread in rate coefficients for H_3^+ measured by stationary and flowing afterglow experiments will be discussed in Chapter 7. It is conceivable that the spread in afterglow results is related to the problem of controlling the vibrational population of H_3^+ (Johnsen 2005).

So far the discussion has concerned measurements at room temperature. It is of course desirable to measure the rate coefficient as a function of electron temperature. The approach taken with the FALP apparatuses is to heat or cool the entire FALP system by using either ohmic heaters or refrigerant liquids. This has the advantage as compared with the stationary afterglow technique that the electron and ion temperatures are the same as the gas temperature, in contrast to the situation in a microwave heated stationary afterglow plasma. The Birmingham FALP allowed measurements in the 95–600 K range, while the one in Georgia has an even wider quoted range, 80–600 K (Adams & Babcock 2005). The high-pressure flowing afterglow (HPFA) in Prague is designed to be suitable for studies of cluster ions, where the high pressure facilitates the formation of cluster ions. The smaller dimensions of the flow tube make it possible to use electric field gradients to raise the electron temperature above that of the ions and the gas. This may seem like a similar technique to microwave heating of the stationary afterglow, but there are two important differences: the temperature range is much smaller and the Langmuir probe is used to measure the electron temperature (Glosík & Plašil 2000).

2.4.2 Measurements of product state distributions and branching ratios

One can distinguish three different approaches to obtaining information about the dissociative recombination products by means of the flowing afterglow technique. An ultraviolet absorption technique was used at the FALP apparatus in Rennes to measure the concentration of recombination product atoms in the afterglow plasma (Quéffelec *et al.* 1985, 1989, Vallée *et al.* 1986, Rowe *et al.* 1988). In Birmingham (Adams, Herd, & Smith 1989), a method based on laser-induced fluorescence was implemented to detect the OH radical, which has an electronic spectrum that can be conveniently probed by a laser. In the first work (Adams, Herd, & Smith 1989), the OH radical was used to measure the H atom concentration via the reaction $H + NO_2 \rightarrow OH + NO$, but in a subsequent work on OH radical formation in dissociative recombination of H_3O^+, Herd, Adams, and Smith (1990) probed the OH reaction products directly. The two techniques, ultraviolet absorption and laser-induced fluorescence, were combined in one paper (Adams *et al.* 1991) addressing the fractional H atom contributions, f_H, to the product distributions for a series of protonated molecules (N_2H^+, HCO^+, N_2OH^+, $OCSH^+$, H_2CN^+, H_3O^+, H_3S^+, NH_4^+, and CH_5^+). In general, good agreement between the two techniques was found. Figure 2.21 shows the Birmingham FALP apparatus with the laser-induced fluorescence technique.

The FALP apparatus in Pittsburgh has also been used in combination with laser-induced fluorescence (Gougousi, Johnsen, & Golde 1997a,b), and is the second to be used with optical emission spectroscopy (Skrzypkowski *et al.* 1998, Rosati, Johnsen, & Golde 2003, 2004). The great advantage with recording the afterglow's emission spectrum is that it gives detailed information about the electronic states of the recombination products. In a study of dissociative recombination of CO_2^+ with

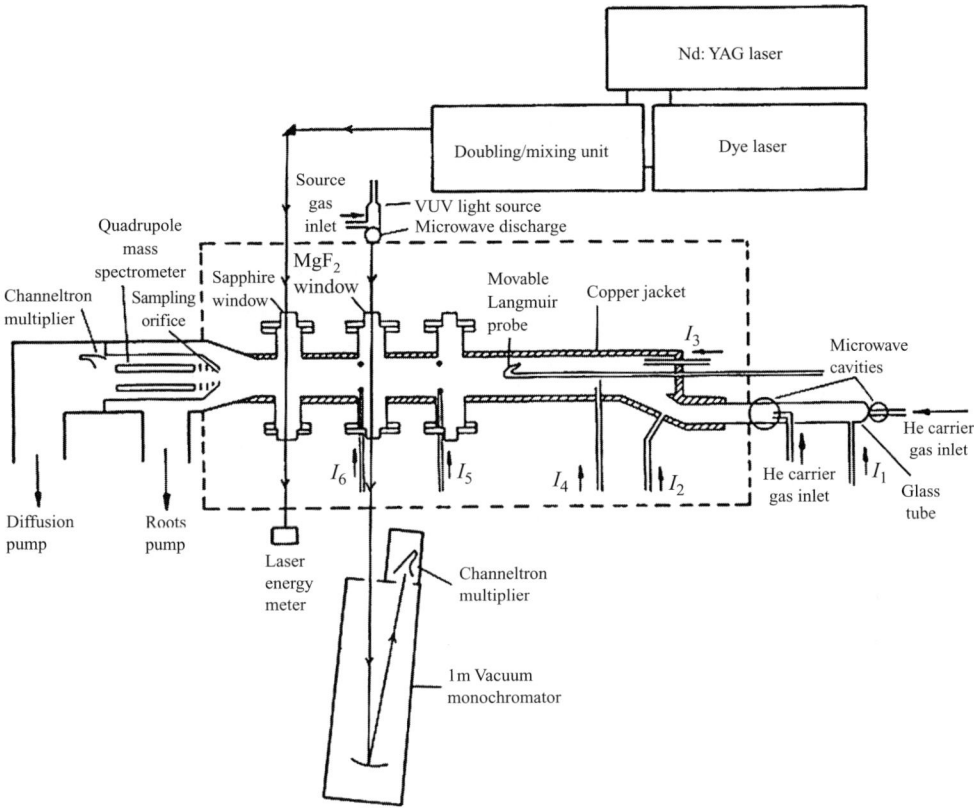

Figure 2.21 The Birmingham FALP apparatus. Laser induced fluorescence (LIF) is used to measure the H-atom density via the LIF spectrum of OH produced in the reaction H + NO$_2$ → OH + NO. (Reused with permission from N. G. Adams, C. R. Herd, M. Geoghegan, D. Smith, A. Canosa, J. C. Gomet, B. R. Rowe, J. L. Queffelec, and M. Morlais, *Journal of Chemical Physics*, **94**, 4852 (1991). Copyright 1991, American Institute of Physics.)

the FALP apparatus in Pittsburgh, not only was the total yield of CO ($a\,^3\Pi$) + O determined to be 0.29 ± 0.10 (Skrzypkowski *et al.* 1998), but it was also possible to determine the contribution to the yield of the $a\,^3\Pi$ state by cascading from higher lying electronic triplet states ($a'\,^3\Sigma^+$, $d\,^3\Delta_i$, $e\,^3\Sigma^-$). For obvious reasons the technique is applicable only to electronically excited states.

In ion storage rings the complete product branching ratios can be measured. A new technique based on flowing afterglows which will achieve the same is presently being developed by Adams and coworkers (Adams & Babcock 2005, Molek *et al.* 2006, McLain *et al.* 2006). The idea is to quench the recombination by adding electron attaching gases such as SF$_6$, block the recombining ions from entering the detection system, and probe the neutral product distribution by electron impact. The

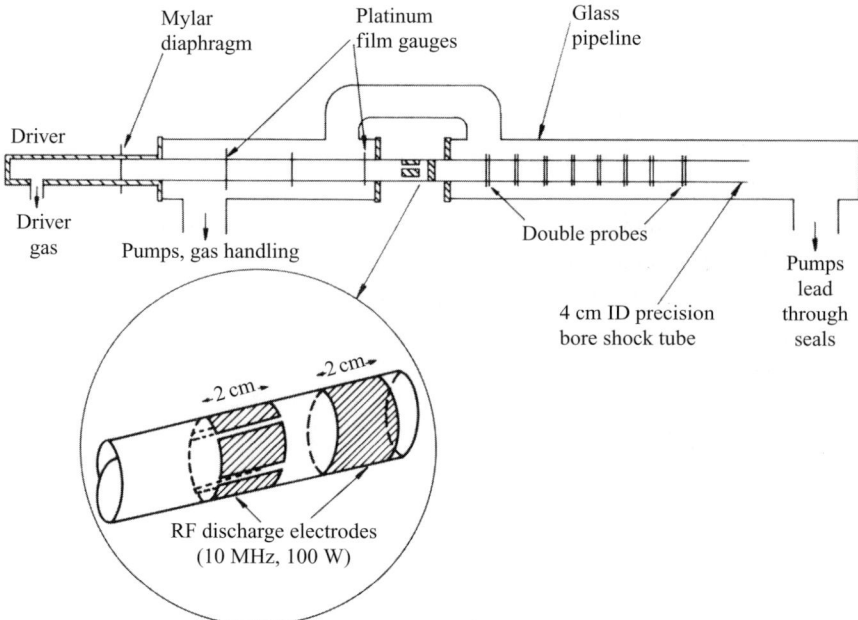

Figure 2.22 Schematic representation of shock tube apparatus. (Reproduced with permission from A. J. Cunningham and R. M. Hobson, "Dissociative recombination at elevated temperatures I. Experimental measurements in krypton afterglows," *J. Phys. B* **5**, pp. 1773–1783, (1972), IOP Publishing Limited.)

system has been tested on CH_5^+ and the preliminary data indicate good agreement with the storage ring result (Semaniak *et al.* 1998).

2.5 Shock-tube technique

The shock-tube technique for the study of dissociative recombination was initiated in order to measure the rate coefficient at elevated temperatures, but in contrast to the microwave heating stationary afterglow method described in Section 2.3.1, the ion and electron temperatures are equal. This can also be achieved with a stationary afterglow apparatus by heating the walls surrounding the gas, but only for a limited temperature range. The shock-tube technique allows a much wider temperature range.

The shock-tube technique for the study of dissociative recombination was introduced by Fox and Hobson (1966) and Cunningham and Hobson (1969). It is described briefly in the review by Bardsley and Biondi (1970) but hardly at all in later reviews. The technique was most used in the early 1970s (Cunningham & Dobson 1972a,b,c, O'Malley, Cunningham, & Hobson 1972), with occasional

2.5 Shock-tube technique

measurements occurring later (Ogram, Chang, & Hobson 1980, Davidson & Hobson 1987, Chang *et al.* 1989, Jiang *et al.* 1989). Figure 2.22 shows the principle of the technique.

The basic principle of the shock-tube technique is similar to that of the flowing afterglow. Ions are flowing in a tube and the decay of the recombining plasma is measured. There are, however, important differences. A cold driver gas at overpressure on the left of the diaphragm in Fig. 2.22 allows shock waves in the range Mach 2 to Mach 6 to be propagated in the tube. This allows translational temperatures in the range 500–3500 K, where the upper temperature limit is set by thermal ionization by collisions. Ionization is induced in the low-pressure gas by the electrodes in the middle of the tube a few hundred microseconds before the shock wave arrives, and the charge density is measured by the eight probes down-stream of the discharge electrodes. The ions and electrons are heated to the translational temperature of the shock wave.

The shock tube used to study dissociative recombination did not include a mass spectrometer, and Bardsley and Biondi (1970) expressed some concerns about the accuracy of the ion density measurements in the early version of the technique (Cunningham & Hobson 1969). Nevertheless, it seems that the limited number of measurements on the rare gas and atmospheric ions carried out with this technique are reliable (see Chapter 6 for discussions of results).

3
Theoretical methods

3.1 Introduction

Significant progress has been made in the theoretical understanding of collisions of atoms, molecules, and electrons. If we consider just the ground state of a polyatomic system, accurate quantum chemistry calculations exist for very large systems. Indeed, standard well-documented quantum chemistry packages exist that are available commercially. When considering the structure of excited states, and the surfaces describing such states, the situation becomes more complicated. However, spectroscopically accurate potential energy curves exist for a number of diatomics, and a smaller yet still significant number of triatomic and polyatomic systems. When considering collisions, large-scale molecular dynamics studies using model potential functions can be routinely carried out for thousands of particles. The current state-of-the-art in such studies of heavy-particle dynamics uses Car–Parinello simulations in which the molecular dynamics following classical equations of motion is driven by quantum forces calculated at each step of the simulation. Even for these systems, algorithms have been developed that allow large-scale simulations involving hundreds of particles. In the area of electron scattering, accurate elastic differential cross sections exist for systems ranging from diatomics such as H_2, to larger polyatomic systems such as ethylene, benzene, and silane. However, accurate calculations of dissociative recombination exist for only a handful of diatomics. Calculations for partial branching ratios are even more rare. There are still discrepancies between theory and experiment for the simplest of diatomics HD^+, and there exists only one accurate *ab initio* calculation in full dimensions on a triatomic, H_3^+.

What are the differences that so complicate the study of dissociative recombination reactions? The basic problem lies in the energy of the process that is being studied. Although the collision energy of the incoming electron can be very low (meV range), when the electron recombines the total energy is at the ionization potential of the neutral. Therefore, a large number of open electronic states exist

3.1 Introduction

in the system. Indeed, there exist an infinite number of Rydberg states converging to each state of the ion. In addition, the states may lie close in energy, leading to breakdown of the Born–Oppenheimer approximation and large nonadiabatic couplings between states. Finally, one must address not only bound electronic states, but also states with a free electron and resonant states that lie embedded in the free electron plus ion continuum.

Dissociative attachment shares the complications of a resonant state embedded in the continuum, but lacks the complications of the Rydberg states. This process can be described as

$$AB + e^- \to (AB^-)^* \to A + B^-. \tag{3.1}$$

Few accurate calculations exist for this process also. Indeed, only a handful of *ab initio* full dimensional studies that treat the entire scattering process into final states have been published, even for diatomics. The case of dissociative attachment will be discussed later in this volume.

We will first give a qualitative description of the process of dissociative recombination, and then describe the theoretical methods that have been used to study it.

In Figs. 3.1 and 3.2 we illustrate the types of dissociative recombination mechanisms. As the electron approaches the ion, it is captured into a temporary neutral state. In the "direct" mechanism this is a resonant state, i.e., a valence state of the neutral that is embedded into the scattering continuum of the system. As opposed to a Rydberg state converging to the ground state of the ion, this state represents a double excitation with respect to the system (ion plus incoming electron), and therefore the electronic coupling is nonzero between the resonant state and the background continuum. In the "direct" case, illustrated in Figure 3.1(a), this state is dissociative. As the neutral molecule dissociates, it can autoionize, leaving the ion in an excited vibrational state. If this autoionization can occur into the dissociative vibrational continuum of the ion it leads to dissociative excitation. At some point, the system crosses the ion curve and becomes stable with respect to autoionization. It then proceeds to products. Often, this state can then cross, in the diabatic picture, the Rydberg states converging to the ground state of the ion curve before the system has reached the asymptotic region. These crossings do not affect the total dissociative recombination cross section, but can dramatically change the final state product distribution.

This simple "direct" dissociative recombination case, which is directly analogous to the case of dissociative attachment, is complicated by the Rydberg states that converge to the ground state of the molecular ion. In the "indirect" mechanism, the electron is captured into an excited vibrational state of a Rydberg state. These

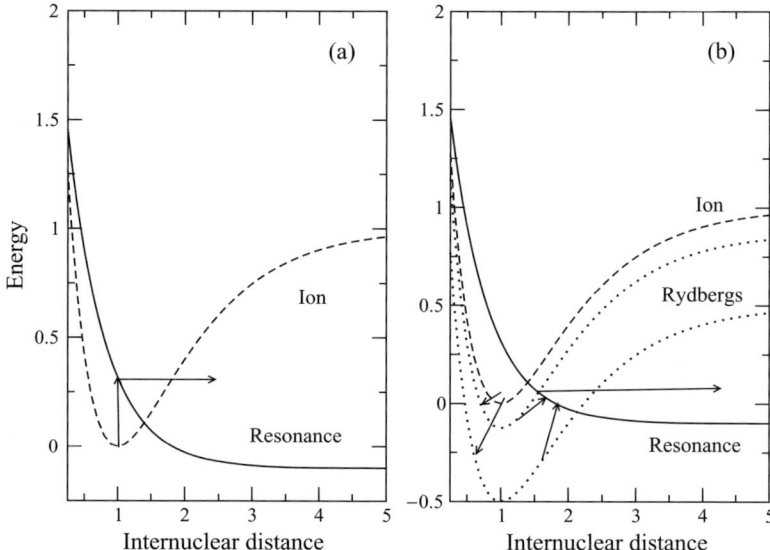

Figure 3.1 Schematic representation of the dissociative recombination process.
(a) Direct process: The resonant state (solid line) is dissociative and crosses the ion state (dashed line) at an internuclear distance larger than the equilibrium bond distance for the ion. The arrows show the path to dissociation, first the electron collision becoming trapped into the resonant state and then dissociating on that potential energy curve.
(b) Indirect process: The Rydberg states converging to the ion are shown as dotted lines. The electron is now first captured into an excited vibrational state of a Rydberg which then couples to the resonant state, followed by dissociation.

high lying Rydberg states, since they parallel the ion, are bound and cannot directly dissociate. However, it is possible for the Rydberg state to couple to the dissociative resonant state. This would then lead to dissociation to products. Again, these states can autoionize, yielding the ion in an excited vibrational state, and curve crossings can occur during the dissociation. This process is illustrated in Fig. 3.1(b). Note that these two processes are indistinguishable paths to the same product. Therefore they interfere. This causes sharp peaks in the total cross section as a function of energy, and similar oscillations in the final state distribution as a function of energy.

In addition, low-lying excited states of the molecular ion may exist. Each of these states will also include a series of Rydberg states, which converge up to each state of the molecular ion. Some, and in some cases many, of these Rydberg states will lie above the ground state of the molecular ion. Although they are Rydberg states, the electronic configuration represents a double excitation from the ground state. These are the so-called "core-excited" Rydberg states, described by Guberman

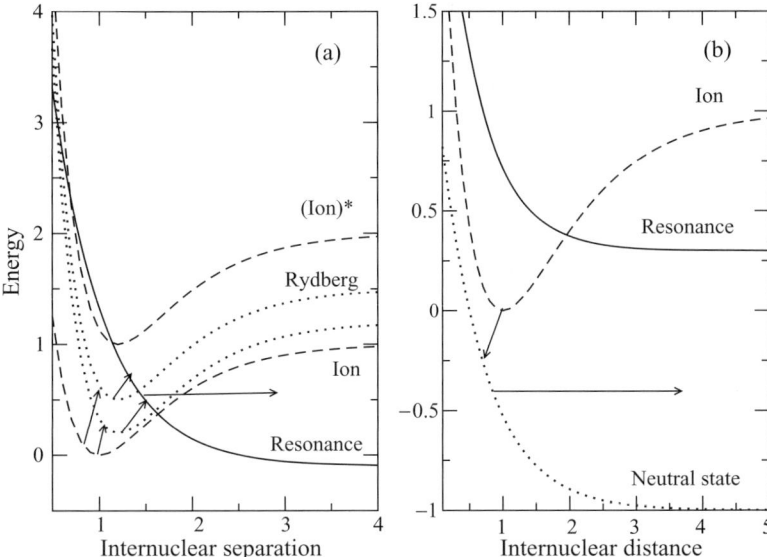

Figure 3.2 Schematic representation of the dissociative recombination process.

a) Core-excited resonances: The ground and low-lying excited state of the ion (dashed lines) are crossed by the dissociative resonance state (solid line). A series of Rydberg states (dotted lines) exist converging to the excited state of the ion. The electron is first captured into a vibrational state of these Rydbergs that then couple to the resonant state leading to dissociation.

b) Direct predissociation: A dissociative neutral state (dotted line) lies close in energy but does not cross the ion potential energy curve (dashed line). The resonant curve lies higher in energy (solid curve). The electron is captured into this low-lying neutral state and then dissociation occurs.

(1995b) in the dissociative recombination of OH^+. During the scattering event, the incoming electron can be captured into one of these states. If the parent state of the Rydberg series is bound, these resonances must couple to a dissociative state in order to lead to neutral products and hence to dissociative recombination. Again, there are multiple indistinguishable paths to products leading to interference structure in the total dissociative recombination cross section. This process is illustrated in Fig. 3.2(a). This has been observed experimentally in the systems OH^+ and CH^+. Theoretical calculations on CH^+ and CD^+ have shown that this mechanism is responsible for the broad resonances appearing at higher energy in the total cross section (see Chapter 5).

There is one further possibility. In some systems, the dissociative recombination proceeds, not by excitation to a resonant state, but by nonadiabatic coupling from the background (ion plus incoming electron) to a high-lying dissociative Rydberg state. This has been called a "direct predissociation" of the (ion + electron) system. This process was first described by Guberman (1994) in his studies of the low-energy dissociative recombination of HeH$^+$. The process is shown in Fig. 3.2(b), where the lower curve represents a dissociative neutral state lying below the ion. This situation was believed to occur only in a handful of cases. However, calculations on systems such as NeH$^+$, H$_3^+$, and HCO$^+$, for example, indicate that it is much more common than supposed. Further work is necessary in more systems to see whether the "direct" resonance case is the exception or the rule.

Therefore, what is required for a calculation of dissociative recombination? Clearly, one must obtain accurate information about all potential energy surfaces involved in the dissociative recombination process. This must include the resonance (or in some cases resonances) involved in the reaction, for not only the region of the surface where the resonance lies in the electronic continuum, but also after it crosses the ion state and becomes bound. If the "indirect" process is important (usually at lower electron impact energies) it is also necessary to describe the Rydberg states, as well as the ground state of the molecular ion, accurately. If just a total dissociative recombination cross section is desired, it is usually only necessary to obtain the surface in and near the Franck–Condon region of the target vibrational state of the ion and to know the asymptotic limits of the state, i.e., if the channel is open or closed. If partial cross sections into final products must be obtained, these surfaces must be known into the asymptotic region. In some treatments this includes the quantum defects for these states.

Second, the couplings between these surfaces must be obtained. Again, this includes both the coupling between the (electron + ion) system and the resonance and Rydbergs, as well as the coupling between the Rydbergs and the resonance. As with the surfaces, this information is only necessary in a restricted area of internuclear distances if just the total cross section is required. However, final state distributions would require any couplings that occur before the final state fragments have reached the asymptotic region.

Finally, the dynamics of the dissociative recombination process must be described. To produce accurate results, whatever procedure is used must treat all processes that occur, for example, indirect vs. direct, on an equal footing to describe correctly the interaction and interference between these processes.

In the sections below we will outline the current state of the art in the calculation of dissociative recombination cross sections, and outline the various methods that have been used to attack these difficult problems. Figure 3.3 shows a flow-chart for the calculation of a dissociative recombination cross section. The section of the chapter associated with each of the major components of the process is indicated, and the reader is directed to that particular section for a detailed discussion of that material.

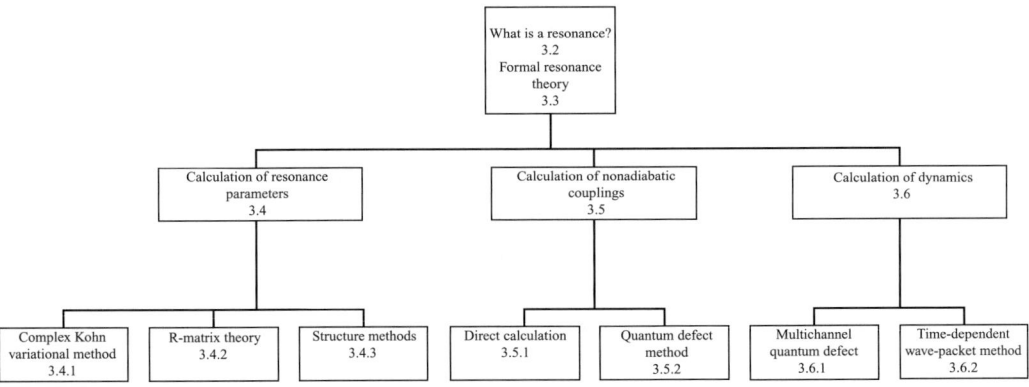

Figure 3.3 "Flow-chart" describing the path of a calculation of dissociation recombination and a layout of this chapter.

3.2 What is a resonance?

In the previous section, we introduced the concept of a resonance that is critical for the dissociative recombination process. What is a resonance? The simple picture is a bound state embedded in the continuum. This concept, although valuable for a qualitative picture of the process, must be put on a firmer mathematical basis to produce quantitative results. First, consider the case of simple potential scattering. If the potential is spherically symmetric, scattering in each partial wave l can be considered separately. For potential scattering, the Jost function is defined by:

$$f_l(k) = 1 + \frac{1}{k} \int_0^\infty h_l^+(r) U(r) \phi_l(r) dr \qquad (3.2)$$

where $U(r)$ is the potential, $h_l^+(r)$ is the Hankel function and $\phi_l(r)$ is the regular wave function. Now consider the behavior of the Jost function near a resonance. The S-matrix can be written as:

$$S_l(k) = \frac{f_l(-k)}{f_l(k)}. \qquad (3.3)$$

This means that $S_l(k)$ has poles where $f_l(k)$ has zeros. Assume $f_l(k) = 0$, and $\text{Im}(k) > 0$, then the wavefunction goes asymptotically as

$$\phi_l(r) \underset{r \to \infty}{\to} f_l(-k) h_l^+(kr), \qquad (3.4)$$

which implies

$$\phi_l(r) \underset{r \to \infty}{\propto} e^{-\text{Im}(k)r}, \qquad (3.5)$$

which is a bound state. However, if $\text{Im}(k) < 0$ then:

$$\phi_l(r) \underset{r \to \infty}{\propto} e^{\text{Im}(k)r}, \tag{3.6}$$

which blows up, so this is a resonance. For illustrative purposes, consider simple potential scattering from the spherically symmetric potential:

$$V(r) = \begin{cases} -V_0, & r < r_0 \\ \dfrac{l(l+1)}{2r^2}, & r > r_0 \end{cases}, \tag{3.7}$$

where V_0 is a constant. The S-matrix for this case can be obtained in closed form for each partial wave l. In this case (Vanroose, McCurdy, & Rescigno 2002)

$$S(k) = e^{-il\pi} \frac{[(l+1) - Kr_0 \cot(Kr_0)]h_l^-(kr_0) - kr_0 h_{l+1}^-(kr_0)}{[(l+1) - Kr_0 \cot(Kr_0)]h_l^+(kr_0) - kr_0 h_{l+1}^+(kr_0)}, \tag{3.8}$$

where $K = \sqrt{k^2 + V_0}$. At a resonance or a bound state, the S-matrix has a pole. As the magnitude of V_0 is increased the resonance moves smoothly into a bound state. Since

$$S(-k^*) = S(k)^* \tag{3.9}$$

the poles appear in pairs. In Figure 3.4 we plot the behavior of these poles as the well is deepened, i.e., the magnitude of V_0 is increased. The poles move towards the negative imaginary axis, eventually becoming pure imaginary. For the case $l = 0$, one pole moves up the axis, becomes a virtual state and then a bound state (see Figure 3.4, left panel). For the case $l \neq 0$, no virtual state is formed, the state moves smoothly into a bound state (Figure 3.4, right panel). Since E is proportional to k^2, when $\text{Re}(k) = 0$, E is less than zero, i.e., a bound state. When $\text{Re}(k) \neq 0$ (and $\text{Im}(k) \neq 0$), the state now has a complex energy with $E = E_{\text{res}} - i\Gamma/2$. This resonant state is not an eigenstate of the system. Consider the probability of the state that can be written as:

$$|\Psi(r,t)|^2 = \left| \Phi(r) e^{-iEt/\hbar} \right|^2 = |\Phi(r)|^2 \tag{3.10}$$

for E real. However, for E complex

$$|\Psi(r,t)|^2 = \left| \Phi(r) e^{-i(E_{\text{res}} - i\Gamma/2)t/\hbar} \right|^2 = |\Phi(r)|^2 e^{-\Gamma t/5\hbar}. \tag{3.11}$$

Therefore the resonant state has a lifetime and is not a stationary state of the Hamiltonian for the system.

Near a resonance if we assume:

$$f_l(\bar{k}) = 0, \tag{3.12}$$

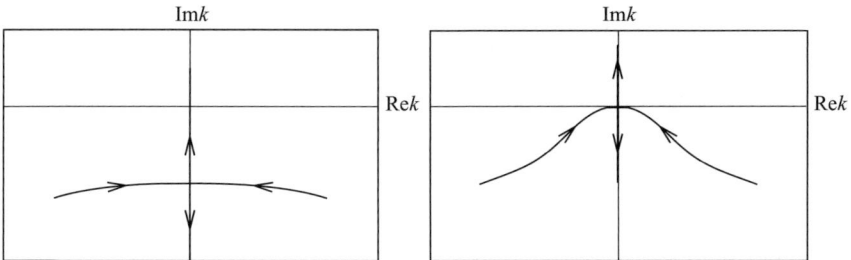

Figure 3.4 The path of the poles of $S(k)$, as a function of increasing well depth, for the potential given by Eq. (3.7): left panel: $l = 0$; right panel: $l \neq 0$. (Reprinted figure with permission from W. Vanroose, C. W. McCurdy, and T. N. Rescigno, *Phys. Rev. A* **66**, pp. 032720-1–10, (2003). Copyright (2003) by the American Physical Society.)

where $\bar{k} = k_r - ik$, we can expand $f_l(k)$ about $\bar{k}$, i.e.,

$$f_l(k) = \frac{df}{dk}\bigg|_{\bar{k}} (k - \bar{k}). \tag{3.13}$$

The phase shift for a fixed value of l, $\delta_l(k)$, is given by

$$\begin{aligned}\delta_l(k) &= -\arg(f_l) = -\arg(\dot{f}_l) - \arg(k - \bar{k}) \\ &\equiv \delta_{\text{bg}} + \delta_{\text{res}},\end{aligned} \tag{3.14}$$

where the phase shift has been split between the slowly varying portion due to the background and that due to the resonance that rapidly varies. This means we can write the resonant portion of the phase shift as

$$\delta_{\text{res}}(k) = -\arg(k - \bar{k}). \tag{3.15}$$

Therefore, as shown in Fig. 3.5(a), as k sweeps past $\bar{k}$, $\delta \to \delta + \pi$. Now instead of plotting the phase shift as a function of k it is plotted as a function of E in Fig. 3.5(b). It can be seen that

$$\sin(\delta_{\text{res}}) = \frac{\Gamma/2}{[(E - E_{\text{res}})^2 + \Gamma^2/4]^{1/2}} \tag{3.16}$$

and since

$$\sigma_l \propto \frac{\sin^2(\delta)}{E} = \frac{\sin^2(\delta_{\text{bg}} + \delta_{\text{res}})}{E} \tag{3.17}$$

if $\delta_{\text{bg}} \approx 0$ then

$$\sigma_l \propto \frac{\Gamma^2/4}{((E - E_{\text{res}})^2 + \Gamma^2/4)}. \tag{3.18}$$

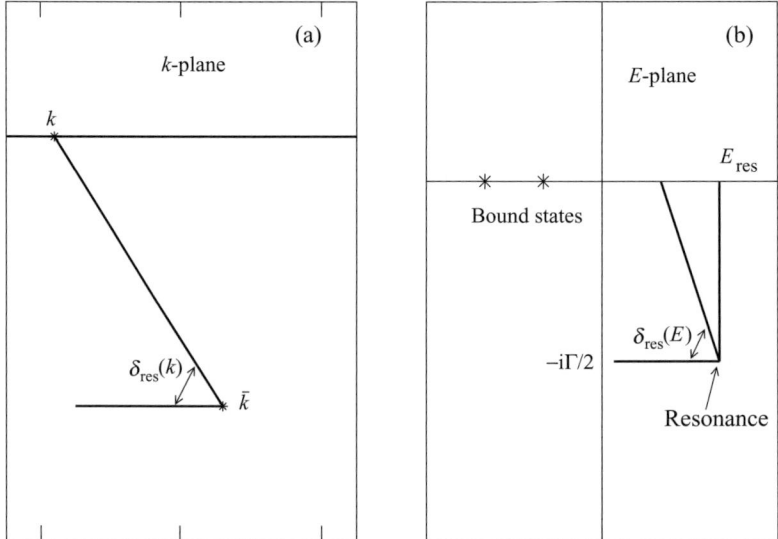

Figure 3.5 Schematic representation of the resonance part of the phase shift
(a) Representation in k-plane showing $\delta_{\text{res}}(k)$
(b) Representation in E-plane with bound states (stars) and displaying $\delta_{\text{res}}(E)$, resonance position is $E = E_{\text{res}} - i\Gamma/2$

This is the Breit–Wigner form for the cross section. The multichannel case is more complicated, but straightforward to derive. In this case the Breit–Wigner form for the T-matrix near a resonance becomes

$$T_{ll'} \propto \frac{\gamma_l \gamma_l'}{((E - E_r)^2 + \Gamma^2/4)}, \qquad (3.19)$$

where γ_l is the partial width.

3.3 Formal resonance theory

Having introduced the concept of a resonance, it must now be related to the dynamics of the dissociative recombination process. Let us first examine the simplest case of dissociative recombination as advanced by Bates (1950b) and following the derivation of O'Malley (1966). In this picture, the ground state of the ion is crossed by an excited state of the neutral which is embedded in the background of the electron + ion system (see Fig. 3.1(a)). It is possible to construct a Q-space projection operator that separates the wavefunction into the background and the resonances. This divides the wavefunction into two subspaces P and Q. These operators are assumed to be orthonormal, i.e.,

$$PP = QQ = 1 \qquad (3.20)$$

3.3 Formal resonance theory

and

$$PQ = QP = 0. \qquad (3.21)$$

Q in the following derivation is taken to be a single dissociative state. The generalization to several resonant states is obvious. Therefore

$$\Psi = P\Psi + Q\Psi \qquad (3.22)$$

and

$$Q = |\Phi\rangle \langle \Phi|, \qquad (3.23)$$

where Φ is the electronic wavefunction of the resonant state, which is a function of both R, the internuclear separation, and q, the electronic coordinates. For simplicity we treat the case of a diatomic, in the general case R becomes a vector. However, the derivation is easily generalized. Note that when Q operates on the Hamiltonian or the wavefunction, the integration is over electronic coordinates. Starting with the time-independent Schrödinger equation

$$H\Psi = E\Psi, \qquad (3.24)$$

substitute Eq. (3.22) to obtain

$$HP\Psi + HQ\Psi = EP\Psi + EQ\Psi. \qquad (3.25)$$

Now operate with P on both sides of the equation

$$PHP\Psi + PHQ\Psi = PEP\Psi + PEQ\Psi. \qquad (3.26)$$

But using Eqs. (3.20), (3.21), and (3.22) one obtains

$$(E - PHP)P\Psi = PHQ\Psi. \qquad (3.27)$$

Similarly by operating with Q the following equation can be derived:

$$(E - QHQ)Q\Psi = QHP\Psi. \qquad (3.28)$$

Formally solving Eq. (3.27), with the boundary condition of incoming flux in the P-space, we obtain:

$$P\Psi_{k,n}^{+} = P\Psi_{k,n}^{0,+} + (E - PHP + i\eta)^{-1} PHQ\Psi_{k,n}^{+}. \qquad (3.29)$$

The states are indexed by k, the incident wave vector of the incoming scattering electron, and the initial internal state n. Note also that in the definition of the Green's function $(E - PHP + i\eta)^{-1}$ we have added the term $i\eta$, as standard for convergence, and obtained the outgoing wave equation now indicated with $\Psi_{k,n}^{+}$ instead of $\Psi_{k,n}$.

We then use this result for $P\Psi_{k,n}^+$ in Eq. (3.28) leading to the following equation for $Q\Psi_{k,n}^+$:

$$(E - QHQ)Q\Psi_{k,n}^+ = QHP\Psi_{k,n}^{0,+} + QHP(E - PHP + i\eta)^{-1}PHQ\Psi_{k,n}^+. \quad (3.30)$$

We will assume that the Born–Oppenheimer approximation is valid and that $Q\Psi_{k,n}^+$ can be represented as a product of the electronic and nuclear wavefunction, i.e., the form:

$$Q\Psi_{k,n}^+ = |\Phi\rangle \, \xi_{k,n}(R). \quad (3.31)$$

where $\xi_{k,n}(R)$ are nuclear wavefunctions depending on the initial rovibrational state of the molecule and the initial momentum. Multiplying by $\langle\Phi|$ and integrating over the electronic coordinates, we derive the equation for $\xi_{k,n}$, the nuclear portion of the $Q\Psi_{k,n}^+$. The operator $(E - QHQ)$ becomes:

$$\langle\Phi|\,(E - QHQ\,|\Phi\rangle\,\xi_{k,n}(R) = E\xi_{k,n}(R) - \langle\Phi|\,H\,|\Phi\rangle\,\xi_{k,n}(R), \quad (3.32)$$

but

$$H = T + V, \quad (3.33)$$

where T is the nuclear kinetic energy operator and V is the potential. Working within the Born–Oppenheimer approximation, and defining

$$\varepsilon(R) = \langle\Phi|\,V\,|\Phi\rangle, \quad (3.34)$$

the "diabatic potential curves" (Estrada & Domcke 1989), the left hand side of Eq. (3.30) is

$$(E - T - \varepsilon(R))\xi_{k,n}(R), \quad (3.35)$$

$QHP\Psi_{k,n}^{0,+}$ becomes $\langle\Phi|HP\Psi_{k,n}^{0,+}\rangle$ and the coupling term $QHP(E - PHP + i\eta)^{-1}PHQ$ becomes

$$\langle\Phi|\,HPG_{PP}^+PH\,|\Phi\rangle\,\xi_{k,n}(R). \quad (3.36)$$

To simplify this term further, expand G_{PP}^+ in a spectral representation so the expression becomes:

$$\langle\Phi|HPG_{PP}^+PH|\Phi\rangle\xi_{k,n}(R)$$
$$= \sum_v P \int dE' \frac{\langle\Phi|H|P\Psi_{k,v}^{0,+}\rangle}{E - E'} \int dR' \langle P\Psi_{k,v}^{0,+}|H|\Phi\rangle\xi_{k,n}(R')$$
$$- i\pi \sum_v \langle\Phi|H|P\Psi_{k,v}^{0,+}\rangle \int dR' \langle P\Psi_{k,v}^{0,+}|H|\Phi\rangle\xi_{k,n}(R'), \quad (3.37)$$

where $P \int dE'$ represents a principal value integral and the two sums over v run over the open vibrational channels. Make the Born–Oppenheimer separation for $P\Psi_{k,n}^{0,+}$, i.e.,

$$P\Psi_{k,n}^{0,+} = |\psi_{k_v}^+\rangle \chi_n(R), \tag{3.38}$$

so

$$i\pi \sum_v \langle\Phi|H|P\Psi_{k,v}^{0,+}\rangle \int dR' \langle P\Psi_{k,v}^{0,+}|H|\Phi\rangle \xi_{k,n}(R')$$
$$= i\pi \int dR' \sum_v |\chi_v(R)\rangle\langle\chi_v(R')|\langle\Phi|H|\psi_{k_v}^+\rangle\langle\psi_{k_v}^+|H|\Phi\rangle \xi_{k,n}(R'). \tag{3.39}$$

At sufficiently high energy that the sum over open vibrational channels becomes complete, one obtains the "boomerang" model (Herzenberg, 1968, Birtwhistle & Herzenberg 1971, Dubé & Herzenberg, 1975), and

$$\sum_v |\chi_v(R)\rangle\langle\chi_v(R')| = \delta(R - R'), \tag{3.40}$$

so

$$i\pi \int dR' \sum_v \langle\Phi|H|\psi_{k_v}^+\rangle\langle\psi_{k_v}^+|H|\Phi\rangle \xi_{k,n}(R') = i\pi |\langle\Phi|H|\psi_{k_v}^+\rangle|^2 \xi_{k,n}(R). \tag{3.41}$$

Define

$$\Gamma(R, k_v) = 2\pi |\langle\Phi|H|\psi_{k_v}^+\rangle|^2. \tag{3.42}$$

Then

$$i\pi \sum_v \langle\Phi|H|P\Psi_{k,v}^{0,+}\rangle \int dR' \langle P\Psi_{k,v}^{0,+}|H|\Phi\rangle \xi_{k,n}(R') = \frac{i\Gamma(R, k_v)}{2} \xi_{k,n}(R) \tag{3.43}$$

and the principal value term is incorporated into a diagonal energy shift. The dependence of the width Γ on energy is usually ignored, using a high-energy argument as before. This leads to the following equation for the nuclear wavefunctions:

$$\left[E - T - \varepsilon(R) + \frac{i\Gamma(R)}{2}\right] \xi_{k,n}(R) = \sqrt{\frac{\Gamma_i(R)}{2\pi}} \chi_{n,k}(R). \tag{3.44}$$

This equation is general and, for example, can be applied to the case of dissociation attachment. A similar expression, using some of the same approximations, that specifically addresses the dissociative recombination process, including the indirect mechanism involving Rydberg states, has been derived by Giusti-Suzor, Bardsley, and Derkits (1983). In this formulation the wavefunction is expanded as:

$$\Psi(q, R) = \sum_v \int dE' c_v(E') \psi_{v,E'}(q, R) \chi_v(R) + \Phi(q, R) \xi_d(R)$$
$$+ \sum_{p,v} d_{v,p} \psi_p(q, R) \chi_v(R). \tag{3.45}$$

The first sum, indexed by v, is summed over vibrational states, the second sum is summed over both vibrational states and the Rydberg states, indexed by p. The first term describes the background scattering, the second term is the resonant state, and the third term describes coupling to the Rydberg states. Note that $\Psi(q, R)$ is given by the Schrödinger equation:

$$[T + H(q, R) - E]\Psi(q, R) = 0. \tag{3.46}$$

The wavefunction for the resonant state is given by $\xi_d(R)$. As in the previous derivation, to solve for this multiply through by $\Phi(q, R)$ and integrate. This leaves:

$$(T + E_{\text{res}} - E)\xi_i(R) = -V_E(R)\left[\chi_i(R) - i\pi \sum_v \langle \chi_v | V_E | \xi_i \rangle \chi_v \right]$$
$$+ \left\{ \sum_{p,v} \rho_p d_{p,v} \left[\chi_v(R) - i\pi \sum_{v'} W_{vv'} \chi_{v'}(R) \right] \right\}, \tag{3.47}$$

where the dissociative wavefunction has been indexed by i, the initial vibrational wavefunction of the ion, ρ_p is the density of Rydberg states, and

$$\rho_p W_{vv'} = \langle \psi_p(q, R)\chi_{v'}(R) | T | \psi_{vE}(q, R)\chi_v(R) \rangle \tag{3.48}$$

is the nonadiabatic coupling between Rydberg states. We have also used Eq. (3.42):

$$V_E(R) = \left[\frac{\Gamma(R, k_v)}{2\pi} \right]^{1/2} = \langle \Phi | H | \psi_{v,E}(q, R) \rangle. \tag{3.49}$$

Note that if coupling to the Rydberg states is ignored, and the sum of v is complete, this equation reduces to Eq. (3.44). Equation (3.47) is the basic equation that must be solved for the dissociative recombination process.

Equation (3.47), or the local approximation to this expression, Eq. (3.44), can be used to obtain information about the angular distribution of fragments produced by dissociative recombination. Early work in this area in the case of dissociative attachment was carried out by O'Malley and Taylor (1968) and then expanded to include the case of dissociative recombination by Guberman (2004). This work assumed the axial recoil limit, i.e., that the dissociation is prompt so the recoil axis does not rotate, sometimes called the "slow rotation limit." Within this approximation, which is valid for most applications, and if the electron capture is dominated by a single partial wave, then the square of that partial wave spherical harmonic is the product angular distribution. For example, if the predominant partial wave is $l = 0, m = 0$ this would lead to an isotropic distribution, whereas a collision where $p\sigma$, $l = 1, m = 0$ dominates would lead to a $\cos^2 \theta$ distribution. An interesting extension of this work to the case where multiple partial waves contribute has been derived for dissociative attachment (Haxton et al. 2006) and applied to the case of H_2O and H_2S. Although a similar method should apply to dissociative recombination, this work has yet to be done.

We will now discuss the calculation of the necessary resonance curves and couplings needed as input before this equation can be solved.

3.4 Resonance parameters and structure

The total dissociative recombination, the final state distributions, and the angular distributions can only be accurately determined if the input to the dynamics calculations is accurate. We will outline below how such parameters are obtained.

As was shown in Section 3.2 it is possible to obtain the electronic resonance parameters, $E_{\text{res}}(R)$, the position of the resonance, and $\Gamma(R)$, the autoionization width (which is related to the coupling), by analyzing the S-matrix from state-of-the-art electron–molecular ion scattering calculations. For single, isolated resonances, the eigenphase sum is fit to a Breit–Wigner form. For more complicated cases, a multichannel generalization can be used (Orel, Kulander, & Rescigno 1995). In addition, as resonances become closer spaced in energy, it is more accurate to fit the time-delay matrix (Stibbe & Tennyson 1996).

The scattering methods used to study these systems must have the flexibility to use the highly accurate configuration interaction wavefunctions that are necessary to study these resonant systems. These calculations extend beyond static exchange and include the effects of polarization, producing accurate reliable cross sections and resonance parameters. Two such methods have been used to obtain the necessary couplings and surfaces for dissociative recombination, the complex Kohn variational method and the R-matrix method. Both methods work with the fixed-nuclei approximation, i.e., that the positions of the nuclei are fixed, and then the electron scattering calculation is carried out.

These methods employ a closed coupling expansion for the wavefunction that, in the case of electronically elastic scattering, is of the form:

$$\Psi = \sum_{\Gamma} A\Phi_0(\mathbf{r}_1 \ldots \mathbf{r}_N) F_{\Gamma}(\mathbf{r}_{N+1}) + \sum_{\mu} d_{\mu} \Theta_{\mu}(\mathbf{r}_1 \ldots \mathbf{r}_{N+1}), \quad (3.50)$$

where $A\Phi_0(\mathbf{r}_1 \ldots \mathbf{r}_N) F_{\Gamma}(\mathbf{r}_{N+1})$ represents the antisymmetrized product of Φ_0, the initial target wavefunction of the ion and the scattered wavefunction $F_{\Gamma}(\mathbf{r}_{N+1})$ and the second sum over μ contains square-integrable, $(N+1)$ electron configuration state functions described further below. The first sum, which we denote as the P-space portion of the wavefunction, runs over the energetically open target states. We denote the second sum (Q-space) as the correlation portion of the wavefunction. This describes polarization and/or correlation effects due to electronically closed channels.

Note that the P- and Q-spaces referred to here and in the literature for electron scattering are similar to, but not exactly the same as, those used in Section 3.3 and in that literature. In both cases, *P*-space refers to the background space of an electron and the molecular ion in a given electronic state. *Q*-space in this case includes the resonant state, but also includes other correlation/polarization terms that are not part of the resonant state. Also at sufficiently high energies, *P*-space may include more than one term if more than one channel is energetically open. For example, in the case of electron scattering from HCN^+ up to energies of ~ 0.9 eV only the $^2\Pi$ state of the ion is an open target state. This state is degenerate so the sum in Eq. (3.48) would run over the two degenerate components of the $^2\Pi$ state. At ~ 0.9 eV the next state of the ion, $^2\Sigma$, is energetically open so now the sum would run over three energetically open target states.

For the correlation part of the wavefunction, two classes of terms are generally included. The first class is the set of all $(N+1)$-electron configuration state functions that can be formed from the active space of target orbitals. These are generally referred to as "penetration terms" (Rescigno *et al.* 1995, Rescigno, Lengsfield, & McCurdy 1995). Since the scattering functions $F(\mathbf{r}_{N+1})$ are constructed from bound and continuum functions which are, by construction, orthogonal to the target orbitals, the penetration terms are needed to relax any constraints implied by this strong orthogonality. In addition to the penetration terms, a second class of "CI relaxation terms" (Rescigno *et al.* 1995, Rescigno, Lengsfield, & McCurdy 1995) is needed. The target ground state is built as a fixed linear combination of a number of electron configuration state functions, say M, from the active space. The target CI calculation can also produce $(M-1)$ excited states, which are presumed to be energetically closed. The CI relaxation terms are constructed as the direct product of these states and the orbitals used to describe the scattered electron. In other words, this class of CI relaxation terms is simply the complement $(1-P)$ of the *P*-space portion of the wavefunction. This complement, combined with the penetration terms, constitutes the correlation part of the wavefunction.

It is at this point that the two methods split, employing different methods to solve the scattering equations. The details for each will be outlined below.

3.4.1 Complex Kohn variational method

The complex Kohn variational method is an algebraic variational technique that has been developed to study both heavy-particle (reactive) collisions (Zhang & Miller 1987) and electron scattering problems. It uses a trial wavefunction that is expanded in terms of square-integrable (Cartesian Gaussian) and continuum basis functions that incorporate the correct asymptotic boundary conditions. Detailed descriptions of the method have been given in several review articles (see, for instance, Rescigno *et al.* (1995), Rescigno, McCurdy, and Lengsfield (1995)), so only a brief summary is included.

3.4 Resonance parameters and structure

In Eq. (3.50), the scattering function, $F_\Gamma(\mathbf{r}_{N+1})$ that represents the wavefunction of the scattered electron is further expanded in a combined basis of linear combination of symmetry-adapted molecular orbitals (Gaussians) ϕ_i and numerical continuum (regular f_l and outgoing g_l^+ Coulomb functions) basis functions:

$$F_\Gamma(\mathbf{r}) = \sum_i c_i \phi_i(\mathbf{r}) + \sum_{l,m} \left[f_l(kr)\delta_{ll_0}\delta_{mm_0} + T_{ll_0mm_0} g_l^+(kr) \frac{Y_{lm}(\hat{r})}{r} \right], \qquad (3.51)$$

where $Y_{lm}(\hat{r})$ are spherical harmonics. The (N+1)-electron electron configuration state functions describe short-range correlations and the effects of closed channels and are critical to striking a proper balance between intra-target electron correlation and correlation between target and scattered electrons. Applying the stationary principle for the T-matrix,

$$T_{\text{stat}} = T_{\text{trial}} - 2\int \Psi(H-E)\Psi, \qquad (3.52)$$

results in a set of linear equations for the coefficients c_i, d_μ and $T_{ll_0mm_0}$. The T-matrix elements are the fundamental dynamical quantities from which all fixed-nuclei cross sections are derived. In solving the variational equations, Feshbach partitioning is used to combine the penetration and relaxation terms into an optical potential. For more details on this subject, we refer the reader to Rescigno *et al.* (1995), Rescigno, McCurdy, and Lengsfield (1995).

3.4.2 R-matrix method

The R-matrix method is also a variational method, and was first used to study electron–atom scattering (Burke, Hibbert & Robb 1971), then diatomics (Schneider & Hay, 1976), and now polyatomic systems (Gillan, Tennyson, & Burke 1995, Morgan, Tennyson, & Gillan 1998). In the R-matrix method, the electron configuration space is divided into two regions. The first is an inner spherical region, consisting of a sphere of radius 10–20 a_0, centered at the center of gravity of the molecule. This radius depends on the application and the spatial dimensions of the wavefunctions involved. It is necessary that the charge density for the target (N-electron) wavefunction be contained within the sphere. In this region, the full interactions of the ($N+1$) system must be described, including exchange. The ($N+1$) problem is solved within the sphere, i.e., all integrals are done on 0 to R (the R-matrix radius). As in the complex Kohn method, the number of configurations is the number of configurations used to describe the target times the number of continuum orbitals (see Eq. (3.50)). Therefore, in both methods there must be a compact target space, usually consisting of no more than a few thousand configurations. In both the R-matrix method and the complex Kohn method for polyatomic molecules, the target wavefunction is expanded in Gaussian-type orbitals. The R-matrix method has the advantage that for diatomics, the more accurate (and more compact) Slater-type orbitals can be employed. The solutions obtained inside the sphere are matched at the boundary to the asymptotic form of the wavefunction. Details of this procedure are reviewed elsewhere (Gillan, Tennyson, & Burke 1995, Morgan, Tennyson, & Gillan, 1998). This matching procedure yields the cross sections and eigenphases that can be fit to obtain resonance information. The R-matrix method has the advantage that the extra computational cost for each electron scattering energy is small, compared to the complex Kohn method, so a fine grid in energies, needed when many resonances are present, is easy to obtain.

3.4.3 Structure methods

Both the R-matrix and the complex Kohn method yield direct information about the resonant states before they cross the ground state of the molecular ion. Standard quantum chemistry methods yield information about the Rydberg states and the resonance states after the crossing. However, it is also possible to use these methods to obtain information about the resonance position and width. Care must be taken. If one carries out a standard quantum chemistry calculation, diagonalization will yield roots that lie above the ground state of the molecular ion. These states represent, for the most part, the molecular ion with an electron in a diffuse orbital. These curves parallel the ion curve and as more diffuse basis functions are added, they drop in energy closer to the ionic curve. True resonant states, representing a double excitation from the electron plus molecular ion, in general do not parallel the ground state ion curve and do not change significantly in energy as the basis becomes more diffuse. This is the basis for the stabilization method of Hazi and Taylor (1970) used in studies of dissociative attachment resonances.

An example of this is shown in Fig. 3.6 for the case of H_2^+. In the first case, a full two-electron configuration interaction (CI) calculation in a basis set of 36 Gaussian-type basis functions for the H_2 system is carried out. The roots of the CI matrix in the case of singlet spin coupling are plotted in Fig. 3.6(a) as a function of the internuclear separation. The ion curve is shown as a dashed line. Note the series of curves that appear above the ion, but which parallel the ion. These are not resonant states. One can track the resonance, which appears as a "ripple" in these states. An approximation to the resonant state, shown as a dotted curve in the figure, is calculated by restricting the CI, requiring the lowest molecular orbital to be unoccupied (an approximation to the exact method described below). As the more diffuse functions are added to the basis, these curves drop in energy and more appear. This can clearly be seen in Fig. 3.6(b) where the results of a similar

3.4 Resonance parameters and structure

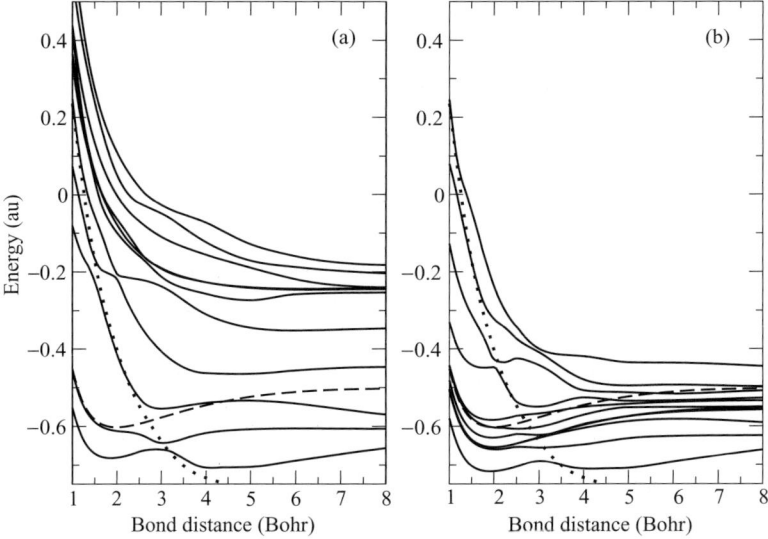

Figure 3.6 Roots of a configuration interaction (CI) calculation. H$_2$ solid curves; H$_2^+$ dashed curve; approximate resonance position obtained from restricted CI dotted curve: (a) full two-electron CI in basis set of 36 functions; (b) full two-electron CI in basis set of 59 functions.

calculation, but using a much larger basis set of 59 basis functions, are shown. Again it is possible to see the diabatic resonant state (an approximation to which is shown by the dotted curve) in its effect on these background states.

A variety of approximate methods have been tried to project out the diabatic resonance curve. This can be done exactly in the case of a one-electron ion (Bottcher 1974, O'Malley & Geltman 1965), but only approximately for a many-electron system. For example, consider the case of H$_2$, for the first (Q_1) set of resonances (Guberman, 1983b). One can define

$$P(1) = |1\sigma_g(1)\rangle\langle 1\sigma_g(1)| \tag{3.53}$$

and

$$Q(1) = 1 - P(1), \tag{3.54}$$

where $1\sigma_g(1)$ is the exact wavefunction for the ground state of H$_2^+$. The resonant states are two-electron states so a two-electron projection operator must be defined as

$$P(1, 2) = P(1) + P(2) - P(1)P(2) \tag{3.55}$$

and

$$Q(1,2) = Q(1)Q(2) = 1 - P(1,2). \tag{3.56}$$

However, as the number of electrons increases, the definition of the P and Q operators and the procedure to implement their application are more complicated. Usually approximations are made, such as the use of a restricted basis that contains no diffuse Rydberg-like basis functions. This will only approximate the diabatic state and in the case of strong mixing can induce errors in the curves.

The resonance width can also be approximated using quantum chemistry techniques. Using Fermi's Golden Rule, the width can be written as:

$$\Gamma(R) = 2\pi\rho \, \langle A\{\Phi_{\text{ion}}(q,R)\phi_n(q,R)\}|\, H\, |\Phi_{\text{d}}(q,R)\rangle^2 \tag{3.57}$$

where ρ is a density of states, and A is the antisymmetrizer operator, which antisymmetrizes the product of the multielectron wavefunction for the ion and the wavefunction for the free electron. H is the Hamiltonian and $|\Phi_{\text{d}}(q,R)\rangle$ is the resonance wavefunction. The integration is over the electronic coordinates q. $\Phi_{\text{ion}}(q,R)$ is a compact valence state, so the matrix element is only nonzero for the region near the molecule over which it has nonzero amplitude. Therefore, only that region of space close to the molecule contributes. In this region, a diffuse Rydberg orbital (with high n) has approximately the same behavior as a free continuum Coulomb orbital. The difference due to the normalization can be taken into account through the density of states factor. Therefore it is possible to carry out the calculation with a diffuse basis and obtain the autoionization width:

$$\langle A\{\Phi_{\text{ion}}(q,R)\phi_E(q,R)\}|\, H\, |\Phi_{\text{d}}(q,R)\rangle$$
$$= \langle A\{\Phi_{\text{ion}}(q,R)\phi_n(q,R)\}|\, H\, |\Phi_{\text{d}}(q,R)\rangle\, n_{\text{eff}}^{1.5}(R), \tag{3.58}$$

with n_{eff}, the effective quantum number, given by:

$$(E_{\text{ion}}(R) - E_n(R)) = \frac{1}{2n_{\text{eff}}^2} = \frac{1}{2(n-\mu)^2}, \tag{3.59}$$

where μ is the quantum defect. If, as is commonly the case in dissociative recombination, the resonance is a member of a Rydberg series converging to an excited state of the ion, this would also imply that the contribution from higher resonances, i.e., with larger effective quantum numbers, is smaller. This allows the treatment of contributions of lower members of the series to be included explicitly, but then summing contributions of the remainder of the (infinite) Rydberg series into an "effective" state (Strömhold et al., 1995).

3.5 Nonadiabatic couplings

As shown in Fig. 3.2(b), dissociative recombination can occur by direct predissociation of the ion itself. It is necessary in this process to deal with a slightly different type of coupling between the (electron + ion) and excited neutral system.

This predissociation is a nonradiative transition to a dissociative state caused by a breakdown in the Born–Oppenheimer approximation. This breakdown induces a coupling between the adiabatic Born–Oppenheimer states. The coupling arises from the terms of the kinetic energy operator neglected in the Born–Oppenheimer approximation. Consider the matrix element of the nuclear kinetic energy for the case of a diatomic. If we label two of these electronic states Ψ_i and Ψ_j, it is given by:

$$\left\langle \Psi_i \left| -\frac{1}{2\mu} \frac{1}{R^2} \frac{\partial}{\partial R} R^2 \frac{\partial}{\partial R} \right| \Psi_j \right\rangle = -\frac{1}{2\mu} \left\langle \Psi_i \left| \frac{2}{R} \frac{\partial}{\partial R} + \frac{\partial^2}{\partial R^2} \right| \Psi_j \right\rangle, \quad (3.60)$$

where μ is the reduced mass and the matrix element represents integration over electronic coordinates only. This leads to nonadiabatic coupling of the form given by Lefebvre-Brion and Field (1986):

$$F^{ij} = F^{(1)}_{ij} + F^{(2)}_{ij}, \quad (3.61)$$

with

$$F^{(1)}_{ij} = -\frac{1}{2\mu} \left\langle \Psi_i \left| \frac{\partial}{\partial R} \right| \Psi_j \right\rangle \frac{\partial}{\partial R} \quad (3.62)$$

and

$$F^{(2)}_{ij} = -\frac{1}{2\mu} \left\langle \Psi_i \left| \frac{\partial^2}{\partial R^2} \right| \Psi_j \right\rangle. \quad (3.63)$$

$F^{(2)}_{ij}$ has, in general, been neglected (Guberman 1994, Lefebvre-Brion & Field 1986). The error in this approximation can be estimated by writing $F^{(2)}_{ij}$ in terms of $F^{(1)}_{ij}$, i.e.,

$$F^{(2)}_{ij} = \sum_{k \neq i,j} \left\langle \Psi_i \left| \frac{\partial}{\partial R} \right| \Psi_k \right\rangle \left\langle \Psi_k \left| \frac{\partial}{\partial R} \right| \Psi_j \right\rangle + \frac{\partial}{\partial R} \left\langle \Psi_i \left| \frac{\partial}{\partial R} \right| \Psi_j \right\rangle. \quad (3.64)$$

The sum over k is over a complete set of states. The corresponding equations for a polyatomic are straightforward, but more complicated (Tennyson & Sutcliffe 1982, Lengsfield & Yarkony 1992).

3.5.1 Direct calculation

Calculation of these necessary coupling elements is complicated. The matrix element is not a simple one-electron property, but requires a change in the electronic wavefunction as nuclear coordinates are changed. There are a number of approximate methods for obtaining these coupling elements. The most common of these is the use of the Hellmann–Feynman theorem (Feynman 1939):

$$(E_j - E_i)\left\langle \Psi_i \left| \frac{\partial}{\partial R} \right| \Psi_j \right\rangle = \left\langle \Psi_i \left| \frac{\partial V}{\partial R} \right| \Psi_j \right\rangle. \quad (3.65)$$

The matrix element of $\partial V/\partial R$ can be expressed as the sum of expectation values of the electric dipole operator (a one-electron property) and therefore easily calculated using standard molecular structure techniques. However, this is approximate, becoming accurate only in the limit of a complete set, and has been found to yield results that are qualitatively correct, but quantitatively inaccurate (Orel & Kulander, 1983). Another approximate method, which has been more successful in producing more quantitatively accurate couplings, was proposed by Sidis (1971) for diatomic systems. Using a series of commutator relationships, again with the assumption of a complete basis, it can be shown that:

$$\left\langle \Psi_i \left| \frac{\partial}{\partial R} \right| \Psi_j \right\rangle = R^{-1} \left[\frac{1}{2}(E_i - E_j)\left\langle \Psi_i \left| \sum_k r_k^2 \right| \Psi_j \right\rangle \right.$$
$$\left. + (E_i - E_j)^{-1}\left\langle \Psi_i \left| T \right| \Psi_j \right\rangle \right], \quad (3.66)$$

where T is the kinetic energy operator. Again these are one-electron properties that are easily evaluated. Finally the derivative can be calculated numerically, i.e., by calculating the wavefunction at a position R and a slightly displaced position $(R + \Delta)$:

$$\left\langle \Psi_i \left| \frac{\partial}{\partial R} \right| \Psi_j \right\rangle = \lim_{\Delta \to 0} \langle \Psi_i(R) | \Psi_j(R + \Delta) \rangle$$
$$= \lim_{\Delta \to 0} \frac{1}{\Delta} \sum_l c_l^i(R) c_l^j(R + \Delta)$$
$$+ \lim_{\Delta \to 0} \frac{1}{\Delta} \sum_{k,l; k \neq l} c_l^i(R) c_k^j(R) \langle \chi_l^i(R) | \chi_k^j(R + \Delta) \rangle, \quad (3.67)$$

where $c_l^i(R)$ is the CI coefficient of the lth Slater determinant, χ_l in the ith electronic state. The sum only runs over Slater determinants that differ by a single orbital. This expression is valid to second order, and can yield accurate results. However, the implementation with large-scale CI wavefunctions is plagued with numerical difficulties and cumbersome to implement, requiring at least two points in the R-grid to calculate a coupling at a single point, thereby doubling the size of the calculation.

It is also possible to calculate these nonadiabatic coupling elements analytically (Lengsfield & Yarkony 1992). Quantum chemistry codes that supply such matrix elements are specialized and not generally available. In addition, such methods require a particular form for the wavefunction. The basis set and the wavefunction calculation itself must be more extensive to produce accurate results. Hence, many calculations use the more approximate methods. Such analytic methods have been used to calculate the coupling elements for predissociation of H_3 (Schneider & Orel 1999) and dissociative recombination of HeH^+ (Larson & Orel 2004) and H_3^+ (Schneider, Suzor-Weiner, & Orel 2000).

The preceding discussion addresses the general problem of nonadiabatic couplings. In the case of dissociative recombination what is needed is the coupling between a given Rydberg state and the system of {ion + free low-energy electron}. The shape of the neutral curves in the Franck–Condon region parallels the ion and the shapes of the electronic wavefunctions for high Rydberg states differ mainly in the asymptotic region. Their energies are well represented as a Rydberg series. Therefore, as in the case of the "direct" coupling elements, the coupling from the ground state to the members of a Rydberg series can be scaled to produce the coupling from a given state, N, to the {ion + free low-energy electron}, i.e.,

$$\left\langle \Psi_{\text{continuum}} \left| \frac{\partial}{\partial R} \right| \Psi_N \right\rangle = \left\langle \Psi_{\text{Rydberg}} \left| \frac{\partial}{\partial R} \right| \Psi_N \right\rangle n_{\text{eff}}^{1.5}(R), \qquad (3.68)$$

where n_{eff} is the effective quantum number given by Eq. (3.59).

As described previously (see Section 3.4.3) the coupling scales with $n_{\text{eff}}^{-1.5}$, so its magnitude decreases for higher members of a Rydberg series. In addition, this form also leads to a more qualitative error estimate in the neglect of the second-order term given in Eq. (3.64). Since each term in that sum scales with $n_{\text{eff}}^{-1.5}$, the product will scale as n_{eff}^{-3}. Therefore, the contribution will be smaller with increasing n.

3.5.2 Quantum defect method

The methods described above represent difficult and tedious calculations. Another method to obtain the nonadiabatic couplings has been proposed that shows promise for more efficient calculation and may be of sufficient accuracy to be useful. This method has its roots in the area of quantum defect theory and relies on the identification of a series of curves that can be identified by the l quantum number. The method was developed by Stolyarov, Pupyshev, and Child (1997) and Kiyoshima et al. (2003). Instead of scaling the matrix element of the nuclear kinetic energy operator directly, the Hellman–Feynman theorem is first applied

$$F^{(1)}_{nl,n'l'}(R) = \frac{\left\langle \Psi_{nl} \left| \frac{\partial V}{\partial R} \right| \Psi_{n'l'} \right\rangle}{E_{n'l'} - E_{nl}} \tag{3.69}$$

and the matrix element in the nominator of Eq. (3.69) is then scaled as in Eq. (3.58) by the quantum defects. These couplings elements will be referred to as the Hellman–Feynman (HF) couplings. In these applications (Stolyarov, Pupyshev, & Child 1997, Kiyoshima et al. 2003) only the interaction between the Rydberg states with the same l quantum number is studied. The nonadiabatic coupling to the electron continuum is then given by

$$F^{(1)}_{nl,\varepsilon l}(R) = -\frac{2\sqrt{v_{nl}}\mu'_l}{1 + 2v^2_{nl}\varepsilon}, \tag{3.70}$$

where $\mu'_l = d\mu_l/dR$. The HF couplings between the bound Rydberg states of same l can be calculated from

$$F^{(1)}_{nl,n'l}(R) = \frac{2\sqrt{v_{nl}v_{n'l}}\mu'_l}{v^2_{nl} - v^2_{n'l}}. \tag{3.71}$$

In order to obtain the HF couplings between states with different values of l, *ab initio* calculated coupling elements may be used to calculate an l-mixing function by using a scaling of Eq. (3.71) to the electron continuum functions. By defining the l-mixing functions as

$$\tilde{f}_{ll'}(R) = \frac{\left(v^2_{n'l'} - v^2_{nl}\right) F_{nl,n'l'}}{2\sqrt{v_{nl}v_{n'l'}}}, \tag{3.72}$$

the continuum HF couplings are then given by

$$F^{(1)}_{nl,\varepsilon l}(R) = \frac{2\sqrt{v_{nl}}\tilde{f}_{ll}}{1 + 2v^2_{nl}\varepsilon} \tag{3.73}$$

and similarly the bound–bound couplings are given by

$$F^{(1)}_{nl,n'l'}(R) = \frac{2\sqrt{v_{nl}v_{n'l'}}\tilde{f}_{ll'}}{v^2_{nl} - v^2_{n'l'}}. \tag{3.74}$$

Note that the HF continuum couplings in Eqs. (3.69) and (3.72) depend on the energy of the continuum electron. However, this dependence on this energy has not been found to be very strong in the cases that have been considered (Larson & Orel 2005).

Another advantage of this method is that it is possible to estimate the second-order couplings that were ignored in the direct calculation case of Section 3.5.1. These can be expressed as:

$$F^{(2)}_{nl,n'l}(R) = \frac{2\sqrt{\upsilon_{nl}\upsilon_{n'l}}}{\upsilon_{nl}^2 - \upsilon_{n'l}^2}\left[\mu_l'' - \frac{(\mu_l')^2}{\upsilon_{nl}^2 - \upsilon_{n'l}^2}\left(3\upsilon_{n'l} + \frac{\upsilon_{nl}^2}{\upsilon_{n'l}}\right)\right], \quad (3.75)$$

which describes the coupling between two Rydberg states with the same l quantum number.

$$F^{(2)}_{nl,\varepsilon l}(R) = \frac{2\sqrt{\upsilon_{nl}}}{1 + 2\varepsilon\upsilon_{nl}^2}\left[\mu_l'' + \frac{(\mu_l')^2}{\upsilon_{nl}}\left(3 - \frac{4}{1 + 2\varepsilon\upsilon_{nl}^2}\right)\right] \quad (3.76)$$

describes the coupling of a given Rydberg state with quantum number l to the continuum (with the same l). It is not possible to obtain the couplings between different l states.

3.6 Calculation of dynamics

3.6.1 Multichannel quantum defect theory

Multichannel quantum defect theory (MQDT) is the most successful and most widely used method of obtaining accurate dissociative recombination cross sections. The method has been applied to a number of diatomics and extended to triatomics by Kokoouline (Kokoouline & Greene 2003a,b). In the MQDT treatment, bound and continuum states are handled simultaneously and on an equal footing. This was the first method that correctly described the direct and indirect dissociative recombination paths and allowed interference between them (Giusti 1980). The method has been reviewed a number of times, for example, Florescu-Mitchell and Mitchell (2006) in which the description is quite detailed. The method will be summarized here.

MQDT interpolates the concept of the quantum defect for bound states (electron energy is negative) to the phase shift (electron energy is positive). The system is divided into channels. Each channel is a set of states with a common ionic core in a specific rovibrational level and an electron of energy E. No restriction is placed on the electron energy. It can be negative or positive, so the state can be bound or free. Therefore, a channel can allow ionization ($AB^+ + e^-$) or dissociation ($A^* + B$). It is characterized by the threshold energy for fragmentation (which determines whether the channel is open or closed) and a set of quantum numbers ($n, v, l, \ldots$). The electron configuration space is then divided into different regions. First, there is the short-range region, the region of strong interactions. (This is similar to the idea used in R-matrix theory of the inner region.) In the short-range region, nothing is done to distinguish whether channels are open or closed, or states are bound or free. In this region one obtains phase shifts and mixing coefficients. The next region is the external region. Here one uses the correct asymptotic wavefunctions, which are linear combinations of Coulomb wavefunctions. It is only at this point that open and closed channels are distinguished. A frame transformation is used to connect the two regions.

First consider a single-channel case, with a short-range potential, $V(r) = 0$ for $r > r_0$, where r_0 is some fixed radius. Clearly the wavefunction for the system, $\phi_E(r)$, must be a linear combination of the regular and irregular solutions for the case where $V(r) = 0$, i.e.,

$$\phi_E(r) = f_E(r)\cos\eta(E) - g_E(r)\sin\eta(E), \tag{3.77}$$

where

$$\tan\eta(E) = -\pi K(E), \tag{3.78}$$

$\eta(E)$ is the phase shift, and $K(E)$ is the electronic reaction matrix. In the multi-channel case this becomes:

$$\Psi_{iE} = \phi_{iE} + \sum_j \pi K_{ji}(E)\phi_{jE} \tag{3.79}$$

and

$$\sum_j \pi K_{ji} U_{j\alpha} = -\tan\eta_\alpha U_{i\alpha}, \tag{3.80}$$

where U is the unitary eigenvector of K. Note that K, U, and η are all functions of E. However, for simplicity in the following derivation, this dependence will not be explicitly stated. Now as in Section 3.3, divide the problem into the two subspaces Q, where Q is the resonant state, and $P = 1 - Q$ is the background. In the following derivation there is assumed to be a single dissociative state. The generalization to several resonant states is obvious. The wavefunction Ψ in the inner region, where short-range interactions are important can be written as

$$\Psi_\alpha = \sum_v U_{v\alpha}\left(\Psi_v^A \cos\eta_\alpha - \overline{\Psi}_v^A \sin\eta_\alpha\right) + U_{d\alpha}\left(\Psi_d \cos\eta_\alpha - \overline{\Psi}_d \sin\eta_\alpha\right). \tag{3.81}$$

When $r > r_0$, the asymptotic wavefunctions are as in Eq. (3.77), with a phase shift that within the Born–Oppenheimer approximation only behaves parametrically on the internuclear separation R, leading to the following expression for the ionization channel wavefunctions:

$$\Psi_\alpha^A \underset{r>r_0}{=} \Phi_{\text{core}}(q^+, R)\chi_v(R)[f_l(v, r)\cos\pi\mu(R) - g_l(v, r)\sin\pi\mu(R)], \tag{3.82}$$

where f_l and g_l are the regular and irregular Coulomb wavefunctions, $\chi_v(R)$ is a bound vibrational wavefunction, and $\Phi_{\text{core}}(q^+, R)$ is the electronic wavefunction

3.6 Calculation of dynamics

for the initial state of the ion with q^+ representing the electronic coordinates in the ion. Similarly within the Born–Oppenheimer approximation, we can write:

$$Q = |\Psi_d^E\rangle\langle\Psi_d^E|, \tag{3.83}$$

with

$$\Psi_d^E(q, R) = F_d(k, R)\Phi_d(q, R), \tag{3.84}$$

where $\Phi_d(q, R)$ is the neutral dissociative electronic state and q are the electronic coordinates in this neutral state. Of course, r is one of these indistinguishable electronic coordinates, which when used shows the asymptotic dependence. Note that $F_d(k, R)$ is the solution to the nuclear wave equation with the dissociative potential $U_d(R)$:

$$\left[-\frac{1}{2\mu}\frac{\partial^2}{\partial R^2} + U_d(R) - E \right] F_d(k, R) = 0. \tag{3.85}$$

With asymptotic behavior

$$F_d(k, R) \underset{R\to\infty}{\sim} (2\mu/k\pi)^{\frac{1}{2}} \sin[kR + \delta(E)]. \tag{3.86}$$

This leads to this final form for the wavefunction:

$$\Psi_\alpha = \sum_v \chi_v(R)\Phi_{\text{core}}(q^+, R)U_{v\alpha}[f_l(v, r)\cos(\pi\mu(R) + \eta_\alpha) \\ - g_l(v, r)\sin(\pi\mu(R) + \eta_\alpha)] \\ + \Phi_d(q, R)U_{d\alpha}[F_d(k, R)\cos\eta_\alpha - G_d(k, R)\sin\eta_\alpha], \tag{3.87}$$

where $G_d(k, R)$ is the solution to Eq. (3.85) with asymptotic behavior:

$$G_d(k, R) \underset{R\to\infty}{\sim} (2\mu/k\pi)^{\frac{1}{2}} \cos[kR + \delta(E)] \tag{3.88}$$

This describes the wavefunction in the inner region. In the outer region, one must account for two different types of decay channels. First, there are the ionization decay channels, which represent a free electron moving away from the molecular ion. This can be represented as

$$\Psi_{v^+}^B = \Phi_{\text{core}}(q^+, R)\chi_{v^+}(R)[c_{v^+} f_l(v^+, r) - d_{v^+} g_l(v^+, r)], \tag{3.89}$$

where $\chi_{v^+}(R)$ are vibrational wavefunctions for the ion and where the coefficients c_{v^+} and d_{v^+} depend only on the vibrational quantum number not on the internuclear separation R, since we assume that as r increases the outgoing electron can no longer exchange energy with the core. The other possible decay channel is a dissociative channel, i.e., the channel leading to dissociative recombination. These are represented by Eq. (3.84). In order to relate the wavefunction in the inner region to that in the outer region, we must match solutions. This is done by expressing the wavefunctions in the inner region (Eq. (3.87)) in terms of a superposition of the wavefunctions in Eqs. (3.84) and (3.87). Therefore, we have:

$$\sum_v \chi_v(R) U_{v\alpha} \cos(\pi\mu(R) + \eta_\alpha) = \sum_{v^+} \chi_{v^+}(R) C_{v^+\alpha} \qquad (3.90)$$

and

$$\sum_v \chi_v(R) U_{v\alpha} \sin(\pi\mu(R) + \eta_\alpha) = \sum_{v^+} \chi_{v^+}(R) S_{v^+\alpha}. \qquad (3.91)$$

Since the expansion is done in the asymptotic region, the coefficients $C_{v^+\alpha}$ and $S_{v^+\alpha}$ are independent of the internuclear separation R. These equations can be solved for the coefficients yielding

$$C_{v\alpha} = \sum_{v^+} U_{v^+\alpha} \langle \chi_{v^+}(R) | \cos(\pi\mu(R) + \eta_\alpha) | \chi_v(R) \rangle \qquad (3.92)$$

and

$$S_{v\alpha} = \sum_{v^+} U_{v^+\alpha} \langle \chi_{v^+}(R) | \sin(\pi\mu(R) + \eta_\alpha) | \chi_v(R) \rangle, \qquad (3.93)$$

where v and v^+ have been interchanged. Matching the dissociative decay channels simply yields:

$$C_{d\alpha} = U_{d\alpha} \cos \eta_\alpha \qquad (3.94)$$

and

$$S_{d\alpha} = U_{d\alpha} \sin \eta_\alpha. \qquad (3.95)$$

Equations (3.89)–(3.95) can be used in Eq. (3.87) to yield:

$$\Psi_\alpha = \sum_v \Phi_{\text{core}}(q^+, R) \chi_v(R) [f_l(v, r) C_{v\alpha} - g_l(v, r) S_{v\alpha}] \\ + \Phi_d(q, R)[F_d(\kappa, R) C_{d\alpha} - G_d(\kappa, R) S_{d\alpha}]. \qquad (3.96)$$

The coefficients in Eqs. (3.92) and (3.93) are often referred to as the channel mixing coefficients. If there is no R dependence in the quantum defect, this is merely an overlap between vibrational wavefunctions for two different states (the ion and a neutral), so the variation in the quantum defect mediates this coupling. Equation (3.96) is the result of a frame transformation from the frame of Eq. (3.87), where the scattering electron is strongly coupled to the core, to the external frame, where the outgoing electron is no longer coupled to the core, leading to an expression that is independent of both the internuclear separation and the vibrational wavefunction.

3.6 Calculation of dynamics

It is now necessary to abstract an S-matrix to obtain a cross section for the dissociative recombination process. It is necessary to eliminate the closed channels in order to efficiently abstract the open-channel S-matrix. This is of the form:

$$S_{oo} = X_{oo} - X_{oc}(X_{cc} - e^{-2i\pi v})^{-1} X_{co}. \quad (3.97)$$

The subscripts o and c refer to open and closed channels respectively, and v is a diagonal matrix of dimension $c \times c$, whose elements are given by:

$$v_v = \frac{1}{[2(E_v - E_T)]^{1/2}} \quad (3.98)$$

where E_T is the threshold energy, and only closed channels ($E_v > E_T$) are included. The cross section for dissociative recombination is then

$$\sigma_v = \left(\frac{\pi}{k^2}\right) \frac{g_f}{g_i} |S_{dv}|^2, \quad (3.99)$$

where g_i and g_f are the degeneracies of the initial and final states, respectively and k, the wave number, is given by

$$k = [2(E_T - E_v)]^{1/2}. \quad (3.100)$$

In order to use this method, one must determine the K-matrix (Eq. (3.79)). This has been done by direct evaluation for the H$_2$ system (Ross & Jungen 1987). In a number of other calculations (Guisti 1980, Giusti-Suzar, Bardsley, & Derkits 1983, Nakashima, Takagi, & Nakamura 1987, Nakamura, Takagi, & Nakashima 1989, Sun & Nakamura 1990) a perturbative approach has been employed. The method starts with the Lippman–Schwinger equation,

$$K = V + VGK. \quad (3.101)$$

In first order only the first term is retained in the expansion of K. Early studies of dissociative recombination using MQDT worked within this approximation. This leads to the expression:

$$V_{lv,d} = \int \chi_v(R) V_{E,l}(R) F_d(E, R) dR = K_{lv,d}. \quad (3.102)$$

This is just the matrix element of the electronic coupling $V_{E,l}$ (see Eq. (3.49)) between the initial vibrational wavefunction and the final dissociative wavefunction. There is no mixing between the various l-components in the interaction and there is no indirect coupling. More recent work has extended this to second order, i.e. where

$$K = V + VGV. \quad (3.103)$$

In a similar fashion to Section 3.3, the P/Q representation of G, the standing wave Green function for the uncoupled channels and the K-matrix, is defined. This is

$$PGQ = QGP = 0.$$

$$PGP = P\left[P\frac{1}{(E - H_0)}P\right] = G_P, \qquad (3.104)$$

$$QGQ = P\left[Q\frac{1}{(E - H_0)}Q\right] = G_Q,$$

where P represents the principal value operator and H_0 excludes the electronic interaction. These can be written in a spectral representation

$$G_P = \sum_v |\chi_v\rangle \langle \chi_v| P \int dE \frac{|\Phi_{\text{core}}(q^+, R)\varphi(E)\rangle \langle \Phi_{\text{core}}(q^+, R)\varphi(E)|}{E_T - (E_v + E)} \qquad (3.105)$$

and

$$G_Q = |\Phi_d(q, R)\rangle \langle \Phi_d(q, R)| P \int dE \frac{F_d(E, R) F_d(E, R')}{E_T - (E_d + E)}. \qquad (3.106)$$

Again, note the similarities to Eq. (3.37). This leads to a form for the K-matrix in second order as:

$$K_{lv,l'v'} = \frac{1}{W} \iint \chi_v(R) V_{E,l}(R) F_d(E, R_<) G_d(E, R_>) V_{E,l'}(R') \chi_{v'}(R') \, dR dR', \qquad (3.107)$$

where W is the Wronskian:

$$W = F_d'(E, R) G_d(E, R) - F_d(E, R) G_d'(E, R). \qquad (3.108)$$

This allows mixing between the l-channels through the indirect interaction. Inclusion of closed channels and an extension beyond first order in the K-matrix leads to the resonance structure and couplings between channels. This has been discussed in the case of H_2^+ by Schneider, Dulieu, and Giusti-Suzor (1991) and for O_2^+ by Guberman and Giusti-Suzor (1991).

Most calculations have been carried out at the first- or second-order level. It has been shown (Ngassam *et al.* 2003) that when the electronic interaction between the dissociative states and the continuum, $V_{E,l}(R)$, is energy independent (or this dependence is small enough to be neglected) convergence occurs at second order. This means the second-order expression:

$$K = V + VGV = V + V\left(\frac{1}{E - H_0}\right)V \qquad (3.109)$$

is exact. Takagi (1996), for the case of H_2^+, carried out a numerical solution of the Lippman–Schwinger equation (Eq. (3.101)) using an iterative grid method, which showed significant differences between first and second order when the energy dependence of $V_{E,l}(R)$ was strong. If the energy dependence is separable in E and R, i.e., if

$$V_{E,l}(R) = c_l(E)V_l(R), \tag{3.110}$$

it has been shown (Pichl, Nakamua, & Horacek 2000) that the solution can be efficiently obtained by solving a system of algebraic equations.

Note that the discussion here has ignored the existence of rotation. The effects for initial rotation and rotational coupling are a topic of current research in the study of dissociative recombination. Takagi (1993) has developed the MQDT method to include rotational effects. The details of the method will not be included here. Rotational effects were also included in the study of dissociative recombination in the H_3^+ system (Kokoouline & Greene 2003a,b). Although only limited systems have been studied (Takagi 1993, 2004, Schneider *et al.* 1997, Kokoouline & Greene 2003a,b), the trend indicated by these calculations is that rotation is important in systems where the direct coupling is small, and can have a strong effect in the position and shape of the resonant structure in the cross section due to the indirect mechanism.

The MQDT treatment has been extended to triatomic systems: for a detailed discussion of the modifications necessary see Kokoouline and Greene (2003a,b). In this treatment the system was studied in hyperspherical coordinates. The hyper-radius Q is fixed and the Schrödinger equation solved in the space of the other two hyperangles. This leads to a set of adiabatic hyperspherical potential curves and the corresponding adiabatic states. The K-matrix is now a function not of R, but of the hyperspherical radius Q. The reaction matrix connects states Λ to Λ', where Λ is the body-frame projection of the electron angular momentum l. The calculation used Siegert pseudostates to allow the dissociating flux to escape. An additional complication occurs in this system, in that Jahn–Teller coupling, which of course cannot exist for a diatomic, plays a critical role in the dissociative recombination process and must be included to describe the process accurately. It has been speculated that other such couplings, for example, Renner–Teller (Mikhailov *et al.* 2006) may be important for other polyatomic systems. This is an area of current research. Indeed, treatments beyond the case of a single active degree of freedom are lacking.

3.6.2 Time-dependent wave packet method

Wave packet methods, i.e., numerical solutions of the time-dependent Schrödinger equation, are standard techniques in the study of the dynamics of chemical reactions, dating back to their introduction in this area by Heller in the 1970s (Heller 1981).

The wave packet picture is physically intuitive, allowing one to watch the evolution of the system dynamics in time, as energy flows through the available modes of the molecule leading to bond breaking and flux between the various states of the system. Since the wave packet is described by its value on a numerical grid of points, there are no serious problems with coupled potential energy surfaces (Orel & Kulander 1988, Heather *et al.* 1988) or a three-body continuum (Leforestier *et al.* 1991), which can be difficult to handle with basis set techniques. Wave packet methods have the additional advantage that the cross section is obtained at all energies for a single calculation.

Early applications were in the area of photodissociation, and were restricted to diatomics or limited dimensionality in larger systems. Increase in computer capabilities and advanced propagation methods, such as the multiconfiguration time-dependent Hartree (MCTDH) method have greatly increased the dimensionality of the systems that can be studied. For example, using MCTDH, the pyrazine molecule has been studied treating all 24 vibrational modes explicitly (Beck *et al.* 2000). The major difficulties are in obtaining the necessary surfaces and couplings in multiple dimensions.

The wave packet approach relies explicitly on the introduction of states. Each state then contains all possible vibrational wavefunctions, including the dissociative vibrational continuum that can be expressed accurately on the grid. In contrast, the MQDT method sets a maximum vibrational state and then contains all Rydberg states that are closed at that vibrational level.

The wave packet method proceeds by the direct integration of the time-dependent Schrödinger equation:

$$i\hbar \frac{\partial}{\partial t} \Psi(\mathbf{r}, t) = H(\mathbf{r})\Psi(\mathbf{r}, t). \tag{3.111}$$

As written, the method is exact. The approximations come in the definition of the Hamiltonian, and the initial conditions for the propagation.

This equation is recast into a time-dependent equation, as first shown by McCurdy and Turner (1983), by formally solving it in terms of the resolvent, or nuclear Green's function:

$$\Psi_i(R) = G(E)V_E \chi_{v_i} = \frac{1}{E-H} V_E \chi_{v_i}. \tag{3.112}$$

In the "local" approximation, described in Section 3.3, Eq. (3.44), the Hamiltonian is given by:

$$H = T + E_{\text{res}}(R) - i\frac{\Gamma(R)}{2}. \tag{3.113}$$

3.6 Calculation of dynamics

The Green's function is then written as the Fourier transform of the propagator for the time-dependent Schrödinger equation producing

$$\Psi_i(R) = -i \int_0^\infty dt e^{iEt} \Psi_i(R,t) = -i \int_0^\infty dt e^{iEt} e^{-iHt} V_E \chi_{v_i}. \quad (3.114)$$

Therefore the problem reduces to solving the time-dependent Schrödinger equation, with the above Hamiltonian and an initial condition of

$$\Psi(R; t=0) = V_E(R)\chi_{v_i}. \quad (3.115)$$

Note that the quantity $V_E(R)\chi_{v_i}$ is often referred to as the "entry amplitude." The cross section is given by

$$\sigma(E) = \frac{2\pi^3}{E} \frac{g_f}{g_i} |T(E)|^2, \quad (3.116)$$

where $T(E)$ for the case of dissociation is given by the projection of the wave packet $\Psi(R,t)$ onto the asymptotic final eigenstates of the system at long times when the wave packet has reached the asymptotic region of the potential and the autoionization loss has gone to zero, i.e.,

$$T(E) = \lim_{t \to \infty} \langle \Phi(E) | \Psi(t) \rangle \quad (3.117)$$

(Kulander & Heller 1978). The equations shown above are for the case of a diatomic, the generalization of these equations to polyatomic systems is obvious.

At very low energies, as discussed in Section 3.3, the "local" approximation is no longer accurate so the action of the Hamiltonian on $\Psi(R)$ in Eq. (3.111) becomes:

$$H(R)\Psi(R) = (T + E_{\text{res}}(R))\Psi(R) + O_{\text{nonloc}} \quad (3.118)$$

where the nonlocal operator $O_{\text{nonloc}}(R)$ is defined by:

$$O_{\text{nonloc}}(R) = -i\pi \sum_j \chi_{v_j}(R) \int_0^\infty dR' \chi_{v_j}(R') V_E(R') \Psi(R'). \quad (3.119)$$

The initial wavefunction and definition of the cross section remain the same. Note that now the operation of H on $\Psi(R)$ involves integration over R, so H becomes a nonlocal operator. The Hamiltonian is also energy-dependent, since as new vibrational channels open new terms must be included in the sum in Eq. (3.119). A similar equation in the case of dissociative attachment has been derived by Domcke and Estrada (1988). This was applied to the case of direct dissociative recombination of HD$^+$ (Orel 2000a) and compared with first- and second-order MQDT calculations on the same system. For low energies (< 0.2 eV) these treatments give identical results. Between 0.2 eV and 0.5 eV, the second-order and wave packet results are identical, but differ slightly from the first-order ones. Finally above 0.5 eV some difference between the second-order and the wave packet results can be observed. This is to be expected, since the wave packet is nonperturbative, and includes the coupling to all orders.

In the case where the coupling is the nonadiabatic coupling described in Section 3.5, and the second derivative coupling $F_{ij}^{(2)}$ (Eq. (3.63)) can be neglected, the initial wavefunction is given by (Krause et al. 1992):

$$\Psi(R, t = 0) = F_{ij}^{(1)} \chi_i(R), \qquad (3.120)$$

where $F_{ij}^{(1)}$ is given in Eq. (3.62) and $\chi_i(R)$ is the initial vibrational wavefunction of the ion. One then propagates this initial wave packet using Eq. (3.111) with the Hamiltonian as just $T + V$, where T is the kinetic energy operator, and V the potential for the neutral potential energy surface. The cross section is obtained as before by projection onto the asymptotic states. When more than one surface is involved, the initial wave packet becomes a vector, and the Hamiltonian becomes a matrix, with the couplings between states appearing in the off-diagonal elements.

The MQDT and the time-dependent wave packet approaches can be viewed as complementary. The time-dependent wave packet method has been applied to just a handful of dissociative recombination and dissociative excitation (see Chapter 9) problems, whereas the MQDT approach has been more widely used. For diatomics, and in particular at low electron impact energies, the MQDT is much more efficient. In particular, in cases where the crossing of the resonance and ion is near the equilibrium of the ion, and the nonlocal form must be employed, the time-dependent wave packet becomes energy dependent, and it is no longer true that the cross section is obtained at all energies for a single calculation, each new threshold involves a new calculation. However, the time-dependent wave packet method is more efficient at high energy where a number of resonant states are involved in the dissociative recombination process and a large number of channels are open, especially when the dissociative excitation becomes dominant. The real power of the wave packet method is in the study of dissociative recombination and dissociative excitation in polyatomic systems where more than one degree of freedom is important. This area has just begun to be explored and holds great promise for the future.

There is one other difference between the two methods. Since the MQDT method does not distinguish states, only channels, it provides no information about the final state distribution of products. With a single dissociative state, this is not an issue. However, as soon as another state opens, for example, when the Rydberg states are energetically open, these states cannot be distinguished. There have been attempts to combine methods to obtain branching methods in several cases. For example, Zajfman et al. (1997) calculated a total dissociative recombination cross section

3.6 Calculation of dynamics

using MQDT in the HD$^+$ system. The Landau–Zener approximation was then used to estimate the distribution of flux into the Rydberg states that cross the $(2p\sigma_u)^2$, $^1\Sigma_g^+$ resonance state along its dissociation path. Carata *et al.* (2000) combined MQDT calculations and a close-coupling calculation on interacting $^2\Pi$ states to study the branching ratios in the dissociative recombination of CH$^+$ and CD$^+$. Another interesting example, described in detail in Chapter 6, is the production of O (^{1}S) in the dissociative recombination of O$_2^+$.

On the other hand, time-dependent wave packet calculations do distinguish states and can provide information about branching ratios. When branching ratios are calculated, it is necessary that the potential energy curves are accurate, not only in the Frank–Condon region, but also in the region where the resonance state becomes electronically bound. In addition, any couplings that occur before the final state fragments have reached the asymptotic region must also be known. Final state distributions have been calculated using the time-dependent wave packet method for HeH$^+$ (Larsson & Orel 1999), HD$^+$(Larsson & Orel 2001), and the rare-gas diatomic ions, He$_2^+$ (Royal & Orel 2005), Ne$_2^+$ (Ngassam & Orel 2006) and Ar$_2^+$ (Royal & Orel 2006).

4
The H_2^+ molecule

The simplest diatomic system is H_2^+. This molecule and its isotopolog HD^+ have been the focus of both theory and experiment, serving as benchmarks between experiments as well as between experiment and theory. H_2^+ is abundant in a variety of environments, such as diffuse interstellar clouds, planetary atmospheres, and at the edges of the plasmas in tokamaks. Dissociative recombination of H_2^+ can produce fast neutral atoms in their ground or excited states, as well as H^- and H^+ through the process of ion-pair formation. This process strongly drives the kinetics of these systems. As will be seen in this chapter, although H_2^+ can be viewed as the simplest diatomic, dissociative recombination in this system is anything but simple, representing a challenge to both theory and experiment.

The relevant potential energy curves for dissociative recombination in H_2^+ are shown in Figs. 4.1 and 4.2. The ground state of H_2^+, a one-electron system, is $1\sigma_g$, leading to a $^2\Sigma_g^+$ state. This state is bound with an equilibrium internuclear separation slightly larger than the neutral. The first state of the ion promotes this one electron into the next lowest lying orbital, $2p\sigma_u$, leading to the $^2\Sigma_u^+$ state. As can be seen in Fig. 4.1, there exists one set of resonances, generally referred to as the Q_1 states, that forms a Rydberg series converging to the first excited state ($^2\Sigma_u^+$) of the ion. The configuration of these states is of the form, $2p\sigma_u nl_u$ or $2p\sigma_u nl_g$ with $l = \sigma, \pi$ or δ. The lowest member of this series, a $^1\Sigma_g^+$ state of H_2 with configuration $2p\sigma_u 2p\sigma_u$ crosses the ground state of the molecular ion just above the energy of the ground vibrational level. Diabatically this state correlates to the ion-pair limit ($H^+ + H^-$) at large internuclear separations. At higher energies, the other members of this Rydberg series, first $2p\sigma_u 2s\sigma_g$ and then higher states, become accessible. Other Rydberg series, converging to higher excited states of the ion (for example, the Q_2 states converging to the $^2\Pi_u$ state) are important at even higher energies. In addition, there exists a Rydberg series converging to the ground state of the ion as shown in Fig. 4.2.

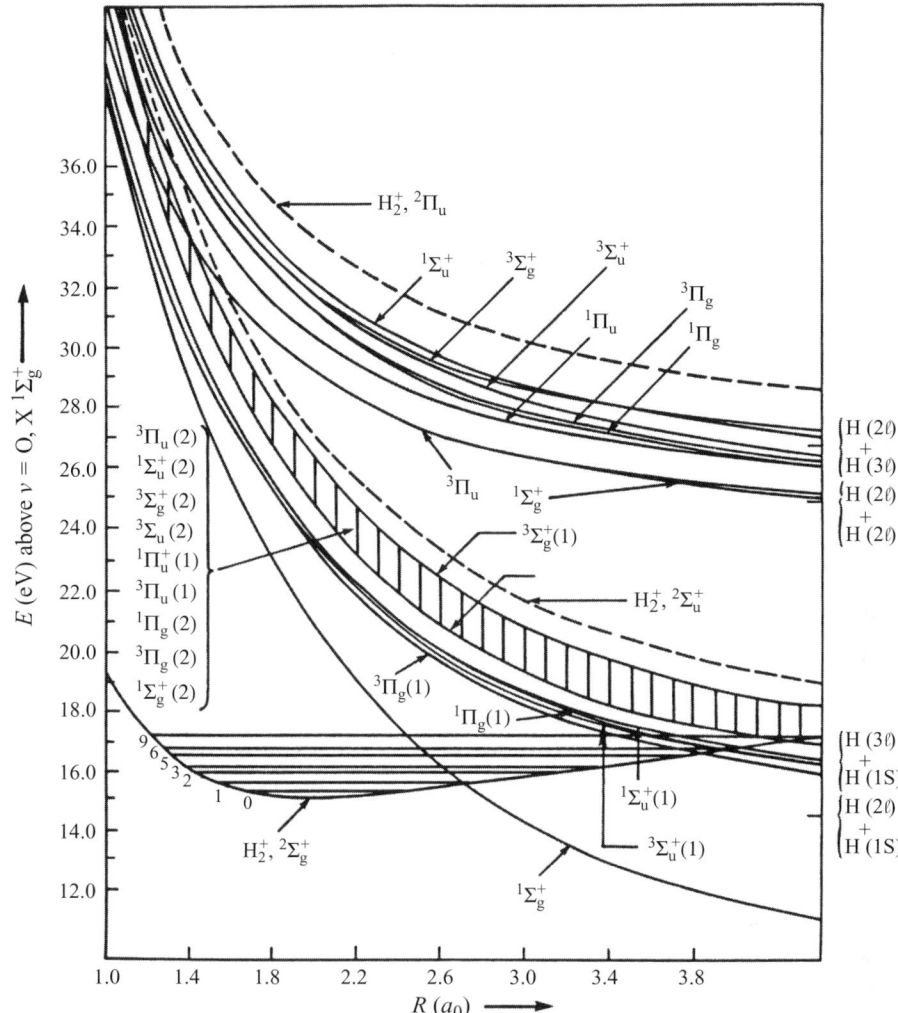

Figure 4.1 The potential curve for the ground state of H_2^+, the electronically excited $^2\Sigma_u^+$ and $^2\Pi_u$ states of H_2^+ (dashed lines), and the potential curves for the Q_1 and Q_2 autoionizing states of H_2. The Q_1 states are labeled by a (1) or (2) indicating the first or second root of that symmetry. For the higher lying Q_2 states only the lowest root of each symmetry is shown. A closely spaced group of Q_1 states bounded by $^3\Sigma_g^+(1)$ and $^1\Pi_u(2)$ is shown with vertical lines. These states are labeled in the bracket on the left of the figure in order of increasing energy at internuclear distance $R = 1.4\ a_0$. (Reused with permission from S. L. Guberman, *Journal of Chemical Physics*, **78**, 1404 (1983). Copyright 1983, American Institute of Physics.)

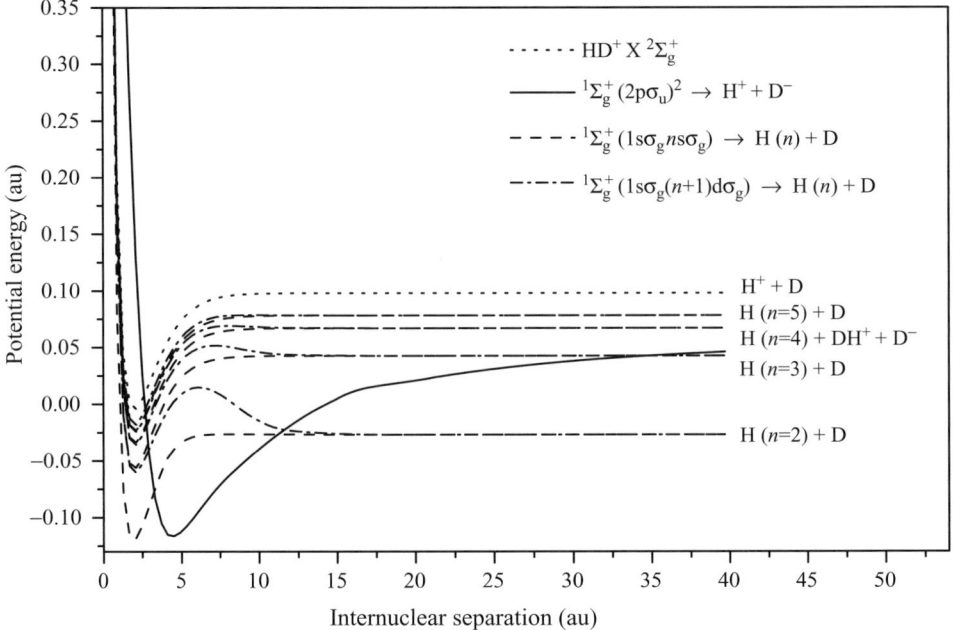

Figure 4.2 Potential energy curves for HD$^+$ and HD. The ground state X $^2\Sigma_g^+$ of the HD$^+$ ion is shown with a dotted line. The energy scale is relative to the zeroth vibrational level of the ion. The lowest dissociative state of the HD molecule is the $^1\Sigma_g^+$ $(2p\sigma_u)^2$ state, which correlates with the ion pair at infinity. The diabatic state, shown with a solid line in the figure, crosses the neutral $^1\Sigma_g^+$ Rydberg states situated below the ionic ground state. (Reprinted figure with permission from Å. Larson and A. E. Orel, *Phys. Rev. A* **64**, pp. 062701-1–8, (2003). Copyright (2003) by the American Physical Society.)

The history of studies of this system involved close interactions between theory and experiment. The development of the stationary afterglow technique, described in Section 2.3, naturally led to experiments with molecular hydrogen. However, H$_2^+$ was found to be very reactive, leading to its conversion to H$^+$, H$_3^+$, and possibly H$_5^+$, thus resulting in an unknown mixture, so the dissociative recombination rate coefficient of H$_2^+$ could not be measured or inferred using these techniques. This difficulty was overcome by the use of merged-beam techniques (Peart & Dolder 1974a) discussed in Chapter 2. This experiment used a beam of H$_2^+$ from an ion source, accelerated to 60 keV and intersected with an electron beam at 10° with the electron beam energy chosen to produce the desired collision energy. The results are shown in Fig. 4.3. These ion sources for H$_2^+$ relied on the formation of the ion via the impact of energetic electrons on H$_2$. As can be seen in Fig. 4.1, the potential minimum of H$_2^+$ lies at larger internuclear separation than in H$_2$. Therefore, this

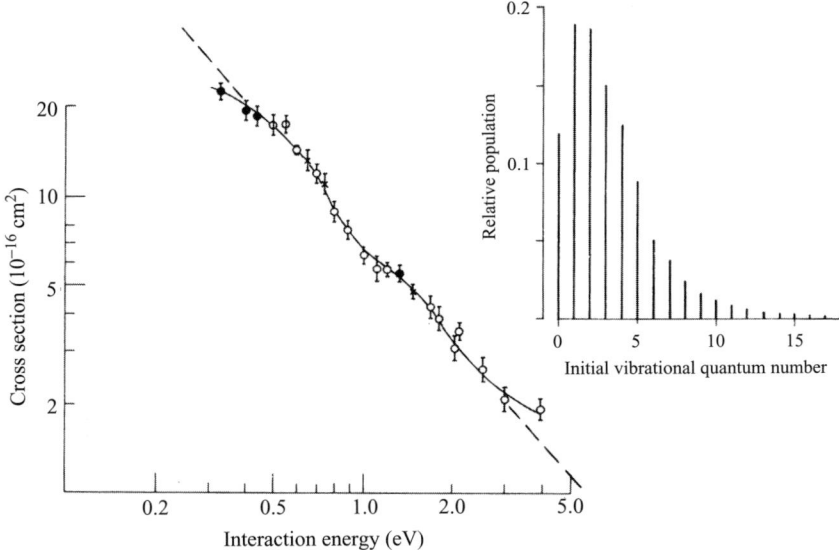

Figure 4.3 Results from measurement of dissociative recombination of H_2^+ by means of the technique with inclined beams (Peart and Dolder 1974a). The cross section for dissociative recombination is plotted against the interaction energy E (see Chapter 2). The solid circles, hollow circles, and crosses, respectively, denote measurements made with ion beam energies of 40, 60, and 50 keV. The dashed line illustrates a cross section propotional to $E^{-1.14}$, which roughly fits the points. The inset shows a population of vibrational levels of H_2^+ which is believed to represent the initial state of the H_2^+ beam in the inclined-beam experiment. (Reproduced with permission from B. Peart and K. T. Dolder, "Collisions between electrons and H_2^+ ions. V. Measurements of cross section for dissociative recombination," J. Phys. B **7**, pp. 236–243, (1974), IOP Publishing Limited.)

method of producing H_2^+ leads to the population of all 19 available vibrational levels generally assumed to be given by a Frank–Condon distribution or von Busch and Dunn distribution (von Busch & Dunn 1972). This vibrational distribution is also shown in Fig. 4.3. The peak of this distribution was found to be in $v = 2$, but with significant population in even higher states. Therefore these early experiments for the dissociative recombination of H_2^+ (Peart & Dolder 1974a) and merged-beam experiments on the isotopolog D_2^+ (Peart & Dolder 1973a) studied dissociative recombination from ions in a range of initial vibrational states. These experiments were followed by others using similar techniques that studied the recombination of electrons with D_2^+ producing specific final states, first D ($n = 4$) (Phaneuf et al. 1975) followed by experiments that monitored Lyman α radiation from the dissociation products, thereby obtaining the cross section into the D (2p) states (Vogler & Dunn 1975). In addition, the first experiments (Peart & Dolder, 1975) were carried out on the ion-pair formation process:

$$e^- + H_2^+ \rightarrow H^+ + H^-, \tag{4.1}$$

which will be further discussed in Chapter 9.

The first calculations on dissociative recombination of H_2^+ (Bauer & Wu 1956) described the process as direct capture of the electron by the molecular ion. This was approximated by treating the electron–electron repulsion as a perturbation between wavefunctions, one having two electrons in H_2^+ eigenstates, and the other with one electron in an H_2^+ eigenstate and the other in an undistorted plane wave to describe the incoming electron. Besides the perturbative treatment, this ignored the significant effect of the long-range Coulomb force. This calculation led to an estimate for the cross section of the order of 10^{-15} cm^2 at 1 eV which was roughly an order of magnitude larger than experiment. These calculations were improved by Wilkins (1966) who used a Coulomb wave for the incoming electron, i.e., treating the molecular ion as H_2^+ and attempted to obtain a capture rate into the $(2p\sigma_u)(2s\sigma_g)\,^3\Sigma_u^+$ state of hydrogen. Note (see Fig. 4.1) this is not the lowest resonant state. This led to an estimate of the capture rate on the order of 10^{-8} cm^3 s^{-1}. This calculation was repeated by Dubrovksy, Ob'edkov, and Janev (1967) and Dubrovsky and Ob'edkov (1967), but in this case considered capture into the lowest resonant state $(2p\sigma_u)(2p\sigma_u)\,^1\Sigma_g^+$ producing an order of magnitude smaller estimate for the rate coefficient. These early calculations made numerous approximations and can only be described as producing estimates for the cross sections.

The first calculations that used the formal resonance theory to calculate the cross section due to the direct mechanism, described in Chapter 3, where the electron is captured into the resonant state and proceeds to dissociation only, were those of Bottcher (1976). In these calculations, the electron was again treated as a Coulomb wave, but a variational treatment was used to obtain the resonance curve and couplings. The calculations were averaged over the vibrational distribution shown in Fig. 4.2 and produced results that were in qualitative agreement with the experiments (Peart & Dolder 1973a, 1974a). This averaging leads to low resolution. In particular, although this work included a discussion about the effect of the indirect mechanism (as described in Chapter 3) that had been predicted to occur in this system (Bardsley 1968b), it only included an approximation of its effect on the thermal rate coefficient.

The lack of a well-characterized initial vibrational state hindered comparison between theory and experiments. One attempt was made, by Auerbach et al. (1977) using a H_2/He mixture in a radio-frequency ion source, to produce a vibrationally cooler source. This resulted in some cooling, believed to produce ions dominated by $v = 0$, 1, and 2, but the residence time in the source was too small to produce cooling to the ground state. This experiment was also carried out with a much higher-energy resolution than previous experiments and showed previously unseen structure in the cross sections as a function of electron energy, which was attributed to the indirect mechanism. Averaging the results of Bottcher (1976) over what was believed to be the experimental vibrational distribution, produced a cross section an order of magnitude larger than experiment. This led to further work to understand the mechanism of dissociative recombination in this system.

Later calculations (Zhdanov & Chibisov 1978, Zhdanov 1980) suggested that although they had a lower capture probability the higher Rydberg states $1\sigma_u nl_u$ would dominate the dissociative recombination due to their lower autoionization rate. Data used in this calculation for the autoionization rate, as well as the treatment of the dynamics of dissociation, were approximate. Derkits, Bardsley, and Wadehra (1979) carried out a calculation within the local approximation (see Chapter 3 for details) with a constant width function which indicated that capture into these higher dissociative states was not important for low-energy electrons and low vibrational excitation in the target. Rai Dastidar and Rai Dastidar (1979) used the Feshbach projection operator method and a single-configuration molecular orbital wavefunction for the resonance state. Their calculations lead to an autoionization width that was an order of magnitude larger than that of Bottcher (1976). These data were used to calculate the dissociative recombination cross section for H_2^+, D_2^+, and HD^+. Comparison to the experiments of Peart and Dolder (1974a) led to results that agreed better with experiment especially at lower incident electron energy. At this time, theory was hampered by a lack of accurate molecular data, especially autoionization widths for this system.

This lack of data was addressed by several groups. The calculation can be divided into two general areas: bound state techniques and electron scattering calculations. In the bound state area, calculations at a single internuclear separation were carried out using the complex coordinate method (Moiseyev & Corcoran 1979) that were larger in both the energy position and width than previous calculations. These were followed by calculations for the position and autoionization width for the first $2p\sigma_u 2p\sigma_u \ ^1\Sigma_g^+$ resonance at a number of internuclear positions using Feshbach projection operators and the Stieltjes imaging technique (Hazi, Derkits, & Bardsley 1983). This calculation used a larger basis set and a more accurate configuration interaction wavefunction. Although the energy position was similar to previous calculations, the autoionization width had a very different behaviour with increasing internuclear separation (Bottcher & Docken 1974). Hazi, Derkits, and Bardsley (1983) also noted the dependence of the autoionization width on the energy of the ejected electron. This "nonlocal" effect was not addressed until much later. These results were confirmed by accurate electron scattering calculations (Collins & Schneider 1983, Takagi & Nakamura 1980). The agreement between position and autoionization width for the $^1\Sigma_g^+$ resonance state obtained by these very different methods was excellent. Guberman (1983b) carried out *ab initio* calculations on the lowest 24 doubly excited autoionization states (see Fig. 4.1). A Feshbach projection operator technique was used to describe these states, and the potential energy curves and autoionization widths for a wide range of internuclear separations were derived.

Also the importance of the second path, the indirect mechanism, involving capture into the autoionizing, vibrationally excited states of the Rydbergs converging to the ground state of the ion (see Fig. 4.2), first postulated in 1968 (Bardsley 1968b) was recognized. It is crucial that these two channels be treated equivalently, since they represent indistinguishable paths to the same product. Calculations using the MQDT method (Giusti-Suzor, Bardsley, & Derkits 1983, Nakashima, Takagi, & Nakamura 1987) indicated that this path interfered destructively with the dissociation directly on the dissociative state leading to a dip in the dissociative recombination cross section. Calculations using the CI method (Hickman 1987) indicated an enhancement of the cross section when the indirect channel was added. Later work by Van der Donk, Yousif, and Mitchell (1991) and Schneider, Dulieu, and Giusti-Suzor (1991), both using the MQDT method but with different treatments of the variation of quantum defect with internuclear separation and the treatment of second-order coupling, confirmed the existence of dips in the experimental cross section due to interference between the direct and indirect paths. Some disagreement as to which Rydberg state caused individual structures in the cross section (Van der Donk, Yousif, & Mitchell 1991, Schneider, Dulieu, & Giusti-Suzor 1992) persisted, but the mechanism itself was confirmed.

On the experimental side, attempts to cool the target more to lower vibrational energy, in order to compare better to theory (and to distinguish between various theories) also continued. Sen, McGowan, and Mitchell (1987) used an inhomogeneous radio-frequency electric field to confine the ions for up to milliseconds, cooling the ions down to a mixture of $v = 0$ and $v = 1$. The ions from this source were used in a merged electron–ion beam apparatus to study dissociative recombination at higher resolutions than previously possible. These experiments clearly showed for the first time, the narrow resonances associated with capture into the Rydberg states (Hus *et al.* 1988, Van der Donk, Yousif, & Mitchell 1991). The agreement of theory with experiment was good. This comparison is shown in Fig. 4.4.

Although there was good qualitative agreement between theory and experiments, further work required an experiment using a well-characterized single initial vibrational state. This was provided by the ion storage ring experiments. Although H_2^+ has no dipole moment, its isotopolog HD^+ is infrared active. Depending on the initial vibrational distribution in the source at injection, the HD^+ was found to cool radiatively down to the ground vibrational states within less than a second. This system was used as the benchmark for measurements, and a coordinated series of experiments was carried out at three ion storage ring facilities, CRYRING, ASTRID, and

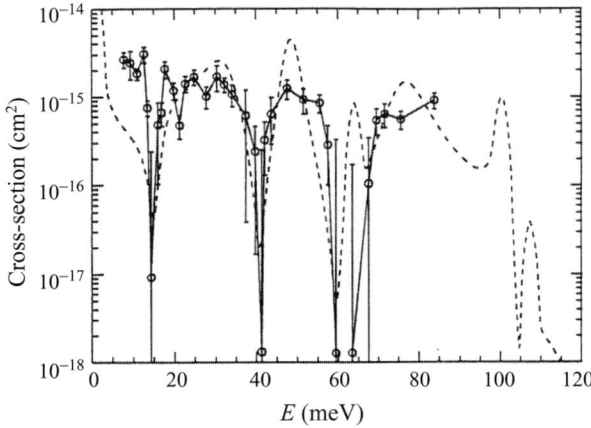

Figure 4.4 Comparison of the merged-beam measurement of the dissociative recombination of H_2^+ (Van der Donk et al. 1991) and MQDT calculations (Schneider, Dulieu, & Giusti-Suzor 1991, 1992). The dashed curve is the theoretical cross section for the intial ion level $v = 0$ convoluted with a 2.5 meV triangular apparatus function (see Chapter 2 for a discussion of the resolution in merged-beam experiments). (Reproduced with permission from I. F. Schneider, O. Dulieu, and A. Giusti-Suzor, "The role of Rydberg states in the H_2^+ dissociative recombination with slow electrons," *J. Phys. B* **7**, pp. 236–243, (1991), IOP Publishing Limited.)

TSR (Al-Khalili et al. 2003). The excellent agreement between these three experiments is shown in Fig. 4.5. As can be seen in the figure, not only is the magnitude of the cross section in agreement in the three experiments, but there is also good agreement between the positions of individual resonance features. These experiments have provided much greater resolution and a much more stringent test of theory.

In order to match the increasing resolution in the experiment, the theory needed to be improved to consider effects that had been averaged out in the previous experiments. The first process considered was the effect of rotation and rotational coupling in calculations. Takagi (1993) modified the MQDT equations to allow for the effect of rotation. In addition, the off-the-energy-shell effect, a higher-order dynamical effect induced in the electronic CI, was also included. This improved MQDT method was applied to HD^+ in the low-energy region and compared with the TARN II storage ring results (Tanabe et al. 1995). The results are shown in Fig. 4.6. Schneider et al. (1997) also included the effect of rotation, but not the off-shell interaction, and compared the results with those from CRYRING shown in Fig. 4.7. Another set of calculations using the integral variant of the MQDT (Ivanov & Golubkov 1984, 1985, Golubkov & Ivanov 1990) were carried out to study dissociation recombination in H_2^+, D_2^+ and HD^+ (Golubkov, Golubkov, & Ivanov 1997). The results were in good agreement with those available from experiment. However,

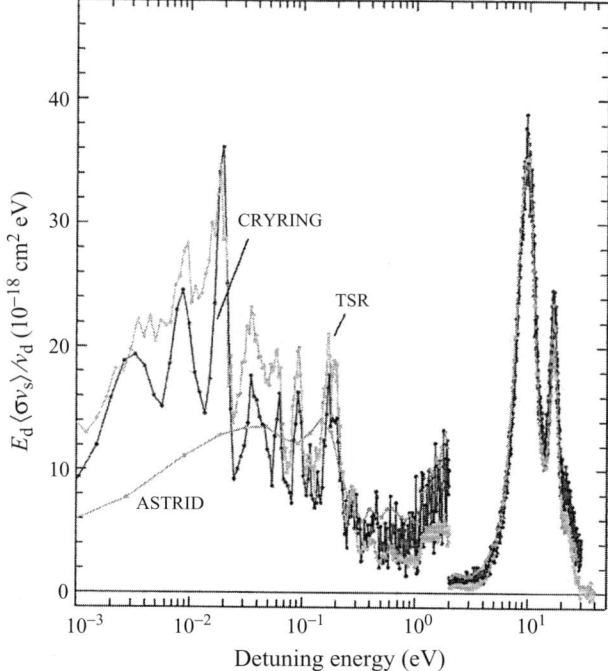

Figure 4.5 Absolute dissociative recombination cross sections, $\langle v_r\sigma\rangle/v_d$, for HD$^+$ ($v=0$) scaled with detuning energy E_d, so that the y-axis is $E_d\langle v_r\sigma\rangle/v_d$ in units of cm^{2}eV. The transverse electron-energy spread, $kT_\perp$, was 2 meV in CRYRING, 4.5 meV in the TSR, and 22 meV in ASTRID. The higher energy spread in the ASTRID electron beam, and hence lower resolution, is the reason why the structure is not really visible below 10^{-2}eV. For the energies above 2 eV, the scaled cross section has been multiplied by 0.1. (Reprinted figure with permission from A. Al-Khalili, W. J. van der Zande, S. Rosén et al., *Phys. Rev. A* **68**, pp. 042702-1–14, (2003). Copyright (2003) by the American Physical Society.)

this calculation highlighted the sensitivity to molecular data of the final cross sections. Any averaging in the experiment over initial states resulted in washing out differences.

Note that to go beyond this level of comparison, the issue of the initial rotational distribution must be addressed. The ion sources for the storage ring produce vibrationally and rotationally excited ions. After the ions are injected into the ring, vibrational cooling occurs until the ions are in the ground vibrational state. At this point, the ions are still rotationally hot. The usual assumption is that the rotational distribution is a Boltzmann distribution characterized by the temperature of the vacuum ring system. Work is continuing to better characterize the rotational excitation in the ring by means of imaging of the recombination products. Figure 4.8 shows the best comparison between data from the TSR and MQDT calculations (Schneider & Wolf 2007, private communication).

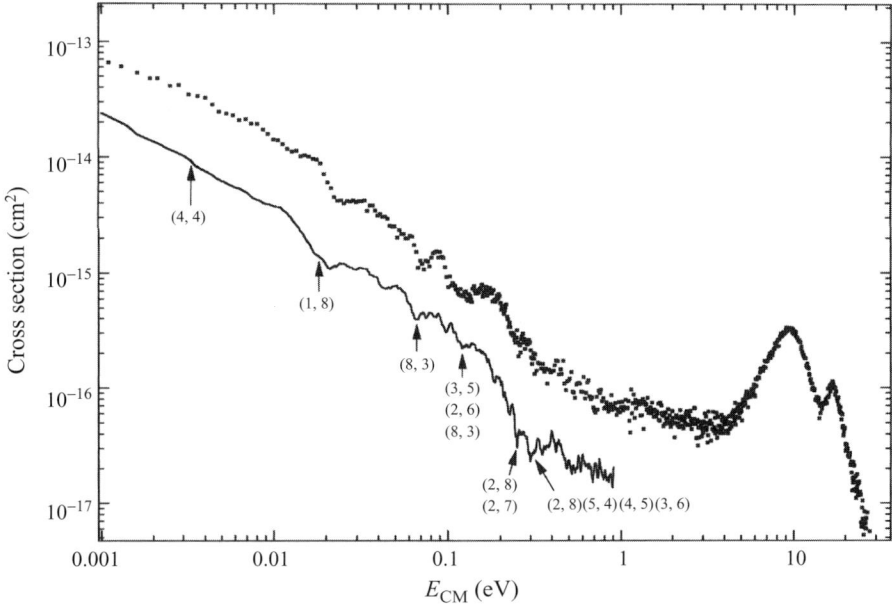

Figure 4.6 Relative dissociative recombination cross sections for HD$^+(v=0)$ measured in the TARN II storage ring as a function of the center-of-mass energy (i.e., the detuning energy). The transverse electron-energy spread in this experiment was 10 meV. The lower curve is theoretical cross section calculated by means of MQDT. The numbers in the parentheses represent the vibrational (v) and principal (n) quantum numbers of the resonance states into which the electron is captured in the indirect process. (Reprinted figure with permission from T. Tanabe, I. Katayama, H. Kamegaya *et al.*, *Phys. Rev. Lett.* **75**, pp. 1066–1069, (1995). Copyright (2003) by the American Physical Society.)

At low energies (below 2.7 eV), only the lowest dissociative channel, the $2p\sigma_u 2p\sigma_u\,^1\Sigma_g^+$ state, as well as the Rydberg states converging to the ground state of the ion are needed to describe the low-energy cross section accurately. As the incident electron energy is increased, or dissociative recombination from higher initial vibrational states is considered, the effect of the indirect mechanism, i.e., the contribution from the low-lying Rydberg states, decreases, but higher resonance states must be included. Three additional ion states, $^1\Pi_g$, $^1\Sigma_u^+$, and $^3\Pi_g$, must be included to reproduce the cross section between 5 and 12 eV. The entire dissociative Rydberg series converging to each new ion state must be considered. These have been included (Strömholm *et al.* 1995) explicitly for $n = 3$–5, with the effect of the rest of the (infinite) series of states included via an effective Rydberg state, with a total width given by

$$\Gamma_{\text{eff}} = \sum_{n=6}^{\infty} \Gamma_n = \left(\sum_{n=6}^{\infty} \frac{2Ry}{(n^*)^3} \overline{\Gamma} \right), \tag{4.2}$$

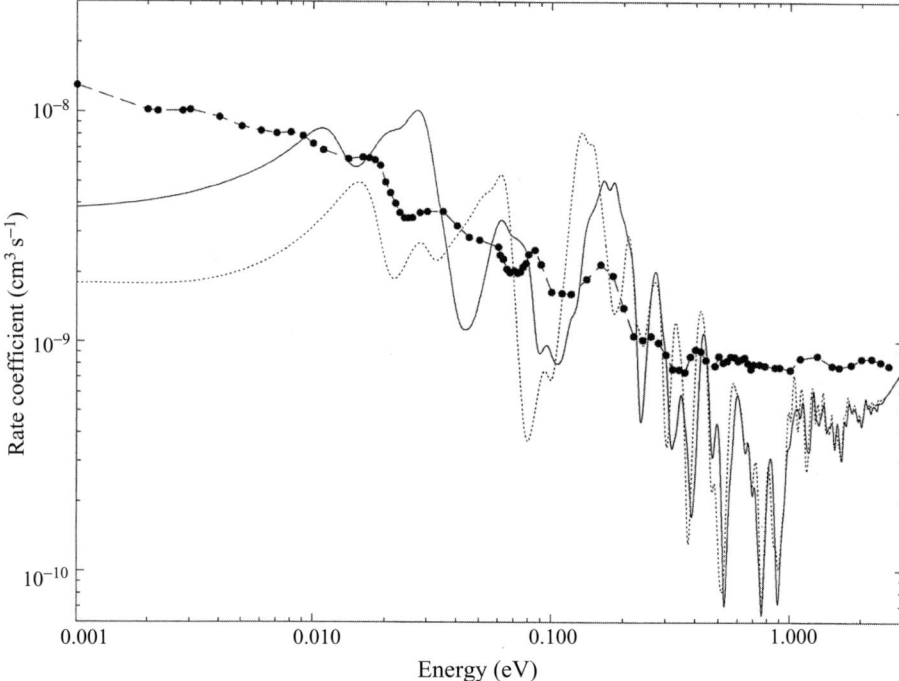

Figure 4.7 Rate coefficient, $\langle v_r \sigma \rangle$, for the dissociative recombination of HD$^+$ ($v = 0$) measured in CRYRING. In this experiment the transverse electron-energy spread was 10 meV. The filled circles are from the experiment by Strömholm *et al.* (1995), the dotted curve is MQDT calculations with rotational effects neglected, and the full drawn curve is with rotational interaction included. In the calculations the ion was assumed to be in $v = 0$ and in its lowest rotational level. The theoretical cross section was convoluted with the energy spread in the electron beam. (Reproduced with permission from I. F. Schneider, C. Strömholm, L. Carat, X. Urbain, M. Larsson, and A. Suzor-Weiner, "Rotational effects in HD$^+$ dissociative recombination: theoretical study of resonant mechanisms and comparison with ion storage ring experiments," *J. Phys. B* **30**, pp. 2687–2705, (1997), IOP Publishing Limited.)

where $\overline{\Gamma}$ is a dimensionless width parameter that is determined from the lower-n values.

Additional states would be required to understand the cross section at higher energy. In addition to the new dissociative resonant states, at these energies autoionization into the dissociative vibrational continuum of the ground ionic states becomes possible, i.e., the process

$$H_2(v) + e^- \rightarrow H_2^{**} \rightarrow H + H^+ + e^-. \tag{4.3}$$

Neglect of this channel leads to an underestimate of the autoionization loss and thereby an increase in the dissociative recombination cross section. This is included

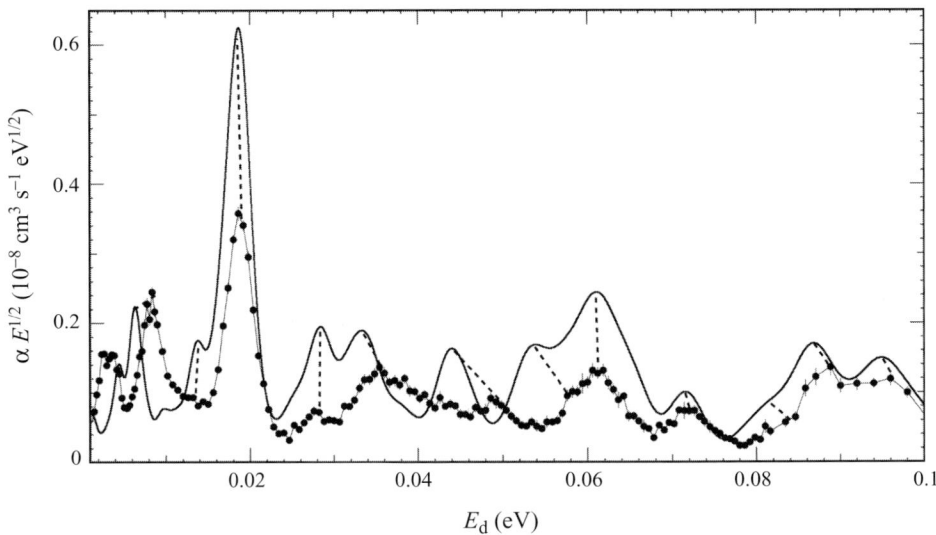

Figure 4.8 Dissociative recombination of HD^+ ($v = 0$). The x-axis shows the detuning energy and the y-axis the product $E_d^{1/2}\langle v_r\sigma\rangle$. The experimental results were measured with the photocathode electron target in the TSR (Orlov et al. 2005) characterized by $kT_\perp = 0.5$ meV. The MQDT results were obtained by convoluting the theoretical cross section with the 0.5 meV energy spread, and by averaging over the calculated results for a rotational distribution of 300 K. (Private communication by I. F. Schneider and A. Wolf 2007.)

in the MQDT treatment by the addition of discretized continuum states. The results of these calculations are compared in Fig. 4.9 to those from CRYRING experiments (Strömholm et al. 1995). In contrast to earlier experiments at the TSR (Forck et al. 1993b) and those at TARN II (Tanabe et al. 1995) these experiments obtained absolute cross sections that could be directly compared with theory. Agreement with experiment in this high-energy region is excellent.

Experiments using the TSR (Amitay et al. 1998, 1999) have opened up another area of investigation. In these experiments the vibrational distribution in the ring is probed as a function of time, from injection until total relaxation. A portion of the stored beam is abstracted and its vibrational energy distribution determined by a Coulomb explosion technique. This is coupled with experiments on the dissociative recombination cross section and product state distribution of the fragments. These data can then be fit to obtain the time evolution of the vibrational state distribution and deconvolved to produce dissociative recombination rate coefficients as a function of the initial vibrational state. Although there is reasonable agreement between theory and experiment, significant differences persist especially for $v = 3$ and $v = 5$, as shown in Fig. 4.10.

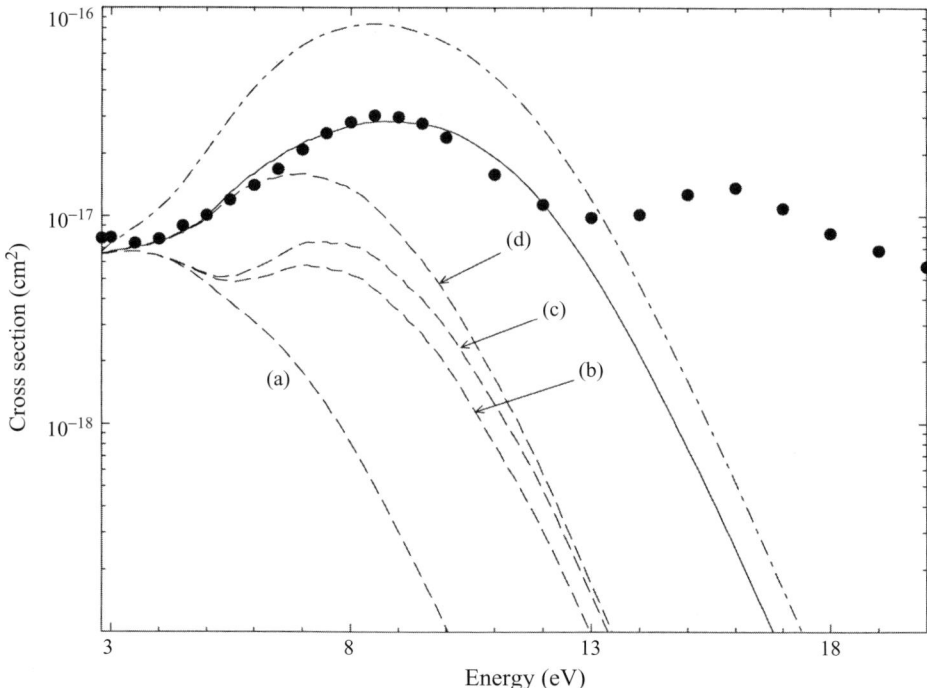

Figure 4.9 Cumulative contributions of dissociative states of HD to the theoretical (MQDT) dissociative recombination cross section of HD$^+$ ($v = 0$) compared with experimental results from CRYRING; (a) theory, contribution of $(2p\sigma_u)^2\,^1\Sigma_g^+$ state only; (b) same as (a) plus contribution of the lowest $^1\Pi_g$ state; (c) same as (b) plus contribution of the lowest $^1\Sigma_u^+$ state; (d) same as (c) plus contribution of the lowest $^3\Pi_g$ state;———, contribution of entire series of Rydberg dissociative states for the 4 symmetries added (i.e., the final theoretical result of $v = 0$, $N = 0$); —·—·— same, without including dissociative autoionization; – – – – – – – –, theoretical result for $v = 1$; •, experimental results from CRYRING. The second peak observed at higher energy is probably due to a second group of dissociative states with a $2p\pi_u$ ion core, which were not included in the calculations. (Reprinted figure by permission from C. Strömholm, I. F. Schneider, G. Sundström et al., "Absolute cross sections for dissociative recombination of HD$^+$: Comparison of experiment and theory," Phys. Rev. A **52**, pp. R4320–4323, (1995). Copyright (1995) by the American Physical Society.)

In addition, these experiments lead to information about the so-called superelastic collision of electrons and HD$^+$

$$e^-(\varepsilon) + H_2^+(v) \to e^-(\varepsilon') + H_2^+(v') \qquad v' < v. \tag{4.4}$$

This process competes with dissociative recombination and occurs via similar mechanisms. Calculations using the R-matrix method (Sarpal & Tennyson 1993) and the MQDT method with first-order perturbative treatment of the electronic

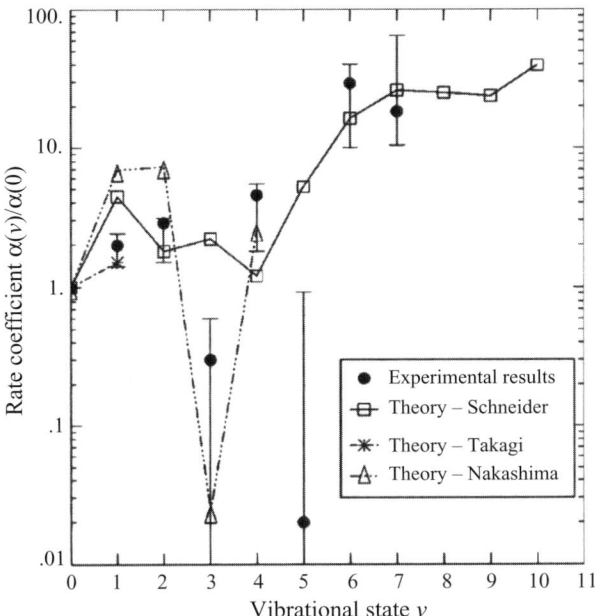

Figure 4.10 Experimental results for dissociative recombination of HD^+. The results were obtained at the TSR at $E_d = 0$ eV as a function of initial vibrational quantum number. The y-axis shows the ratio of the rate coefficient, $\alpha = \langle v_r \sigma \rangle$, for a particular vibration level v and the rate coefficient for $v = 0$. Except for $v = 5$, where the result is only for dissociation to the $n = 2$ asymptote, the displayed values have been summed over all possible final states. Theoretical calculations were performed by Schneider and Suzor-Weiner (Amitay et al. 1998) and the results are shown as □. Also shown are theoretical results from Nakashima, Takagi, and Nakamura (1987) and Takagi (1996). (From Z. Amitay, A. Baer, M. Dahan et al. 1998, "Dissociative recombination of HD^+ in selected vibrational quantum states," *Science* **281**, pp. 75–78. Reprinted with permission from AAAS.)

coupling (Nakashima, Takagi, & Nakamura 1987) were employed to study this process and yielded similar cross sections. This work was repeated by Ngassam et al. (2003) using a second-order MQDT treatment. Although rate coefficients for dissociative recombination as a function of vibrational state are in good agreement with those obtained on the TSR storage ring, the rates for the superelastic process agree with the previous calculations and lie an order of magnitude smaller that the experimental results. At this time there is no explanation for the discrepancy, although work is continuing.

Dissociative recombination of HD^+ leads to the formation of H (*nl*) and D (*n'l'*). Zajfman et al. (1995) showed how the imaging technique can be used to determine the recombination products. At $E_d = 0$, only two asymptotic limits are energetically available: H (1s) + D (1s) and H (1s) + D ($n = 2$)/H ($n = 2$) + D (1s). Based

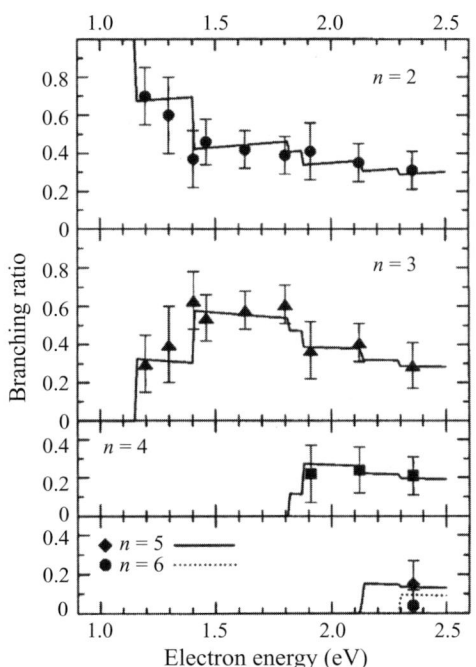

Figure 4.11 Branching ratios for the dissociative recombination of HD$^+$ ($v = 0$) as a function of electron energy for the final states $n = 2, n = 3, n = 4$, and $n = 5, 6$, measured in the TSR. The symbols are the experimental results and the lines the results of the Landau–Zener calculations. (Reprinted figure by permission from D. Zajfman, Z. Amitay, M. Lange *et al.*, "Curve crossing and branching ratios in the dissociative recombination of of HD$^+$," *Phys. Rev. Lett.* **79**, pp. 1829–1832, (1997). Copyright (1997) by the American Physical Society.)

on the potential curves shown in Fig. 4.1 and the recombination mechanism, one would anticipate the H (1s) + D ($n = 2$)/H ($n = 2$) + D (1s) limit to be dominant, and this is also precisely what Zajfman *et al.* (1995) found. In a subsequent paper (Zajfman *et al.* 1997) they did a careful study of the recombination products as a function of electron energy. The measured branching ratios are shown in Fig. 4.11. Zajfman *et al.* (1997) applied a Landau–Zener formulation of the dissociation along the $(2p\sigma_u)^2\ ^1\Sigma_g^+$ state. As can be seen from Fig. 4.11, this approach was very successful.

5
Diatomic hydride ions

Molecular ions XH$^+$, where X is an atom with at least two electrons, come next in complexity after H$_2^+$ and its isotopologs. When Bardsley and Biondi (1970) reviewed dissociative recombination there were no results on this type of ion, and 20 years later (Mitchell 1990a) results for only four XH$^+$ ions had been reported. More recently they have been the subjects of much experimental and theoretical work, and a separate chapter is needed to describe these systems.

5.1 HeH$^+$

The HeH$^+$ ion is composed of the two most abundant elements in the universe. It was discovered by Hogness and Lunn (1925) using mass spectrometry. The first experimental information about rovibrational levels in HeH$^+$ came not from spectroscopy but from ion-beam experiments by Schopman, Fournier, and Los (1973) in which they detected protons from 10 keV HeH$^+$ decomposing either spontaneously or as a result of collisional breakup. They concluded that HeH$^+$ populated rovibrational levels in the $X\ ^1\Sigma^+$ state close to the dissociation limit either as a result of their formation in the ion source or by collision-induced excitation. These rovibrational levels were trapped behind a centrifugal potential barrier and predissociated to He + H$^+$. By measuring the energy of the proton, Schopman, Fournier, and Los (1973) were able to infer information about the energy levels. Their work inspired Dabrowski and Herzberg (1977) to consider the reverse process:

$$\text{He} + \text{H}^+ \rightarrow \text{HeH}^{+*} \rightarrow \text{HeH}^+ + h\nu, \tag{5.1}$$

which they labeled radiative recombination (radiative association is a more appropriate name). Dabrowski and Herzberg (1977) calculated rovibrational levels in the ground state and made predictions of an infrared spectrum. They also pointed out that process (5.1) could lead to the formation of HeH$^+$ in the interstellar medium. Tolliver, Kyrala, and Wing (1979) used a fast ion beam merged with a laser beam

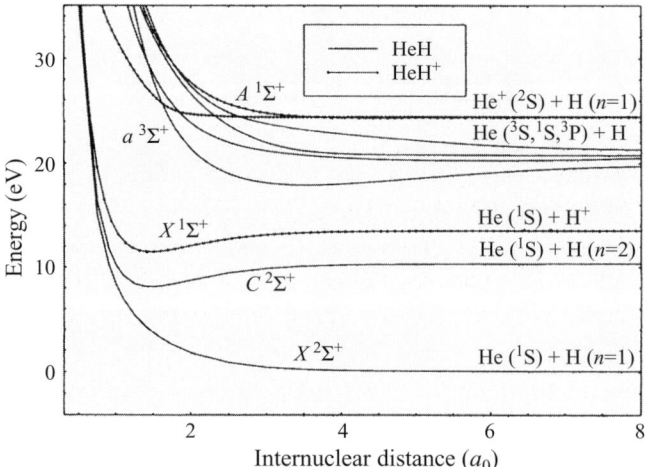

Figure 5.1 Potential curves for HeH$^+$ and HeH. (Reprinted figure by permission from C. Strömholm, J. Semaniak, S. Rosén, *et al.*, "Dissociative recombination and dissociative excitation of ^{4}HeH$^+$: Absolute cross sections and mechanisms," *Phys. Rev. A* **54**, pp. 3086–3094, (1996). Copyright (1996) by the American Physical Society.)

at an angle $\theta = 11$ mrad to observe the first vibration–rotation spectrum of HeH$^+$. It is interesting to note that all the important experimental steps quoted here were done at accelerators. Observations of infrared transitions by means of laser spectroscopy with discharge cells came somewhat later (Bernath & Amano 1982, Liu, Ho, & Oka 1987, Crofton *et al.* 1989). Being isoelectronic with H$_2$, and thus a molecule of fundamental significance, HeH$^+$ has been a subject of great theoretical interest. The first theory dates back to the early 1930s (Glockler & Fuller 1933), and had by the late 1970s reached spectroscopic accuracy (Bishop & Cheung 1979; this paper also includes a review of earlier work). The quasibound levels close to the dissociation limit in the HeH$^+$, first probed by Schopman, Fournier, and Los (1973), have been studied at high accuracy with a fast-ion-beam laser apparatus by Carrington *et al.* (1983). This work also contains theoretical predictions of all bound and quasibound levels of ^{4}HeH$^+$ and its three isotopologs. Figure 5.1 shows potential curves for HeH$^+$ and HeH.

Black (1978), Flower and Roueff (1979), and Roberge and Dalgarno (1982) discussed possible mechanisms for the formation and destruction of HeH$^+$ in astrophysical plasmas. Black (1978) assumed the dissociative recombination rate coefficient of HeH$^+$ to be 10^{-8} cm^3 s^{-1}, a number that had been used since 1970 (Hunten 1969, Prasad & Capone 1971). Prasad and Capone (1971) admitted, however, that the number was speculative. Flower and Roueff (1979) were sceptical and argued that the rate coefficient must be smaller because of the lack of a neutral state curve-crossing of the ion ground state. Roberge and Dalgarno (1982) conjectured that dissociative recombination of HeH$^+$ is negligible for the same reason. Adams and Smith (1989) made the first attempt to measure the dissociative recombination of HeH$^+$, using the FALP technique; they found the rate to be comparable to that of He$^+$ (see Section 7.2.1 for a discussion of this experiment in connection with H$_3^+$) and concluded that HeH$^+$ recombines with a rate coefficient of 10^{-11} cm^3 s^{-1}, thus corroborating the conjecture by Roberge and Dalgarno (1982). This is, as far as we know, the only attempt to measure the recombination of HeH$^+$ with an afterglow technique; all subsequent studies have been done using beam techniques. Yousif and Mitchell (1989) used the merged-beam technique to study the recombination of ^{4}HeH$^+$, found a structured cross section, and used it to estimate the rate coefficient to be 10^{-8} cm at 300 K. This is the same value as had been guessed by Hunten (1969) and Prasad and Capone (1971) almost 20 years earlier, but apparently Yousif and Mitchell were unaware of these previous papers. Yousif and Mitchell used dissociative excitation to infer that under optimal ion source conditions the ^{4}HeH$^+$ ions were predominantly in the ground vibrational level. They also criticized the FALP experiment and argued that the sensitivity in a FALP experiment is not sufficient to measure a rate coefficient as small as 10^{-11} cm^3 s^{-1}. This is a justified criticism.

Yousif and Mitchell (1989) found sharp resonances in the dissociative excitation cross section for the channel HeH$^+$ + e$^-$ → He$^+$ + H at 20 eV and 26 eV, and interpreted these peaks as being evidence for their having vibrationally cold HeH$^+$ in the beam. However, this did not explain why the peaks were so sharp. This problem was amplified by the calculations of Orel, Rescigno, and Lengsfield (1991), which gave an almost flat cross section in the region 21.5–26 eV, with no sharp structures. Strömholm et al. (1996) measured a flat cross section in the same region, but their result was about 60% larger than the theoretical cross section. Yousif and Mitchell (1989) did not report the observation of a He + H$^+$ channel, which would have revealed resonant dissociative excitation (see Section 9.1). The sharp peaks in Yousif and Mitchell's experiment are even more difficult to understand in view of what happened five years after the initial publication. Mitchell and coworkers (Yousif et al. 1994) used the same single-pass merged-beam apparatus as in the original experiment, but this time they also added oxygen to the ion source. The effect of making this addition was to render the cross section immeasurably small. They interpreted the originally measured cross section (Yousif & Mitchell 1989) as due to the presence of the electronically excited, metastable, $a\ ^3\Sigma^+$ state, and proposed that the vanishing cross section was a result of quenching of the recombining $a\ ^3\Sigma^+$ state in the ion source. They also calculated the vibrational level spacing in the $a\ ^3\Sigma^+$ state and found that resonances in the cross section in the absence of oxygen quenching matched these level spacings. This implies that the electronic ground state of HeH$^+$ recombines so slowly that it did not give rise to a signal in the merged-beam experiment (Yousif et al. 1994), a result in agreement with the findings of Adams and Smith (1989). So where does that leave the dissociative excitation resonances Yousif and Mitchell (1989) observed? Yousif et al. (1994) suggested that these resonances were indeed due to the ground state, but the obvious test of searching for them when oxygen was added to the ion source was apparently never performed. They also pointed out that Michels (1989) was the first to suggest that the dissociative recombination of HeH$^+$ was due to a metastable triplet state rather than the electronic ground state. This is not entirely correct. The suggestion of HeH$^+$ recombining through an excited state actually derives from Mitchell's group, which is also pointed out by Michels (1989). It is mentioned in a very brief note by Yousif and Mitchell (1988, p. 1010) that "Recent measurements of $e +$ HeH$^+$ ions have indicated that excited ions can display large recombination cross sections which rapidly decrease as the excited states are quenched." It is not clear from this quotation whether they were referring to vibrationally and/or electronically excited states, but in Mitchell and Yousif (1989) they mention electronically excited states as one possibility.

Datz and Larsson (1992) used HeH$^+$ as a case study ion to address the question of whether there were advantages in using an ion storage ring to study dissociative recombination. They found that this was indeed the case, and HeH$^+$ was chosen for the first dissociative recombination experiment at TARN II (Tanabe et al. 1993). Tanabe and coworkers found a high-energy peak at 20 eV and a second peak when they varied the detuning energy across 0 eV. Their measurement was not an absolute one, but they noted that the peak centered on 0 eV decreased with respect to the 20 eV peak as a function of storage time. This time dependence of the peak ratio was interpreted as vibrational distribution of the HeH$^+$ ions, which had a larger effect on the 0 eV peak than on the 20 eV peak. Tanabe et al. (1994) investigated the isotope effect and found that the 0 eV peak for HeD$^+$ was smaller relative to the 20 eV peak than with HeH$^+$, as shown in Fig. 5.2. They also put a 1-mm wire in front of the surface barrier detector and inferred from the modification of the pulse height spectrum that recombination at low electron energy gave a kinetic energy release of about 1.0 eV, which suggests that low-energy recombination leads to formation of He $+$ H ($n = 2$). The addition of oxygen to the ion source was tested and found to have no effect on the cross section; however, Yousif et al. (1994) in a note added in proof pointed out that Tanabe et al. (1994) did not add sufficient amounts of oxygen to cause triplet state quenching. Yousif et al. also estimated the lifetime of the $a\,^3\Sigma^+$ state to be extremely long, which would suggest that storage in a ring for a few seconds would not be sufficient to remove the triplet state. Sundström et al. (1994a) performed the first absolute cross section measurement of a HeH$^+$ isotopolog in an ion storage ring when they used CRYRING to study ^{3}HeH$^+$. The addition of oxygen to a 30% level in the ion source gave no effect on the cross section. Orel, Kulander,

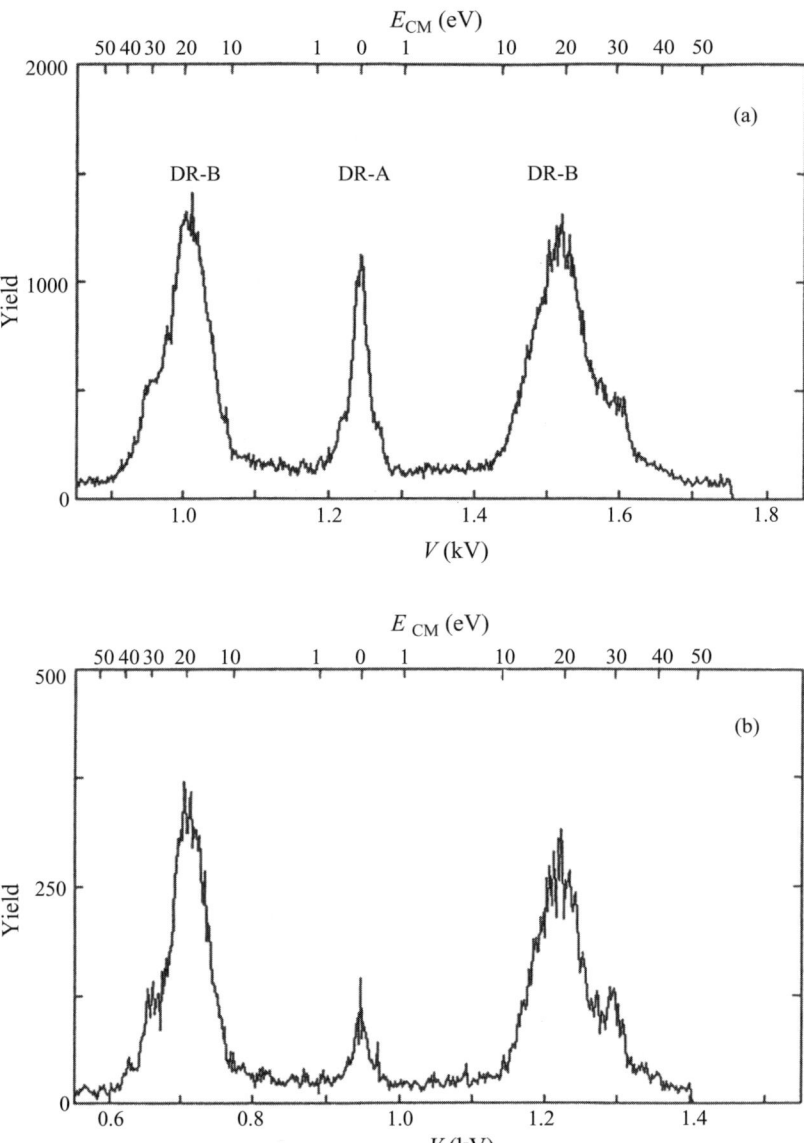

Figure 5.2 The yield of formation of a) He + H and b) He + D in HeH(D)$^+$ + e collisions in the TARN II storage ring. The voltage applied to the electron cooler cathode is shown on the abscissa, and on top the center-of-mass energy is shown. (Reprinted figure by permission from T. Tanabe, I. Katayama, N. Inoue, *et al.*, "Origin of the low-energy component and isotope effect on dissociative recombination of HeH$^+$ and HeD$^+$," *Phys. Rev. A* **49**, pp. R1531–R1534, (1994). Copyright (1994) by the American Physical Society.)

and Rescigno (1995) applied the complex Kohn variational method for the electron scattering problem and a wave packet approach for the dissociation dynamics to obtain theoretical results for the 20 eV peak. They showed that this peak arises from electron capture into Rydberg states converging to the $a\,^3\Sigma^+$ and $A\,^1\Sigma^+$ states. By including eight resonance states in the calculations, four $^2\Sigma$ states and one $^2\Pi$ state converging to the $a\,^3\Sigma^+$ state $((1\sigma 2\sigma^2), (1\sigma 2\sigma)^3 3\sigma, (1\sigma 2\sigma)^3 4\sigma, (1\sigma 2\sigma)^3 5\sigma$, and $(1\sigma 2\sigma)^3 1\pi$, where the superscript 3 is used to label triplet coupling), the two lowest Rydberg states associated with the $A\,^1\Sigma^+$ state $((1\sigma 2\sigma)^1 3\sigma$ and $(1\sigma 2\sigma)^1 1\pi)$, and the $(1\sigma 3\sigma^2)$ state, they obtained a cross section slightly larger than the absolute cross section measured in CRYRING (Sundström et al. 1994a). Larson and Orel (1999) also calculated the cross section for ^{4}HeH$^+$ in the high-energy region and found a cross section about 60% higher than the one measured by Strömholm et al. (1996) (there is a misprint in the caption of Figure 5.8 in Larson & Orel (1999); the experimental data for ^{4}HeH$^+$ are from Strömholm et al. (1996) and for ^{3}HeH$^+$ from Sundström et al. (1994a)). However, it was later found (Orel 2000b) that there is a factor of 2 error in the definition of the cross section given by Orel, Kulander, and Rescigno (1995) and Larson and Orel (1999). After correction, the theoretical cross sections are lower than the experimental ones; the agreement is still good.

Simultaneously with the experimental work, theoretical work on HeH$^+$ was being pursued by Guberman (1994) and by Tennyson's group (Sarpal, Tennyson, & Morgan 1994). This work represents the first *ab initio* calculations of dissociative recombination with low-energy electrons in the absence of a curve-crossing. Guberman performed the calculations for ^{3}HeH$^+$ while Tennyson's group chose to study ^{4}HeH$^+$. The results from CRYRING (Sundström et al. 1994a) had been reported at ICPEAC in Aarhus in 1993 (Larsson et al. 1993b) and were known to Guberman, and Tennyson's group used the results of Yousif and Mitchell (1989) for comparison. Ironically, the new results from Mitchell's group (Yousif et al. 1994) were published at about the same time that Sarpal, Tennyson, and Morgan (1994) submitted their paper for publication.

Guberman (1994) performed accurate electronic structure calculations of the relevant potential curves of HeH$^+$ and HeH, and instead of constructing diabatic potential curves, which is the usual procedure for curve-crossing recombination, Guberman used adiabatic potential curves. In the adiabatic representation the electronic Hamiltonian is diagonal and the coupling between different adiabatic states occurs through the nuclear kinetic-energy operator:

$$T_N = -\frac{\hbar^2}{2\mu}\frac{1}{R^2}\frac{\partial}{\partial R}R^2\frac{\partial}{\partial R}, \qquad (5.2)$$

where R is the internuclear distance and μ the reduced mass. Let us label the wavefunctions of the HeH$^+$ $X\,^1\Sigma^+$ state $\psi_v\chi_v$, where v is the vibrational quantum number, ψ_v is the many-electron wavefunction including the free-electron orbital, and χ_v is the vibrational wavefunction. The dissociative states of HeH are not repulsive states crossing through the $X\,^1\Sigma^+$ state, but Rydberg states converging to $X\,^1\Sigma^+$. The dissociation limit of these Rydberg states is below the lowest vibrational level of the $X\,^1\Sigma^+$ state; Fig. 5.1 shows the $C\,^2\Sigma^+$ state of HeH dissociating to He (^{1}S) + H ($n = 2$). If the wavefunctions for the Rydberg states are labeled $\psi_d \chi_d$, the interaction matrix elements take the form (Guberman 1994; the mechanism is further described in Chapter 3):

$$\langle \psi_d \chi_d | T_N | \psi_v \chi_v \rangle = -\frac{\hbar^2}{2\mu} \langle \chi_d | B(R) + 2A(R) \partial/\partial R | \chi_v \rangle, \quad (5.3)$$

where

$$B(R) = \langle \psi_d | \partial^2 / \partial R^2 | \psi_v \rangle \quad (5.4)$$

and

$$A(R) = \langle \psi_d | \partial / \partial R | \psi_v \rangle. \quad (5.5)$$

The matrix elements $B(R)$ and $A(R)$ are of comparable magnitude (Guberman 1994), but in Eq. (5.3) the element $A(R)$ is followed by the $\partial/\partial R$ operator. This means that if the derivative of the vibrational wavefunction χ_v is large when $A(R)$ is not vanishingly small, the matrix element in Eq. (5.3) can become quite large. The operator $\partial/\partial R$ couples states in regions where they change character. Expressed in terms of electronic wavefunctions expanded in a linear combination of configuration state functions, this corresponds to a region of internuclear distances where sizable components in the expansion of one electronic state switch to become sizable components in the expansion of the other electronic states, i.e., in the region of an avoided crossing. Kubach et al. (1987) have shown that the $X\,^2\Sigma^+$ ($1\sigma^2 2\sigma$) and $A\,^2\Sigma^+$ ($1\sigma^2 3\sigma$) states have an avoided crossing around 0.8 a_0 as a result of an avoided crossing between the 2σ and 3σ molecular orbitals. A similar study by van Hemert and Peyerimhoff (1991) revealed a similar strong avoided crossing between the $C\,^2\Sigma^+$ and $D\,^2\Sigma^+$ states, also around $R = 0.8\,a_0$. Figure 5.3 shows the potential curves of the HeH$^+$ ground state and the relevant HeH states, and the product $2A(R)\partial\chi_{v=0}/\partial R = 2A(R)\mathrm{d}\chi_0/\mathrm{d}R$, where $A(R)$ is that between the $C\,^2\Sigma^+$ and $D\,^2\Sigma^+$ states. This product reaches its maximum value at an internuclear distance smaller than the left turning point of the $v = 0$ level of the ion ground state. The operator $\partial/\partial R$ has the effect of increasing the range of internuclear distances over which dissociative recombination can occur, and it is this effect which makes recombination possible even in the absence of a curve crossing. Because the maximum occurs in the region to the left of the classical turning of $v = 0$, tunneling

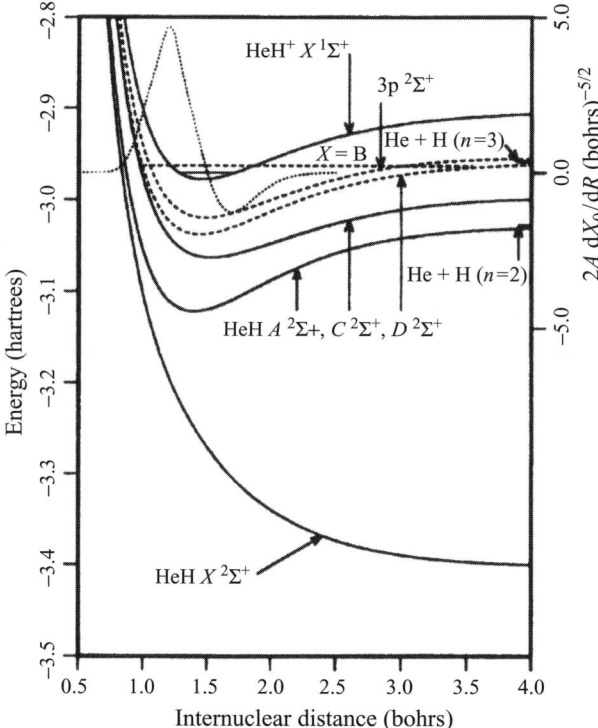

Figure 5.3 The calculated HeH dissociative potential energy curves (solid lines), the $n=3$ Rydberg states (dashed lines), and the HeH$^+$ electronic ground state (solid line) with its lowest vibrational level. The $v=8$ resonance level of the $D\,^2\Sigma^+$ state is shown as a dashed line. The dotted line shows the product $2A(R)\mathrm{d}\chi_0/\mathrm{d}R$, with its ordinate axis on the right. (Reprinted figure by permission from S.L. Guberman, "Dissociative recombination without a curve crossing," *Phys. Rev. A* **49**, pp. 4277–4280, (1994). Copyright (1994) by the American Physical Society.)

plays an important role, so important that Bates (1994) named it tunneling mode recombination.

Guberman (1994) performed MQDT calculations using the potential curves and coupling elements discussed above. He found that the cross section is dominated by the indirect mechanism and that the totally dominating decay route is the $C\,^2\Sigma^+$ state:

$$\begin{aligned}\mathrm{HeH}^+\,(X^2\,\Sigma^+(v=0)) + \mathrm{e}^- &\to \mathrm{HeH}\,(\mathrm{Rydberg}) \to \mathrm{HeH}\,(C^2\,\Sigma^+)\\ &\to \mathrm{He} + \mathrm{H}\,(n=2).\end{aligned} \quad (5.6)$$

The prediction of dissociation to the H ($n=2$) state was in agreement with Tanabe *et al.*'s (1994) data, and as we shall see later, this has been verified convincingly by the use of imaging techniques. Guberman's calculated cross section for ^{3}HeH$^+$

was in very good agreement with results from CRYRING (Larsson et al. 1993b, Sundström et al. 1994a).

Sarpal, Tennyson, and Morgan (1994) adopted the R-matrix method to the HeH$^+$ case where the indirect mechanism dominates and where there is no crossing. Basically they identified and treated the same mechanism as Guberman, i.e., nonadiabatic coupling at small internuclear distances, and their cross section for ^{4}HeH$^+$ was in good agreement with the first single-pass experiment (Yousif & Mitchell 1989). They pointed out that the mechanism, relying on tunneling, would give the isotope effect shown in Fig. 5.2, at least qualitatively. Sarpal, Tennyson, and Morgan (1994), however, differed from Guberman in the prediction of the dominant decay channel. At low electron energy they found the dissociation flux going predominantly to the He + H ($n = 1$) channel. Guberman (1995a) also calculated the cross section for ^{4}HeH$^+$, but did not make a comparison with the results of the Tennyson group. Such a comparison was, however, made by Strömholm et al. (1996), and the agreement between the two theoretical calculations is fair; the cross section of Sarpal, Tennyson, and Morgan (1994) is larger than Guberman's below 0.06 eV and smaller at larger electron energies.

The measurement in CRYRING (Sundström et al. 1994a) gave a cross section with very little structure, in contrast to Guberman's calculated cross section (Guberman 1994), which was highly structured. The reason for the lack of structure in the CRYRING data was that the experiment was carried out with an electron beam with $kT_\perp = 0.1$ eV. Shortly after the CRYRING experiment, the cathode and transport magnets of the electron cooler were modified (Danared et al. 1994) so that an electron beam with $kT_\perp = 0.01$ eV was obtained. A new experiment on ^{3}HeH$^+$ with the colder electron beam was performed (Mowat et al. 1995), and this time the structure in the cross section was revealed. Figure 5.4 shows a comparison of the CRYRING result (Mowat et al. 1995) with the theoretical results of Guberman (1994). The theoretical cross section was folded with the electron-velocity distribution of the CRYRING electron beam, as described in Chapter 2. The agreement is quite good and the CRYRING experiment represents a confirmation of the noncrossing mechanism found by Guberman (1994) and Sarpal, Tennyson, and Morgan (1994).

The imaging technique described in Section 2.2.1 was applied to different isotopologs of HeH$^+$ at CRYRING (Strömholm et al. 1996, Semaniak et al. 1996) and the TSR (Semaniak et al. 1996). These experiments provided firm evidence that low-energy recombination of HeH$^+$ results in He + H ($n = 2$). Figure 5.5 shows distance distributions from three different measurements (Semaniak et al. 1996). Figures 5.5(a) and 5.5(b) show that dissociative recombination leading to He + H ($n = 1$) has such a small branching ratio that it does not give a signal at the expected distance. Figure 5.5(c) was recorded with the imaging system at the TSR for

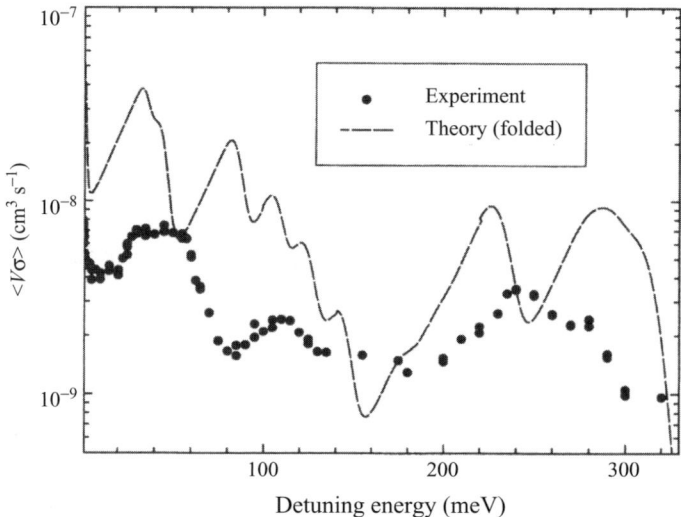

Figure 5.4 The energy resolved rate coefficient, $\alpha(v_d)$ as defined in Eq. (2.11), measured in CRYRING (Mowat et al. 1995) compared with the theoretical results of Guberman (1994). The theoretical cross section was folded with an electron-velocity distribution given in Eq. (2.9) and with $kT_\perp = 0.01$ eV. (Reprinted figure by permission from J. R. Mowat, H. Danared, G. Sundström, et al., "High-resolution, low-energy dissociative recombination spectrum of ^{3}HeH$^+$," *Phys. Rev. Lett.* **74**, pp. 50–53, (1995). Copyright (1995) by the American Physical Society.)

$E_d = 0.577$ eV, which is 0.244 eV above the threshold for He + H ($n = 3$). An analysis of the peak gave a kinetic energy release of 0.243 eV, in perfect agreement with the predicted value. The peak corresponding to He + H ($n = 2$) has disappeared, and it seems that dissociation occurs so that the kinetic energy release is minimized.

The most detailed cross section comparison of experiment and theory was done by the Japanese group of Tanabe et al. (1998). The theoretical calculations were only briefly described in this paper; a much more detailed description of the theoretical calculations was given in a separate paper by Takagi (2004). Tanabe et al. (1998) measured the relative dissociative cross section for ^{3,4}He^{1,2}H$^+$ using an ultracold electron beam with $kT_\perp \approx 1$ meV that had recently been installed at TARN II. Almost simultaneously, relative measurements on ^{3}HeH$^+$ were performed at CRYRING with an equally cold electron beam (Al-Khalili et al. 1998). The structures in the cross sections measured in the two storage rings agree very well for ^{3}HeH$^+$ though a comparison of the absolute levels is of course impossible. Tanabe et al. (1998) compared the experimental results with theoretical calculations, with the caveat that the relative experimental cross sections were positioned to agree with the theoretical results, thus emphasizing a comparison of the structures. Figure 5.6 shows this comparison of experiment and theory.

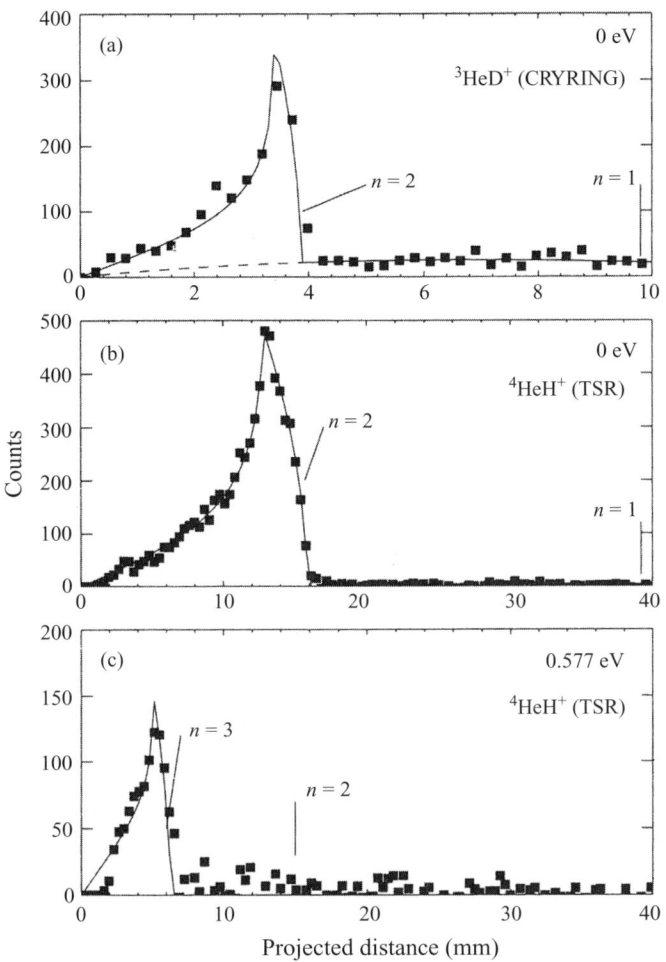

Figure 5.5 Projected distance distributions for ^{3}HeD$^+$ at $E_d = 0$ eV (upper), ^{4}HeH$^+$ at $E_d = 0$ eV (middle), and ^{4}HeH$^+$ at $E_d = 0.577$ eV. The measurement for ^{3}HeD$^+$ was done at CRYRING and the two other measurements at the TSR. The CCD camera at the TSR had more pixels, as can be seen from the larger number of filled squares. (Reprinted figure by permission from J. Semaniak, S. Rosén, G. Sundström, et al., "Product-state distributions in the dissociative recombination of ^{3}HeD$^+$ and ^{4}HeH$^+$," Phys. Rev. A **54**, pp. R4617–R4620, (1996). Copyright (1996) by the American Physical Society.)

Takagi (Tanabe et al. 1998) introduced three improvements as compared with Guberman's calculations (Guberman 1994, 1995a); (i) he allowed for rotation of the molecule; (ii) he treated the Rydberg states in a uniform way, which allowed him to extend the calculations to higher electron energy; (iii) he included the closed dissociation channels. The agreement with the TARN II data is good with the exception of the ^{3}HeD$^+$ isotopolog. The explanation for the poor agreement for ^{3}HeD$^+$

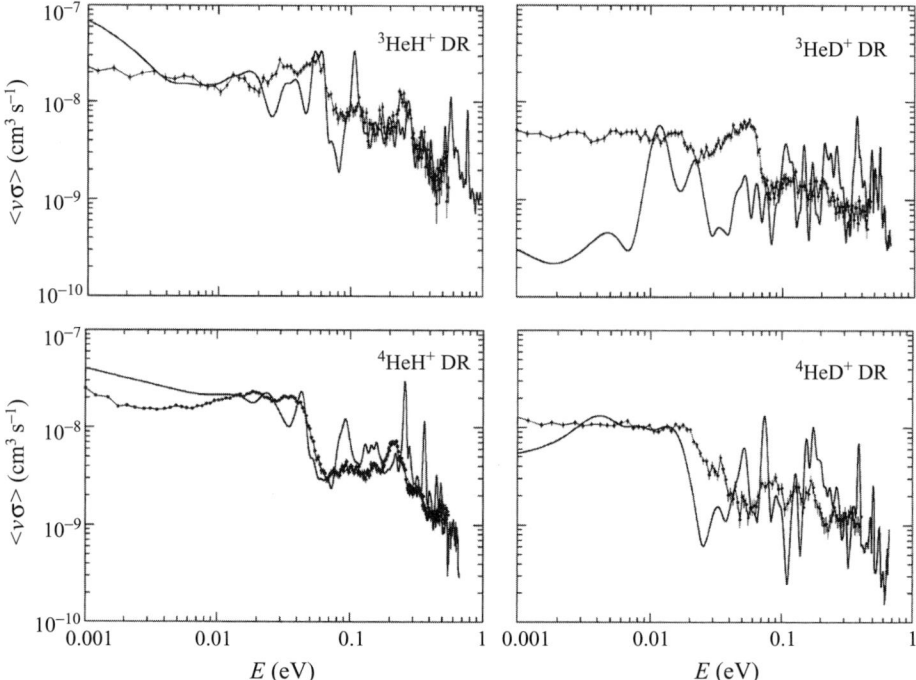

Figure 5.6 Experimental relative rate coefficients (dots) measured in TARN II at a resolution of $kT_\perp \approx 1$ meV and compared with theoretical results (solid lines). The error bars indicate statistical errors. (Reproduced with permission from T. Tanabe, I. Katayama, S. Ono, et al., "Dissociative recombination of HeH$^+$ isotopes with an ultracold electron beam from a superconducting electron cooler in a storage ring," *J. Phys. B* **31**, pp. L297–L303, (1998), IOP Publishing Limited.)

is not known. Tanabe *et al.* (1998) speculated that it may be related to the assumed rotational distribution in ^{3}HeD$^+$, which may be higher for this isotopolog because of its lower dipole moment. This does not seem to be a very plausible explanation; unfortunately, Takagi (2004) does not discuss the problem in his longer paper.

The comparison in Fig. 5.6 is not complete in that the experimental data are not absolute. Figure 5.7 shows a comparison for ^{3}HeH$^+$ of the experimental absolute rate coefficients from CRYRING (Mowat *et al.* 1995). The agreement is very good and clearly better than the experiment/theory agreement in Fig. 5.4.

Strömholm *et al.* (1996) used the measured cross section to calculate the thermal rate coefficient for ^{4}HeH$^+$ and displayed the result in a graph. At 300 K they found the rate coefficient to be 3×10^{-8} cm^3 s^{-1}. This is quite close to the very first guesses by Hunten (1969) and Prasad and Capone (1971). In an interstellar cloud ^{4}HeH$^+$ would be removed by electron recombination with a rate coefficient of

5.1 HeH$^+$

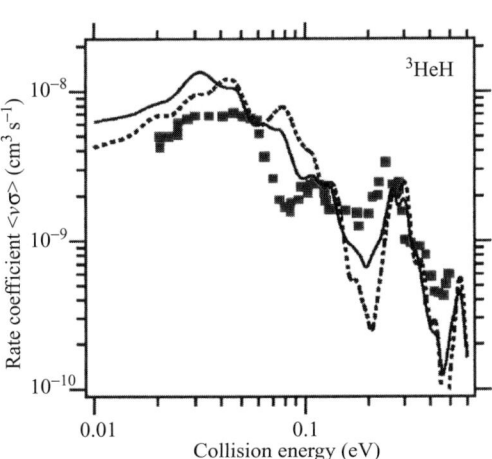

Figure 5.7 Comparison of the experimental rate coefficient for ^{3}HeH$^+$ + e$^-$ → ^{3}He + H measured in CRYRING (Mowat *et al.* 1995) and the theoretically calculated results of Takagi (2004) The experimental result is shown by squares, the solid line is the theoretical result convoluted with an electron-energy distribution of $T_\perp = 10$ meV and $T_\parallel = 0.5$ meV, and assuming a rotational temperature of 800 K; the dotted curve show the same but with a rotational temperature of 300 K. (Reprinted figure by permission from H. Takagi, "Theoretical study of dissociative recombination of HeH$^+$," *Phys. Rev. A* **70**, pp. 022709-1–10, (2004). Copyright (2004) by the American Physical Society.)

about 10^{-7} cm^3 s^{-1}, which is much faster than assumed by Roberge and Dalgarno (1982); HeH$^+$ has not yet been discovered in the interstellar medium.

The product distributions in Fig. 5.5 display the characteristic shape of isotropic distributions. Semaniak *et al.* (1996) also performed distribution measurements for ^{3}HeD$^+$ at electron energies corresponding to the high-energy peak and found evidence of $\cos^2\theta$ distributions. Larson and Orel (1999) addressed this issue in their theoretical study of the recombination of HeH$^+$. The good agreement in terms of the absolute cross section for the high-energy peak found for ^{3}HeH$^+$ and ^{4}HeH$^+$ as discussed above did not carry across to the product distributions. While the experiment (Semaniak *et al.* 1996) gave a clear signature of dissociation to an excited ^{3}He atom and ground state D, this channel does not appear in the theoretical calculations. The reason for this is not understood.

HeH$^+$ and its isotopologs are by now some of the best studied molecular ions with respect to dissociative recombination. HeH$^+$ has provided a paradigm in the understanding of the mechanisms for dissociative recombination by demonstrating, both theoretically and experimentally, that favorable curve crossing is not mandatory for recombination to occur at a reasonable rate. This insight was also very important to the efforts to understand recombination of H$_3^+$ (see Chapter 7). The first experiment to reveal recombination of HeH$^+$ was that of Yousif and Mitchell (1989). The subsequent work by Mitchell's group (Yousif *et al.* 1994), which claims that ground state HeH$^+$ does not recombine, is difficult to understand, and definitely wrong. The same holds for the FALP experiment (Adams & Smith 1989). Although there are details in the dissociative recombination of HeH$^+$ which are not fully understood, the overall picture is very clear. The argument that a molecule must recombine slowly because it lacks suitable curve crossings is obsolete and should not be used again.

5.2 NeH$^+$, ArH$^+$, KrH$^+$, and XeH$^+$

Less work has been invested in understanding the recombination of the other rare gas hydrides, but results are now available for all rare gas hydride ions. The heavier two have been investigated by means of the FALP technique (Geoghean, Adams, & Smith 1991, Le Padellec *et al.* 1997b), and the lighter two with the merged-beam ion storage ring technique (Mitchell *et al.* 2005a,b).

Geoghean, Adams, and Smith (1991) did not observe recombination of KrH$^+$ and XeH$^+$ at a significant rate, and gave only upper limits of 2×10^{-8} and 4×10^{-8} cm^3 s^{-1}. The results of the FALP-MS experiments of Le Padellec *et al.* (1997b) were in agreement with those of Geoghean, Adams, and Smith (1991) for KrH$^+$, but for XeH$^+$ they measured a significantly larger rate coefficient of 8.3×10^{-8} cm^3 s^{-1}. They had no explanation for why their result differed from that of Geoghean, Adams, and Smith (1991) despite the fact that basically the same technique was used. The higher rate coefficient is consistent with theoretical calculations, which show that the ground state of XeH$^+$ is crossed by a neutral $^2\Sigma$ state near $v = 2$ (Mitchell, Lipson, & Sarpal 2003). The rate coefficient for XeH$^+$ has been confirmed by a flowing afterglow experiment by Glosík's group (Korolov *et al.* 2006), and the same group also measured the rate coefficient for KrH$^+$ to be 2×10^{-8} cm^3 s^{-1} at 250 K (Korolov *et al.* 2006).

Just like HeH$^+$, NeH$^+$ lacks a curve-crossing. Recombination of NeH$^+$ has been studied in ASTRID, but the ion beam was too weak to allow an absolute measurement of the cross section (Mitchell *et al.* 2005b). However, an order of magnitude estimate showed that the cross section is comparable to that of HeH$^+$ in the low-energy regime. Two peaks were found in the cross section at electron energies larger than 7 eV. Theoretical work on this system was discussed at the 2007 Dissociative Recombination meeting in Ameland, The Netherlands. ASTRID was also used for experiments with ArH$^+$, and the result was surprising (Mitchell *et al.* 2005a). No recombination signal was found in the low-energy regime, while peaks were found at 7.5 eV, 16 eV, and 26 eV. Theoretical work on this ion is in progress. The lack of recombination of ArH$^+$ with thermal energy electrons is clearly in contrast to the

5.3 CH$^+$

general assumption that it recombines with a rate of 10^{-7} cm^3 s^{-1} (Meulenbroeks et al. 1994).

5.3 CH$^+$

The CH$^+$ ion has received more attention with regard to dissociative recombination than any diatomic hydride, except HeH$^+$. This is related to a long-standing problem in molecular astrophysics, namely the inability of models to account for the abundance of CH$^+$ in diffuse clouds (Herbst 2005). CH$^+$ was first identified in diffuse clouds by Douglas and Herzberg (1941). Bates and Spitzer (1951) discussed the formation and destruction of interstellar molecules, and proposed that if the electronic ground state of CH$^+$ was crossed by a neutral state, the ion would be destroyed by dissociative recombination according to the mechanism that Bates (1950b) had just laid out. Frisch (1972) and Solomon and Klemperer (1972) discussed the formation of interstellar CH$^+$, and Solomon and Klemperer (1972) also considered destruction mechanisms of CH$^+$. They assumed that CH$^+$ is destroyed equally rapidly by dissociative recombination as by dielectronic recombination (see Section 2.1.1), which gave a slow dissociative recombination rate. Theoretical calculations of relevant potential curves of CH$^+$ and CH led Bardsley and Junker (1973) and Krauss and Julienne (1973) to suggest that the dissociative recombination rate be three orders of magnitude larger than that of dielectronic recombination. Giusti-Suzor and Lefebvre-Brion (1977) also calculated potential curves of CH$^+$ and CH and were unable to find a crossing of the zeroth vibrational level of CH$^+$ by a neutral state potential curve, which made them propose a rate similar to that of Solomon and Klemperer (1972). Giusti (1980) used CH$^+$ as a test system when she developed the multichannel quantum defect theory to treat dissociative recombination, but she never attempted to calculate a thermal rate coefficient, and it was not until the work of Takagi, Kosugi, and Le Dourneuf (1991) that MQDT produced a value of 1.12×10^{-7} cm^3 s^{-1} at 120 K, very close to the value of 10^{-7} cm^3 s^{-1} guessed by Bardsley and Junker (1973).

The first measurement of dissociative recombination of CH$^+$ was made by means of the single-pass merged-beam technique (Mitchell & McGowan 1978). It is clear from the title of their 1978 paper that Mitchell and McGowan felt that they had sufficiently compelling evidence that their result was valid for CH$^+$ in the lowest vibrational level of its electronic ground state. They measured a cross section of $(5.12 \pm 0.5) \times 10^{-14}$ cm^2 at an electron energy of 0.01 eV (uncorrected for the factor of 2 error, see Chapter 2), and reported a thermal rate coefficient of $(3.05 \pm 0.3) \times 10^{-7}$ cm^3 s^{-1} at 100 K (also uncorrected). A few years later a slightly larger value of 5×10^{-7} cm^3 s^{-1} (uncorrected) was reported from the same group (Mul et al. 1981). The corrected thermal rate coefficient from the MEIBE experiments,

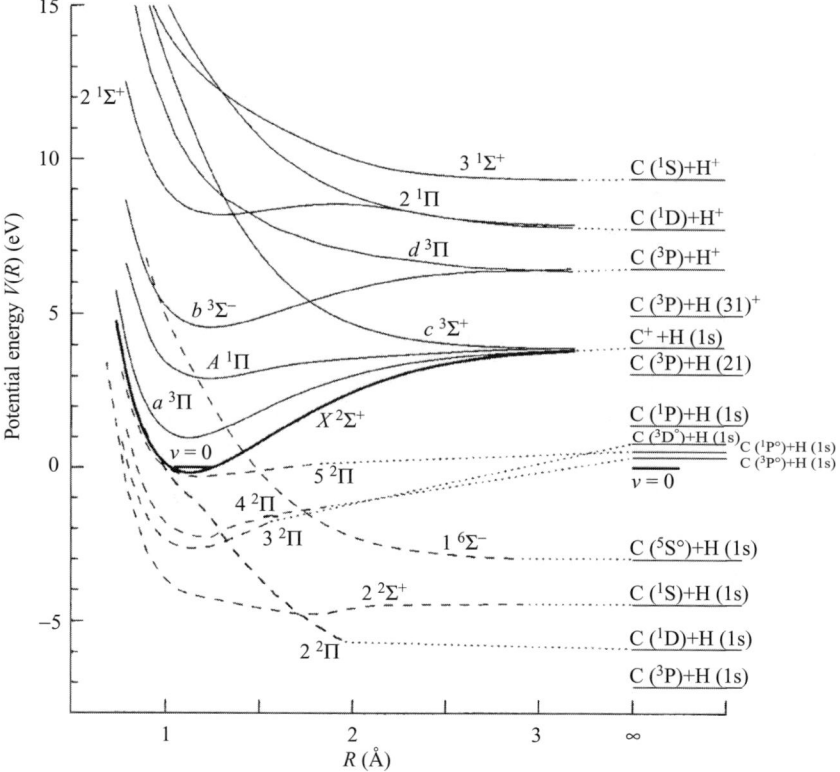

Figure 5.8 Potential energy curves of CH$^+$ (solid lines) and CH (dashed lines). The dotted lines show how the different electronic states correlate with the different dissociation limits. (Reprinted figure by permission from Z. Amitay, D. Zajfman, P. Forck, et al., "Dissociative recombination of CH$^+$: Cross section and final states," *Phys. Rev. A* **54**, pp. 4032–4050, (1996). Copyright (1996) by the American Physical Society.)

as given by Mitchell (1990a), is $1.5 \times 10^{-7}(300/T_e)^{0.42}$ cm^3 s^{-1}, close to the theoretical value of Takagi, Kosugi, and Le Dourneuf (1991).

Forck *et al.* (1994) studied dissociative recombination of CD$^+$ in the TSR, and found resonances in the cross section at 0.8 eV, 8.6 eV, and 11.7 eV. The work on CD$^+$ was followed by a very detailed study of CH$^+$ (Amitay *et al.* 1996b), including cross sections, product branching ratios, and angular distributions. Figure 5.8 shows the potential energy curves of CH$^+$ and CH given in the paper of Amitay *et al.* (1996b).

The $2\,^2\Pi$ state of CH, with the main configuration $1\sigma^2 2\sigma^2 3\sigma 1\pi 4\sigma$ and correlating with the C (^{1}D) + H (1s) limit, is responsible for recombination of CH$^+$ in its $X\,^1\Sigma^+$ state (Takagi, Kosugi, & Le Dourneuf 1991). There is a low-lying electronically excited triplet state, $a\,^3\Pi$ with the dominant configuration $1\sigma^2 2\sigma^2 3\sigma 1\pi$, and there is no reason a priori to assume that this state is not populated in an

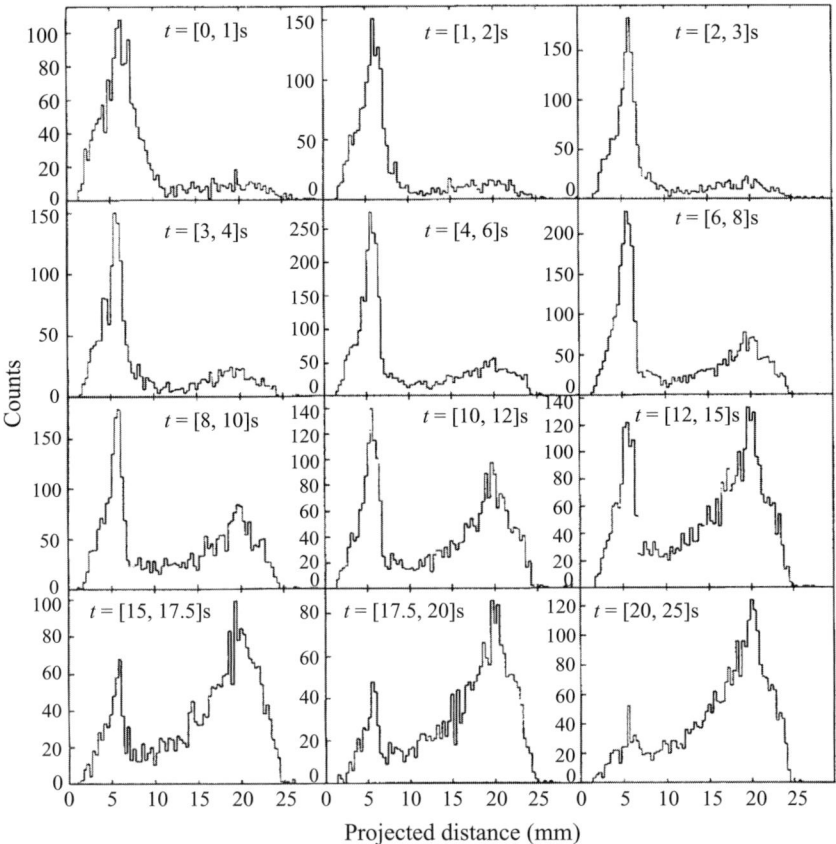

Figure 5.9 Projected distance spectra at $E_d = 0$ eV obtained from $CH^+(X\,^1\Sigma^+(v), a\,^3\Pi\,(v)) + e \rightarrow C + H$ in the TSR at different time intervals after ion production. The left peak is due to recombination of CH^+ in the $a\,^3\Pi\,(v=0)$ state and the right one to recombination of the CH^+ ground state. The dominant asymptotic limit in recombination of the $a\,^3\Pi$ state is $C(^3D) + H(1s)$, which gives a kinetic energy release of 0.44 eV, whereas recombination of the $X\,^1\Sigma^+\,(v=0)$ state predominantly leads to $C(^1D) + H(1s)$ with a kinetic energy release of 5.92 eV. The narrowing of the left peak during the first three seconds occurs because of the vibrational relaxation in the $a\,^3\Pi$ state, which is considerably faster than the radiative transition $a\,^3\Pi \rightarrow X\,^1\Sigma^+$. (Reprinted figure by permission from Z. Amitay, D. Zajfman, P. Forck, et al., "Dissociative recombination of CH^+: Cross section and final states," Phys. Rev. A **54**, pp. 4032–4050, (1996). Copyright (1996) by the American Physical Society.)

ion source. Mul et al. (1981) attempted to use buffer gases to quench the $a\,^3\Pi$ state and vibrationally excited levels in the $X\,^1\Sigma^+$ state, but lacked unambiguous diagnostics to ascertain that they had succeeded. Amitay et al. (1996b) produced CH^+ by charge stripping CH^- and used particle imaging of recombination events to determine the composition of electronic states in their CH^+ beam. Figure 5.9 shows projected distances on the imaging detector plane as a function of storage time in the TSR. The populations of the $X\,^1\Sigma^+$ and $a\,^3\Pi$ states are shown very clearly by the right and left hand peaks, and there is a distinct time dependence reflecting the depopulation of the $a\,^3\Pi$ state and the simultaneous population of the ground state. Amitay et al. (1996b) used the time dependence of the left hand peak to determine the radiative lifetime of the $a\,^3\Pi$ state to be about 7 s. This means that the data obtained by MEIBE (Mitchell & McGowan 1978, Mul et al. 1981) most likely concerned recombination of CH^+ in the two lowest electronic states. This has been discussed further in connection with a study of CH^+ by Sheehan and St.-Maurice (2004a) using MEIBE.

The cross section for dissociative recombination of CH^+ in $X\,^1\Sigma^+$ ($v = 0$) was measured in the TSR by Amitay et al. (1996b) by making cross section measurements in different time windows and by making use of the measured radiative decay of the $a\,^3\Pi$ state. Figure 5.10 shows the result and a comparison with the MEIBE and MQDT results. The agreement between the TSR result and theory is very good below about 0.2 eV, while the resonances at 0.08 eV and 0.33 eV are absent in the theoretical curve. The cross section measured by Mul et al. (1981) was most likely enhanced by contamination of the $a\,^3\Pi$ state. Amitay et al. (1996b) did not give a thermal rate coefficient, but Sheehan and St.-Maurice estimated it to be 1.0×10^{-7} cm^3 s$^{-1}(T_e/300)^{-0.37}$ cm^3 s^{-1}, which is in very good agreement with the theoretical result of Takagi, Kosugi, and Le Dourneuf (1991).

The resonances at 0.08, 0.33, 0.50, 0.59, 0.96, 1.11, 1.55, 8.6, and 11.7 eV in Fig. 5.10 are not present in or are outside the range of earlier experiments. The high-energy resonances are similar in character to that in HeH^+ at 20 eV, i.e., electron capture into dissociative Rydberg states converging to electronically excited states. Dissociative recombination at low energy proceeds through the $2\,^2\Pi$ state, either directly or by the indirect process, which gives rise to narrow resonances appearing as dips in the cross section (Takagi, Kosugi, & Le Dourneuf 1991); these resonances were too narrow to be resolved in the TSR experiment. We must finally discuss the seven resonances between 0.08 eV and 1.55 eV. The TSR team observed resonances of this type in the work on CD^+ (Forck et al. 1994), at 0.8 eV and also, although weaker, at 0.4 eV and 1.4 eV, and attributed them to a new type of indirect mechanism, first conceived of by Guberman (1989, 1995a). In the indirect mechanism introduced by Bardsley (1968b), the energy of the free electron is removed by vibrational excitation of the ion core, thus allowing the electron to enter a Rydberg orbital. The Rydberg state is then predissociated by the resonant state. Forck et al. (1994) proposed that the excess energy of the free electron can also be liberated by an electronic excitation of the ion core, so that the electron can enter a Rydberg orbital. The electronic excitation and simultaneous electron capture leads to the formation of a Rydberg state converging to a bound electronically excited ion state. The capture is stabilized by predissociation of the Rydberg state by a

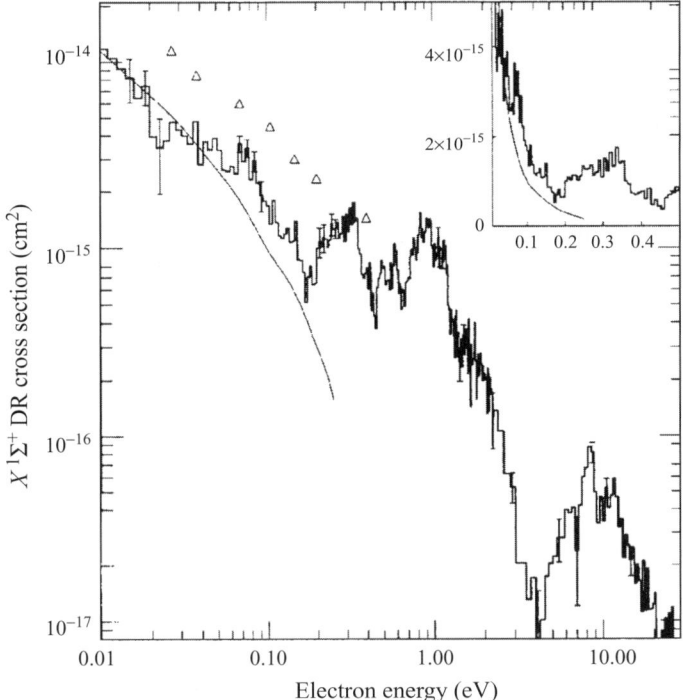

Figure 5.10 The cross section for $CH^+(X\,^1\Sigma^+(v=0)) + e^- \to C + H$ measured in the TSR (Amitay et al. 1996b). The systematic error was estimated to ±50%. The triangles show the factor of 2 corrected result of Mul et al. (1981), and the smooth line (also in the inset) shows the theoretical result of Takagi, Kosugi, and Le Dourneuf (1991) convoluted with the energy spread in the TSR electron beam (17 meV in the transverse direction). (Reprinted figure by permission from Z. Amitay, D. Zajfman, P. Forck, et al., "Dissociative recombination of CH^+: Cross section and final states," Phys. Rev. A **54**, pp. 4032–4050, (1996). Copyright (1996) by the American Physical Society.)

repulsive state. This mechanism was not included in the calculations of Takagi, Kosugi, and Le Dourneuf (1991), which would explain the discrepancy between experiment and theory above 0.2 eV in Fig. 5.10.

Carata et al. (1997) performed exploratory MQDT calculations addressing the ion–core indirect mechanism, and this work was followed by extensive calculations that were published a few years later (Carata et al. 2000). Electron scattering calculations were carried out to determine the resonance positions and autoionization widths. Adiabatic potential energy curves were calculated by a MCSCF + CI procedure, and Rydberg series converging to the three lowest electronic states were identified. Using a representation by dominant configurations, these Rydberg series were $(X\,^1\Sigma^+)np\pi$, $(a\,^3\Pi)ns\sigma$, $(a\,^3\Pi)np\sigma$, $(A\,^1\Pi)ns\sigma$, and $(A\,^1\Pi)np\sigma$. The adiabatic states arising from the CI calculations showed avoided crossings, and quantum

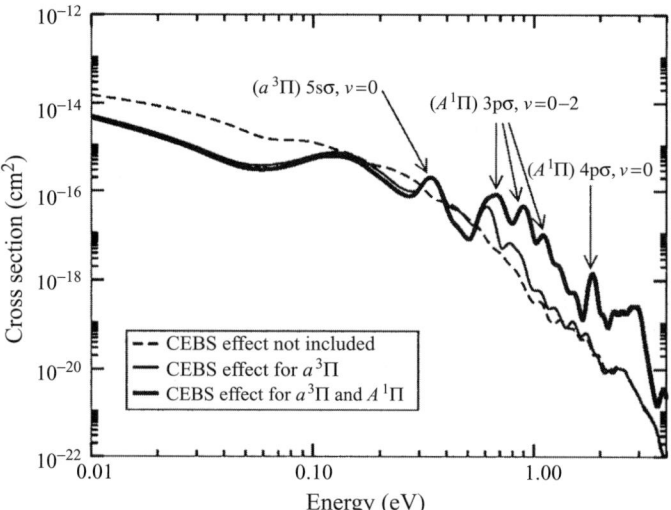

Figure 5.11 The calculated cross section for dissociative recombination of CH$^+$ with and without inclusion of core-excited bound Rydberg states (CEBS). The theoretical cross section has been convoluted with the electron-velocity distribution in the TSR electron beam. (Reprinted figure by permission from L. Carata, A.E. Orel, M. Raoult, I. F. Schneider, and A. Suzor-Weiner, "Core-excited resonances in the dissociative recombination of CH$^+$ and CD$^+$", *Phys. Rev. A* **62**, pp. 052711-1−10, (2000). Copyright (2000) by the American Physical Society.)

defects that varied rapidly at these avoided crossings. Carata *et al.* (2000) used a subset of configurations to construct quasidiabatic Rydberg states that did not mix with valence states. The interaction strength between the Rydberg states and the dissociative state $2\,^2\Pi$ was calculated by adding the configuration $1\sigma^2 2\sigma^2 3\sigma 1\pi 4\sigma$ (i.e., the dominant configuration for the $2\,^2\Pi$ state in a quasidiabatic representation) to the subset of configurations describing the quasiadiabatic Rydberg states, and by performing a new CI calculation. The interaction strength was then determined by identifying the shift in the Rydberg states caused by the Rydberg–valence interaction. Finally, MQDT calculations were performed, including ionization channels built on electronically excited states of CH$^+$ ($a\,^3\Pi$ and $A\,^1\Pi$). Figure 5.11 shows the calculated dissociative recombination cross section with and without the core-excited indirect mechanism (Carata *et al.* 2000). The inclusion of the core-excited indirect mechanism gives the peaks in the cross section that were observed in the TSR experiments (Forck *et al.* 1994, Amitay *et al.* 1996b).

Amitay *et al.* (1996b) used particle imaging to measure the kinetic energy release at $E_d = 0, 0.11, 0.28, 0.93, 1.04, 1.18$, and 9.04 eV, and derived product branching ratios from these data sets; the results of these measurements are collected in Table I in Amitay *et al.* (1996b). At $E_d = 0$ eV, recombination of CH$^+$ in its $X\,^1\Sigma^+$ state was found to produce C (^{1}D) + H (1s) in 79% of the events and C (^{1}S) + H (1s) in 21%

of the events. The numbers were obtained under the assumption that the angular distribution was isotropic, but it was also verified that this was compatible with allowing the anisotropy coefficients (Amitay et al. 1996b) to be free parameters; the branching ratios did not change and the best fit was obtained when the anisotropy coefficients were zero. The branching ratios remained essentially unchanged when the electron energy was tuned to $E_d = 0.11$ eV (structureless part of the cross section) and $E_d = 0.28$ eV (resonance peak), whereas only the flux to the C (^{1}D) + H (1s) limit showed an isotropic distribution. The observation of a substantial flux to the C (^{1}S) + H (1s) limit is interesting since only the $2\,^2\Sigma^+$ state correlates with this limit, and the $2\,^2\Sigma^+$ state is too far from the ground state of CH$^+$ to cause recombination (see Fig. 5.8). Amitay et al. (1996b) suggested that the C (^{1}S) + H (1s) limit is reached by a radiationless transition from the $2\,^2\Pi$ to the $2\,^2\Sigma^+$ state at 3.2 a_0. Carata et al. (2000) considered the possibility that rotational coupling caused this transition, but found it to be too small to do so. Guberman (2004, 2005) treated angular distributions in dissociative recombination, and made a specific study of CH$^+$. He concluded that spin–orbit coupling between the $2\,^2\Sigma^+$ and $2\,^2\Pi$ states is too small to cause the desired transition needed to redirect flux to the C (^{1}S) + H (1s) limit, and, based on quantitative calculations, rejected the suggestion by Carata et al. (2000) that the $2\,^2\Sigma^+$ state could also contribute to dissociative recombination of CH$^+$ ($X\,^1\Sigma^+$ ($v = 0$)) by the same radial non-Born–Oppenheimer interaction that drives recombination of HeH$^+$. Having excluded any possibility of the formation of C (^{1}S) + H (1s), Guberman (2005) went on to investigate the effect of an anisotropic dissociation entirely along $2\,^2\Pi$ at $E_d = 0$ eV, and found that he could obtain a good fit to the experimental data by allowing the angular distribution to be slightly anisotropic. It is too early to tell which interpretation is correct.

The fast rate by which CH$^+$ recombines contributes to the problem of explaining the abundance of this ion in diffuse interstellar clouds (Herbst 2005). CH$^+$ is believed to be formed by the reaction

$$C^+ + H_2 \rightarrow CH^+ + H, \tag{5.7}$$

which is endothermic and only occurs at temperatures above 1000 K or at non-thermal collision energies. It is destroyed by reactions with atoms and molecules and by recombination with electrons. And we know now that this latter reaction is much more effective than dielectronic recombination.

5.4 NH$^+$ and OH$^+$

There has been very little work done on the recombination of NH$^+$. The cross section was reported in connection with a broader survey of the energy dependence of dissociative recombination (McGowan et al. 1979). The rate coefficient given by McGowan et al. (1979) concerns 120 K and includes the factor of 2 error. The corrected temperature-dependent rate coefficient is given in Mitchell's review (Mitchell 1990a).

More attention has been given to OH^+. A single-pass merged-beam experiment (Mul *et al.* 1983) showed that OH^+ recombines slower than CH^+. Subsequent MQDT calculations gave an even slower recombination rate (Guberman 1995b), which suggested that the merged-beam experiment was carried out with OH^+ ions occupying vibrationally/electronically excited states. Guberman showed that the $X\,^3\Sigma^+$ ground state of OH^+, with leading configuration $1\sigma^2 2\sigma^2 3\sigma^2 1\pi^2$, recombines primarily by the direct mechanism through the $2\,^2\Pi$ state of OH, and he calculated the thermal rate coefficient to be $(6.3 \pm 0.7) \times 10^{-9} \times (T_e/300)^{-0.48}$ cm^3 s^{-1} (10 K $< T_e <$ 1000 K). The calculated cross section was about a factor of 6 smaller than that measured by Mul *et al.* (1983). The most reasonable explanation for this large difference between experiment and theory is that the single-pass experiment was contaminated by excited states. The slow recombination of OH^+ is not a result of a poor vibrational overlap between the $X\,^3\Sigma^+$ and $2\,^2\Pi$ states, but is rather due to a small electronic capture width and a slow variation of the quantum defect with internuclear distance (Guberman 1995b). The relative cross section was measured at the TSR by Amitay *et al.* (1996a), and they found resonances deriving from the same ion–core indirect mechanism as discussed in Section 5.3 for CH^+.

The $2\,^2\Pi$ state correlates with the O (^{1}D) + H (1s) limit, which results in 6.1 eV being converted to kinetic energy when ground state OH^+ recombines with low-energy electrons (Guberman 1995b). Particle imaging spectroscopy was used at CRYRING by Strömholm *et al.* (1997) to study dissociative recombination of OH^+, and, surprisingly, a channel was found to be opened at $E_d = 1.64$ eV for which the kinetic energy release was only 120 meV. Two possible origins of this very-low-release channel were discussed by Strömholm *et al.* (1997). First, the possibility of recombination of the low lying $a\,^1\Delta$ state was considered. Strömholm *et al.* calculated the radiative lifetime of this state and found it to be 0.03 s, which means that it would have decayed by the time the recombination experiment in CRYRING started. Second, dissociative recombination of OH^+ $X\,^3\Sigma^-$ was considered, and it was suggested that part of the dissociating flux goes to the O (^{3}P) + H ($n = 2$) limit for $E_d = 1.64 - 2.03$ eV. This would occur if the flux in this interval followed Rydberg states of $^2\Pi$ character and dissociated to O (^{3}P) + H ($n = 2$). Detailed calculations are needed to verify this conjecture.

5.5 LiH$^+$

The chemistry in the early universe was confined to the formation of diatomic molecules from hydrogen, helium, and lithium. Lepp and Shull (1984) calculated the formation rates of H_2, HD, HeH^+, and LiH by gas-phase reactions. Molecules

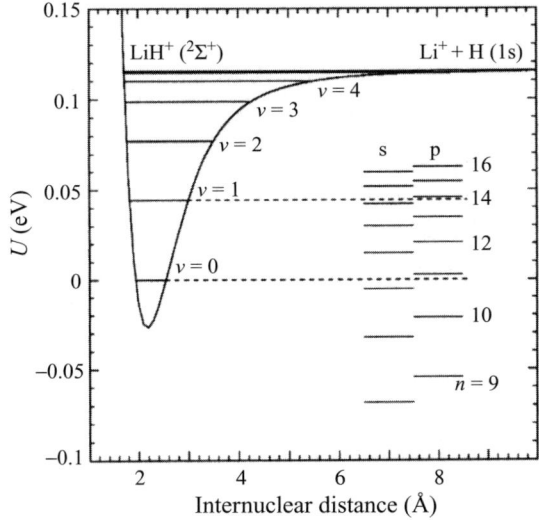

Figure 5.12 Potential energy curve for the $X\,^2\Sigma^+$ state of LiH$^+$, with the vibrational energy levels and the H(1s) + Li(nl), $l=0$ and 1, energy levels displayed. (Reprinted figure by permission from S. Krohn, M. Lange, M. Grieser, *et al.*, "Rate coefficients and final states for the dissociative recombination of LiH$^+$," *Phys. Rev. Lett.* **86**, pp. 4005–4008, (2001). Copyright (2001) by the American Physical Society.)

with a dipole moment played an important role in the early universe since they provided the cooling that was necessary for the primordial gas to collapse and form the first stars. In their analysis of the lithium chemistry, Stancil, Lepp, and Dalgarno (1996) included the formation of LiH$^+$ by radiative association and its destruction by dissociative recombination. In the absence of measurements or theoretical calculations, Stancil, Lepp, and Dalgarno estimated the rate coefficient to be 2.6×10^{-8} cm^3 s^{-1}, and this value was also adopted by Galli and Palla (1998). This is an interesting estimate in that it is much higher than those of more than a decade earlier for HeH$^+$ (see Section 5.1). Of course, by 1996 it was known that despite the absence of a curve-crossing HeH$^+$ recombined at a nonnegligible rate, and it was assumed that LiH$^+$, also lacking a curve-crossing, would recombine at about the same rate as HeH$^+$. The result of the only experimental determination of the rate coefficient for LiH$^+$ recombination with electrons (Krohn *et al.* 2001, Zajfman, Schwalm, & Wolf 2003) therefore came as a surprise – with a thermal rate coefficient of $(3.8 \pm 1.4) \times 10^{-7}$ cm^3 s^{-1} it was clear that LiH$^+$ recombines as fast as molecular ions that have a favorable curve-crossing.

Figure 5.12 shows the potential energy curve for the $X\,^2\Sigma^+$ ground state of LiH$^+$. It is weakly bound and in combination with the low ionization potential of atomic lithium (5.39 eV) this leads to a large number of energetically allowed final states even in recombination with thermal energy electrons.

The experiment of Krohn *et al.* (2001) was carried out at the TSR. Because of the formation of Li atoms in high Rydberg states as a result of the recombination of LiH^+, the normal means of detecting the reaction products failed; the Li atoms were field ionized by the dipole magnet following the electron cooler section. This did not prevent the rate coefficient from being determined. The lifetime of the stored ion beam was dependent on how rapidly LiH^+ were destroyed in the interaction region, and by measuring the beam lifetime with and without the electron beam, the rate coefficient could be extracted.

Attempts were made to identify a mechanism for dissociative recombination of LiH^+ (Florescu *et al.* 2004), but the calculated value fell short of the experimental result by an order of magnitude. Čurík and Greene (2007a,b) applied MQDT in combination with Siegert states (see Section 3.6.1) and obtained quantitative agreement with experiment. This work represents a very interesting illustration of how the indirect mechanism alone can generate a large recombination cross section. The electron is initially captured into a Rydberg state by excitation of one quantum of vibration, and Čurík and Greene labeled this the doorway state. If this state is energetically isolated, the electron system autoionizes and dissociative recombination is inefficient. In the $LiH^+ + e^-$ system, however, the initial Rydberg state is overlapped by higher-v/lower-n perturbing resonance levels, and the interaction of the initial state with this manifold of resonances leads to a resonant Rydberg complex with a significantly increased dissociative recombination rate. There is no reason why this mechanism should not also be important in other systems. The notion that a favorable curve-crossing is needed for efficient dissociative recombination is obsolete.

6
Diatomic ions

The efforts to study dissociative recombination of diatomic molecular ions were focused early on the rare-gas dimer ions and the atmospheric diatomics. Interest in the rare-gas dimer ions as suitable benchmark molecules for the stationary afterglow technique was discussed in Section 2.3.1, and interest in the atmospheric ions dates back to Kaplan's (1931) discussion of the airglow.

6.1 Rare-gas dimer ions: He_2^+, Ne_2^+, Ar_2^+, Kr_2^+, Xe_2^+

The rare-gas ion, He_2^+, was the subject of the earliest studies of dissociative recombination. In 1949, Biondi and Brown (1949b) used a microwave discharge to study electron–ion recombination in helium, leading to a value for the rate coefficient of 1.7×10^{-8} cm^3 s^{-1} at 300 K. This surprisingly high value led Bates (1950a) to propose that dissociative recombination, i.e.,

$$He_2^+ + e^- \rightarrow He + He, \tag{6.1}$$

rather than processes involving the atomic ion, such as radiative recombination

$$He^+ + e^- \rightarrow He^* + h\upsilon \tag{6.2}$$

or three-body recombination

$$He^+ + e^- + He \rightarrow He + He, \tag{6.3}$$

was responsible for the observed decay of the electron density. Earlier, Bates and Massey (1947) had proposed a similar reaction for recombination in the ionosphere as a possibility to explain electron removal in the nocturnal ionosphere. For He_2^+, however, this high rate coefficient was not found in later experiments. These (Oskam & Mittelstadt 1963, Collins & Robertson 1965, Chen, Leiby, & Goldstein 1961) could only place upper bounds on the rate, ranging from 2×10^{-11} cm^3 s^{-1} to 4.0×10^{-9} cm^3 s^{-1}. Rogers and Biondi (1964) found line broadening in the late part of the afterglow plasma and interpreted this as dissociative recombination. Ferguson, Fehsenfeld, and Schmeltekopf (1965) analyzed previous experiments, pointing out the importance of collisional radiative recombination,

$$\mathrm{He}^+ + e^- + e^- \rightarrow \mathrm{He}^* + e^-, \tag{6.4}$$

and concluded that the rate of dissociative recombination from the ground vibrational state was extremely small ($< 3 \times 10^{-10}$ cm^3 s^{-1}). This conclusion was supported by early calculations of the potential energy curves of He$_2^*$ states by Mulliken (1964) that indicated there was no favorable crossing and that dissociative recombination of electrons was unlikely except from higher vibrational states. Deloche et al. (1976) carried out a careful series of experiments on a high-pressure helium afterglow at room temperature, then constructed a model taking into account the variety of processes that can occur. This included not only processes involving electrons and the ground atomic and molecular states, but also excited vibrational states and metastable states. The model led to an upper bound to the rate coefficient of 5×10^{-10} cm^3 s^{-1}. Further calculations (Cohen 1976, Guberman & Goddard 1975) agreed with the early conclusion of Mulliken that there were no dissociative resonance curves that crossed in the Frank–Condon region of the ground vibrational state of He$_2^+$.

As is the case for H$_2^+$ the most common isotopolog, ^{4}He^{4}He$^+$, has no dipole moment. However, the mixed isotopolog ^{3}He^{4}He$^+$ does. It was demonstrated by Urbain et al. (2000) that just like HD$^+$, ^{3}He^{4}He$^+$ cools vibrationally in a storage ring.

Calculations were carried out (Carata et al. 1999) in order to quantify the low-energy cross section. These used quantum chemistry calculations to determine the potential energy curves and quantum defects for the bound states, and electron scattering calculations using the complex Kohn variational method to determine the resonance curves and couplings. These data were then used in an MQDT calculation, and the cross section was studied as a function of energy and initial vibrational state. Both the ground state $(1\sigma_g^2 1\sigma_u) X\,^2\Sigma_u^+$ and the dissociative first excited state $(1\sigma_g 1\sigma_u^2) A\,^2\Sigma_g^+$ of the molecular ion dissociate to the lowest limit, He$^+$ (1s, ^{2}S) + He (1s^2, ^{1}S). As first discussed by Guberman and Goddard (1975) several dissociative pathways are possible. The lowest dissociative resonant state is the $^3\Sigma_g^+$ state with the dominant configuration, $(1\sigma_g 1\sigma_u^2 2\sigma_g)$. This state is the lowest member of the Rydberg series converging to the first excited state of the ion. This state was found to dominate the low-energy cross section (Carata et al. 1999). Two other states, $^1\Sigma_g^+$ and the $^3\Pi_u$, were included in this study. An example of the final results is shown in Fig. 6.1. These calculations led to a value for the rate coefficient in the temperature range of 250–350 K of

6.1 Rare-gas dimer ions: He_2^+, Ne_2^+, Ar_2^+, Kr_2^+, Xe_2^+

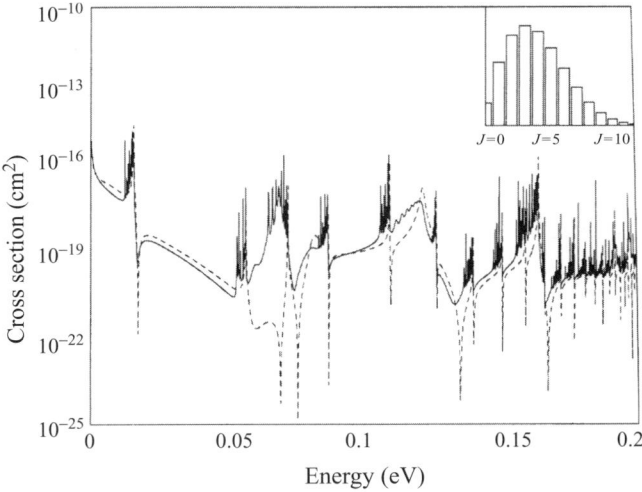

Figure 6.1 Calculated dissociative recombination cross section of $^3He^4He^+$ (contribution of the $^3\Sigma_g^+$ dissociative state only). The calculation was performed for $^3He^4He^+$ in $v=0$ and with a Boltzmann distribution of rotational levels at 300 K. The inset shows the corresponding Boltzmann distribution of initial rotational levels. The dashed line represents the $J=0$ cross section for comparison. (Reprinted figure by permission from L. Carata, A. E. Orel, and A. Suzor-Weiner, "Dissociative recombination of He_2^+ molecular ions," *Phys. Rev. A* **59**, pp. 2804–2812, (1999). Copyright (1999) by the American Physical Society.)

$$\alpha = 1.04 \times 10^{-8} T^{-0.9} \text{ cm}^3 \text{ s}^{-1},$$

which corresponds to a rate at 300 K of 6.1×10^{-11} cm^3 s^{-1}, which is consistent with the experimental data.

Experiments on ASTRID (Urbain *et al.* 2005) and TSR (Pedersen *et al.* 2005) extended the range of electron impact energies to 15 eV. The ASTRID experiments compared the cross section for $^4He^4He^+$ with that of $^3He^4He^+$ as a function of the storage time in the ring. There was no significant change in the $^4He^4He^+$ cross section. The $^3He^4He^+$ isotopolog, on the other hand, showed a significant drop in magnitude of the cross section for electron collision energies of less than 1 eV, indicating cooling of excited vibrational states. The data were modeled to abstract the low-energy rate coefficient leading to an upper limit of 6.0×10^{-10} cm^3 s^{-1}, in agreement with previous experiments, but somewhat higher than that given by theory (Carata *et al.* 1999). The TSR experiments used $^3He^4He^+$ and Coulomb explosion imaging techniques to monitor the vibrational population of the stored ions. In addition, in the TSR ring time-dependent measurements of the dissociative recombination rate coefficients and the fragment distributions were carried out.

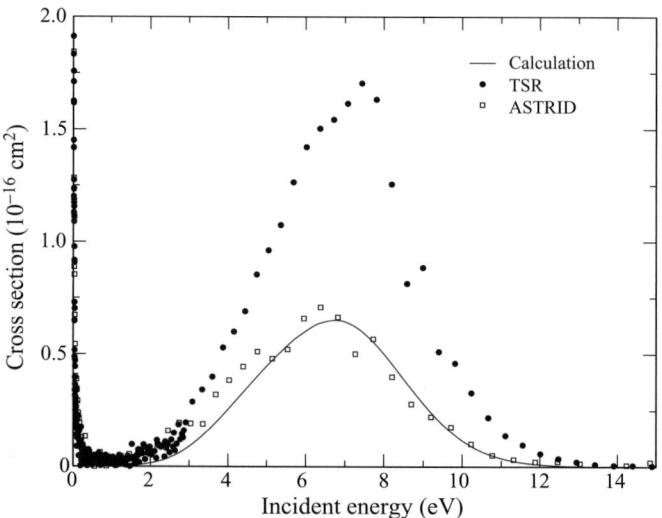

Figure 6.2 Total cross section for dissociative recombination of He_2^+. The solid line represents the cross section calculated by Royal and Orel (2005), the solid circles show the experimental results from the TSR (Pedersen et al. 2005), and the unfilled squares show the experimental results from ASTRID (Urbain et al. 2005). (Reprinted figure by permission from J. Royal and A. E. Orel, "Dissociative recombination of He_2^+," *Phys. Rev. A* **72**, pp. 022719-1–8, (2005). Copyright (2005) by the American Physical Society.)

This led to a rate coefficient of $(3.3 \pm 0.9) \times 10^{-10}$ cm^3 s^{-1}, somewhat larger than the ASTRID results and theory.

In the higher-energy (>1 eV) region the cross section is dominated by a series of resonances, Rydberg states converging to the lowest state of the ion. These states are all dissociative and cross the ground state of the ion at large internuclear distances. In this region the direct mechanism dominates. Using the complex Kohn variational method to describe electron scattering from the ground state of the ion as a function of internuclear distance, the potential energy curves and autoionization widths for 24 of these resonance states were obtained (Royal & Orel 2005). Time-dependent wave-packet calculations were used to determine the dissociative recombination cross section. The result is shown in Figure 6.2, together with the results from the TSR and ASTRID experiments. The theory agrees well with the ASTRID results. The TSR results are larger by a factor of 2.6. This factor is constant over the energy range reported in both experiments. The reason for the discrepancy between the two measurements is unknown and is a current issue of research. Calculations and experiments studying different isotopologs and the dissociative excitation channel are continuing.

The theoretical calculations also predicted the branching ratios into the final atomic states. The results are shown in Fig. 6.3. At low energy, the $n = 2$ states

6.1 Rare-gas dimer ions: He_2^+, Ne_2^+, Ar_2^+, Kr_2^+, Xe_2^+ 147

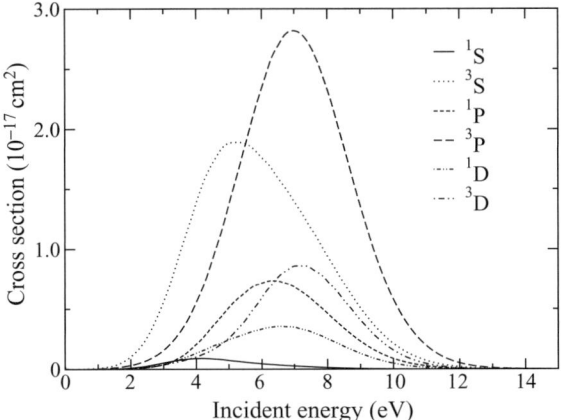

Figure 6.3 Calculated partial cross sections based on asymptotic atomic state symmetry for dissociative recombination of He_2^+. (Reprinted figure by permission from J. Royal and A. E. Orel, "Dissociative recombination of He_2^+," *Phys. Rev. A* **72**, pp. 022719-1–8, (2005). Copyright (2005) by the American Physical Society.)

were found to dominate. As the energy increased, the branching into $n = 3$ states increases, finally dominating at collision energies above 6 eV. There are no experimental measurements of the branching ratios in this energy region. At low energy, experiments show all the fragments are in $n = 2$, with fragmentation into He (1s2p, ^{3}P) dominating $(58.6 \pm 5.2)\%$ followed by He (1s2s, ^{1}S) at $(37.4 \pm 4.0)\%$ (Pedersen *et al.* 2005).

All the experimental studies of Ne_2^+ dissociative recombination have been performed using afterglow techniques described in Chapter 2. In these types of experiments, the recombination rate coefficients are generally determined from the decay of the electron density with time following the removal of the external ionizing source. These experiments as summarized by Frommhold, Biondi, and Mehr (1968) led to consistent values of the rate coefficient at 300 K on the order of 1.7×10^{-7} cm^3 s^{-1}. However, there was not consistency in the variation of the rate coefficient with temperature. Most experiments (Frommhold, Biondi, & Mehr 1968, Kasner 1968) were performed over a restricted range of gas (ion) temperatures ($T_{gas} < 500$ K). These experiments found the rate coefficients for dissociative recombination of Ne_2^+ with a temperature dependence close to $T_e^{-0.5}$ over a range of electron temperatures between 300 K and 10^4 K. More recently, Chang *et al.* (1989) measured the rate coefficients for dissociative recombination of Ne_2^+ at elevated gas and electron temperatures using a shock-heated pulsed-flow radio-frequency-discharge-pulsed afterglow. The variation of the rate coefficients was shown to be between $T_e^{-0.6}$ and $T_e^{-1.0}$ for elevated gas temperatures of 2200 K and 2700 K, respectively. The temperature behavior then changed to approximately $T_e^{-1.5}$ when the electron and gas temperature were kept equal but above 1500 K. This rather unusual temperature dependence was supported by the earlier works of Cunningham and coworkers (Cunningham & Hobson 1969, Cunningham, O'Malley, & Hobson 1981), which predicted a change in the slope of the rate coefficient for dissociative recombination of rare-gas diatomic ions (around 900 K for Ne_2^+) as a function of electron energy leading to $T_e^{-1.5}$ in the high-temperature region. This temperature behavior was first attributed to the dominance of the indirect mechanism. Bardsley (1968b) showed that the indirect process gives rate coefficients that decrease as $T_e^{-1.5}$ and therefore the actual temperature dependence of the rate coefficients should depend on the relative strength of the two mechanisms. His treatment considered the two processes independently, thus neglecting interferences between them. Another attempt to explain the $T_e^{-1.5}$ behavior of the rate coefficients was made by O'Malley and coworkers (O'Malley, Cunningham, & Hobson 1972) with a "low-vibrational-state model." However, this model is based on a single resonant state description of dissociative recombination, with the resonant state crossing the ion so that the cross section is maximum for a low vibrational state $v = 0, 1, 2$.

In order to explain the temperature dependence more accurate calculations were necessary. The ground state of Ne_2^+ has overall $^2\Sigma_u^+$ symmetry with the electronic configuration

$$1\sigma_g^2 1\sigma_u^2 2\sigma_g^2 2\sigma_u^2 1\pi_u^4 1\pi_g^4 3\sigma_g^2 3\sigma_u.$$

The first excited state of the ion is dissociative with the configuration

$$1\sigma_g^2 1\sigma_u^2 2\sigma_g^2 2\sigma_u^2 1\pi_u^4 1\pi_g^4 3\sigma_g 3\sigma_u^2.$$

As in the case of He_2^+, there exists a Rydberg series with configuration

$$\left(1\sigma_g^2 1\sigma_u^2 2\sigma_g^2 2\sigma_u^2 1\pi_u^4 1\pi_g^4 3\sigma_g 3\sigma_u^2\right)^{1,3} n\lambda_{g,u,}$$

where λ can be σ or π, converging to the first excited states. In contrast to He_2^+, these states cross the ground state of the ion in the Franck–Condon region. Extensive electron scattering calculations using the complex Kohn variational method were carried out to determine the resonance curves and autoionization widths (Ngassam & Orel 2006). This information was then used as the input in MQDT to determine the dissociative recombination rate coefficient. In order to compare these results with existing measurements, the theoretical cross sections were converted into rate coefficients after a convolution with an isotropic Maxwell–Boltzmann distribution of electron energies and averaged over the initial vibrational states populated at a given gas/ion temperature. Figure 6.4 displays the rate coefficients at room temperature (300 K) for various initial vibrational levels. The rate coefficients fluctuate with the initial vibrational energy level. This behavior can be explained by the

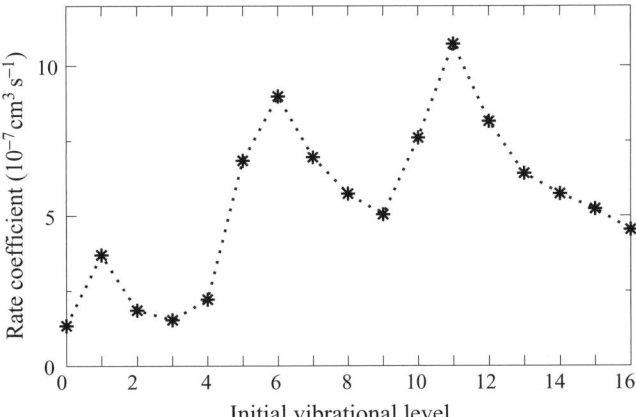

Figure 6.4 Rate coefficients for dissociative recombination of Ne_2^+ at fixed electron temperature of 300 K as a function of the initial vibrational level of the ion. (Reprinted figure by permission from V. Ngassam and A. E. Orel, "Dissociative recombination of Ne_2^+ molecular ions," *Phys. Rev. A* **73**, pp. 032720-1–10, (2006). Copyright (2006) by the American Physical Society.)

relative contribution from various resonant states for different initial vibrational levels. The peak at $v = 1$ is due to a good overlap of the wavefunction of the first vibrational excited state ($v = 1$) of the ion with the lowest $^{1,3}\Sigma_g^+$ resonant states. Then, the contribution to the process from these lowest $^{1,3}\Sigma_g^+$ states drops dramatically and around $v = 6$, the contributions to the process of the lowest $^{1,3}\Pi_u$ and $^{1,3}\Sigma_u^+$ states become more important, once more due to a good overlap of the wavefunctions. The last peak is due to the opening of new routes for dissociative recombination (lowest $^{1,3}\Pi_g$ and the second resonant states of the other symmetries).

The absolute rate coefficients for dissociative recombination, at fixed gas temperature (300 K, 10 000 K) and for a gas temperature that is always equal to electron temperature, as a function of electron temperature are displayed in Fig. 6.5. Also shown are the lines representing slopes of $T_e^{-0.5}$ and $T_e^{-1.5}$. At 300 K, the rate coefficients decrease as $T_e^{-0.5}$ for low electron temperatures. This occurs because the process is dominated by a single resonance. As the electron temperature increases, new routes for dissociative recombination open causing a deviation from the $T_e^{-0.5}$ behavior around 700 K. At elevated gas temperatures (10 000 K), the rate coefficients are higher at low electron energy since many resonant states contribute to the process even at low electron temperatures and the high vibrational states dissociate faster. At these temperatures, the rate also decreases faster than $T_e^{-0.5}$ even at low energy. When the electron temperature is kept equal to the gas temperature, the rate decreases as $T_e^{-0.5}$ at low energy with a break in the slope after 2000 K. Then the

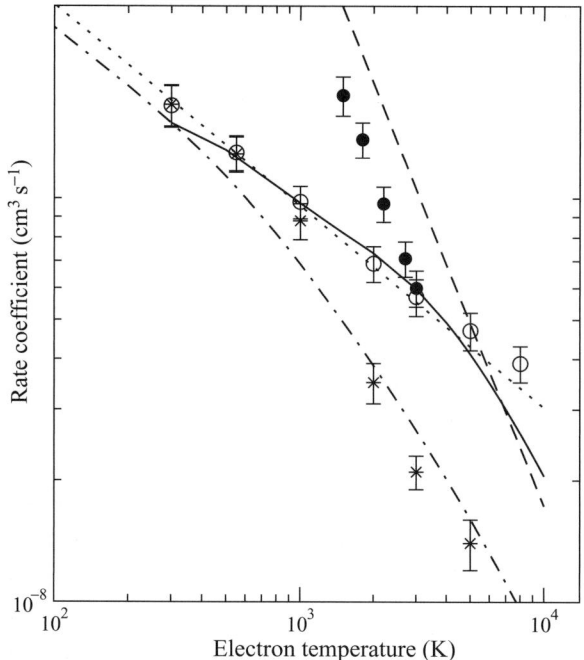

Figure 6.5 Comparison of calculated rate coefficients for dissociative recombination of Ne_2^+ with various experiments. The solid line represent the theoretical rate coefficients for $T_{gas} = T_e$ calculated by Ngassam and Orel (2006). The open circles are measurements of Frommhold, Biondi, and Mehr (1968) at 300 K. The stars represent the result of Cunningham, O'Malley, and Hobson (1981), while the solid circles are measurements of Chang *et al.* (1989) both performed with $T_{gas} = T_e$. The other lines represent theoretical results from Ngassam and Orel (2006): $-\cdot-\cdot-\ T_{gas} = 300$ K; $---\ T_e^{-1.5}$; $\cdots\cdots\ T_e^{-0.5}$.

rate coefficient decreases faster and eventually as approximately $T_e^{-1.5}$. In Fig. 6.5, the calculated rate coefficients are compared with those from various experiments. The break observed in the rate coefficients as a function of electron and gas temperature is clearly displayed in the experimental data of Cunningham and coworkers (Cunningham, O'Malley, & Hobson 1981). There is good agreement between theory and experiment both for the magnitude and temperature dependence of the rate coefficient.

The earliest measurements of the dissociative recombination of Ar_2^+ were those of Biondi and Brown (1949b) who reported a rate coefficient of 3×10^{-7} cm^3 s^{-1}. Issues with gas purity, with the most probable contaminant being N_2^+, caused this rate to become suspect. Further experiments (Biondi 1951, Biondi & Holstein 1951) followed by a reanalysis (Biondi 1963) led to a value at 300 K of 7×10^{-7} cm^3 s^{-1}, while Redfield and Holt (1951), using a similar pulsed microwave discharge, derived a value of 1.07×10^{-6} cm^3 s^{-1}. Work by Sexton and Graggs (1958) found a

6.1 Rare-gas dimer ions: He_2^+, Ne_2^+, Ar_2^+, Kr_2^+, Xe_2^+

range of values from 4×10^{-7} cm^3 s^{-1} to 8×10^{-7} cm^3 s^{-1}, but the results increased with decreasing gas pressure, possibly due to an experimental artifact. Oskam and Mittelstadt (1963) produced a rate of $(6.7 \pm 0.5) \times 10^{-7}$ cm^3 s^{-1} that was independent of pressure. Subsequent work (Mehr & Biondi 1968, Shiu & Biondi 1978) produced values of $(8.5 \pm 0.8) \times 10^{-7}$ cm^3 s^{-1} and $(9.1 \pm 0.9) \times 10^{-7}$ cm^3 s^{-1}. Although the results of Cooper, van Sonsbeek, and Bhave (1993) produced even higher values, $(1.07 \pm 0.19) \times 10^{-6}$ cm^3 s^{-1} at 295 K, the more recent experiments of Okada and Sugawara (1993), gave the somewhat lower value of $(8.2 \pm 0.5) \times 10^{-7}$ cm^3 s^{-1}. However, considering the range of the more recent experiment, the higher values are more favored.

The variation of the rate coefficient with temperature was studied in parallel with the attempt to determine its value. As in the case of Ne_2^+, there are two classes of experiments. In the afterglow experiments, the gas temperature remained constant, but the electron collision energy changed. This led to a variation of $T_e^{-0.66}$ (Mehr & Biondi 1968). In shock-tube experiments, both the gas temperature and the electron collision energy are changed simultaneously, $T_e = T_{gas}$. In this case, the rate coefficient at high temperature was found to vary as $T_e^{-1.5}$ (Fox & Hobson 1966), while at low temperature the variation was found to be $T_e^{-0.5}$ (Cunningham & Hobson 1969). As in the case of Ne_2^+, this was explained by O'Malley and coworkers (O'Malley, Cunningham, & Hobson 1972) with a "ground vibrational state (GVS) model," based on a single resonant state with the assumption that the ground vibrational state has a much larger cross section than the higher vibrational states.

Extensive electron scattering calculations using the complex Kohn variational method were carried out to determine the resonance curves and autoionization widths (Royal & Orel 2006). As was seen in Ne_2^+, a number of resonances cross the ground ion curve near the Franck–Condon region. However, the crossing occurs for small internuclear separation, less than the equilibrium bond distance. These data were used as the input to a calculation of the dissociative dynamics, using a nonlocal complex potential to obtain cross sections for dissociative recombination for each resonant state. The results when the theoretical cross sections were convoluted with an isotropic Maxwell–Boltzmann distribution of electron energies, averaged over the initial vibrational states populated at a given gas/ion temperature, and converted into rate coefficients at room temperature (300 K) are shown in Fig. 6.6 for various initial vibrational levels. Note that in this case, since $v = 1$ and $v = 2$ have higher rates than $v = 0$, the conditions for O'Malley's GVS model (O'Malley, Cunningham, & Hobson 1972) are not met, and the extension to the "low-vibrational-state model" is a better approximation. The theoretical results are compared with experiment in Fig. 6.7. The temperature dependence of the rate coefficients is in good agreement with experiment. Although the rate coefficient is

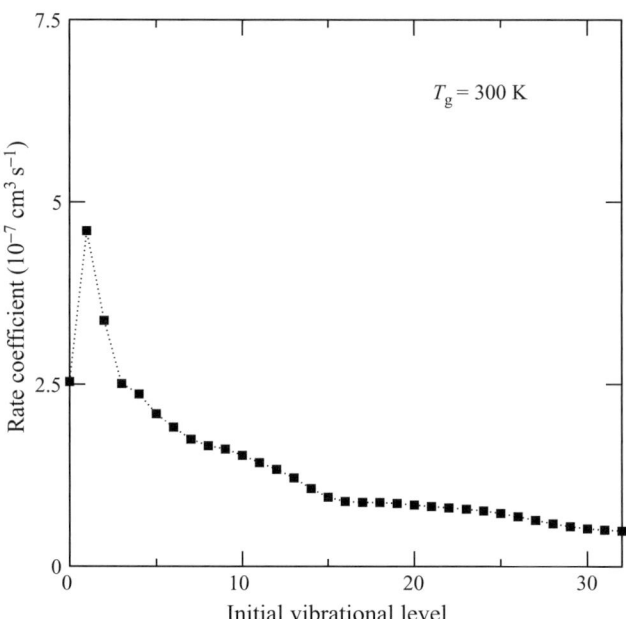

Figure 6.6 The rate coefficient for dissociative recombination of Ar_2^+ at $T_e = 300$ K as a function of vibrational level. The $v = 1$ and 2 vibrational levels have the largest rate coefficient. (Reprinted figure by permission from J. Royal and A. E. Orel, "Dissociative recombination of Ar_2^+," *Phys. Rev. A* **73**, pp. 042706-1–12, (2006). Copyright (2006) by the American Physical Society.)

within the range of the experimental data, it lies about a factor of 2 below the more recent experiments.

It had been thought that dissociative recombination of rare-gas ions heavier than He_2^+ would yield one atom in the np^6 atomic ground state and the second atom in the $np^5(n + 1)p$ excited state (Huestis 1982). This belief was due to optical measurements which showed that the $np^5(n + 1)p$ radiation to the lower $np^5(n + 1)s$ states dominated over radiation from higher states. However, no radiation from these $np^5(n + 1)s$ states could be observed. This could be due to the states being either metastable or radiating in the ultraviolet, which was not detectable in earlier experiments (Ramos *et al.* 1995a,b). The calculations (Royal & Orel 2006) showed that the resonance states dissociated to either Ar ($3s^2 3p^6$) + Ar ($3s^2 3p^5 4s$) or Ar ($3s^2 3p^6$) + Ar ($3s^2 3p^5 4p$) atomic limits. For gas temperatures of 300 K and 2000 K, over a range of electron temperatures, the Ar (4s) state was the primary contribution resulting in 67–83% for $T_{gas} = 300$ K and 72–86% for $T_{gas} = 2000$ K of the final atomic fragments over the electron collision temperatures studied (300–6000 K). The remaining 17–33% for $T_{gas} = 300$ K and 14–28% for $T_{gas} = 2000$ K is attributed to the Ar (4p) atomic fragment. Product states from the

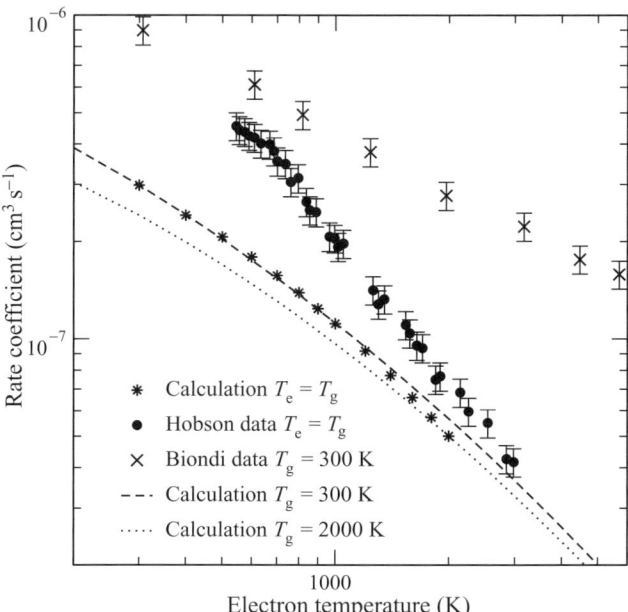

Figure 6.7 Rate coefficients for dissociative recombination of Ar_2^+. The calculated results are from Royal and Orel (2006), and the experimental data are from Cunningham and Hobson (1969) and Shiu and Biondi (1978). (Reprinted figure, with some small modifications, by permission from J. Royal and A. E. Orel, "Dissociative recombination of Ar_2^+," *Phys. Rev. A* **73**, pp. 042706-1–12, (2006). Copyright (2006) by the American Physical Society.)

dissociative recombination of rare gases have been produced experimentally, using time-of-flight spectra of metastable fragments emerging from discharges (Barrios *et al*. 1992, Ramos *et al*. 1995a,b). These measurements showed products, but there was no information about the relative yield of each product. The products observed were also strongly influenced by changes in the discharge conditions. Ramos *et al*. (1995b) reported that the Ar (4s) atomic fragments dominate the dissociative recombination products. In addition, these experiments detected ground-state fragments, Ar $(3s^2 3p^6)$ + Ar $(3s^2 3p^6)$ (Ramos *et al*. 1995a). The calculations (Royal & Orel 2006) showed no evidence of these products, nor is there any other experimental evidence for these states. If ground-state products are produced, they must originate from a different mechanism.

Very little work has been done on the other rare-gas ions. In the case of Kr_2^+, Oskam and Mittelstadt (1963) reported a rate coefficient of 1.2×10^{-6} cm^3 s^{-1} at room temperature. Shock-tube results (Cunningham & Hobson 1972a) yielded a value of 2.2×10^{-7} cm^3 s^{-1} at 1430 K with a variation with temperature of $T_e^{-1.5}$ keeping $T_{gas} = T_e$ as discussed above. Using this slope to extrapolate to room temperature yields results in agreement with the previous experiment. Shiu and Biondi (1977) obtained a value of 1.6×10^{-6} cm^3 s^{-1} at 300 K and a temperature dependence of $T_e^{-0.5}$ with the gas temperature fixed at 300 K and only the electron temperature allowed to vary. These temperature dependences are similar to those seen in Ne$_2^+$ and Ar$_2^+$ and are presumed to have the same mechanism. In the case of Xe$_2^+$, Shiu, Biondi, and Sipler (1977) reported a rate coefficient of 2.3×10^{-6} cm^3 s^{-1} at room temperature, which is in agreement with the measurement of Lennon and Sexton (1959), but about 60% higher than that reported by Oskam and Mittelstadt (1963). This high value was explained by Bates (1991a) as "super-dissociative-recombination," but no other calculations have been carried out to test this mechanism. The results of Hu, Mitchell, and Lipson (2000), who studied the products produced by dissociative recombination in Xe$_2^+$, indicate that lower states are favored over higher Rydberg states, which may weaken the case for the "super-dissociative-recombination" mechanism.

6.2 The atmospheric ions: O_2^+, N_2^+, and NO$^+$

Before discussing the individual atmospheric ions, we would like to point to the very useful review of the dissociative recombination of O_2^+, N_2^+, and NO$^+$ by Sheehan and St.-Maurice (2004b).

6.2.1 O_2^+

The formation and destruction of ions in the terrestrial ionosphere were discussed in Chapter 1, and ionospheric ions played an important role in the conceptual development of processes such as dissociative recombination. Being the dominant molecular ions in the ionosphere, O_2^+, N_2^+, and NO$^+$ are usually termed the atmospheric ions.

O_2^+ occupies the role of a benchmark ion for experimental dissociative recombination techniques. One reason for this is that O_2^+ is easy to deactivate vibrationally, which means that almost all experiments have been carried out with ions predominantly in the lowest vibrational level. Another reason is that the ion is easy to produce and to perform measurements on, which means that it can be used as a calibration ion when a new experiment is being tested.

Electron–ion volume recombination in oxygen was studied by Biondi and Brown (1949b), but it was then almost two decades before measurements of the electron-temperature dependence of the rate coefficient were performed with mass spectrometric monitoring of the recombining plasma. It is probably fair to say that early measurements by Sayers and Kerr (1957), Holt (1959), Kasner, Rogers, and Biondi (1961), Anisimov, Vinogradov, and Goland (1964), Mentzoni (1965), and Gunton

(1967) were quite reliable, with rate coefficients falling in the range (2–3) × 10^{-7} cm^3 s^{-1}. The brief note by Kasner and Biondi (1967) was the lead-up to the more accurate measurements by Kasner and Biondi (1968) and Smith and Goodall (1968). The recombination measurements of Smith and Goodall (1968) were some of the first for which a Langmuir probe was used to measure the electron density. Mehr and Biondi (1969) performed the measurement which has been regarded as the definitive measurement, and the one to which all subsequent measurements have been compared. The result from this measurement is given in Mitchell's review of 1990 (Mitchell 1990a), as the first value in the review by Florescu-Mitchell and Mitchell (2006), and as the earliest experiment in the critical survey of Sheehan and St.-Maurice (2004a). Mehr and Biondi (1969) measured the dissociative recombination rate coefficient of O_2^+ over a wide range of electron temperatures while keeping the gas and ion temperatures constant at room temperature. The results of this and subsequent measurements are given in Table 6.1.

The results in Table 6.1 represent the most consistent set of data on dissociative recombination for any molecular ion. The ion storage ring result of Peverall (2001), for which the cross section was used to derive the rate coefficient, is in excellent agreement with the afterglow results, which are all internally consistent. The only result that is inconsistent is the one from the single-pass merged-beam experiment of Mul and McGowan (1979a). The reason for this is that the single-pass experiment was done with O_2^+ ions containing vibrational excitations. In a study of dissociative recombination of O_2^+ with different controlled and measured vibrational populations, Petrignani *et al.* (2005c) were able to deduce vibrationally resolved cross sections. They found that the cross sections for $v = 1$ and 2 were smaller than that for $v = 0$, as can be seen in Table 6.1. This explains qualitatively the smaller rate coefficient measured by Mul and McGowan (1979a).

With the rate coefficient for O_2^+ ($v = 0$) so firmly established, it may be surprising that there is no theoretical calculation of this quantity. The reason is that the theoretical efforts (by means of MQDT) have mainly been devoted to the production rate of O (^{1}S):

$$O_2^+ (v) + e^- \rightarrow O + O(^1S). \tag{6.5}$$

These efforts were mainly due to Guberman, starting with calculations of the relevant potential energy curves (Guberman 1979, 1983a) and model calculations (Guberman 1986b), followed by a sequence of papers dealing with the formation of O (^{1}S) and O (^{1}D) (Guberman 1987, 1988, 1989, Guberman & Giusti-Suzor 1991, Guberman 1997). The formation of O (^{1}D) has also been addressed by Seong & Sun (1996) using an approach similar to that of Guberman.

Figure 6.8 gives a good idea as to why a calculation of the total rate coefficient would be a daunting task; the lower part of the ground-state potential energy curve

Table 6.1. O_2^+ dissociative recombination rate coefficients

Technique	Reference	Rate coefficient (cm^3 s^{-1})	Temperature (K) Vibrational level[a]
Stationary afterglow	Kasner & Biondi 1968	$(2.2 \pm 0.2) \times 10^{-7} (T_e/295)^{-1}$	$205 \leq T_e \leq 690$ $v = 0$
Stationary afterglow	Mehr & Biondi 1969[b]	$(1.95 \pm 0.20) \times 10^{-7} (T_e/300)^{-0.70}$	$300 \leq T_e \leq 1200$ $v = 0$
Ion storage ring	Peverall et al. 2001	$(2.40 \pm 0.48) \times 10^{-7} (T_e/300)^{-0.70}$	$200 \leq T_e \leq 1000$ $v = 0$
Ion storage ring	Petrignani et al. 2005c[c]	$(0.7 \pm 0.3) \times 10^{-7}$	$v = 1, T_e = 300$
		$(1.25 \pm 0.4) \times 10^{-7}$	$v = 2, T_e = 300$
FALP	Mahdavi, Hasted, & Nakshbandi 1971	$(1.9 \pm 0.5) \times 10^{-7}$	$T_e \approx 300$ $v = 0$
FALP	Alge, Adams, & Smith 1983[d]	$1.95 \times 10^{-7} (T_e/300)^{-0.70}$	$200 \leq T_e \leq 600$ $v = 0$
FALP	Španěl, Dittrichová, & Smith 1993	$2.0 \times 10^{-7} (T_e/300)^{-0.65}$	$300 \leq T_e \leq 2000$ $v = 0$
FALP	Gougousi, Golde, & Johnsen 1997	$(1.9 \pm 0.2) \times 10^{-7}$	$T_e = 300$ $v = 0$
FALP-MS	Mostefaoui et al. 1999	1.9×10^{-7}	$T_e = 300$ $v = 0$
FALP	McLain et al. 2004	$(2.24 \pm 0.3) \times 10^{-7} (T_e/300)^{-0.70}$	$100 \leq T_e \leq 500$ $v = 0$
Ion trap	Walls & Dunn 1974[e]	$1.95 \times 10^{-7} (T_e/300)^{-0.66}$	$100 \leq T_e \leq 5000$ $v = 0$
Shock tube	Cunningham & Hobson 1972b	$1.82 \times 10^{-7} (T_e/300)^{-0.63}$	$600 \leq T_e \leq 2500$ $v = 0$
Merged beams	Mul & McGowan 1979a; Sheehan & St.-Maurice 2004b[f]	$(0.90 \pm 0.14) \times 10^{-7} (T_e/300)^{-0.49}$	$T_e \leq 1200$ $v \geq 0$

[a] In the ion storage ring experiment by Peverall et al. (2001), the vibrational distribution was measured and found to be completely dominated by $v = 0$. For all other experiments except that of Petrignani et al. (2005c) the vibrational distribution was inferred.
[b] The variation between 1200 K and 5000 K was determined to be $T_e^{-0.56}$. The analysis of the data by Mehr and Biondi was not flawless, as discussed by Dulaney, Biondi, and Johnsen (1987); however, the error at the highest temperature was estimated to be no more than 9%.
[c] Assuming the rate coefficient for $v = 0$ to be 2.4×10^{-7} cm^3 s^{-1}.
[d] The same group, using stationary afterglow techniques, measured the rate coefficient as $(2.1 \pm 0.3) \times 10^{-7}$ cm^3 s^{-1} (Smith & Goodall 1968), $(2.1 \pm 0.5) \times 10^{-7}$ cm^3 s^{-1} (Smith et al. 1970), and $(2.0 \pm 0.2) \times 10^{-7}$ cm^3 s^{-1} (Plumb, Smith, & Adams 1972) at room temperature, but for some reason these works were not referenced by Alge, Adams, and Smith (1983). A value of $(2.1 \pm 0.2) \times 10^{-7}$ cm^3 s^{-1} is given in the first FALP paper from the group (Smith et al. 1975).
[e] Walls and Dunn did not give an analytic expression for the rate coefficient. The expression given here was taken from McLain et al. (2004).
[f] For the range 1200 K < T_e < 5000 K the electron-temperature dependence was $T_e^{-0.51}$ (Sheehan and St.-Maurice 2004b).

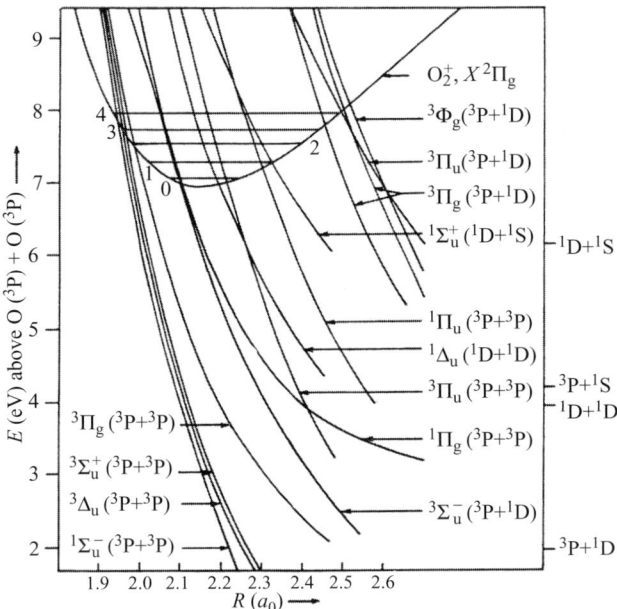

Figure 6.8 Calculated diabatic potential energy curves of O_2 by Guberman (1979) providing routes for dissociative recombination of O_2^+, and the ground state $X\ ^2\Pi_g$ potential energy curve of O_2^+. (*International Journal of Quantum Chemistry: Quantum Chemistry Symposium No. 13*, 1979, pp. 531–540. Copyright (1979, John Wiley & Sons, Inc.). Reprinted with permission from John Wiley & Sons, Inc.)

of O_2^+ is crossed by no less than 14 repulsive curves of O_2. Guberman chose a different strategy. Since there was consensus in the experimental community on the total recombination rate coefficient, a far more interesting problem was that of the formation of O (^{1}S) from dissociative recombination of O_2^+. Guberman (1979, 1983a) identified the $^1\Sigma_u^+$ state as the only possible route to O (^{1}S) and then proceeded to calculate the partial rate coefficient for recombination of O_2^+ through the $^1\Sigma_u^+$ state, and by making use of the experimental rate coefficient he was able derive the O (^{1}S) quantum yield (Guberman 1987), i.e., the number of O (^{1}S) atoms formed per O_2^+ recombination event. In a subsequent paper, Guberman (1988) calculated the O (^{1}D) quantum yield, O (^{1}D) being the initial state in the red airglow at 6330 Å.

The O (^{1}S) state is discussed in Chapter 1 in connection with the auroral green line and the green airglow, which arise from the O (^{1}S $\rightarrow\ ^1$D) emission at 5577 Å. Kaplan had suggested in 1931 that the auroral green line could be due to dissociative recombination of O_2^+. Nicolet (1954) was the first to propose dissociative recombination of O_2^+ as the production mechanism of O (^{1}S), and hence the source of the green airglow, in the E and F regions of the terrestrial ionosphere. Donahue *et al.* (1968) analyzed photometric and electron flux sounding rocket data, and proposed based on this analysis that dissociative recombination of O_2^+ is the dominant excitation mechanism of the 5577 Å line. Measurements of the emission line profiles by means of a Fabry–Perot spectrometer provided conclusive evidence that dissociative recombination of O_2^+ is indeed the formation mechanism (Hernandez 1971) (see e.g. the review by Bates (1982)).

The dissociative recombination of O_2^+ was the source of the green airglow stimulated laboratory experiments by Zipf (1970, 1980b), and an O (^{1}S) quantum yield of 0.1 was obtained by optical measurements of an afterglow plasma (Zipf 1970). A caveat in this experiment, however, was that the vibrational state distribution was not well known. Guberman's theoretical result for the zeroth vibrational level was 0.0016 at an electron temperature of 300 K. The crossing of the $^1\Sigma_u^+$ state curve and the ground-state curve of O_2^+ occurs between $v = 1$ and 2, and Guberman (1987) as expected obtained much higher quantum yields for $v = 1$ and 2 than for $v = 0$. Observations of the terrestrial ionosphere by means of sounding rockets and ground-based instruments gave O (^{1}S) quantum yields from the dissociative recombination of O_2^+ in the range 0.02–0.23 (Bates 1982, Hays & Sharp 1973, Frederick *et al.* 1976, O'Neil, Lee, & Huppi 1979, Abreu *et al.* 1988, Yee, Abreu, & Colwell 1989, Takahashi *et al.* 1990). Zipf (1980b) repeated his experiment of 10 years earlier, the reason being the potential energy curves of Guberman (1979), which led him to believe that the high quantum yields of O (^{1}S) observed in the ionosphere could not have only dissociative recombination of O_2^+ as their source: it would require an implausible amount of vibrationally excited O_2^+ in the night-time ionosphere (Zipf 1979). The new and more extensive experiment by Zipf (1980b) showed that the principal source of O (^{1}S) was dissociative recombination of O_2^+ in $v = 4-12$. Yee and Killeen (1986) analyzed satellite Fabry–Perot interferometer data of the 5577 Å line and found that the best agreement was obtained when the O_2^+ ions were assumed to be in $v = 1$ and 2. But Zipf (1988) pointed out that increasing the population in vibrationally excited levels of the O_2^+ $X\,^2\Pi_g$ ground state could lead to unphysical conditions which did not take due account of the vibrational deactivation of the ground state. There was also a need to account for the quantum yield of O (^{1}D); according to Guberman's (1988) calculations, the quantum yield of O (^{1}D) decreases with increasing v. Zipf (1988) combined the theoretical quantum yields of Guberman (1987) with the vibrational distribution of O_2^+ in the ionosphere estimated by Fox (1986) and obtained quantum yields falling far short of those observed by aeronomers.

Bates and Zipf (1980) speculated that one way out of the dilemma could be that the quantum yield of O (^{1}S) would increase as a function of electron temperature; in the F region the electron temperature can be considerably higher than the gas temperature. Zipf (1970, 1980a) had performed the experiments at room temperature (i.e. $T_e = T_{gas} \approx 300$ K); in a new experiment, Zipf (1988) used microwave heating to vary the electron temperature between 300 K and 3500 K. The results did not show the required increase in quantum yield for O (^{1}S) (Zipf 1988). Deprived of other alternatives Zipf concluded that (Zipf 1988, p. 627): "If the in-situ measurements of O_2^+ (v) verify the predictions of Fox (1986) and if future laboratory experiments continue to confirm Guberman's theoretical work, then an additional source of O (^{1}S) atoms in the nocturnal F-region that mimics process (1) [i.e., dissociative recombination of O_2^+] will probably have to be found."

Bates (1990) tentatively suggested that inclusion of the indirect mechanism could have a significant effect on the O (^{1}S) quantum yield. This proposal was tested in a large MQDT calculation by Guberman and Giusti-Suzor (1991), who showed that the indirect mechanism had essentially no effect on the quantum yield. They obtained a quantum yield of 0.0012, which is slightly lower than when the indirect mechanism was neglected. Guéffelec et al. (1989) applied the FALP-MS technique with optical detection and tried to extract the quantum yield for O_2^+ ($v = 0$), but were unable to obtain an unambiguous determination of the quantum yields for different vibrational levels. By the early 1990s it was clear that the problems with the green airglow would disappear if the quantum yield for O (^{1}S) were higher for dissociative recombination of O_2^+ in its zeroth vibrational level. A charge transfer experiment by Helm et al. (1996), in which a beam of O_2^+ ions was passed through a Cs cell in order to generate high Rydberg states, provided a first hint. Helm et al. (1996) used particle imaging of field-induced dissociating Rydberg states to measure the quantum yield of O (^{1}S) to be 0.033. This was not dissociative recombination, but involved Rydberg states close to those involved in recombination.

Kella et al. (1997) provided the first experimental evidence that dissociative recombination of O_2^+ ($v = 0$) gives a much higher production of O (^{1}S) than Guberman's (1987) calculation suggested. In an experiment using ASTRID, with an O_2^+ beam populating vibrational levels up to $v = 5$, Kella et al. (1997) applied the particle imaging technique to measure the quantum yield of O (^{1}S) to be 0.05 when the electron-beam detuning energy was 0 eV. Figure 6.9 shows the distribution of projected distances for dissociative recombination of $^{18}O^{16}O^+$ (Kella et al. 1997). They obtained an O (^{1}S) quantum yield of 0.05 ± 0.02 for recombination of O_2^+ in $v = 0$. (The use of a mixed $^{18}O^{16}O^+$ ion was in order to add a small dipole moment, which gave some vibrational cooling during the storage phase in ASTRID.)

Guberman had presented a possible mechanism at the American Physical Society in Washington DC in April 1997, a few months before the Kella et al. paper was published. Guberman's poster presentation is given as reference 27 in Kella et al.'s paper, but with an unfortunate misprint, later corrected in a "Corrections and clarifications" section of *Science* (see Kella et al. (1997)). The

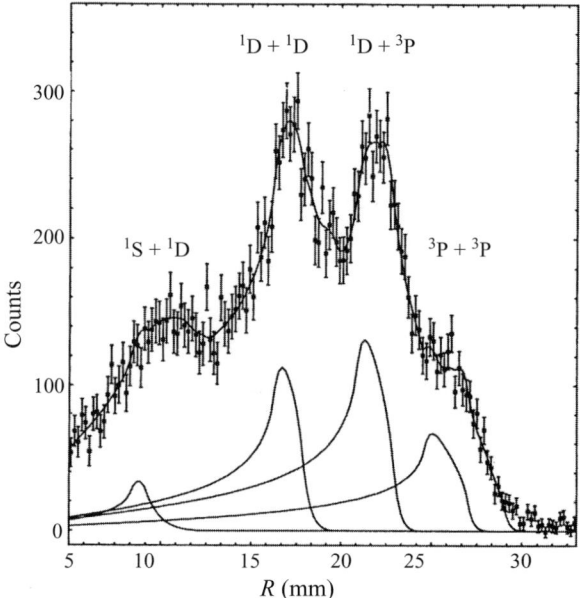

Figure 6.9 The distribution of projected distances for $^{18}O^{16}O^+$ measured in ASTRID with zero detuning of the electron beam. The small peak at $R = 9$ mm is due to O_2^+ $(v = 0) + e^- \rightarrow O(^1S) + O(^1D)$ and obtained by peak fitting. The experimental data, with error bars, are due to O_2^+ $(v = 0-\text{"5"}) + e^- \rightarrow O + O$, where "5" is used to label the fact that the true vibrational distribution was unknown; the peak fitting included six vibrational levels. The peak fitting included a parameter that was equal to the population of a specific vibrational level multiplied by the recombination rate coefficient for this same level. The data did not allow a separation of level population and the rate coefficient. (From D. Kella, L. Vejby-Christensen, P. J. Johnson, H. B. Pedersen, and L. H. Andersen 1997, "The source of the green light emission determined from a heavy-ion storage ring experiment," *Science* **276**, pp. 1530–1533. Reprinted with permission from AAAS.)

surviving written account of Guberman's poster, available as a short abstract at the web page of the American Physical Society (APS/AAPT Joint Meeting, April 18–21, 1997),[1] does not mention a mechanism for O (^{1}S) production. One can assume that the mechanism was added to the poster after the abstract had been submitted.

Guberman (1997) published the full account of his theoretical work on the formation of O (^{1}S) atoms about six months after the meeting in Washington DC. Figure 6.10 shows the potential energy curves involved in recombination of O_2^+ leading to the formation of O (^{1}S). Guberman included for the first time spin–orbit coupling between two resonant states in a calculation of a dissociative recombination cross section. The ground state of O_2^+ has the leading configuration $1\sigma_g^2$

---

[1] Guberman, S. L. 1997, "O (^{1}D) from dissociative recombination of O_2^+," D15.56, Session D15 – Poster Session.

6.2 The atmospheric ions: O_2^+, N_2^+, and NO^+

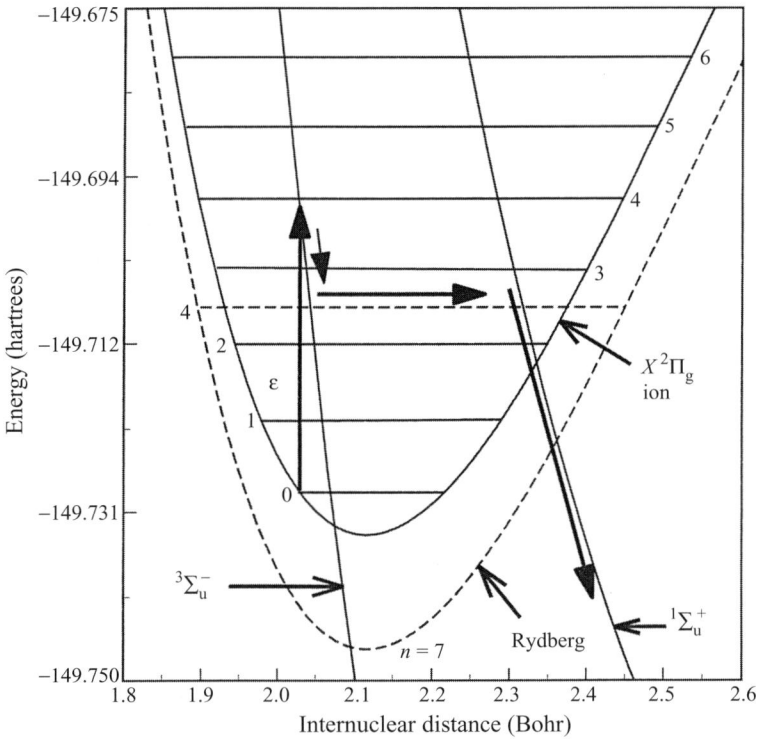

Figure 6.10 The spin-orbit coupling mechanism for the dissociative recombination of O_2^+ ($v=0$). The electron is initially captured into the $^3\Sigma_u^-$ state, where the vertical arrow labelled ε illustrates the kinetic energy of the electron. The Rydberg state is of mixed $^3\Sigma_u^-/^1\Sigma_u^+$ character. The molecule starts to dissociate along the $^3\Sigma_u^-$ state (which does not lead to formation of O (^{1}S)), switches over to the Rydberg state of mixed character. The molecule is now residing in a Rydberg state above the ionization potential. Just like in the indirect mechanism, the Rydberg state is predissociated by the $^1\Sigma_u^+$ state, which dissociates to O (^{1}S) + O (^{1}D). (From S. L. Guberman 1997, "Mechanism for the green glow of the upper ionosphere," *Science* **278**, pp. 1276–1278. Reprinted with permission from AAAS.)

$1\sigma_u^2\, 2\sigma_g^2\, 2\sigma_u^2\, 3\sigma_g^2\, 1\pi_u^4\, 1\pi_g$, and the $^3\Sigma_u^-$ and $^1\Sigma_u^+$ Rydberg states are built on this core with an electron in a diffuse $n\pi_u$ Rydberg orbital. Guberman calculated the spin–orbit interaction between the $^3\Sigma_u^-$ and $^1\Sigma_u^+$ Rydberg states and included this coupling in the MQDT calculations. Since the $^3\Sigma_u^-$ resonance state has a favorable crossing with respect to the $v=0$ level of the $X\,^2\Pi_g$ ion ground state, electron capture into the $^3\Sigma_u^-$ state is effective. The doubly excited O_2 molecule is then transferred to a Rydberg state of mixed $^3\Sigma_u^-/^1\Sigma_u^+$ and finally transferred to the $^1\Sigma_u^+$ resonance state, which dissociates to O (^{1}S) + O (^{1}D). Inclusion of the spin–orbit driven mechanism gave a quantum yield of 0.016 for $v=0$ and $T_e = 300$ K (Guberman 1997). This result is compatible even with the smallest yields coming

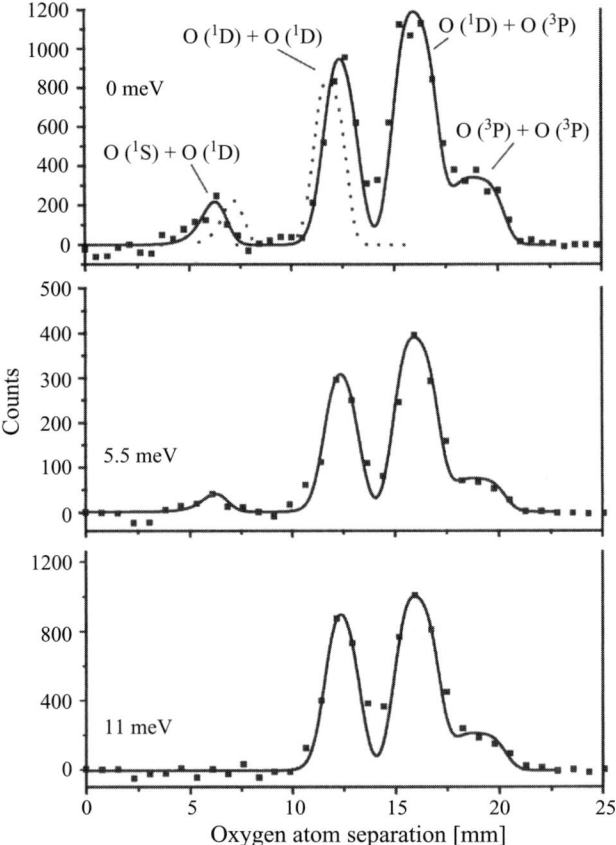

Figure 6.11 Distance distributions measured by Peverall *et al.* (2000, 2001) in CRYRING using a three-dimensional imaging detector. The solid lines are best fits to the data (filled squares). The four dissociation limits observed are identified in the zero energy distribution. (Reused with permission from R. Peverall, S. Rosén, J. R. Peterson, *et al.*, *Journal of Chemical Physics*, **114**, 6679 (2001). Copyright 2001, American Institute of Physics.)

from aeronomy observation, and in agreement with the result of Kella *et al.* (1997).

Peverall *et al.* (2000, 2001) used CRYRING and a hollow cathode ion source operated at > 0.1 Torr, which was sufficient to quench vibrationally excited O_2^+ ions collisionally before they were extracted from the source. This made possible by means of particle imaging a direct measure of the O (^{1}S) quantum yield as a function of the electron energy. Figure 6.11 shows the results at four different electron energies. Most strikingly, recombination leading to O (^{1}S) + O (^{1}D) shows a strong variation with electron energy. At 11 meV O (^{1}S) + O (^{1}D) has completely

6.2 *The atmospheric ions*: O_2^+, N_2^+, *and* NO^+

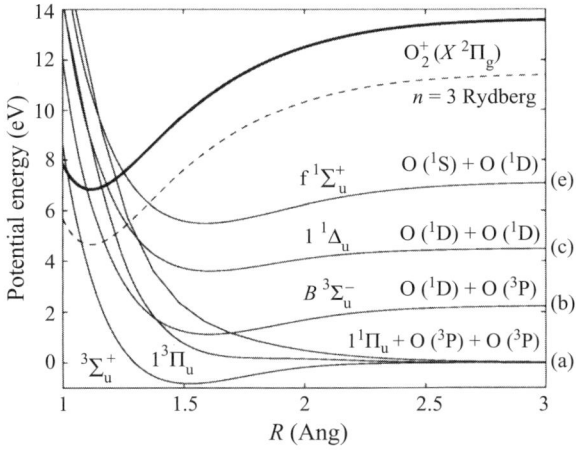

Figure 6.12 Schematic potential energy curves relevant to the dissociative recombination of O_2^+. (Reused with permission from A. Petrignani, F. Hellberg, R. D. Thomas, M. Larsson, P. C. Cosby, and W. J. van der Zande, *Journal of Chemical Physics*, **122**, 234311 (2005). Copyright 2005, American Institute of Physics.)

disappeared. The experimental variation is in qualitative agreement with theoretical calculations by Guberman, included in Peverall *et al.* (2000).

It follows from Fig. 6.4 that four sets of reaction products were identified. They arise from:

$$O_2^+ \left(X\,^2\Pi_g, v=0 \right) + e^- \to O\left(^3P\right) + O\left(^3P\right) + 6.95\,\text{eV}, \quad (6.6a)$$

$$O_2^+ \left(X\,^2\Pi_g, v=0 \right) + e^- \to O\left(^1D\right) + O\left(^3P\right) + 4.99\,\text{eV}, \quad (6.6b)$$

$$O_2^+ \left(X\,^2\Pi_g, v=0 \right) + e^- \to O\left(^1D\right) + O\left(^1D\right) + 3.02\,\text{eV}, \quad (6.6c)$$

$$O_2^+ \left(X\,^2\Pi_g, v=0 \right) + e^- \to O\left(^1S\right) + O\left(^1D\right) + 0.80\,\text{eV}. \quad (6.6d)$$

The energies are given for nominally 0 eV electrons. The channel $O\,(^1S) + O\,(^3P)$ is energetically open by 2.77 eV but was not observed (Petrignani *et al.* 2005c). Figure 6.12 shows potential energy curves of O_2 and O_2^+ of relevance to the dissociative recombination leading to the four channels (6.6a)–(6.6d). The channels leading to the formation of $O\,(^1D)$ are the major sources of hot $O\,(^1D)$ atoms in the thermosphere above 160 km (Fox & Hać 1997, Kharchenko, Dalgarno, & Fox 2005).

Petrignani *et al.* (2005b) extended the measurements of Peverall *et al.* (2000, 2001) to include a broader range of electron energies. Figure 6.13 shows the quantum yields for $O\,(^1D)$, $O\,(^3P)$, and $O\,(^1S)$ as a function of electron energy. The quantum yield for $O\,(^1D)$ atoms is close to unity, something which had already been found by Zipf (1970). The $O\,(^1S)$ quantum yield, which Peverall *et al.* (2000,

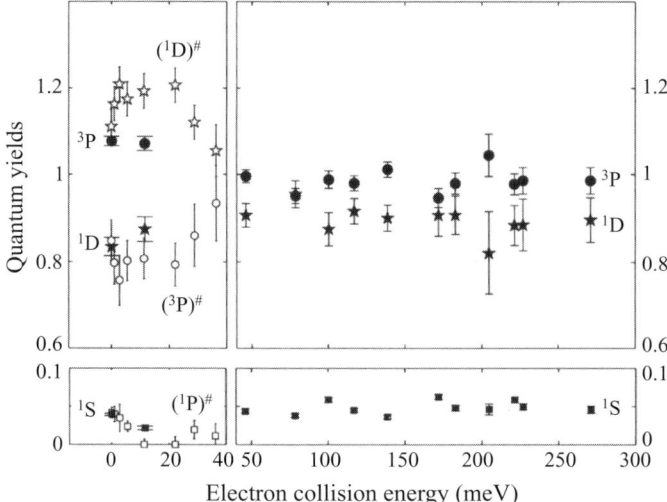

Figure 6.13 Quantum yields for O (^{1}D), O (^{3}P), and O (^{1}S) measured at CRYRING in two different experiments. The unfilled symbols are from Peverall *et al.* (2000, 2001) and the filled symbols from Petrignani *et al.* (2005b). (Reused with permission from A. Petrignani, F. Hellberg, R. D. Thomas, M. Larsson, P. C. Cosby, and W. J. van der Zande, *Journal of Chemical Physics*, **122**, 234311 (2005). Copyright 2005, American Institute of Physics.)

2001) found to disappear at 11 meV, recovers and then shows an oscillatory behavior which seems to have no counterpart in the O (^{1}D) and O (^{3}P) yields.

The story is not complete without addressing the question of whether the O (^{1}S) quantum yield depends on the O_2^+ vibrational level. That this is the case has been assumed at least since Guberman (1979) published potential energy curves for O_2 relevant to dissociative recombination of O_2^+. The first experimental measurements of vibrationally resolved O (^{1}S) quantum yields were performed at CRYRING by Petrignani *et al.* (2005c), who found the yields for $v = 1$ and 2 to be 0.14 and 0.21, respectively. Combining the recombination rate coefficients for $v = 1$ and 2 with the partial rate coefficients for dissociation into channel (6.6d) calculated by Guberman (1987) gives O (^{1}S) quantum yields of 0.11 for $v = 1$ and 0.2 for $v = 2$. This close agreement with the experimental results suggests that the spin–orbit coupling mechanism, obviously not included in Guberman's calculation in 1987, is of less importance for $v = 1$ and 2 than for $v = 0$.

There is still no *ab initio* calculation of the total rate coefficient of O_2^+–electron recombination. Given the complete experimental consensus, this would be a very interesting and worthwhile project, particularly since vibrationally state resolved cross sections are now available.

6.2.2 N_2^+

Comparing the dissociative recombination of N_2^+ and O_2^+, a few points stand out. The experimentally measured rate coefficients are not as consistent for N_2^+ as for O_2^+; there is no merged-beam experiment, single-pass or storage ring multipass, on vibrationally relaxed N_2^+ ions; there is a theoretically calculated rate coefficient for N_2^+; there is no long-standing "green line" problem for N_2^+.

Measurements on N_2^+ follow basically the same pattern as for O_2^+, but with a larger spread in the early measurements that lacked mass spectrometric identification of the recombining ions. The reason is that the formation of cluster ions, such as N_4^+, clouded the measurements. For example, Faire et al. (1958) reported a rate coefficient of 1.2×10^{-6} cm^3 s^{-1} in the afterglow of a pulsed microwave discharge (stationary afterglow), and an even larger rate, 2.0×10^{-6} cm^3 s^{-1}, was reported by Hackham (1965), also from a stationary afterglow experiment, whereas Faire and Champion (1959) obtained 4.0×10^{-7} cm^3 s^{-1} at 400 K, and Mentzoni (1963) an even lower value (1.2×10^{-7} cm^3 s^{-1}). Other results have fallen in between, such as 8.5×10^{-7} cm^3 s^{-1} by Bialecke and Dougal (1958) and 5.9×10^{-7} cm^3 s^{-1} by Kasner, Rogers, and Biondi (1961). However, in most cases strong pressure dependences were found, and the quoted results are those given in the abstracts. By 1965 there was considerable confusion, but the problem had been clearly identified and the remedy was obvious. Kasner and Biondi (1965), Kasner (1967), and Mehr and Biondi (1969) used mass spectrometric techniques to ensure that N_2^+ was the only significant afterglow ion. The results of the Mehr and Biondi (1969) experiment and subsequent measurements, and theoretical results, are given in Table 6.2 and displayed in Figure 6.14.

Recombination of N_2^+ has also been treated theoretically. Guberman (1989) reported preliminary potential curves for electronic states in N_2 involved in the dissociative recombination of N_2^+. This work was followed by a series of publications (Guberman 1991, 1993, 2003c) which presented theoretical rate coefficients for recombination of the zeroth vibrational level in the ion ground state, as shown in Table 6.2, and also for vibrationally excited levels (Guberman 2003c). Figure 6.15 shows potential energy curves for N_2^+ and N_2 taken from Guberman's work.

A visual inspection of the curves in Fig. 6.15 does not give any particular grounds for assuming that the rate coefficient should have a strong vibrational dependence. The curves were not available when Orsini et al. (1977) proposed that $v = 1$ recombines 10 times faster than $v = 0$. The suggestion was an attempt to explain aeronomic data from the Atmosphere Explorer mission. This was met with immediate opposition by Biondi (1978), who was able to use preliminary, unpublished results from Zipf to make his case. Torr and Orsini (1978) found other means to explain the data. When Zipf's (1980a) results were published, it was the first time that vibrationally

Table 6.2. N_2^+ dissociative recombination rate coefficients

Technique	Reference	Rate coefficient (cm^3 s^{-1})	Temperature (K) Vibrational level[a]
Stationary afterglow	Mehr & Biondi 1969	$(1.8 \pm 0.40) \times 10^{-7}(T_e/300)^{-0.39}$	$300 \leq T_e \leq 5000$ $v = 0$
Stationary afterglow/LIF	Zipf 1980a[b]	2.2×10^{-7}	$v = 0$ $T_e = 300$ $T_{vib} = 1500$ K
Ion storage ring	Peterson et al. 1998[c,d]	$(1.75 \pm 0.09) \times 10^{-7}(T_e/300)^{-0.30}$	$80 \leq T_e \leq 1000$ $v = 0\text{--}3$
Merged beams	Sheehan & St.-Maurice 2004b	$(1.50 \pm 0.23) \times 10^{-7}(T_e/300)^{-0.39}$ $(1.88 \pm 0.28) \times 10^{-7}(T_e/300)^{-0.56}$	$100 \leq T_e \leq 1200$ $1200 < T_e < 4000$ $v = 0\text{--}6$
FALP	Mahdavi, Hasted, & Nakshbandi 1971	$(2.2 \pm 0.4) \times 10^{-7}$	$T_e \approx 300$ $v = 0$
FALP	Geoghegan, Adams, & Smith 1991	2.0×10^{-7}	$T_e = 300$ $v = 0$
FALP	Canosa et al. 1991a	2.6×10^{-7}	$T_e = 300$ $v = 0$
Shock tube	Cunningham & Hobson 1972c	$1.78 \times 10^{-7}(T_e/300)^{-0.37}$	$300 \leq T_e \leq 2700$ $v = 0$
Merged beams	Mul & McGowan 1979a[e]	$1.8 \times 10^{-7}(T_e/300)^{-0.50}$	$100 \leq T_e \leq 20000$ $v \geq 0$

Theory; MQDT	Guberman 1993	$1.6 \times 10^{-7}(T_e/300)^{-0.37}$	$v = 0$
		$2.1 \times 10^{-7}(T_e/300)^{-0.20}$	$v = 0$
		$1.9 \times 10^{-7}(T_e/300)^{-0.87}$	$v = 1$
		$1.3 \times 10^{-7}(T_e/300)^{-0.49}$	$v = 2$
Theory; MQDT	Guberman 2003c[f]	$1.6 \times 10^{-7}(T_e/800)^{-0.37}$	$v = 0$
		$0.81 \times 10^{-7}(T_e/800)^{-0.78}$	$v = 1$
		$0.96 \times 10^{-7}(T_e/800)^{-0.29}$	$v = 2$

[a] The vibrational distribution is estimated except for the experiments by Zipf (1980a) and Peterson et al. (1998).

[b] Zipf (1980a) obtained state resolved rate coefficients: $\alpha(v = 0) = 2.15 \times 10^{-7}(T_e/300)^{-0.38}$ cm^3 s^{-1}, $\alpha(v = 1) = 2.42 \times 10^{-7}(T_e/300)^{-0.38}$ cm^3 s^{-1}, $\alpha(v = 2) = 2.7 \times 10^{-7}(T_e/300)^{-0.38}$ cm^3 s^{-1}. Zipf's interpretation of his data has been criticized by Johnsen (1987, 1989), who pointed out that the vibrationally excited ions in Zipf's experiment could not have survived for milliseconds without being collisionally de-excited were it not for the fact that they were kept vibrationally hot by collisions with N$_2$ molecules with internal temperature of 1500 K. Bates and Mitchell (1991) attempted a reanalysis of Zipf's data and obtained $\alpha(v = 0) = 2.6 \times 10^{-7}(T_e/300)^{-0.38}$ cm^3 s^{-1}. The result given here is the rate coefficient for N$_2^+$ ions at a vibrational temperature of about 1500 K.

[c] Sheehan and St.-Maurice (2004b) have used the cross section of Peterson et al. (1998) to derive the rate coefficient $(1.50 \pm 0.23) \times 10^{-7}(T_e/300)^{-0.38}$ cm^3 s^{-1}, in perfect agreement with their single-pass merged-beam result. Sheehan and St.-Maurice have also given expressions for the rate coefficients from their own experiment and from Peterson et al. (1998) for 1200 K $\leq T_e \leq$ 4000 K.

[d] The measured distribution was: $v = 0$ (46%), $v = 1$ (27%), $v = 2$ (10%), $v = 3$ (16%).

[e] We do not list the merged-beam result of Noren, Yousif, and Mitchell (1989), which was possibly affected by a calibration error (Florescu-Mitchell & Mitchell 2006).

[f] The values given here do not agree with those given in the reference. These are the corrected results given at www.sci.org/Kluwer.N2+.pdf

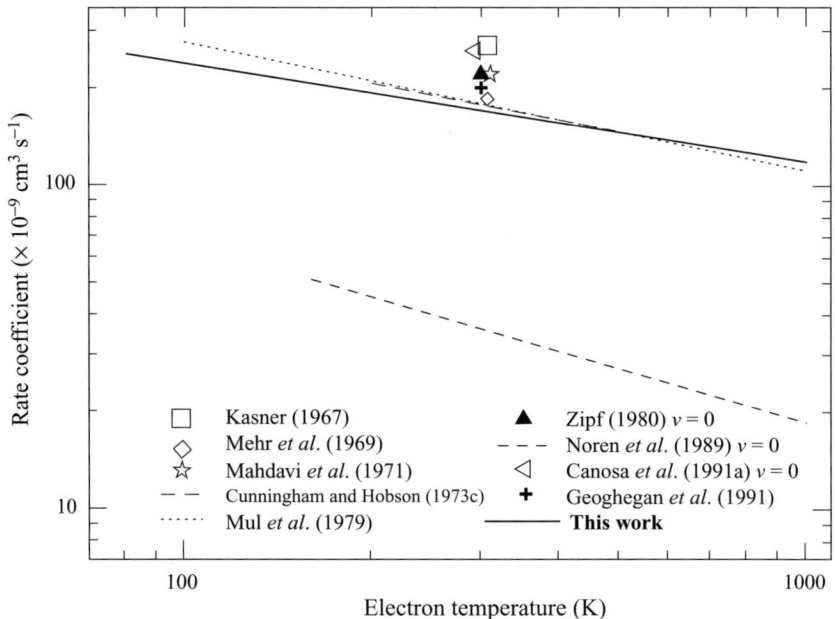

Figure 6.14 Electron-temperature dependent rate coefficients as deduced from the CRYRING experiment (Peterson *et al.* 1998), and results from the literature (see Table 6.2). (Reused with permission from J. R. Peterson, A. Le Padellec, H. Danared, *et al.*, *Journal of Chemical Physics*, **108**, 1978 (1998). Copyright 1998, American Institute of Physics.)

state resolved rate coefficients became available for a molecular ion. The interpretation of the data has been criticized, and rightly so, by Johnsen (1987, 1989), and the data have been reanalyzed by Bates and Mitchell (1991). Nevertheless, Zipf performed a pioneering experiment which at least ruled out a strong vibrational dependence of the rate coefficient. Evidence for a weak vibrational dependence came from the CRYRING experiment of Peterson *et al.* (1998), who used two different ion sources as injectors to CRYRING and measured the vibrational state distributions by particle imaging. Despite markedly different vibrational distributions, the cross sections were very similar. The beam experiments (Mul & McGowan 1979a, Peterson *et al.* 1998, Sheehan & St.-Maurice 2004b), which were carried out on vibrationally excited N_2^+, gave consistently lower rates than did the afterglow experiments on ground-state ions. This provides evidence that the rate coefficient for N_2^+ ($v > 0$) is lower than that for N_2^+ ($v = 0$), and is in agreement with Guberman's calculated results. Sheehan and St.-Maurice (2004b) even went as far as to quantify the ratio $\alpha(N_2^+ (v = 0))/\alpha(N_2^+ (v > 0))$ and suggested it was 2.8. This is at least partly in disagreement with Guberman's (2003c) calculated rate coefficients for $v = 1$ and 2. The final answer must await the larger calculations being carried

6.2 The atmospheric ions: O_2^+, N_2^+, and NO^+

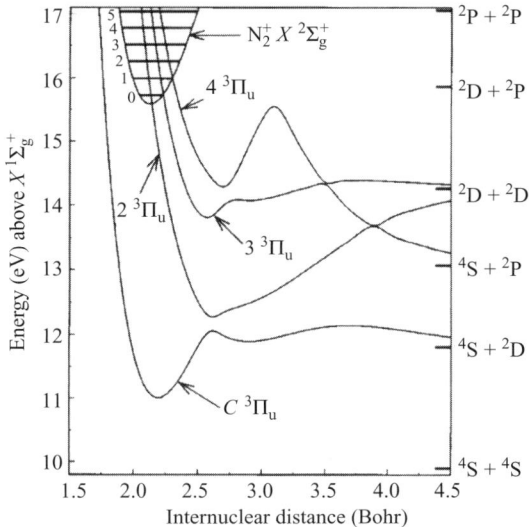

Figure 6.15 Potential energy curves of molecular nitrogen and its ion calculated by Guberman (1993). The ion ground state is shown together with a selection of doubly excited repulsive neutral valence states. It is interesting to note that none of these states correlates with the ground state products N (^{4}S) + N (^{4}S). (Reused with permission from J. R. Peterson, A. Le Padellec, H. Danared, *et al.*, *Journal of Chemical Physics*, **108**, 1978 (1998). Copyright 1998, American Institute of Physics.)

out by Guberman (mentioned in Guberman (2003c)), and experiments of the type carried out for O_2^+ in CRYRING (Petrignani *et al.* 2005c).

Figure 6.15 shows that N_2^+ recombining with thermal-energy electrons can give rise to many different combinations of ground- and excited-state N atoms. The dissociation limits and the associated total kinetic energy of the fragments are:

$$N_2^+(X\ ^2\Sigma_g^+, v=0) + e^- \rightarrow N(^4S) + N(^4S) + 5.82\,\text{eV}, \tag{6.7a}$$
$$N_2^+(X\ ^2\Sigma_g^+, v=0) + e^- \rightarrow N(^4S) + N(^2D) + 3.44\,\text{eV}, \tag{6.7b}$$
$$N_2^+(X\ ^2\Sigma_g^+, v=0) + e^- \rightarrow N(^4S) + N(^2P) + 2.25\,\text{eV}, \tag{6.7c}$$
$$N_2^+(X\ ^2\Sigma_g^+, v=0) + e^- \rightarrow N(^2D) + N(^2D) + 1.06\,\text{eV}, \tag{6.7d}$$
$$N_2^+(X\ ^2\Sigma_g^+, v=0) + e^- \rightarrow N(^4S) + N(^4S) - 0.13\,\text{eV}$$
$$\text{(accessible only for } v > 0\text{)}. \tag{6.7e}$$

According to the calculations by Guberman (1991, 1993), the $2\ ^2\Pi_u$ state plays an important role in the dissociative recombination of N_2^+. The many avoided crossings between 2.5 and 4.0 a_0 (see Figure 6.15) make it less obvious which of channels (6.7a)–(6.7b) will be dominant. The potential curves, however, strongly suggest that channel (6.7a) is negligible.

Table 6.2. *Branching ratios for dissociative recombination of* $N_2^+(v=0)$

Channels	Kella et al. (1996)[a]	Peterson et al. (1998)	Quéffelec et al. (1985)
N (^{4}S) + N (^{4}S)	0.0	0.0	
N (^{4}S) + N (^{2}D)	0.46 ± 0.08	0.37 ± 0.08	0.15
N (^{4}S) + N (^{2}P)	0.08 ± 0.06	0.11 ± 0.06	
N (^{2}D) + N (^{2}D)	0.46 ± 0.06	0.52 ± 0.04	0.85

[a] These results were obtained with ^{15}N^{14}N$^+$. Guberman (2003c) found that the rate coefficient for ^{15}N^{14}N$^+$ was 26% larger than the one for ^{14}N$_2^+$ at $T_e = 300$ K. Thus, it cannot be excluded that the difference between the storage ring results is due to an isotope effect.

The kinetic energies of the N atoms produced by dissociative recombination of N_2^+ are far in excess of thermal energies. Recombination of ^{15}N^{14}N$^+$ would lead to production of ^{15}N atoms that are slower than the lighter ^{14}N. The possibility of dissociative recombination of N_2^+ leading to an isotope fractionation resulting from the selective escape of ^{14}N from the atmosphere of Mars has been pointed out (Wallis 1978, Fox 1993a): this would require channel (6.7b) to have a reasonable branching ratio. Lammer and Bauer (1993) studied the atmospheric mass loss from Titan and found that dissociative recombination of N_2^+ in this case played a minor role. The isotope fractionation in the Mars atmosphere stimulated two experiments at ion storage rings with the aim of measuring the product distributions in the recombination of N_2^+. The first was performed at ASTRID (Kella et al. 1996) and the second at CRYRING a few years later (Peterson et al. 1998). Figure 6.16 shows the results for CRYRING using a hollow cathode ion source as injector. The results of the storage ring experiments are given in Table 6.2, and compared with the only available additional experimental data, those from a FALP experiment. The agreement between the storage ring results is good, whereas the FALP experiment gave a higher value for the N (^{2}D) + N (^{2}D) channel.

Guberman (1993) predicted N (^{4}S) + N (^{2}D) to dominate (branching 0.88) by means of a Landau–Zener calculation that considered only the $2\,^3\Pi_u$ state. If the initial electron capture takes place into the $2\,^3\Pi_u$ state, the N (^{4}S) + N (^{2}D) channel is reached if the $2\,^3\Pi_u$ state crosses the avoided crossing with the $C\,^3\Pi_u$ state at 2.6 a_0. Guberman included more $^3\Pi_u$ states in a later calculation, quoted by Peterson et al. (1998) as a private communication from Guberman. Inclusion of more $^3\Pi_u$ states reduced the branching to 0.70, which is still larger than the experimental values. In this calculation he also found 0.27 for the N (^{2}D) + N (^{2}D) channel, which is a little lower than the experimental results.

The experimental data suggest that the $2\,^3\Pi_u$ state cannot dominate dissociative recombination of N_2^+. If it did, the rate coefficient would depend more strongly on

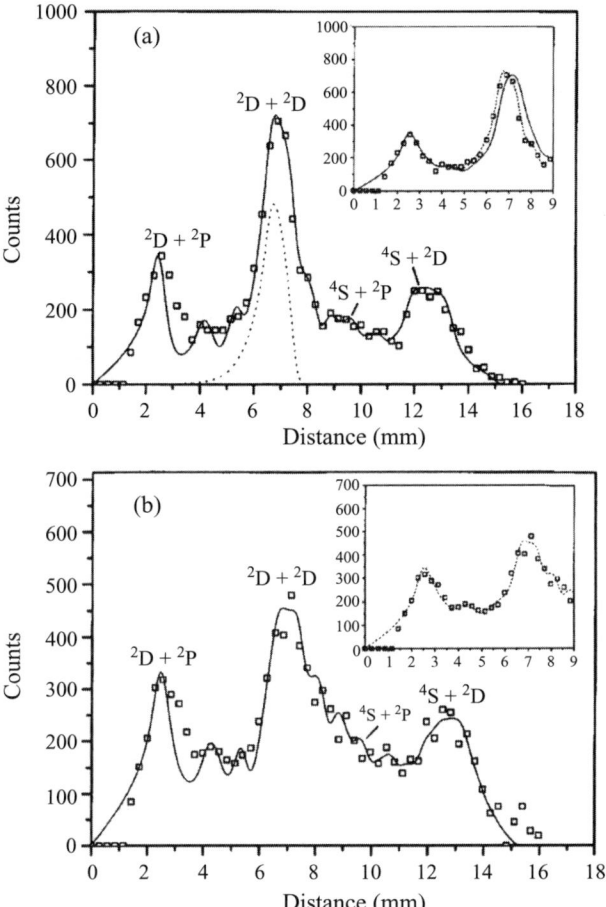

Figure 6.16 Distance distribution spectrum for the products of dissociative recombination of N_2^+. The dotted line is the model spectrum of a single vibrational feature, $N_2^+ (v = 0) \rightarrow N\,(^2D) + N\,(^2D)$ at a rotational temperature of 300 K. The inset shows a fit with a rotational temperature of 1400 K, which was the temperature used by the ASTRID group (Kella et al. 1996). (Reused with permission from J. R. Peterson, A. Le Padellec, H. Danared, et al., Journal of Chemical Physics, **108**, 1978 (1998). Copyright 1998, American Institute of Physics.)

the vibrational level, and the $N\,(^4S) + N\,(^2D)$ product channel would dominate. This is not in agreement with observations.

6.2.3 NO^+

Although the neutral molecule NO is a minor species in the upper atmosphere, its low ionization potential (9.26 eV) means that NO^+ is an important atmospheric ion. Since it cannot be neutralized in collisions with molecular oxygen and nitrogen, it must wait to recombine with a free electron in order to be neutralized. The first measurement of dissociative recombination of NO^+, in which the ion in the decaying plasma was identified as NO^+, was performed by Gunton and Shaw (1965). They used a pulsed hydrogen lamp to ionize NO and measured the electron density with a microwave technique. Young and St. John (1966) measured the ion density in a plasma created by chemi-ionization, but they did not use mass spectrometric identification. Even earlier than the work by Gunton and Shaw (1965) is the measurement of a recombination coefficient in shock-heated air (Stein *et al.* 1964) in which it was assumed that electrons were removed primarily by recombination with NO^+. In Table 6.4 we list as the first reliable measurement, that of Weller and Biondi (1968) using the stationary afterglow technique.

The rate coefficient for $v = 0$ at 300 K is consistently around 4×10^{-7} cm^3 s^{-1} in all experiments with ground-state ions. Equally consistent is that the rate coefficient drops when NO^+ contains vibrational excitations. For a while, there was confusion about the temperature dependence of the rate coefficient. Huang, Biondi, and Johnsen (1975) derived a $T_e^{-0.37}$ dependence using a stationary afterglow with microwave heating, which was difficult to reconcile with the steeper dependence measured by Walls and Dunn (1974) using an ion trap technique. When Torr, St.-Maurice, and Torr (1977) obtained a $T_e^{-0.85}$ dependence based on the Atmosphere Explorer satellite data, the NO^+ chemistry in the upper atmosphere came into question. The temperature dependence measured by Huang, Biondi, and Johnsen (1975) gave a much higher rate coefficient at 2000 K than did the dependence measured by Walls and Dunn (1974), and Torr and Torr (1979) tried to identify an additional source of atmospheric NO^+, in particular associative ionization of $N(^2D) + O$, as had been suggested by Zipf (1978). They were unsuccessful because the cross section for associative ionization was too low, and the problem remained. The temperature dependence measured by Huang, Biondi, and Johnsen (1975) was not without support. Theoretical calculations by Michels (1980) gave a $T_e^{-0.39}$ dependence. There is also an interesting remark in Michels's technical report. In a private communication from Walls and Dunn, they mentioned a problem with their data point at 0.04 eV, which was the lowest electron energy for which they measured a cross section. When the merged-beam technique was used to access collisions at lower energies, a $T_e^{-0.50}$ dependence was obtained by Mul and McGowan (1979a). Thus, around 1980, the situation with the temperature dependence was unclear. Later, Dulaney, Biondi, and Johnsen (1987) remeasured the electron-temperature dependence by means of the microwave-heated afterglow technique and found that some erroneous assumptions had been made in Huang, Biondi, and Johnsen (1975), and the new dependence of $T_e^{-0.75}$ (Dulaney, Biondi, & Johnsen 1987) was in good agreement with the results of Walls and Dunn (1974) and Torr, St.-Maurice, and Torr (1977), and the FALP and shock-tube experiments (see Table 6.4). The

Table 6.3. NO^+ dissociative recombination rate coefficients

Technique	Reference	Rate coefficient ($cm^3\ s^{-1}$)	Temperature (K) Vibrational level[a]
Stationary afterglow	Weller & Biondi 1969[b]	$(4.1 \pm 0.3) \times 10^{-7}(T_e/300)^{-1.0}$	$200 \leq T_e \leq 450$, $v = 0$
Stationary afterglow	Dulaney, Biondi, & Johnsen 1987[c]	$(4.2 \pm 0.2) \times 10^{-7}(T_e/300)^{-0.75}$	$295 \leq T_e \leq 4500$, $v = 0$
FALP	Mahdavi, Hasted, & Nakshbandi 1971	$(3.4 \pm 0.6) \times 10^{-7}$	$T_e \approx 300$, $v = 0$
FALP	Alge, Adams, & Smith 1983	$4.0 \times 10^{-7}(T_e/300)^{-0.9}$	$200 \leq T_e \leq 600$, $v = 0$
FALP	Španěl, Dittrichová, & Smith 1993	$4.0 \times 10^{-7}(T_e/300)^{-0.85}$	$300 \leq T_e \leq 5000$, $v = 0$
FALP	Mostefaoui et al. 1999	$(3.9 \pm 1.0) \times 10^{-7}$ $(1.6 \pm 1.0) \times 10^{-7}$	$v = 0, T_e = 300$ $v \geq 0, T_e = 300$
Shock tube	Davidson & Hobson 1987[d]	$(4.3 \pm 0.8) \times 10^{-7}(T_e/300)^{-0.7}$	$300 \leq T_e \leq 4500$, $v = 0$
Merged beams	Mul & McGowan 1979a	$1.15 \times 10^{-7}(T_e/300)^{-0.50}$	$100 \leq T_e \leq 10\,000$, $v \geq 0$
Merged beams	Sheehan & St.-Maurice 2004b	$(1.0 \pm 0.2) \times 10^{-7}(T_e/300)^{-0.48}$ $(1.14 \pm 0.17) \times 10^{-7}(T_e/300)^{-0.62}$	$100 \leq T_e \leq 1200$ $1200 < T_e < 4000$, $v \geq 0$

(cont.)

Table 6.4. (cont.)

Technique	Reference	Rate coefficient (cm^3 s^{-1})	Temperature (K) Vibrational level[a]
Ion storage ring	Vejby-Christensen et al. 1998[e]	$(4.0 \pm 2.0) \times 10^{-7}$	$T_e = 300$ $v = 0$
Ion trap	Walls & Dunn 1974[f]	4.3×10^{-7}	$T_e = 300$ $v = 0$
Atmospheric Explorer satellite	Torr, St.-Maurice, & Torr 1977	$4.2 \times 10^{-7} (T_e/300)^{-0.85}$	$300 \le T_e \le 2700$ $v = 0$
Flame	Burdett & Hayhurst 1978	$2.3 \times 10^{-7} - 1.4 \times 10^{-7}$	$1820 \le T_e \le 2650$ $v = 0$
Theory; MQDT	Schneider et al. 2000b	$3.7 \times 10^{-7} (T_e/300)^{-0.60}$	$v = 0$

[a] Estimated except for the storage ring experiment.
[b] Weller and Biondi (1968) hesitated to express the temperature dependence using a power law. They wrote (p. 205): "Actually, the recombination coefficient apparently varies more rapidly with temperature between 200 and 300 K than between 300 and 450 K."
[c] An earlier measurement by Huang, Biondi, and Johnsen (1975) gave an incorrect temperature dependence.
[d] The measurement was done in the electron-temperature range 2800–4500 K.
[e] Sheehan and St.-Maurice have used the cross section published by Vejby-Christensen et al. (1998) to derive an electron-temperature dependence of $T_e^{-0.69}$.
[f] The ion trap measurement by Walls and Dunn (1974) was the first of the NO$^+$ recombination cross section. The cross section was used to calculate electron-temperature-dependent rate coefficients. No analytic expressions for these dependences were given, but the rate coefficient was displayed graphically in the range 100 K $\le T_e \le$ 40 000 K. Torr, St.-Maurice, and Torr (1977) derived the following expression based on the cross section data of Walls and Dunn (1974): $\alpha(T_e) = (4.3 \pm 1) \times 10^{-7} (T_e/300)^{-(0.83+0.16-0.08)}$ cm^3 s^{-1}.

6.2 The atmospheric ions: O_2^+, N_2^+, and NO^+

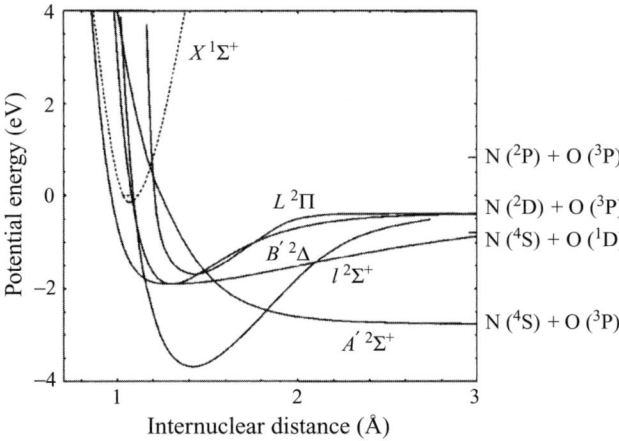

Figure 6.17 Potential energy curves for NO^+. (Reprinted figure by permission from L. Vejby-Christensen, D. Kella, H. B. Pedersen, and L. H. Andersen, "Dissociative recombination of NO^+", *Phys. Rev. A* **57**, pp. 3627–3634, (1998). Copyright (1998) by the American Physical Society.)

afterglow result (Dulaney, Biondi, & Johnsen 1987) is also in good agreement with the dependence derived by Sheehan and St.-Maurice (2004b) from the ion storage ring cross section measurement (Vejby-Christensen *et al.* 1998).

NO^+ has played an important role in the development of the theory of dissociative recombination. Bardsley (1968b) chose NO^+ for an attempt at quantitative calculation of the recombination coefficient because of the wealth of spectroscopic results available (Miescher 1966), which allowed Bardsley to extract parameters such as the electron capture width by extrapolating valence–Rydberg coupling matrix elements for NO states below the ionization limit. The $X\,^1\Sigma^+$ ground state of NO^+ has the configuration $(1\sigma)^2(2\sigma)^2(3\sigma)^2(1\pi)^4(4\sigma)^2(5\sigma)^2$. The ground state of NO is obtained by adding a 2π electron to NO^+. Bardsley included two states in NO, $B'\,^2\Delta$ with configuration $\ldots(5\sigma)(2\pi)^2$ and $B\,^2\Pi$ with configuration $\ldots(1\pi)^3(4\sigma)^2(5\sigma)^2(2\pi)^2$. Figure 6.17 shows the relevant potential energy curves involved in dissociative recombination of NO^+. In addition to the two just mentioned, the $L\,^2\Pi$, $A'\,^2\Sigma^+$, and $I'\,^2\Sigma^+$ states are also shown. Bardlesy (1968b) calculated a rate coefficient of 2.6×10^{-7} cm^3 s^{-1}, which is quite close to the experimental results in Table 6.4. Only the direct mechanism was included and only recombination through the $B'\,^2\Delta$ and $B\,^2\Pi$ states was considered. Bardsley (1983) had access to better molecular data and more potential energy curves (Michels 1981) and included in addition to the $B'\,^2\Delta$ and $B\,^2\Pi$ states also the $L\,^2\Pi$, $A'\,^2\Sigma^+$, and $I'\,^2\Sigma^+$ states. He obtained rate coefficients of 4.3×10^{-7} cm^3 s^{-1}, 4.6×10^{-7} cm^3 s^{-1}, and 1.9×10^{-7} cm^3 s^{-1} for the ionic states $v = 0$, 1, and 2, respectively.

Lee (1977) applied a multichannel treatment that included both the direct and indirect mechanisms to the recombination of NO$^+$, and of the states mentioned above neglected only $L\,^2\Pi$. The rate coefficient came out a factor of 2 larger than the one measured by Walls and Dunn (1974).

Theoretical work on NO$^+$ was done by Michels (1974, 1975, 1980), but this appeared only as technical reports, which are not easily accessible outside the USA, and even inside the USA require some efforts to obtain. The final report of Michels (1980) has, to the best of our knowledge, never been referenced, whereas the 1974 and 1975 reports are referenced in some papers. Michels (1980) obtained a rate coefficient of 1.2×10^{-7} cm^3 s^{-1} for $v = 0$ at 300 K. He also found that the most important contribution to this rate came from the $A'\,^2\Sigma^+$ state, in contrast to Bardsley (1968b) and Lee (1977). The temperature dependence determined by Michels (1980) supported the one measured by Huang, Biondi, and Johnsen (1975).

The MQDT calculation by Sun and Nakamura (1990) was the first to include all five states and the full theoretical apparatus. Sun and Nakamura (1990) also had at their disposal more accurate R-dependent quantum defects than those previously available (Nakashima et al. 1989). The temperature dependence of the rate coefficient, $T_e^{-0.82}$, was in good agreement with the experiments, but the total rate coefficient at 300 K was low, 1.6×10^{-7} cm^3 s^{-1}. Whereas Bardsley (1968b) had found the $B\,^2\Pi$ and $B'\,^2\Delta$ states to be almost equally important in recombination of NO$^+$, with contributions of about 60% and 40%, respectively, Sun and Nakamura (1990) found the recombination cross section via the $B\,^2\Pi$ state to be an order of magnitude more important than the second most important state, $B'\,^2\Delta$. They noted that this made the calculations sensitively dependent on the accuracy of the potential energy curve of the $B\,^2\Pi$ state. Vâlcu et al. (1998) investigated the effect of rotational interaction in the recombination of NO$^+$ and found that it can be neglected.

When Schneider et al. (2000b) applied MQDT to the problem of NO$^+$ recombination, they made use of extensive new molecular data obtained in R-matrix calculations by Rabadán and Tennyson (1996, 1997). One more state, $3\,^2\Pi$, was included. Figure 6.18 shows the calculated cross section and a comparison with the theoretical results of Sun and Nakamura (1990), and the experimental results from ASTRID (Vejby-Christensen et al. 1998). Schneider et al. (2000b) found the $B\,^2\Pi$ and $B'\,^2\Delta$ states to be the dominant recombination pathways, just as Bardsley (1968b) had done. At higher electron energy, the $3\,^2\Pi$ state also becomes important and gives rise to the peak in the cross section between 5 and 6 eV. For these three states, Schneider et al. found the effect of the indirect process to be minor (of the order of 20%) and mainly destructive. On the other hand, for a state such as $L\,^2\Pi$, for which the direct process is very weak, they found that the cross section was increased by several orders of magnitude by the inclusion of the indirect

6.2 The atmospheric ions: O_2^+, N_2^+, and NO^+

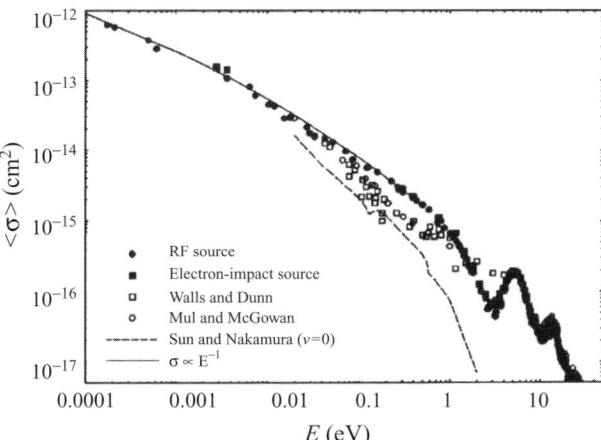

Figure 6.18 The effective cross section, $\langle v_r \sigma \rangle / v_d$ (see Section 2.1.1), for dissociative recombination of NO^+ ($X\ ^1\Sigma^+$, $v = 0$). The calculation of $\langle v_r \sigma \rangle / v_d$ was performed with the electron-velocity distribution parameters of the ASTRID electron beam. The experimental results are from ASTRID (•), Walls and Dunn (1974) (□), and Mul and McGowan (1979a) (○). The grey curve is the theoretical result of Sun and Nakamura (1990). (Reprinted figure by permission from L. Vejby-Christensen, D. Kella, H. B. Pedersen, and L. H. Andersen, "Dissociative recombination of NO^+," *Phys. Rev. A* **57**, pp. 3627–3634, (1998). Copyright (1998) by the American Physical Society.)

mechanism, just as was found by Guberman (1994) for HeH^+ and by Orel, Schneider, and Suzor-Weiner (2000) and Schneider, Suzor-Weiner, and Orel (2000) for H_3^+. The rate coefficient obtained by integrating the cross section (Schneider *et al.* 2000b) is given in Table 6.4 and must be considered to be in very good agreement with the experimental results. It should be noted that the theoretical result is not fully *ab initio*; it relies on a calibration of the relevant potential energy curves against spectroscopic data.

In a sequel to the work of Schneider *et al.* (2000b), Motapon *et al.* (2006) used MQDT to calculate vibrationally state resolved cross sections and rate coefficients, along with other collisional properties of NO^+. They showed that the rate coefficient decreases for vibrationally excited levels and found very good agreement with the result of Mostefaoui *et al.* (1999) obtained for vibrationally excited NO^+.

Dissociative recombination of NO^+ with zero-energy electrons leads to the following products:

$$NO^+\left(X\ ^1\Sigma^+, v = 0\right) + e^- \rightarrow N\left(^4S\right) + O\left(^3P\right) + 2.77\,\text{eV}, \quad (6.8a)$$
$$NO^+\left(X\ ^1\Sigma^+, v = 0\right) + e^- \rightarrow N\left(^4S\right) + O\left(^1D\right) + 0.80\,\text{eV}, \quad (6.8b)$$
$$NO^+\left(X\ ^1\Sigma^+, v = 0\right) + e^- \rightarrow N\left(^2D\right) + O\left(^3P\right) + 0.38\,\text{eV}, \quad (6.8c)$$

$$\text{NO}^+ \left(X\,^1\Sigma^+, \nu = 0\right) + e^- \to \text{N}\,(^2\text{P}) + \text{O}\,(^3\text{P}) - 0.81\,\text{eV}, \qquad (6.8\text{d})$$

$$\text{NO}^+ \left(X\,^1\Sigma^+, \nu = 0\right) + e^- \to \text{N}\,(^4\text{S}) + \text{O}\,(^1\text{S}) - 1.42\,\text{eV}, \qquad (6.8\text{e})$$

$$\text{NO}^+ \left(X\,^1\Sigma^+, \nu = 0\right) + e^- \to \text{N}\,(^2\text{D}) + \text{O}\,(^1\text{D}) - 1.59\,\text{eV}, \qquad (6.8\text{f})$$

$$\text{NO}^+ \left(X\,^1\Sigma^+, \nu = 0\right) + e^- \to \text{N}\,(^2\text{P}) + \text{O}\,(^1\text{D}) - 2.78\,\text{eV}, \qquad (6.8\text{g})$$

$$\text{NO}^+ \left(X\,^1\Sigma^+, \nu = 0\right) + e^- \to \text{N}\,(^2\text{D}) + \text{O}\,(^1\text{S}) - 3.81\,\text{eV}, \qquad (6.8\text{h})$$

$$\text{NO}^+ \left(X\,^1\Sigma^+, \nu = 0\right) + e^- \to \text{N}\,(^2\text{P}) + \text{O}\,(^1\text{S}) - 5.00\,\text{eV}, \qquad (6.8\text{i})$$

$$\text{NO}^+ \left(X\,^1\Sigma^+, \nu = 0\right) + e^- \to \text{N}\,(^4\text{S}) + \text{O}\,(^5\text{P}) - 6.38\,\text{eV}, \qquad (6.8\text{j})$$

where the minus sign for (6.8d)–(6.8j) is used to show that these channels are closed for low-energy electrons. Of particular interest is the channel leading to the formation of N (^{2}D). In the terrestrial atmosphere the reaction N (^{2}D) + O$_2$ → NO + O is the most important source of NO, and in their discussion of sources of atmospheric N (^{2}D), Oran, Julienne, and Strobel (1975) considered dissociative recombination of NO$^+$. They concluded that (p. 3070) "In summary, neither experiment nor theory currently provides us with a quantitative branching ratio of NO$^+$ dissociative recombination into N (^{2}D). A qualitative picture based on Bardlsey's theory and published spectroscopic information indicates a branching ratio appreciably less than unity but in excess of 0.5." Lawrence, Kley, and Stone (1977) using an atomic absorption technique measured the branching ratio for channel (6.8c) to be 0.76 ± 0.06. This is in good agreement with the storage ring result of 0.85 ± 0.06 (Vejby-Christensen et al. 1998), obtained by the application of particle imaging. Vejby-Christensen et al. (1998) also recorded imaging spectra at electron energies of 0.75 eV and 1.35 eV. At 0.75 eV no new channels are opened, so there were no new peaks in the distance spectrum. However, they found that a better fit to the data was obtained by assuming a $\sin^2\theta$ angular distribution. When the electron was increased to 1.35 eV, the N (^{2}P) + O (^{3}P) channel opened. This gave rise to a small peak in the distance spectrum (Vejby-Christensen et al. 1998), although they did not explicitly identify the peak as coming from the N (^{2}P) + O (^{3}P) channel. Also at 1.35 eV the distance spectrum was best fitted with a $\sin^2\theta$ angular distribution, and the branching ratio for N (^{2}D) + O (^{3}P) was determined to be 0.65 at both electron energies.

Hellberg et al. (2003) performed a detailed study of NO$^+$ using CRYRING. Figure 6.19 shows a distance spectrum recorded at 0 eV electron energy with an imaging detector. The branching ratios extracted from this distance spectrum are given in Table 6.5.

The spin-forbidden channel N (^{4}S) + O (^{1}D) has a zero branching ratio, but this was only found if the toroidal correction was included; without this correction the branching ratio was 0.015. A complicating factor in the analysis for both Vejby-Christensen et al. (1998) and Hellberg et al. (2003) was also the presence of the

6.2 The atmospheric ions: O_2^+, N_2^+, and NO^+

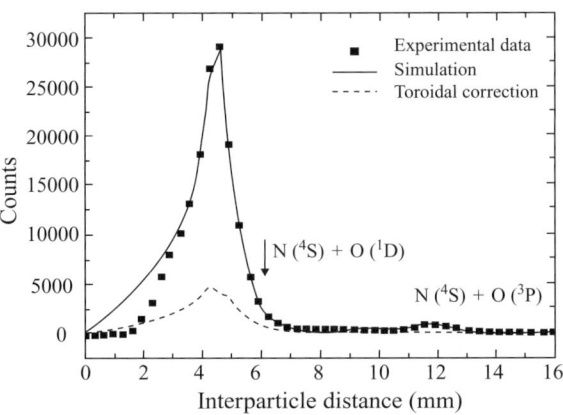

Figure 6.19 Histograms of distances between N-atoms and O-atoms deriving from 0 eV dissociative recombination of NO^+ in CRYRING. The three open dissociation channels are labeled. The bold curve is a fit from an optimized simulation spectrum using the branching ratios given in Table 6.5. The thinner curve is the magnitude of the toroidal correction, which was required in order to determine the small branching ratios. A rotational temperature of 1300 K was applied in the fitting (similar as in Vejby-Christensen et al. 1998). This is higher than one would have expected based on the ion source temperature of about 900 K, and suggests that there could have been an additional source of rotational excitation in the ion source. (Reused with permission from F. Hellberg, S. Rosén, R. Thomas, et al., *Journal of Chemical Physics*, **118**, 6250 (2003). Copyright 2003, American Institute of Physics.)

metastable $a\,^3\Sigma^+$ state. Hellberg et al. (2003) turned this into an advantage, and by analyzing the distance spectrum of recombining NO^+ ($a\,^3\Sigma^+$) ions at different storage times, they measured the radiative lifetime of the $a\,^3\Sigma^+$ state to be 730 ± 50 ms.

Hellberg et al. (2003) measured distance spectra at electron energies of 1.25 eV and 5.6 eV. At 1.25 eV they were able to confirm Vejby-Christensen et al.'s observation that the angular distribution was $\sin^2\theta$, implying that there was a preference for recombining NO^+ to break up perpendicular to the electron-beam axis. The distance spectrum at 5.6 eV, which is at the peak in the cross section in Fig. 6.19, is shown in Fig. 6.20. Schneider et al. (2000b) calculated the contributions to this peak and found that the $3\,^2\Pi$ state made the main contribution, but that the $A'\,^2\Sigma^+$ state also made a contribution at the 10% level. Since the $A'\,^2\Sigma^+$ state correlates diabatically with the ground-state limit $N(^4S) + O(^3P)$ this suggests that there should be an imprint of this limit in the distance spectrum. Such an imprint is visible in the spectrum shown in Fig. 6.20, but it is smaller than 10%.

Hellberg et al. (2003) applied a statistical model to predict the product branching ratios for dissociative recombination of NO^+ (Van der Zande 2000). In this model

Table 6.5. *Branching ratios of dissociative recombination of $NO^+(v=0)$*

Channels	Lawrence, Key, & Stone (1977)	Vejby-Christensen et al. (1998)	Hellberg et al. (2003)
N (^{4}S) + O (^{3}P)			0.05 ± 0.02
N (^{4}S) + O (^{1}D)			0.00 ± 0.02
N (^{2}D) + O (^{3}P)	0.76 ± 0.06	0.85 ± 0.06	0.95 ± 0.03

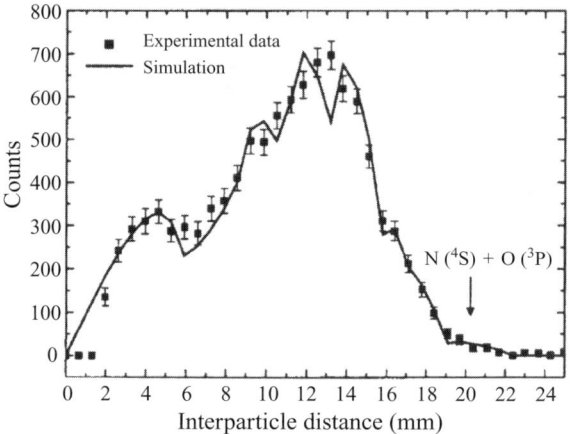

Figure 6.20 Histogram of distances between N-atoms and O-atoms deriving from 5.6 eV dissociative recombination of NO^+ in CRYRING. The arrow indicates the maximum distance for ground-state atoms. The curve is the result of a simulation including all channels except the spin-forbidden ones, N (^{4}S) + O (^{1}D) and N (^{4}S) + O (^{1}S). The angular distributions were assumed to be isotropic, except for the N (^{2}P) + O (^{1}S) channel, for which a $\cos^2\theta$ distribution was used. (Reused with permission from F. Hellberg, S. Rosén, R. Thomas, et al., *Journal of Chemical Physics*, **118**, 6250 (2003). Copyright 2003, American Institute of Physics.)

the spin multiplicity and orbital angular momentum degeneracy are multiplied together; N (^{4}S) + O (^{3}P) gives $4 \times 1 \times 3 \times 3 = 36$, and so on. Figure 6.21 shows the experimental results for the $X\,^1\Sigma^+$ state at electron energies 0 eV, 1.25 eV, and 5.6 eV, and for the $a\,^3\Sigma^+$ state at 0 eV. Note how the outgoing flux in recombination of NO^+ ($a\,^3\Sigma^+$) is distributed over many more channels than is the case for NO^+ ($X\,^1\Sigma^+$) at 0 eV. The statistical model works very well for NO^+ judging by the good agreement between the experimental and model branching ratios.

6.3 Other diatomic ions

There are surprisingly few ions that do not fall into any of the categories in this chapter or in Chapters 4 and 5. Most of them contain at least one carbon atom.

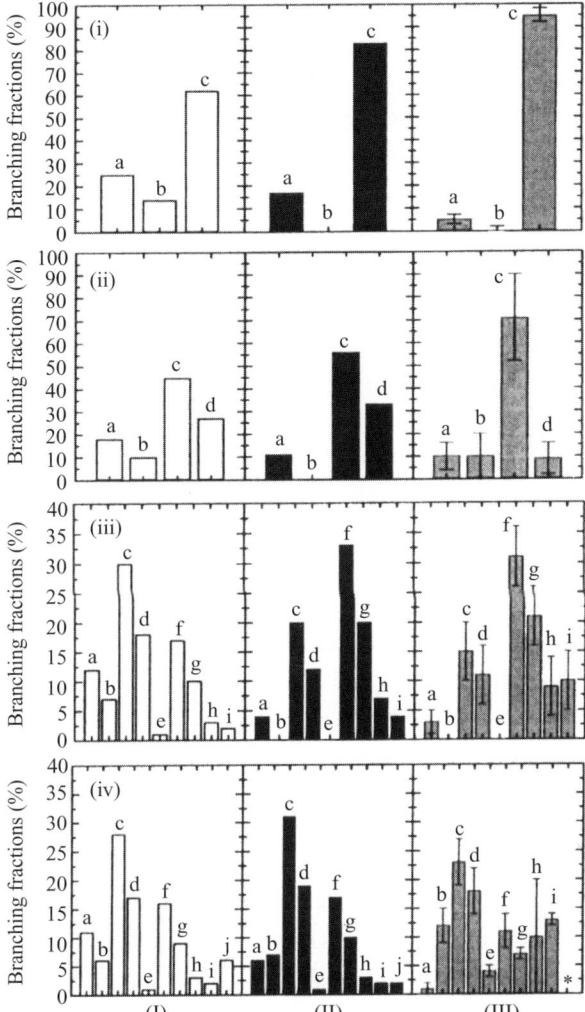

Figure 6.21 Experimental and model dissociative recombination branching ratios for the $X\,^1\Sigma^+$ state of NO^+ at (i) 0 eV, (ii) 1.25 eV, (iii) 5.6 eV collision energy, and (iv) the $a\,^3\Sigma^+$ state of NO^+ at 0 eV. In each case, three data sets are plotted. (I) shows the branching ratios determined from the multiplicity of available states. (II) shows branching ratios determined from the multiplicity of available states, but with the branching ratios for the N (^4S) + O (^1D) and N (^4S) + O (^1S) channels set equal to zero. (III) shows the experimental data. The histogram bars (a)–(j) correspond to the channels (6.8a)–(6.8j). (Reused with permission from F. Hellberg, S. Rosén, R. Thomas, et al., Journal of Chemical Physics, **118**, 6250 (2003). Copyright 2003, American Institute of Physics.)

Two dimer ions have been studied. Jog and Biondi (1981) used the microwave-induced stationary afterglow technique and measured the recombination rate coefficient of Hg_2^+ to be $(4.2 \pm 0.5) \times 10^{-7}(T_e/340)^{-1.1}$ cm^3 s^{-1} over the electron-temperature range $340 \leq T_e \leq 5600$ K. Using optical spectroscopy, Jog and Biondi found that at both $T_e = 340$ K and $T_e = 4900$ K, the 7s and 6d states were strongly populated. Momozaki and El-Genk (2002) analyzed the relaxation time of a decaying Cs plasma, and found that dissociative recombination of Cs_2^+ occurred with a rate in the range 10^{-6}–10^{-7} cm^3 s^{-1}.

Mul and McGowan (1980) in their study of $C_2H_{0\leq n\leq 3}^+$ ions determined the rate coefficient for C_2^+ to be $3.0 \times 10^{-7}(T_e/300)^{-0.5}$ cm^3 s^{-1}, a result not given in the original article but in Mitchell's review (Mitchell, 1990a). Le Padellec et al. (1999) found a very similar result for CN$^+$, with a rate coefficient of 3.4×10^{-7} cm^3 s^{-1} and a $T_e^{-0.55}$ temperature dependence over the range 20–2000 K. The product states at zero and elevated electron energies were also measured in this study.

CO$^+$ is used to illustrate cross section measurements in Section 2.1.2. The single-pass experiment with merged beams performed by Mitchell and Hus (1985) gave, after the factor of 2 correction, a rate coefficient $1.0 \times 10^{-7}(T_e/300)^{-0.46}$ cm^3 s^{-1} (Mitchell 1990a), but there are some question marks about the result, related to the detector efficiency in the single-pass experiment (Laubé et al. 1998a). Flowing afterglow experiments at 300 K (Laubé et al. 1998a, Geoghegan, Adams, & Smith 1991) have given rates of $(1.85 \pm 0.5) \times 10^{-7}$ cm^3 s^{-1} and 1.6×10^{-7} cm^3 s^{-1}, respectively. This is somewhat lower than the storage ring result, $(2.75 \pm 0.4) \times 10^{-7}(T_e/300)^{-0.55}$ cm^3 s^{-1} (Rosén et al. 1998a), which, however, was obtained with ^{13}CO$^+$. Guberman's (2003c) work on N$_2^+$ showed a large isotope effect, so it cannot be excluded that part of the difference is due to an isotope effect. It is noteworthy that recombination with thermal-energy electrons gave a clear dominance for ground-state products, C (^{3}P) + O (^{3}P), with a branching ratio of 0.76 (Rosén et al. 1998a). This is radically different from N$_2^+$ and NO$^+$, where none or a very small number of the recombination events give ground-state products. Recombination of the dication of carbon monoxide, CO^{2+}, has been studied in ASTRID (Safvan et al. 1999). The two channels C + O$^+$ and O + C$^+$ were found to have about equal branching ratios for electron energies below 2 eV. The rate of recombination for CO^{2+} was found to be only marginally larger than for CO$^+$ (Seiersen et al. 2003a). Potential energy curves of relevance to the dissociative recombination of CO^{2+} have been calculated by Vinci and Tennyson (2004). In contrast to carbon monoxide, the dication of molecular nitrogen recombines about three times faster than the cation (Seiersen et al. 2003a).

A surprising result emerged from a combined ion storage ring effort on CF$^+$. A rate coefficient of $(5.2 \pm 1.0) \times 10^{-8}(T_e/300)^{-0.8}$ cm^3 s^{-1} was measured in

ASTRID and CRYRING (Novotný et al. 2005a). This is not only a rate which is almost a factor of 10 lower than for other comparable diatomics, but also is unexpected when it is considered that the CF^+ is crossed by neutral resonance states. Since the result was obtained independently at two different storage rings, it is unlikely to be due to a calibration error. More insight from theoretical calculations would be valuable.

7
The H_3^+ molecule

There are two reasons to devote a chapter of this book to the dissociative recombination of H_3^+. First, it is the single most important molecular ion in astrophysics and, second, the dissociative recombination of H_3^+ has been the subject of much controversy for many years, although much progress, experimental and theoretical, has been made during the last 2–3 years. It should be stressed that the bewildering situation which has prevailed at times, and to some extent still prevails, has nothing to do with trivial experimental errors. It is the H_3^+ ion itself which, for reasons to be discussed in this chapter, represents such an outstanding challenge. The following sections are very detailed; readers who would like to avoid such details can go directly to Section 7.2.14.

7.1 History of H_3^+

H_3^+ was discovered by Thomson using the newly developed technique of mass spectrometry (Thomson 1911). In a subsequent paper he wrote "There can, I think, be little doubt that this line is due to H_3, and not to the carbon atom with four charges" (Thomson 1912, p. 241). In fact, Thomson had some doubts, and in his monograph on positive rays (Thomson 1913, 1921), H_3 is referred to as X_3. After the discovery of the deuteron, Thomson was inclined to believe that his discovery 20 years earlier was due to HD^+ (Thomson 1934). But a few years later, in his autobiography, Thomson acknowledged "Again, one of the first things discovered by the photographic method was the existence of H_3^+" (Thomson 1937, p. 363). (See also Oka (2006b) for more details of this history.)

Thomson's work was followed by an experiment by Duane and Wendt (1917) in which they irradiated hydrogen with α-particles and found an increased reactivity of the molecular hydrogen gas. They admitted that their work did not provide direct evidence for the existence of H_3, but speculated that the reactivity could be due to the formation of this molecule. Using a higher gas pressure and lower voltage than Thomson did, and by varying the pressure, Dempster (1916) showed that the ratio between H^+, H_2^+, and H_3^+ in the mass spectrum could be changed, and, in particular, that the H_3^+ peak was stronger than the other peaks at a high pressure of H_2. This was the first direct evidence that H_3^+ is formed in gas-phase reactions by ionization of molecular hydrogen. Dempster did not, however, identify the formation process, and it was not until the work by Hogness and Lunn (1925) that the famous exothermic ion–molecule reaction was formulated:

$$H_2^+ + H_2 \rightarrow H_3^+ + H. \tag{7.1}$$

Interestingly, the work by Thomson (1913) and Duane and Wendt (1916) inspired Bohr (1919) to publish a model of the triatomic hydrogen molecule using the old quantum mechanics. This attempt failed completely, and it was Coulson (1935) who first predicted that H_3^+ has the structure of an equilateral triangle, with each side in the triangle having a length of about 0.85 Å. Coulson (1935) used molecular orbital theory and his predicted bond length came close to the modern value of 0.87 Å. Coulson did not work out all the required integrals for the mutual repulsion of the electrons, and his work is also not free from errors; one can note in passing that the H_3^+ ion is not included in his classic book "Valence" (Coulson 1951). Hirschfelder (1938) succeeded in calculating the energy of both linear and triangular H_3^+, and found the electronic ground state to be an isosceles triangle while the electronically excited state was linear. But Hirschfelder (1938) was well aware that his calculation contained some approximations and pointed out that the real test of whether H_3^+ is an equilateral triangle or not would be an observation of its infrared spectrum. The first modern *ab initio* calculations of H_3^+ were done by Christoffersen, Hagstrom, and Prosser (1964), and by Conroy (1964), and these works established the equilateral triangular structure of H_3^+. Naturally, many theoretical papers have been published since then, e.g. Porter (1982) and Tennyson *et al.* (2000) and references therein, and spectroscopic accuracy is achieved for energy levels up to 10 000 cm^{-1}.

With no bound excited singlet electronic states (only one excited triplet state is weakly bound), infrared spectroscopy of the electronic ground state was the only viable option for observing a spectrum of H_3^+. Because of the low number density of molecular ions even in plasmas, the infrared spectrum of H_3^+ was not observed until 1980 by Oka (1980) after several years of painstaking work. The first observation of an infrared spectrum of D_3^+ was acquired almost simultaneously by means of a fast-ion-beam laser-spectroscopic method (Shy *et al.* 1980), and the two experimental papers were published back-to-back with a theoretical paper by Carney and Porter (1980). The present status of the spectroscopy of H_3^+ has been reviewed by Lindsay and McCall (2001).

During the 1980s and 1990s, H_3^+ was discovered in a variety of astrophysical objects, such as the auroral regions of Jupiter (Trafton, Lester, & Thompson 1989, Drossart, Maillard, & Caldwell 1989), dark molecular clouds (Geballe & Oka 1996), the diffuse interstellar medium (McCall *et al.* 1998), the Galactic center (Geballe *et al.* 1999), and possibly in supernova SN1987A (Miller *et al.* 1992).

In 2000, Oka organized a Royal Society Discussion Meeting in London on the "Astronomy, physics and chemistry of H_3^+" (Herbst *et al.* 2000), and another 6 years later (Oka 2006a; see also Geballe & Oka 2006). H_3^+ has also been the subject of several review articles (Oka 1983, 1992, 2006c, Dalgarno 1994, Tennyson & Miller 1994, 2000, Tennyson 1995, McNab 1995), and specific reviews of the dissociative recombination of H_3^+ have also appeared (Larsson 2000b, Oka 2000, 2003a, Wolf, Lammich, & Schmelcher 2004, Johnsen 2005); also several of the original research articles to be discussed in Section 7.2 contain reviews of the dissociative recombination of H_3^+.

One can finally note as a curiosity that in the early days H_3 was called hyzone (Wendt & Landauer 1920) in analogy with ozone. In contrast to ozone, however, the name hyzone was quickly forgotten (Larsson 2006).

7.2 The dissociative recombination of H_3^+

7.2.1 Experiment and theory until 1988

The development of the stationary afterglow technique, described in Section 2.3, naturally led to experiments with molecular hydrogen. The results emerging from these experiments concerned an unknown mixture of H_2^+, H_3^+, and possibly H_5^+, and although this is discussed in the review by Bardsley and Biondi (1970), they do not list H_3^+ as a molecular ion for which a dissociative recombination rate coefficient had been measured. The first measurement for which it is unambiguous that the recombining ion in the stationary afterglow plasma was H_3^+ was performed by Leu, Biondi, and Johnsen (1973b). They obtained a rate coefficient at 300 K of $(2.3 \pm 0.3) \times 10^{-7}$ cm^3 s^{-1}, which is similar to the rate coefficients for the atmospheric ions O_2^+, N_2^+, and NO^+. The value was an order of magnitude larger than the rate coefficient used in the modeling of the upper atmosphere of Jupiter (Gross & Rasool 1964), in which the model calculations are based on a rate coefficient estimated by Dalgarno in 1963 and communicated privately to Gross and Rasool. The measurement by Leu, Biondi, and Johnsen (1973b) was timely because it coincided with the important papers of Herbst and Klemperer (1973) and Watson (1973) in which they for the first time constructed models for the chemistry of dense interstellar clouds.

The stationary afterglow measurement was repeated 10 years later but then covered a much larger range of electron temperatures (Macdonald, Biondi, & Johnsen

1984). They found a slightly lower rate coefficient at 300 K and explained this by problems of detecting heavier-mass ions in the earlier experiment. The mass spectrometer used by Leu, Biondi, and Johnsen (1973b) was designed to detect light ions in the mass range 1–10 amu, and may have had a reduced transmission for heavier ions leading to an underestimate of the corrections needed for heavier impurity ions. This was later confirmed by Johnsen (1987), who found that CH_5^+ was a persistent impurity in the hydrogen plasma. Since CH_5^+ recombines very effectively with electrons, Johnsen (1987) concluded that the measurements of the H_3^+ rate coefficient (Leu, Biondi, & Johnsen 1973b) and its temperature dependence (Macdonald, Biondi, & Johnsen 1984) must be regarded as questionable (but as we shall see later, the stationary afterglow results must be regarded as credible).

Between the two afterglow experiments, results from inclined-beam (Peart & Dolder 1974c) and merged-beam (Auerbach *et al.* 1977, McGowan *et al.* 1979) experiments were published and found to be consistent with a rate coefficient larger than 10^{-7} cm^3 s^{-1}, as was the result of an ion-trap experiment (Mathur, Kahn, & Hasted 1978).

Kulander and Guest (1979) published the first paper that captured part of the physics involved in dissociative recombination of H_3^+. They focused on recombination at electron energies larger than 1 eV, but concluded also, based on their calculated potential energy curves, that Rydberg states in H_3 must play an important role in low-energy recombination of H_3^+.

Shortly after Macdonald, Biondi, and Johnsen (1984) published the second stationary afterglow study of H_3^+, three papers appeared which radically changed the whole scene. Adams and Smith in Birmingham used their recently developed FALP technique to measure the recombination rate coefficient for a number of ions of astrophysical importance, among them H_3^+ (Adams, Smith, & Alge 1984, Smith & Adams 1984). The H_3^+ ions were produced by Penning ionization of H_2 by helium metastables, followed by formation of H_3^+ by Reaction (7.1). The resulting rate coefficient for H_3^+ was much smaller than earlier results, and Adams, Smith, & Alge (1984) could only set an upper limit that was less than 2×10^{-8} cm^3 s^{-1}. Shortly after these results appeared, Michels and Hobbs (1984) published a theoretical study that supported the results from the Birmingham group. This paper became very influential and was sometimes said to provide a theoretical rate coefficient, which it did not. What Michels and Hobbs (1984) did was to calculate potential energy curves in C_{2v} geometry, which was basically very similar to what Kulander and Guest (1979) had done 5 years earlier. Figure 7.1 shows the calculated curves by Kulander and Guest (1979) and Michels and Hobbs (1984). The potential curves of Michels and Hobbs (1984) immediately became a very strong argument for the Birmingham group not only in support of their own result, but also in explaining why some of the earlier results were flawed. The neutral resonance state, labeled

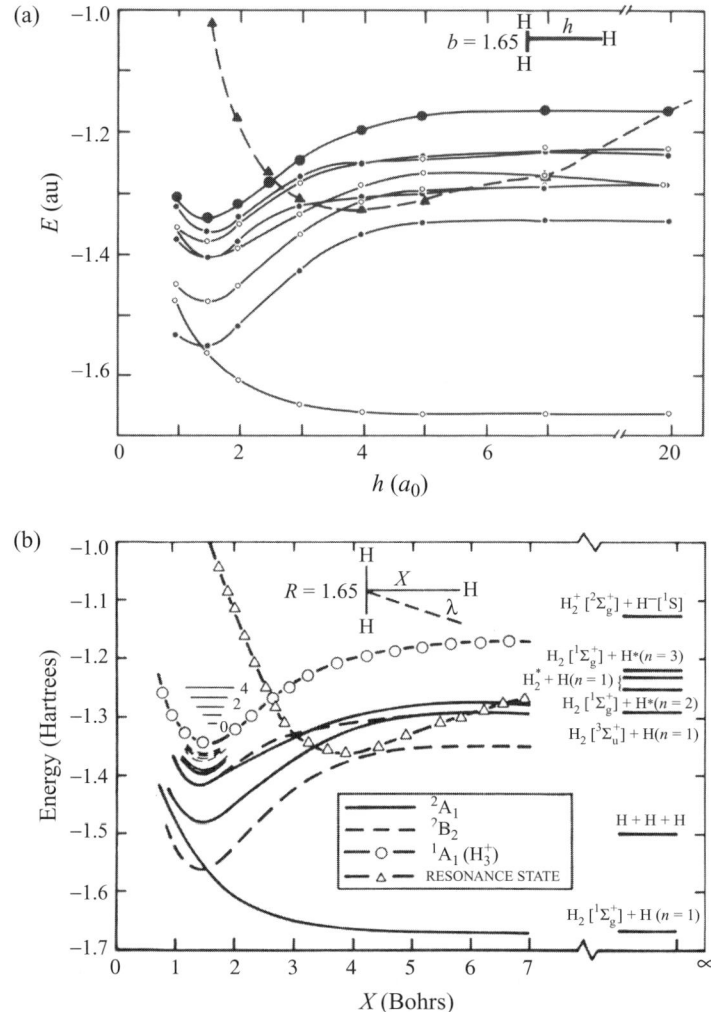

Figure 7.1 (a) Potential energy curves for H_3^+ and H_3. (Reproduced with permission from K. C. Kulander and M. F. Guest, "Excited electronic states of H_3 and their role in the dissociative recombination of H_3^+," *J. Phys. B.* **12**, pp. L501–L504 (1979), IOP Publishing Limited.) (b) Potential curves for H_3^+ and H_3. (Reproduced by permission of the AAS from H. H. Michels and R. H. Hobbs, "Low-temperature dissociative recombination of e + H_3^+," *Astrophys. J. Lett.* **286**, pp. L27–L29, (1984).)

with triangles in Fig. 7.1(b), crosses the potential curve of the H_3^+ ground state at approximately the location of $v = 3$. The vibrational overlap between $v = 0$ and the resonance state is very small. Thus, an experiment in which only $v = 0$ is populated would yield a very low recombination rate, whereas an experiment in which vibrationally excited levels are populated would yield a much larger rate. This is exactly

7.2 The dissociative recombination of H_3^+

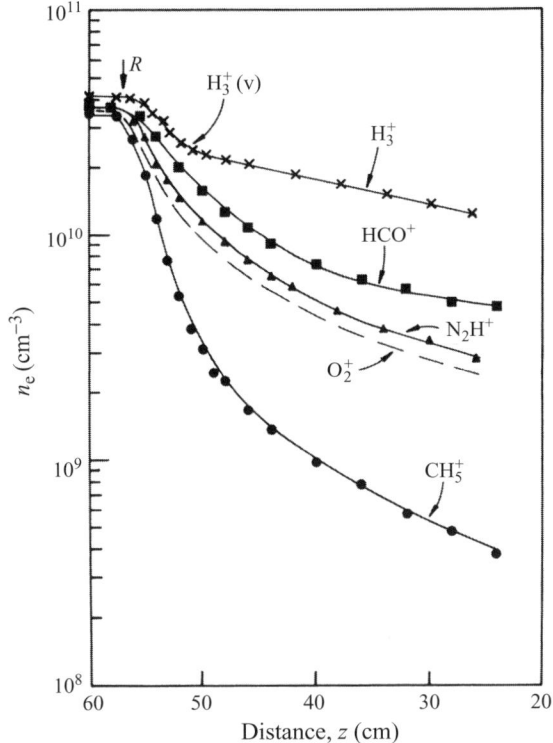

Figure 7.2 Semilogarithmic plots of electron density, n_e, in the afterglow against z, the distance along the afterglow column as measured from the downstream mass spectrometer sampling orifice in the FALP apparatus (see Fig. 2.21). (Reused with permission from N. G. Adams, D. Smith, and E. Alge, *Journal of Chemical Physics* **81**, 1778, (1984). Copyright 1984, American Institute of Physics.)

the argument put forward by Adams, Smith, and Alge (1984). Figure 7.2 shows their measured electron density as a function of distance from the point where the ions were created (see also Fig. 2.21). The electron density initially decays rapidly, and then shows a much slower decay due to ambipolar diffusion. The initial rapid decay was explained by vibrationally excited H_3^+ ions being present in the early part of the afterglow. Once they had been removed either by dissociative recombination or by collision-induced vibrational relaxation, the H_3^+ ($v = 0$) ions essentially did not contribute to the decay of the electron density, according to Adams, Smith, and Alge (1984). The high recombination rate of vibrationally excited H_3^+ was underscored by a merged-beam experiment by Mitchell *et al.* (1984) using a radio-frequency ion source in which the level of vibrational excitation was estimated to be about 65%.

The argument of the presence of vibrationally excited H_3^+ was not used to explain the discrepancy between the FALP (Adams, Smith, & Alge 1984) and the stationary afterglow experiments (Leu, Biondi, & Johnsen 1973b, Macdonald, Biondi, &

Johnsen 1984); instead it was suggested that the stationary afterglow experiments could have been contaminated by a small fraction of H_5^+ ions.

Following the initial H_3^+ publications from the Birmingham group, several papers appeared in which the upper limit was lowered to 10^{-11} cm^3 s^{-1} (Adams & Smith 1987, 1988b, 1989, Smith & Adams 1987). For some reason, these publications, which mostly appeared in conference proceedings, do not contain a detailed account of how this upper limit was obtained. The most detailed is the one published in the proceedings from the first dissociative recombination conference (Mitchell & Guberman 1989). Adams and Smith (1989) had devised a method by which they could compare the recombination rate of different ions by observing the relative mass-spectrometer signals for those ions. Comparing He^+, HeH^+, and H_3^+ they concluded (p. 134) "We have found no measurable differences between the loss of He^+, HeH^+ and H_3^+ ions for very large changes in n_e [electron density] and therefore conclude that $\alpha(He^+) \approx \alpha(HeH^+) \approx \alpha(H_3^+)$ within the limits of accuracy of measuring relative peak heights of the mass spectrometer signal." Adams and Smith (1989) concluded that H_3^+ recombines with the same rate as He^+, i.e., radiatively with a rate coefficient of 10^{-11} cm^3 s^{-1}. The same type of argument comes across in the proceedings from the International Astronomical Union Symposium on Astrochemistry in Goa, India, in December 1985. In the discussion part of the proceedings, Smith in answering the question of whether the limit of Adams and Smith is not close to the radiative recombination rate, concludes that "Our upper limit to the $H_3^+ + e^-$ recombination coefficient is 10^{-11} cm^3 s^{-1} and is indeed close to the radiative rate coefficient" (Adams & Smith 1987, p. 18). Finally, in a book chapter from 1990 they raise the upper limit to 10^{-10} cm^3 s^{-1} (Smith, Adams, & Ferguson 1990).

Further support for a low recombination rate came from a merged-beam experiment by Mitchell's group (Hus *et al.* 1988). They used an ion source in which the H_3^+ ions were confined for a sufficiently long time to undergo collisional vibrational relaxation. Measurements of dissociative excitation thresholds (see Fig. 2.11) convinced them that H_3^+ populated only the two lowest vibrational levels, and their result of 2×10^{-8} cm^3 s^{-1} (rate coefficient deduced from cross section measurement) seemed in agreement with Adams, Smith, and Alge (1984), although not with the papers emerging from Birmingham during 1987–9. Since the 10^{-11} cm^3 s^{-1} rate coefficient was inferred rather than measured, and such a small rate coefficient is well below the sensitivity level of a FALP experiment, Hus *et al.* (1988) felt that their result was in agreement with the more moderate interpretation of the FALP experiment (Adams, Smith, & Alge 1984).

As of 1988, it was argued that a low recombination rate coefficient of H_3^+ was well established, and all experiments giving a higher rate could be explained by either vibrational excitations or ion impurities. An early stationary afterglow experiment with no mass analysis (Persson & Brown 1955), which gave an upper limit of 3×10^{-8} cm^3 s^{-1}, was cited as support of a low rate (Adams & Smith 1989), although it is clear from the review by Bardsley and Biondi (1970) that H$^+$ was the principal ion in the afterglow. A rate coefficient of 10^{-10} cm^3 s^{-1} at $T = 40$ K, essentially the radiative rate coefficient at that temperature, was applied in the hugely influential paper on diffuse cloud chemistry by van Dishoeck and Black (1986).

7.2.2 Experiment and theory 1988–2000

Amano (1988) used a different technique to monitor decaying hydrogen plasma. Instead of measuring the electron density either as a function of the time after the discharge had been fired, as in the stationary afterglow technique, or as a function of distance after ion creation, as in the flowing afterglow technique, Amano measured the concentration of H$_3^+$ ions by infrared laser absorption spectroscopy. He determined the rate coefficient for removal of H$_3^+$ ($v = 0$) in the plasma to be $(1.8 \pm 0.2) \times 10^{-7}$ cm^3 s^{-1} at 210 K, a result that contradicts the FALP findings. There can be no question about the ion being removed, and there can be no question of its quantum state; Amano monitored the $J = 3$, $K = 3$ rotational level of the lowest vibrational level. One can, however, ask whether the H$_3^+$ ions were removed from the afterglow plasma solely by dissociative recombination, or whether other loss processes were also contributing. Adams and Smith (1989) argued that collisional-radiative recombination must be present in Amano's plasma because of the high ion concentration, and hence presumably high electron density. Collisional-radiative recombination is a name introduced by Bates, Kingston, and McWhirter (1962) in a paper in which they developed a statistical theory for general loss mechanisms in a plasma. The collisional-radiative recombination rate increases with the electron density, and the effect is especially marked if the electron temperature is low. In the context of Amano's experiment, collisional-radiative recombination would involve removal of H$_3^+$ by three-body H$_3^+$ + e$^-$ + e$^-$ collisions. The electron density was not given by Amano (1988), and the ion density given in Fig. 3 of Amano's paper contained a factor of 10 error (it was 3×10^{11} cm^{-3}, not 3×10^{12} cm^{-3} as given in Amano (1988); see Amano (1990)). Thus, Adams and Smith's (1989) criticism was justified based on the knowledge they had at the time. Amano (1990) returned with a longer paper in which much more detail of the experiment and data analysis was revealed, and he addressed explicitly the question of collisional-radiative recombination. Amano (1988, 1990) found a linear relation between $1/n_i(H_3^+)$ and the time after termination of discharge, and he interpreted this as evidence of the H$_3^+$ being lost by dissociative recombination. This interpretation was criticized by Bates, Guest, and Kendall (1993), who pointed out that there are circumstances in which the $1/n_i(H_3^+)$ vs. time plot can be linear even when the contribution from collisional-radiative recombination is significant. But they also estimated that the contribution from collisional-radiative recombination at 273 K (the highest temperature used by Amano (1990)) would be 1.2×10^{-8} cm^3 s^{-1}, much smaller than the rate coefficient of H_3^+ loss, which was measured to be 1.8×10^{-7} cm^3 s^{-1} at 273 K. Fehér, Rohrbacher, and Maier (1994) repeated Amano's experiment with two important amendments; they used a Langmuir probe to measure the electron density and they used a mass spectrometer to measure the ion composition in the hydrogen plasma. They found that the electron density was higher than the H_3^+ ($v = 0$) concentration, indicating the presence of vibrationally excited H_3^+ and other positive ions in the plasma, and that it decayed more slowly than the H_3^+ ($v = 0$) concentration. Using the mass spectrometer they could identify the impurity ions (apart from H^+ ions) to be H_2O^+ and H_3O^+, but they were unable to operate the mass spectrometer in a time-resolved mode in order to sample the composition of the afterglow plasma. Lacking quantitative information to perform a full analysis, and not knowing the precise conditions in Amano's experiments, Fehér, Rohrbacher, and Maier (1994) concluded that Amano's rate coefficient (Amano 1990) should be regarded as an upper limit to the dissociative recombination rate coefficient of H_3^+ ($v = 0$).

The Rennes group used their FALP apparatus with a movable mass spectrometer to study recombination of H_3^+. In a letter to the editor of *Astronomy and Astrophysics*, Canosa *et al.* (1991b) reported results in conspicuous disagreement with the FALP results from the Birmingham group (Adams & Smith 1987, 1988b, 1989, Smith & Adams 1987, Smith, Adams, & Ferguson 1990). This preliminary work was followed by a longer paper (Canosa *et al.* 1992) which gave $\alpha(650 \text{ K}) = 1.1 \times 10^{-7}$ cm^3 s^{-1} for H_3^+ ($v = 0$), in good agreement with Amano (1990). At this point it was difficult for the Birmingham group to maintain support for their extremely low rate, which is illustrated by the fact that Smith (1992) in his comprehensive review of interstellar chemistry refers only to the original article (Adams, Smith, & Alge 1984) reporting a more "moderate" rate, and by Glosík's omission of any reference to the extremely low rate despite a section about H_3^+ in his review (Glosík 1992; Smith is acknowledged for useful help with the manuscript).

The Birmingham FALP apparatus was moved to Innsbruck and used in a new series of measurements of dissociative recombination of H_3^+ (Smith & Španěl 1993a,b). This time the FALP apparatus was supplied with an improved mass spectrometer and Langmuir probe. Smith and Španěl (1993a,b) put much more experimental detail into these papers than in the previous publications reporting a rate coefficient $< 2 \times 10^{-8}$ cm^3 s^{-1} (Adams, Smith, & Alge 1984, Smith & Adams 1984), not to mention the papers reporting an immeasurably small rate (Adams & Smith 1987, 1988b, 1989, Smith & Adams 1987, Smith, Adams, & Ferguson 1990). Firstly, they acknowledged that the rate coefficient 10^{-11} cm^3 s^{-1} was incorrectly deduced from the experimental data. Secondly, they pointed out a number of circumstances that conspire to make a determination of $\alpha(H_3^+)$ difficult. Basically, Smith & Španěl (1993a,b) found the same decay curve for the electron density as shown in Fig. 7.2, but they were also able to quantify the rate coefficient, which was found to be $(1-2) \times 10^{-8}$ cm^3 s^{-1} for H_3^+ ($v = 0$) at 300 K. They also discussed previous results and concluded that the FALP results from the Rennes group (Canosa et al. 1991b,1992) concerned vibrationally excited H_3^+, Amano's result (Amano 1990) was much too high because of his assumption of an electron density equalling the H_3^+ density, and the stationary afterglow results (Macdonald, Biondi, & Johnsen 1984) could have been affected by the presence of impurity ions. Mitchell's group published a paper (Yousif et al. 1991) which claimed that their result of a few years earlier, $\alpha(100\ K) = 2 \times 10^{-8}$ cm^3 s^{-1} (Hus et al. 1988), was indeed for H_3^+ in its zeroth vibrational level. The Mitchell group also repeated their study of H_3^+ (Mitchell et al. 1993) with the radio-frequency source rebuilt to allow much higher pressures than in the earlier experiment (Mitchell et al. 1984). It was shown that the low rate of Hus et al. (1988) could be reproduced if the pressure was sufficiently high, but they also made the peculiar observation that by increasing the pressure even further, the rate increased again. They used a measurement of the ion-pair formation cross section to argue that their experiment involved ions only in the lowest vibrational level (Mitchell et al. 1993). As will become clear in Section 7.2.4, this turned out to be an incorrect assumption.

The titles of the two papers by Smith and Španěl (1993a,b), claiming "a controversy resolved" and "experiment and theory reconciled" appear a bit puzzling, in particular in view of a new experiment for which results had just been published when Smith and Španěl (1993a,b) conducted their studies. (In a brief review in 1995 Španěl and Smith admitted the discrepancy with the storage ring result.) The ion storage ring CRYRING had just been commissioned, and one of its very first experiments concerned H_3^+ (Larsson et al. 1993a). Storing H_3^+ for 10 s before the recombination experiment was started allowed the ions to relax vibrationally, and the cross section showed a direct signature of this relaxation. Figure 7.3 shows the cross section measured in CRYRING for detuning energies from 1 meV to 30 eV. Smith and Španěl (1993a,b) used the cross section in Fig. 7.3 to infer a rate coefficient of $(4-5) \times 10^{-8}$ cm^3 s^{-1}, closer to their own result than the results of Canosa et al. (1992) and Amano (1990). This estimate was made by taking the cross section at 0.04 eV and converting it to a rate coefficient by multiplication with the electron velocity. This is not the correct way of obtaining a thermal rate coefficient, and when complementary data had been taken in CRYRING, and a thermal rate coefficient extracted from the cross section, it turned out to be 1.15×10^{-7} cm^3 s^{-1}, in clear disagreement with Smith and Španěl (1993a,b). The peak at 9.5 eV in the cross section in Fig. 7.3 reflects recombination from H_3^+ ($v = 0$) through the resonance

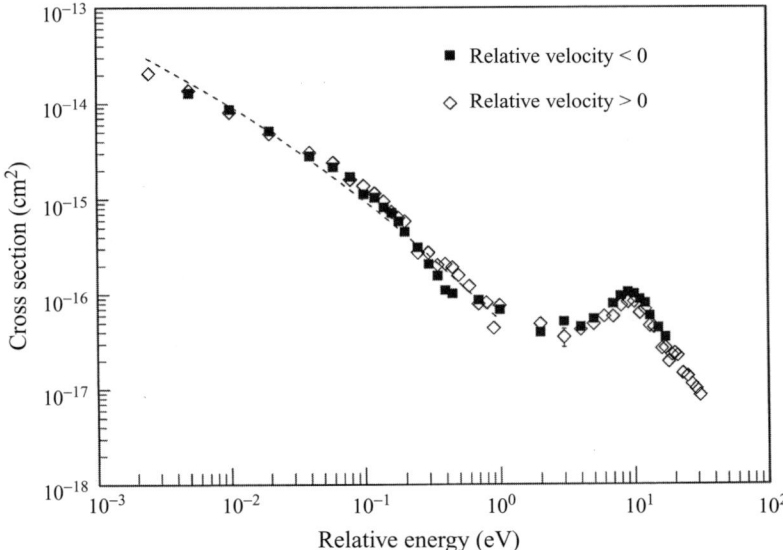

Figure 7.3 Total (effective) cross section for $H_3^+ + e \rightarrow H + H + H$ or $H + H_2$ as a function of detuning energy (labeled "Relative energy" in the figure). Diamonds display data for $v_e > v_i$ and filled squares data for $v_e < v_i$. The term "effective" in parenthesis is used to emphasize that what is plotted on the y-axis is ($<v_r\sigma>/v_d$) (see Section 2.2.1). (Reprinted figure by permission from M. Larsson, H. Danared, J. R. Mowat, "Direct high-energy neutral-channel dissociative recombination of cold H_3^+ in an ion storage ring," *Phys. Rev. Lett.* **70**, pp. 430–433, (1993). Copyright (1993) by the American Physical Society.)

state of H_3 shown in Fig. 7.1. Any contamination of vibrationally excited ions in the beam would distort the peak, which basically reflects the vibrational wavefunction of $v = 0$ (Orel & Kulander 1993). Thus, by about 1994 attemps to resolve discrepancies between results from different experiments had reached deadlock, a deadlock that still has not been completely resolved.

The Rennes group made a great effort to rebut the claim of Smith and Španěl (1993a,b) that their result (Canosa *et al.* 1992) concerned vibrationally excited H_3^+ and hence the rate was expected to be high. They used KrH^+ as the precursor ion to produce H_3^+ by the slightly endothermic reaction $KrH^+ + H_2 \rightarrow H_3^+ + Kr$, which supposedly gave H_3^+ ($v = 0$). They obtained $\alpha(H_3^+, 300 \text{ K}) = (7.8 \pm 2.3) \times 10^{-8}$ cm³ s⁻¹ (Laubé *et al.* 1998b), in disagreement with Smith and Španěl.

Gougousi, Golde, and Johnsen (1994) used the new FALP apparatus in Pittsburgh to study H_3^+. They reported a preliminary recombination rate coefficient of $(1.8 \pm 0.2) \times 10^{-7}$ cm³ s⁻¹ at 300 K; in a more detailed study (Gougousi, Johnsen, & Golde 1995) they found the electron density displayed a similar two-component decay to that in Fig. 7.2. However, their interpretation was different from that of

Adams, Smith, and Alge (1984), the reason being that they reckoned that only H_3^+ in $v = 0$ and 1 would be present in the early afterglow while H_3^+ ($v \geq 2$) would be rapidly destroyed by proton transfer to argon. Instead they attributed the initial fast decay of the electron density to three-body recombination in which the capture of an electron into an H_3 Rydberg state is stabilized by collisions with either an electron or a hydrogen molecule carrying the energy liberated by the capture. In the late afterglow, three-body recombination would not be present and H_3^+ in its lowest vibrational levels would remain unharmed by the electrons. Thus, they concluded that it is possible and even likely that binary dissociative recombination of H_3^+ is a very slow process. The H_3 Rydberg states in the early afterglow could decay by photon emission to lower Rydberg states. However, when Johnsen et al. (2000) searched for an emission spectrum of H_3, they did not find one. A different interpretation has been proposed (Johnsen 2005), based on new theoretical results; this will be discussed in Section 7.2.12. The first observation of an optical emission of H_3 was made by Herzberg (1979) using a hollow cathode discharge tube. The H_3 molecules in this experiment probably arose from dissociative recombination of H_5^+ (Adams, Smith, & Alge 1984).

The agreement with the FALP result of Smith and Španěl (1993a,b) and the single-pass merged-beam result of Mitchell's group (Hus et al. 1988) faded when new experiments were performed in which the voltage across the ion deflector (see Fig. 2.8) was changed from 3 kV cm^{-1} to 200 V cm^{-1} (Mitchell 1994, Yousif, Rogelstadt, & Mitchell 1995). In these studies it was found that the recombination signals increased by a factor of 5 when the voltage was decreased from 3 kV cm^{-1} to 200 V cm^{-1}. This was interpreted as due to formation of long-lived H_3 Rydberg states in the interaction region, which were field ionized in the ion deflector region. When the electric field was lowered, fewer H_3 Rydberg states were ionized. The result at 200 V cm^{-1} was in good agreement with the storage ring result (Larsson et al. 1993a). As an additional check, measurements were performed at 8.88 eV, where direct capture into the resonance state should dominate and the Rydberg states should play an insignificant role, and here no effect of changing the electric field was found (Yousif, Rogelstadt, & Mitchell 1995). The difficulty, however, in reconciling this result with the storage ring result was that even higher electric fields are created when the neutrals go through the dipole magnet following the interaction region in CRYRING (see Fig. 2.12). Yousif, Rogelstadt, and Mitchell (1995) proposed that the dissociative recombination rate in CRYRING is increased because of electric fields *in* the interaction region. This point will be discussed in Section 7.2.4.

In late 2000 a paper was published which once again disturbed the balance between the high- and low-rate camps, this time in the favor of the low rate. Using the AISA apparatus, briefly mentioned in Chapter 2, Glosík et al. (2000) measured a rate coefficient of less than 1.3×10^{-8} cm^3 s^{-1} at 270 K. The oldest (Persson & Brown 1955) and newest stationary afterglow results now agreed with each other, but not with the results from the 1970s (Leu, Biondi, & Johnsen 1973b) and 1980s (Mcdonald, Biondi & Johnsen 1984).

The later 1990s also saw progress in theory. Attempts had been made to formulate semiclassical theories of the dissociative recombination of H$_3^+$ (Bates, Guest, & Kendall 1993, Flannery 1995), but no full scale *ab initio* calculations of the low-energy recombination of H$_3^+$ had been performed until the work of Orel, Schneider, and Suzor-Weiner (2000) and Schneider, Orel, and Suzor-Weiner (2000), which, however, was confined to two dimensions. They obtained a result two orders of magnitude smaller than the storage ring result (Larsson *et al.* 1993a).

Thus, by the end of 2000, the situation was indeed confusing. We will now leave the chronological exposé and present results from ion storage rings and theoretical calculations in a systematic way under different headings. The reason for doing this is that the ion storage ring is the only technique for measuring dissociative recombination of H$_3^+$ that has generated invariably consistent results. This will make the presentation easier to follow, and forms the platform on which results from the other techniques can be discussed.

7.2.3 Ion storage rings: the high-energy peak

The total cross section in Fig. 7.3 peaks at 9.5 eV as a result of an electron capture into the resonance state, which then rapidly carries the system out of the autoionizing region so that the capture is stabilized. Referring to Fig. 7.1, the capture involves a vertical transition from the lowest vibrational level in H$_3^+$ to the resonance state. The resonance state dissociates diabatically to the ion-pair limit H$^-$ + H$_2^+$, but despite the considerable amount of kinetic energy that goes into the nuclear coordinates, the dissociation proceeds to neutral products in about 98% of the events (as follows by comparison with the ion-pair formation rate discussed in Section 7.2.4).

Theoretical calculations of the high-energy recombination process are simpler than those in the low-energy region, and were performed by Orel and Kulander (1993) shortly after the publication of Larsson *et al.* (1993a). In the ground state, H$_3^+$ is an equilateral triangle with D_{3h} symmetry and the electronic configuration $1a_1'^2$. The resonance state is formed by the capture of an electron into the low-lying doubly degenerate e orbital while simultaneously one of the electrons in the $1a_1'$ orbital is excited:

$$\text{H}_3^+ \left(1a_1'^2\right) + e^- \rightarrow \text{H}_3 \left(1a_1 e^2\right) \quad (7.2)$$

The double degeneracy of the e orbital results in four distinct electronic states. If the D_{3h} symmetry is relaxed to C_{2v}, so that only two bond distances are equal, the four electronic states will have the configurations $1a_1 2a_1^2$ and $1a_1 2b_2^2$ of symmetry

$^2A_1$ and $1a_1(2a_1 1b_2)^1$ and $1a_1(2a_1 1b_2)^3$ of symmetry $^2B_2$ (the superscripts 1 and 3 represent the spin coupling of the electrons in the parentheses). Michels and Hobbs (1984) calculated only the lowest $^2A_1$ dissociating to $H^- + H_2^+$, whereas Orel, Kulander, and Lengsfield (1994) calculated all four states. These calculations served as the basis for a complex Kohn electron scattering calculation of resonance energies and widths, and a wave-packet calculation of the dissociation dynamics which gave very good agreement with the measured cross section in the high-energy region (Orel & Kulander 1993). This was the first time that experimental and theoretical dissociative recombination cross sections for H_3^+ were compared, and even though it was later discovered that the theoretical result was a factor of 2 too large (Orel 2000b), the outcome was still encouraging.

7.2.4 Ion storage rings: ion-pair formation

The ion-pair formation process $H_3^+ + e^- \rightarrow H^- + H_2^+$ was first studied experimentally by Peart, Foster, and Dolder (1979) using inclined beams. They used a duoplasmatron ion source operated under such low pressure that they fully realized that their beam of H_3^+ ions could not be vibrationally relaxed. This was inferred from the measured threshold at about 3 eV as compared with the expected 5.4 eV if only $v = 0$ was present in the beam. The reason for using this ion source was that it gave a strong beam of H_3^+, which was needed to measure the small ion-pair production cross section. Their measured cross section was used by Larsson et al. (1993a) to estimate the 98% branching ratio into neutral channels mentioned in Section 7.2.3. Yousif, Van der Donk, and Mitchell (1993) repeated the ion-pair study using their single-pass merged-beam machine and a high-pressure radio-frequency ion source. They were convinced that they could produce vibrationally cold H_3^+ by operating the source at sufficiently high pressure to cause collisional deactivation of the H_3^+ ions. It is possible that they were misled by an erroneous argument as to the position of the threshold for ion-pair production for H_3^+ ($v = 1$). Inspection of the potential energy curves in Fig. 7.1(b) shows that it is certainly true that capture into the resonance state can occur at a much lower energy from $v = 1$ than from $v = 0$ if one also takes into account the shape of the vibrational wavefunctions. Thus, Yousif, Van der Donk, and Mitchell (1993) estimated the threshold for $v = 1$ to be at 3 eV. However, 3 eV is still below the energetic threshold for ion-pair formation by about 2 eV.

The third ion-pair experiment was carried out in CRYRING by Kalhori et al. (2004) using an ion source capable of producing rotationally cold ions (see Section 7.2.9). The results from this experiment together with those from previous experiments are shown in Fig. 7.4. These results are very illustrative. To begin with it is very satisfactory that these three absolute cross section measurements agree so well. The data points for the most vibrationally excited ions, those of Yousif, Van der Donk, and Mitchell when they operated their source to avoid vibrational

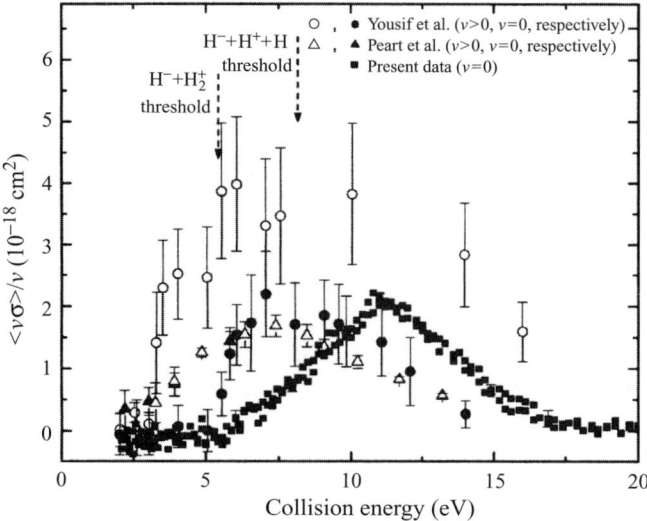

Figure 7.4 Comparison of the cross section for ion-pair formation in $H_3^+ + e$ collisions. The CRYRING data are shown as solid squares (Kalhori et al. 2004), whereas filled and open triangles are used for the data of Peart, Foster, and Dolder (1979), and the filled and open circles for the data of Yousif, Van der Donk, and Mitchell (1993). The filled triangles and circles, respectively, represent the data at high pressure conditions; however, the duoplasmatron ion source used by Peart, Foster, and Dolder (1979) could not be operated at very high pressure, and the caption up in the right corner stating $v = 0$ for the filled triangles should not be taken literally; they were aware that the source could not deliver vibrationally cold ions. (Reprinted figure by permission from S. Kalhori, R. Thomas, A. Al-Khalili, et al., "Resonant ion-pair formation in electron collisions with rovibrationally cold H_3^+," Phys. Rev. A **69**, pp. 022713-1–11, (2004). Copyright (2004) by the American Physical Society.)

de-excitations, were higher, but as soon as they tuned it into the de-excitation mode (i.e., increased H_2 pressure), the cross section dropped, and the only difference between the data sets in Fig. 7.4 is that the peak position of the cross section is shifted and the shape is modified. The peak measured in CRYRING has the shape one would expect for H_3^+ in the lowest vibrational level. The peak position is at a slightly higher energy than the peak in Fig. 7.3, but this is probably due to the facts that the early data (Larsson et al. 1993a) were not corrected for the toroidal effect, and also that the data were obtained with rotationally hot ions. The neutral channel peak was remeasured by Kalhori et al. (2004), and from Fig. 5 in that paper it follows that the peak positions almost coincide. It is obvious that the two other experiments were performed with vibrationally excited ions, as was the single-pass merged-beam experiment in the high-pressure mode. This shows how difficult it is to produce vibrationally cold H_3^+ ions from an ion source.

Theoretical wave-packet calculations of ion-pair formation were carried out in connection with the CRYRING work (Kalhori et al. 2004, Larson, Roos, & Orel 2006). The most advanced of these calculations, two-dimensional in C_{2v}, gave good agreement with experiment, but since the $H + H^+ + H^-$ channel (threshold 8.1 eV) was not included, complete agreement could not be expected. The calculations were also carried out for vibrationally excited ions and show convincingly that the differences between the data sets in Fig. 7.4 are due to vibrational excitations.

The process of ion-pair formation does not involve Rydberg states and the only effect of vibrational excitations is to shift the peak position of the cross section and modify the shape of the peak. The fact that three independent beam experiments gave basically the same result as far as the maximum value of the cross section is concerned is very satisfactory, and shows that differences in the dissociative recombination cross section measured at low energy with different beam techniques are more likely to do with the inherent properties of H_3^+ in different vibrational levels.

7.2.5 Ion storage rings: cross sections

An important question to pose is whether the cross section measured in CRYRING, shown in Fig. 7.3, could be reproduced in other storage rings. In the CRYRING experiments (Larsson et al. 1993a, Sundström et al. 1994b), no effort was made to produce rotationally cold H_3^+, and as we shall see in Section 7.2.9, there are reasons to believe that the ions were rotationally hot, maybe a few thousand kelvin, when data were taken; rotational excitations are difficult to remove even by storing for 10 s. Experiments in two other storage rings have been performed with H_3^+ ($v = 0$) with rotationally excited ions. The first was performed in the TARN II storage ring by Tanabe et al. (2000). They used a SQUID developed by themselves in order to measure the circulating current accurately, and their results were in excellent agreement with the CRYRING result, with one exception: the region around 1 eV in Fig. 7.3 has not been corrected for the toroidal effect (see Section 2.2.1), and hence the TARN II data are lower in that region. Of course, the cross section in this region has no influence on the thermal rate coefficient. Tanabe et al. (2000) did not deduce a rate coefficient from the cross section data, but from the excellent agreement with the CRYRING result, a value of 10^{-7} cm^3 s^{-1} would be expected.

Jensen et al. (2001) measured the cross section for H_3^+ ($v = 0$) and a rotational temperature estimated to be 1000–3000 K using the ASTRID storage ring. They derived a rate coefficient of $(1.0 \pm 0.2) \times 10^{-7}$ cm^3 s^{-1}, in excellent agreement with the two other ring results.

Thus, three independent measurements in three different storage rings, and with a roughly similar internal state distribution ($v = 0$ and a rotational temperature > 1000 K) gave very similar results.

7.2.6 Ion storage rings: electric field effects

It was pointed out by Kulander and Guest (1979) that dissociative recombination of H_3^+ in its lowest vibrational level with thermal energy electrons must involve Rydberg states. As was mentioned in Section 7.2.2, Yousif, Rogelstadt, and Mitchell (1995) proposed that dissociative recombination of H_3^+ in the CRYRING electron–ion interaction region is enhanced by the presence of electric fields.

The electron beam in the CRYRING cooler is guided by a magnetic field of 0.03 T. Even if the ion beam is translationally cooled, it can have a slight divergence so that the ions cross the magnetic field lines resulting in a motional $\mathbf{v} \times \mathbf{B}$ electric field. It is difficult to estimate the size of the field and it is impossible to completely avoid it. The strategy chosen at CRYRING (Larsson et al. 1997, Le Padellec, Sheehan, & Mitchell 1998) was instead to apply an electric field of 30 V cm^{-1}, which is much larger than the remnant electric field. This field is introduced by a slight misalignment of the electron and ion beams. The electron beam follows the magnetic field lines, which means that a magnetic field component perpendicular to the ion beam is introduced by the misalignment. Technically, the misalignment was achieved by means of two pairs of correction coils in the CRYRING electron cooler. These coils are normally used for aligning the electron beam with respect to the ion beam and are mounted symmetrically along the solenoid to the electron cooling region (i.e., the interaction region). This gives a magnetic field perpendicular to the ion beam that is almost constant in the interaction region, and that can be precisely controlled. The magnetic field gives rise to a motional electric field, which Larsson et al. (1997) set to be 30 V cm^{-1} in their experiment on D_3^+. D_3^+ was used to provide also a study of isotope effects, see Section 7.2.7. The addition of an electric field had no measurable effect on the cross section. The most likely explanation is that an electric field of 30 V cm^{-1} is too small to influence the dissociative recombination of D_3^+.

7.2.7 Ion storage rings: isotope effects

The isotope effect measured in ion storage rings is totally unambiguous; however, the same cannot be said for the effect measured by other techniques. In fact, different experiments led to contradictory results. We start by discussing the storage ring measurements, and then continue with measurements by other techniques.

The first isotopolog of H_3^+ to be studied in a storage ring was H_2D^+ (Datz et al. 1995a, Larsson et al. 1996); this study, performed in CRYRING, was discussed in connection with beam cooling in Section 2.2.1. H_2D^+ is the most difficult of the isotopologs to study and it shows the storage ring technique at the height of its powers. The notorious contamination of D_2^+ in the H_2D^+ beam cannot be avoided,

7.2 The dissociative recombination of H_3^+

but, as described in Section 2.2.1, the degree of impurity can be measured by using electron cooling to force the two ions to circulate with slightly different frequencies and a Schottky detector to measure the degree of impurity. But this is not enough. One must also differentiate the signal in the detector. Recombining H_2D^+ and D_2^+ give rise to a single peak in the pulse-height spectrum. In order to measure the cross section, this signal must be decomposed into the contributions from the two ions. The grid technique alone, described in Section 2.2.1, cannot accomplish this. In addition to the grid, a 50-μm-thick Cu absorber was inserted between the grid and the surface barrier detector. The stopping powers (dE/dx) of H and D are the same at identical velocities. Thus, if they pass through a very thin foil they lose the same amount of energy. After the thin foil, however, the velocities are no longer identical and if they pass through a second thin foil, they will lose different amounts of energy. For a number of thin foils, the effect propagates, and the signals for H_2D^+ and D_2^+ can be separated (see Fig. 7 in Datz *et al.* (1995a)). This detector assembly (grid + foil + surface barrier detector) was used to measure the dissociative recombination cross section, and, as will be discussed in Section 7.2.10, product branching ratios. The thermal rate coefficient at 300 K deduced from the cross section was determined to be 6×10^{-8} cm^3 s^{-1}, clearly smaller than for H_3^+. For D_3^+ the measurement did not require any of the procedures used for H_2D^+, and a result of 2.7×10^{-8} cm^3 s^{-1} was obtained (Le Padellec, Sheehan, & Mitchell 1998). Thus, it is clear that substitution of H with D reduces the rate coefficient (D_2H^+ has not been studied in CRYRING). The TARN II ring was used to study D_3^+ and D_2H^+ (Tanabe *et al.* 1996), but only relative cross sections were obtained in that study. At the TSR, extensive studies of D_2H^+ have been carried out (Lammich *et al.* 2003b, 2005), but these have been more concerned with the effect of rotational excitations (see Section 7.2.9), and no thermal rate coefficient has been published. The absolute rate coefficient measured in the electron cooler of the TSR (Lammich *et al.* 2003b, 2005) is, however, larger than would have been anticipated by a comparison with the CRYRING results for H_2D^+.

The development of a new theory for the dissociative recombination will be discussed in Section 7.2.11. Suffice it to note here that this theory (Kokoouline & Greene 2003b, 2005a,b, Greene & Kokoouline 2006) predicts with very good agreement the trend in the rate coefficient for the isotopologs H_3^+, D_3^+, and H_2D^+. This was a very important step forward in the understanding of dissociative recombination of H_3^+.

Isotope effects studied by other methods give more scattered results. Mitchell *et al.* (1984) measured the cross section for H_3^+, D_2H^+, and D_3^+ (an attempt was made also to measure H_2D^+, but the data were not published because of the D_2^+ contamination problem) and found the cross section for H_3^+ to be about a factor of 2 larger than the one for D_3^+, with that for D_2H^+ falling somewhere in between.

These experiments were done with vibrationally excited ions, and when Mitchell's group succeeded in measuring the cross section for vibrationally de-excited H_3^+ (Hus *et al.* 1988) and D_3^+ (Van der Donk, Yousif, & Mitchell 1991) they found the cross sections to be similar.

Adams, Smith, and Alge (1984) studied both H_3^+ and D_3^+ in their FALP apparatus and stated (p. 1781): "The data obtained for D_3^+ at both 95 and 300 K were insignificantly different from those of H_3^+."

Smith and Španěl (1993a) found a small difference in their late afterglow plasma between ions they labeled H_3^+ (0) and D_3^+ (0). This labeling was not intended to suggest that the ions were fully relaxed, but rather that they occupied only low vibrational levels. They found the rate coefficient for H_3^+ (0) to be $(3 \pm 1) \times 10^{-8}$ cm^3 s^{-1} and that of D_3^+ (0) to be $(2 \pm 1) \times 10^{-8}$ cm^3 s^{-1}, and concluded that $\alpha(D_3^+ (v=0))$ is probably $\leq 1 \times 10^{-8}$ cm^3 s^{-1}. In the early part of the afterglow, where the electron density decayed rapidly (see Fig. 7.2), they found that the rate coefficient for H_3^+ was 1.6 larger than that for D_3^+.

Gougousi, Johnsen, and Golde (1995) also found the recombination rate of D_3^+ to be slower than that of H_3^+ in the early part of the afterglow, where recombination (or de-ionization) occurred rapidly (i.e., corresponding to the fast decay component in Fig. 7.2); judging from the numbers they give, H_3^+ recombined a factor of 2 faster than D_3^+. As we shall see in Section 7.2.12, where a different interpretation of the early afterglow will be discussed, this is consistent with the storage ring results.

The Rennes group used their FALP-MS apparatus to study the isotope effect (Laubé *et al.* 1998b) and, surprisingly, found no effect at all. The rate coefficients measured for H_3^+ and D_3^+ were almost identical and clearly within the quoted error.

Glosík's group in Prague has also made simultaneous studies of the dissociative recombination of H_3^+ and D_3^+ (Plašil *et al.* 2002, 2003, Poterya *et al.* 2002) using the AISA apparatus. They found the binary recombination coefficient of H_3^+ to be $< 3 \times 10^{-9}$ cm^3 s^{-1} (at 260 K) while that of D_3^+ was $< 6 \times 10^{-9}$ cm^3 s^{-1} (at 230 K). However, this difference is too small to claim an inverse isotope effect.

7.2.8 Ion storage rings: vibrational excitations

The vibrational excitation of H_3^+ has consistently been discussed in the dissociative recombination papers as a key parameter, and extensive effort has been put into showing that any particular experiment was performed with H_3^+ in low vibrational levels, the zeroth vibrational level, or in a mixture of excited vibrational levels. Only two experiments discussed so far (afterglow experiments after 2000 will be discussed in Section 7.2.12) have been definitive in the identification of the vibrational level for which recombination was studied, namely those of Amano (1988, 1990) and Fehér, Rohrbacher, and Maier (1994). In experiments by means

7.2 *The dissociative recombination of* H_3^+ 203

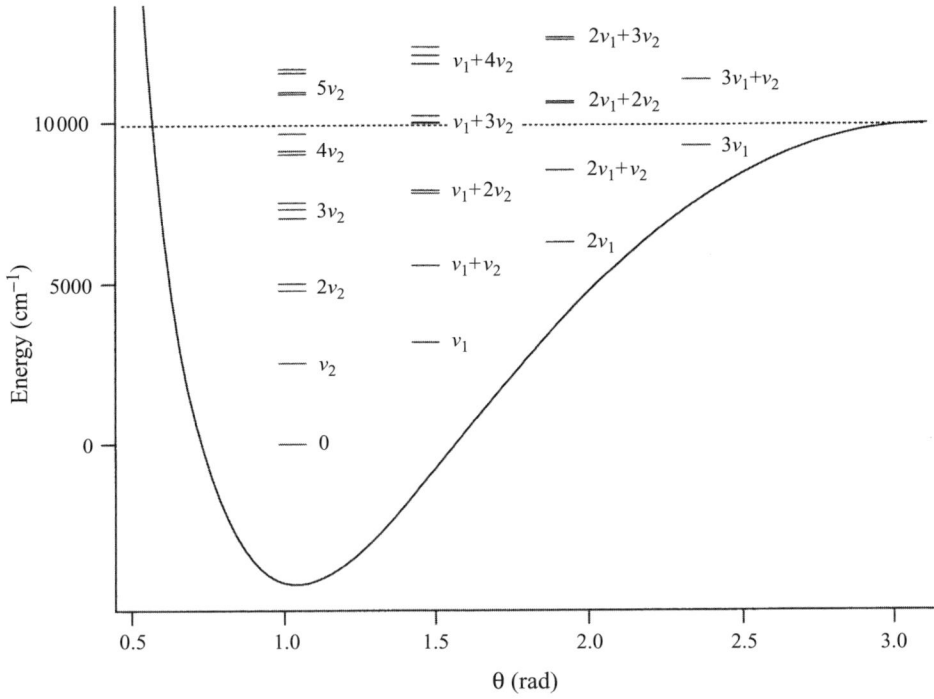

Figure 7.5 Vibrational energy level structure of H_3^+. (Reproduced with permission from McCall 2001.)

of other techniques a range of indirect evidence and arguments has been used to claim the state of excitations of the H_3^+ ions.

A very important contribution to the ion storage ring studies of dissociative recombination was made at the TSR by Kreckel *et al.* (2002). As mentioned in Section 2.2.1, part of the stored ion beam in the TSR can be extracted and sent into a beam line containing a CEI detector, which allows a reconstruction of the vibrational modes occupied by the extracted ions. Since ions are usually injected at full beam energy into the TSR, the acceleration phase that in CRYRING takes between 1 and 2 seconds is bypassed, and the time dependence of vibrational modes can be studied a few milliseconds after production in the ion source.

Before describing the experiment by Kreckel *et al.* (2002) in some more detail, it is necessary to extend the simplified picture of the vibrational levels in H_3^+ that follows from Fig. 7.1, and which has been used so far. Being a triatomic molecular ion, H_3^+ has three vibrational modes, the totally symmetric (breathing) mode v_1 and the doubly degenerate antisymmetric stretch mode v_2. Figure 7.5 shows the vibrational energy level structure of H_3^+.

In the TSR experiment (Kreckel *et al.* 2002), H_3^+ ions were produced in a gas discharge source, accelerated to 1.43 MeV and injected into the ring and stored. The ring was operated in a "slow extraction" mode, so that 10^3 ions per second were extracted from the circulating beam and sent into the CEI beam line. Three-dimensional images of the Coulomb exploding protons were taken with a multihit imaging detector in 1 ms time bins from 2–3 ms to 6 s of storage. The measured asymptotic velocity distributions gave a direct measure of the vibrational wavefunctions, and were compared with a Monte-Carlo simulation procedure fed by theoretical vibrational wavefunctions. A strong time dependence was found during the first second of storage, then after about 2 s of storage the time dependence was gone and a stable shape of proton distributions was measured. A decay component of 25 ms was identified as $v_1 = 2$ and a slower 500 ms component as $v_1 = 1$. These lifetimes are shorter than predicted theoretically for rotationless H_3^+ (Dinelli, Miller, & Tennyson 1992). By making use of a comprehensive line list of infrared transitions in H_3^+ (Neale, Miller, & Tennyson 1996) and a comprehensive list of assignments (Dinelli *et al.* 1997), Kreckel *et al.* (2002, 2004) found excellent agreement between the measured and calculated lifetime of $v_1 = 1$, and reasonable agreement for $v_1 = 2$. But in order to do this, they had to assume that H_3^+ carried 0.3 eV of rotational excitation. Moreover, their study clearly showed that the rotational excitation remained after tens of seconds of storage, and this conclusion was corroborated by model calculations.

The experiment by Kreckel *et al.* (2002) showed that H_3^+ ions that are stored for more than 2 s relax vibrationally to the zeroth vibrational level ($v_1 = v_2 = 0$). Thus, the results from ion storage rings are unambiguously valid for vibrationally relaxed ions. But their work also pointed out a potential problem with the ion storage ring experiments: rotational excitations were present, and had to be addressed seriously as a possible explanation for the conflicting results in the literature.

7.2.9 Ion storage rings: rotational excitations

The experiment by Kreckel *et al.* (2002) was not the only one for which effects of rotational excitations were found. A rotational temperature between 1000 K and 3000 K had been estimated in the ASTRID experiment (Jensen *et al.* 2001), Strasser *et al.* (2001, 2002) found evidence of rotational excitations in their studies of breakup dynamics (Section 7.2.10), Larsson *et al.* (2003) found that the recombination cross section varied depending on whether a low-pressure discharge source or a high-pressure hollow cathode source was used as the injector to CRYRING, and on how long the ions were stored in the ring, and Lammich *et al.* (2003b) in a systematic study of the recombination of D_2H^+ in the TSR noted that decreasing rotational excitation led to a decreased recombination rate. The symmetry is broken

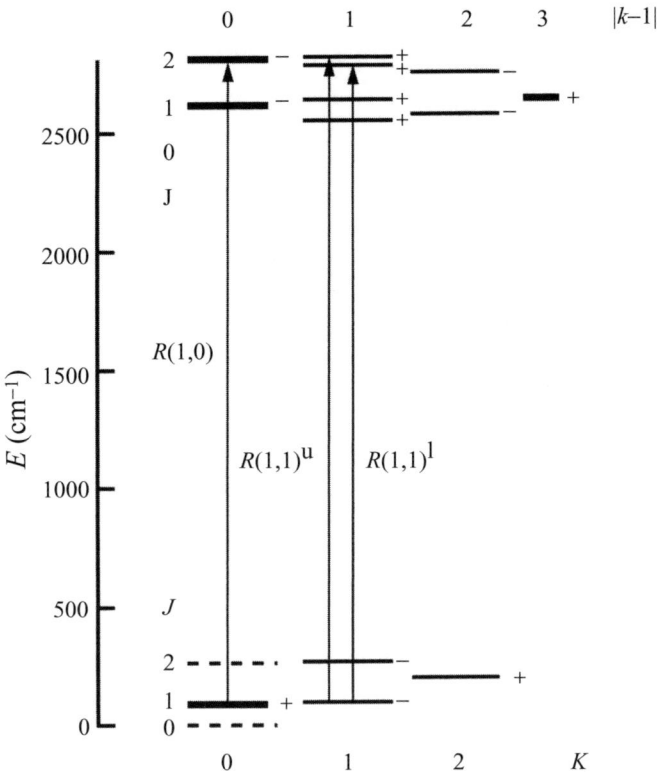

Figure 7.6 Rovibrational energy level diagram of H_3^+ showing transitions observed in interstellar clouds. The rotational levels at 2500 cm^{-1} belongs to the $v_2 = 1$ vibrational level (the superscript u stands for "upper" and l for "lower"). The $J = 0$, $K = 0$ and $J = 2$, $K = 0$ levels (dashed) are forbidden by the Pauli principle. Only $J = 1$, $K = 0$ (*ortho*) and $J = 1$, $K = 1$ (*para*) are populated in cold clouds. (Reproduced with permission from McCall 2001.)

in D_2H^+, which produces an effective dipole moment and leads to rotational relaxation ($T_{\text{rot}} < 0.05$ eV) on time scales comparable to the ion storage time (Wolf *et al.* 2004). Three different schemes were adopted to address the rotational excitation problem, one at CRYRING and two at the TSR.

Figure 7.6 is an energy level diagram of H_3^+ in which the transitions observed in interstellar clouds are shown. The very slow rotational relaxation in the lowest vibrational level can be used to advantage; if H_3^+ is produced rotationally cold and stored in a ring, the blackbody radiation from the beam tube will not heat the ions.

The approach adopted at CRYRING grew out of a collaboration between Stockholm and Berkeley. A dc discharge pinhole supersonic jet source was built at Berkeley, characterized by cavity ring-down spectroscopy, shipped to Stockholm and used as the injector to CRYRING in a preliminary experiment, sent back to

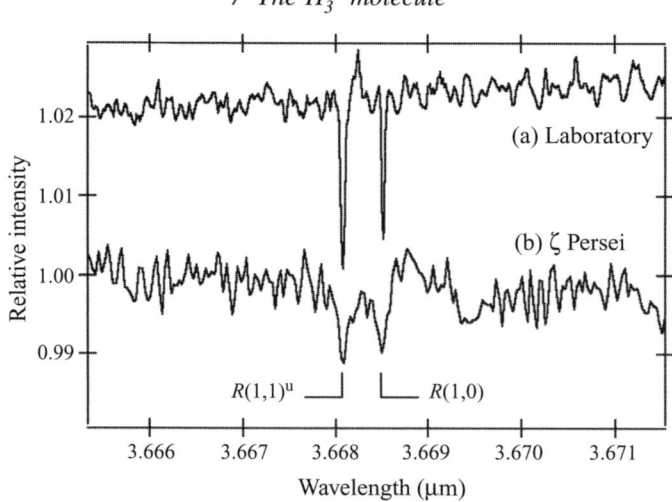

Figure 7.7 Spectra of two H_3^+ transitions arising from the two lowest rotational levels, $J = 1$, $K = 0$ and 1, which are the only levels with significant population in diffuse clouds. R(1,0) is inherently slightly stronger than $R(1,1)^u$, and the $J = 1$, $K = 0$ level is somewhat less populated than the $J = 1$, $K = 1$ level by the Boltzmann factor, so the two absorption lines appear with nearly equal strength. Top trace shows the cavity ring-down absorption spectrum of the supersonic expansion source. In this source, H_3^+ number densities of $\sim 10^{11}$ cm^{-3} were produced in the plasma downstream of a 500-µm pinhole through which hydrogen gas at 2.5 atm expanded supersonically into the vacuum. Based on relative intensity measurements of the two lines, the rotational temperature of the plasma was estimated to be 20–60 K depending on conditions. The bottom trace shows a spectrum of a diffuse cloud towards ζ Persei obtained with the CGS4 infrared spectrometer at the Infrared United Kingdom Telescope (UKIRT). (Reproduced with permission from B. J. McCall, A. J. Huneycutt, R. J. Saykally, et al. 2003.)

Berkeley for modification and further characterizations, and then finally returned to Stockholm for the experiment (McCall et al. 2003, 2004, 2005).

Figure 7.7 shows a cavity ring-down spectrum obtained with the supersonic discharge source. The rotational temperature of the plasma was 20–60 K depending on the conditions, which was sufficiently low to prevent rotational levels above $J = 1$ from being populated. No lines originating from $J = 2$ were observed. The source was not expected to produce vibrationally cold H_3^+, but this was not a prerequisite since the ions cool vibrationally in the ring. The rotational temperature in the CRYRING experiment was estimated to be about 30 K. Figure 7.8 shows the result from the CRYRING experiment in a $<v_r\sigma>$ vs. E_d plot (see Section 2.2.1). It is interesting to compare the results in Figs. 7.3 and 7.8. There are two reasons why the structures at low energy are so clearly revealed in Fig. 7.8, whereas the cross section in Fig. 7.3 is structureless. First, the cross section ($<v_r\sigma>/v_d$) was measured with a transverse electron temperature of 100 meV, whereas the new data

7.2 The dissociative recombination of H_3^+

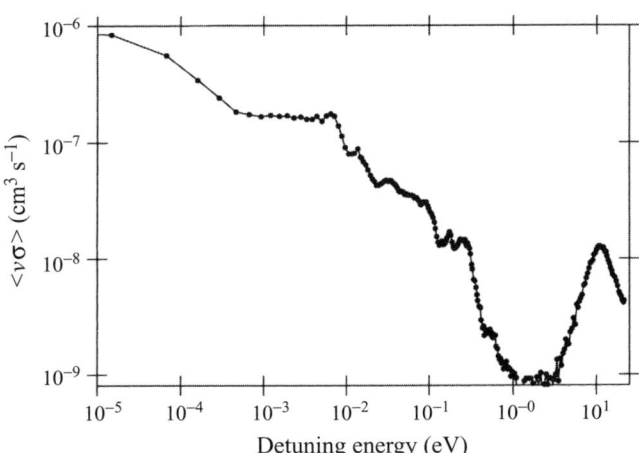

Figure 7.8 Measured dissociative recombination rate coefficient of rotationally cold H_3^+ in CRYRING as a function of detuning energy. (Reproduced with permission from B. J. McCall, A. J. Huneycutt, R. J. Saykally, et al. 2003.)

were obtained with a 2 meV electron beam. Second, the measurement using the supersonic source concerned H_3^+ occupying essentially two quantum states: $v_1 = v_2 = 0$, $J = 1$, $K = 1$ and 0.

The thermal rate coefficient deduced from the results shown in Fig. 7.8 was 6.8×10^{-8} cm^3 s^{-1} ($T_e = 300$ K; $T_{rot} \approx 30$ K). Thus, the reduction of the rotational temperature from "hot" (~several thousands kelvin (Sundström et al. 1994)) to 30 K gave a reduction in the rate coefficient of 40%.

Two schemes were used at the TSR to cool H_3^+ rotationally. The first scheme employed a hollow cathode ion source cooled to 80 K, which was combined with the use of superelastic collisions in the electron beam (see Section 2.2.1). The second was very ambitious, and involved a cryogenic 22-pole radio-frequency ion trap in which the H_3^+ ions were cooled by a 10 K He buffer gas (Kreckel et al. 2005b, Wolf et al. 2006).

Evidence of a subthermal rotational population in D_2H^+ (Lammich et al. 2003b) was used to conduct a similar study of H_3^+ (Wolf et al. 2004). Starting with the cooled hollow cathode source as the injector, H_3^+ that was distinctly colder than that used in previous TSR experiments (Strasser et al. 2001, 2002a, Kreckel et al. 2002) could be obtained, and these ions were subjected to long storage times and further rotational cooling by interaction with the electron beam. A notable difference between measured $<v_r\sigma>$ vs. E_d was noted for storage times of 5–20 s and > 30 s. The rotational temperature for storage times >30-s was estimated to be somewhere between 150 K and 600 K. The absolute cross section was not measured; instead the high-energy peak measured absolutely in CRYRING was used for calibration.

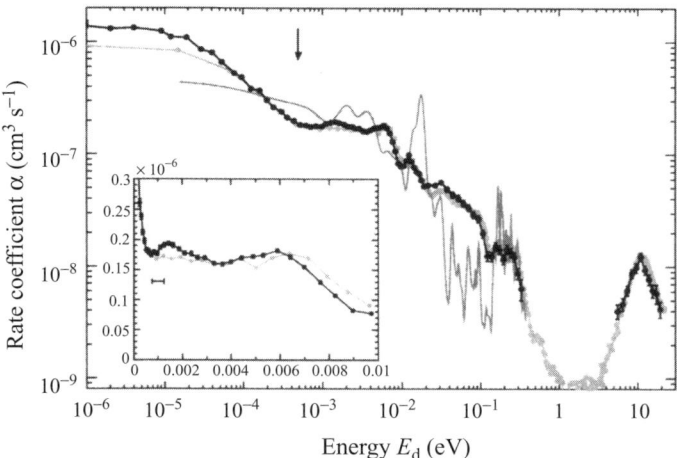

Figure 7.9 Comparison of results for dissociative recombination of H_3^+. The interconnected black dots display the results from TSR (Kreckel *et al.* 2005), the interconnected gray dots display the results from CRYRING (McCall *et al.* 2003; see also Fig. 7.8), and the continuous gray line shows the theoretical results of Kokoouline and Greene (2003a,b). The arrow and the horizontal bar in the inset mark the limiting energy spread $kT_\perp$ at TSR. The theoretical results were convoluted for $kT_\perp = 500$ μeV and $kT_\parallel = 25$ μeV. The small difference between the experimental data of CRYRING and TSR is due to the colder electron beam used in the TSR experiment. The theoretical result does not include averaging over the longtitudinal electron-velocity distribution and the toroidal correction. When this is done, the agreement with the experimental results is better (Kokoouline & Greene 2005b). (Reprinted figure by permission from H. Kreckel, M. Motsch, J. Mikosch, *et al.*, "High resolution dissociative recombination of cold H_3^+ and first evidence for nuclear spin effects," *Phys. Rev. Lett.* **95**, pp. 263201-1−4, (2005). Copyright (2005) by the American Physical Society.)

The TSR data were lower than the CRYRING data over a fairly wide range of detuning energies, ∼0.003−0.1 eV, which was somewhat surprising considering that the TSR beam should have been rotationally warmer than the CRYRING beam.

Any question related to the TSR/CRYRING data was removed when the cryogenic 22-pole radio-frequency ion trap cooled to 10 K was used as the injector to TSR, and the new 0.5 meV electron beam was used as the target (Kreckel *et al.* 2005b, Wolf *et al.* 2006). The agreement between the results from the TSR and CRYRING is absolutely perfect, as shown in Fig. 7.9. Once again normalizing to the high-energy peak, every single structure occurring in the CRYRING data set is precisely reproduced by the TSR experiment (at very low detuning energies, the colder electron beam in the TSR leads to somewhat more pronounced structures). The fact that rotationally cold H_3^+ was produced in two entirely different ion sources is very strong evidence that the experiments were performed with ions only populating the lowest rotational levels, $J = 1$, $K = 1$ and 0. The difference between the TSR data of Wolf *et al.* (2004) and those of CRYRING by McCall *et al.* (2003, 2004) and the new TSR data (Kreckel *et al.* 2005b, Wolf *et al.* 2006) is most likely caused by the difference in rotational temperature. It is easy to assume a priori that the recombination rate should scale linearly with rotational excitation, and for rotational temperatures in excess of room temperature this seems to be the case, but there is no rationale for assuming that this trend continues to low temperatures; rather, the data sets just discussed suggest that this is not the case.

Based on Amano's experiment (1990), Sundström *et al.* (1994b) assumed that the recombination rate should be essentially independent of rotational excitation, at least for $J \leq 4$. Amano (1990) used different rotational levels to probe the rate and found no rotational dependence. However, the exchange of angular momenta in the plasma conditions of the experiment of Amano was so fast so that no rotational dependence could be expected. The efforts at CRYRING and TSR have shown that even if the recombination rate has a rotational dependence, it is too small to explain the large variations in results coming from different techniques. The ion storage rings have also provided a rate coefficient for interstellar conditions.

7.2.10 Ion storage rings: product branching ratios and breakup dynamics

Dissociative recombination of H_3^+ can lead to the following neutral products:

$$H_3^+ + e^- \rightarrow H + H + H \quad (\alpha),$$
$$H_3^+ + e^- \rightarrow H + H_2 \quad (\beta), \quad (7.3)$$
$$H_3^+ + e^- \rightarrow H_3 \text{ (Rydberg)} \quad (\gamma),$$

where H_3 (Rydberg) is used to label the formation of a metastable Rydberg state that potentially can be detected in an experiment. Of course, there are no stable H_3 states.

Mitchell *et al.* (1983) used the grid technique (see Section 2.2.1) to determine the partial cross section for channels α and β; they argued that channel γ was assumed to be very slow since it would proceed through radiative recombination and it was therefore neglected. It is difficult to quantify the results of Mitchell *et al.* (1983) since they display their data in a graph in which the ratio between the cross sections for channels α and β varies quite a lot as a function of electron energy between 0.01 eV and 0.5 eV, however, the three-body channel is always larger; at 0.01 eV the branching ratios are approximately 0.85 for the three-body channel (α) and 0.15 for the two-body channel (β). They estimated that about 65% of the H_3^+ ions were vibrationally excited.

Mitchell and Yousif (1989a) repeated the experiment using the trap source which supposedly delivered vibrationally relaxed ions (Teloy & Gerlich 1974, Hus *et al.* 1988, Yousif *et al.* 1991). The current of H_3^+ ($v = 0,1$) that could be drawn from this source was only about 0.01 nA (Mitchell & Yousif 1989a), and the signal-to-noise problems that were nonnegligible in the experiment with the vibrationally excited ions (Mitchell *et al.* 1983) must have made the experiment an extremely difficult one. This time H_3 Rydberg products were observed, but the results of the experiment were not quantified. A full description of this experiment was not published in a regular publication, but some results can be found in a conference proceeding (Mitchell & Yousif 1989b) and were also given in Mitchell's review in 1990. The branching ratio was 0.52 for the three-body channel, 0.40 for the two-body channel, and, surprisingly, 0.08 (error ±47%) for the one-body channel, i.e., the channel for which metastable Rydberg states are formed (in Mitchell's experiment this amounts to lifetimes of 100 ns). The fact that H_3 Rydberg molecules were observed to survive the 100 ns transport to the detector had a considerable influence on how the dissociative recombination of H_3^+ was viewed at this time. This comes across very clearly in the papers by Gougousi, Johnsen, and Golde (1995) and Canosa *et al.* (1992), for instance.

The experiment by Datz *et al.* (1995b) addressing the same issue using CRYRING is quite remarkable in that the signal was measured against a zero background. The reason for this is that, at the high beam energy of 6.5 MeV amu^{-1} used in the experiment, the products were not stopped by the grid, as described in Section 2.2.1. Instead they passed through the grid with a consequent energy loss of 1.8 MeV per atom. In contrast to the normal use of the grid technique, where the signal is lowered by the insertion of the grid, no particles were stopped by the grid and the signal from the dissociative recombination was completely separated from the peaks derived from collisions with residual gas molecules. At 6.5 MeV amu^{-1}, electron capture from residual gas molecules is completely negligible, which means that the grid-modified pulse-height spectrum could be measured from "0" eV to 20 eV against a zero background. The three-body breakup was found to be even more dominant than in the work by Mitchell (1990a), and the branching ratio was determined to be 0.75 at low electron energy, whereas the two-body channel was found to be 0.25. The Rydberg channel observed by Mitchell and Yousif (1989a,b) was negligible at all electron energies in the CRYRING experiment.

Peterson *et al.* (1992) studied a related aspect of the dissociative recombination of H_3^+, namely predissociation of H_3 in its two lowest Rydberg states, 2s $^2A_1'$ and 2p $^2A_2''$. These states are located about 3.7 eV below H_3^+ but 1 eV above the H + H + H limit, and consequently they can predissociate to both the three- and two-body channels. The Rydberg states were formed by passing an H_3^+ beam through a Cs

7.2 The dissociative recombination of H_3^+

charge-exchange cell, and Peterson *et al.* (1992) used different ion sources in order to manipulate the internal state distributions of the H_3^+ precursor. They found that the two-body channel dominated but also that vibrationally excited precursors enhanced the three-body branching. Mitchell *et al.* (1983) using vibrationally excited H_3^+ found a larger branching ratio for the three-body recombination than the CRYRING experiment (Datz *et al.* 1995b), for which vibrationally relaxed ions were used. Experiments in the TARN II also showed that vibrational excitation in the H_3^+ led to an increase of the three-body channel (Tanabe *et al.* 1996). Returning to the discussion in Section 7.2.9, one might ask whether the branching ratios measured by Datz *et al.* (1995b) were affected by rotational excitations, which could have been as large as 0.2–0.3 eV. McCall *et al.* (2004) therefore remeasured the branching ratios using rotationally cold H_3^+. The beam energy in this experiment was lower, excluding the possibility of using the trick with the translucent grid, and the results $N_\alpha = 0.64 \pm 0.05$ and $N_\beta = 0.36 \pm 0.05$ are somewhat less accurate than the earlier results, but support the conjecture put forward by Peterson *et al.* (1992) that internal excitation increases the production of $H + H + H$. No trace of metastable Rydberg states was found despite the somewhat lower electric field when the neutrals exited the ring through the dipole magnet.

Johnsen *et al.* (2000) used laser-induced fluorescence to probe H atoms in their FALP apparatus. They found that on average 2.2 ± 0.3 H atoms are produced in each recombination event. This value is consistent with the storage ring results; it is obtained at rotationally thermalized conditions and falls between the "hot" and cold storage ring results (Datz *et al.* 1995b, McCall *et al.* 2004).

Strasser *et al.* (2002b, 2003) developed a statistical model in which dissociative recombination was treated as a two-step process, and applied it to H_3^+. They succeeded in obtaining a quite remarkable agreement with the measured results from CRYRING. Figure 7.10 shows a schematic energy level diagram for H_3^+ and the lowest asymptotic limits of H_3, and Fig. 7.11 shows the comparison of the measured branching ratios (Datz *et al.* 1995b) and the statistical model of Strasser *et al.* (2002b, 2003).

The branching ratios for recombination of H_2D^+ were measured by Datz *et al.* (1995a): this experimentally is more difficult than the corresponding measurement on H_3^+. The grid + foil technique mentioned in Section 7.2.7 was used; the combined two-body channels $H + HD$ and $D + H_2$ showed a trend similar to the one for H_3^+ depicted in Fig. 7.11. Statistically, the $H + HD$ channel should be twice as large as the $D + H_2$ channel. In the CRYRING experiment this ratio was measured to be ~ 2.5 over the entire range of electron energies (1 meV to 20 eV). Ejection of an H atom is favored as compared with ejection of a D atom in the two-body channel.

The TSR group has used the imaging technique described in Section 2.2.1 to study the breakup dynamics of H_3^+ and all its deuterated isotopologs (Strasser *et al.*

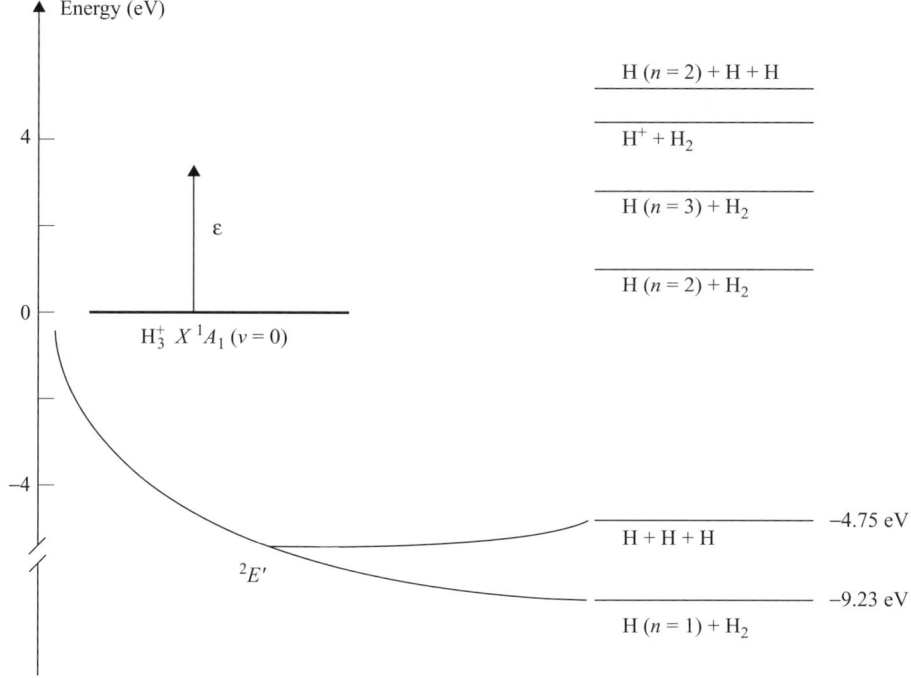

Figure 7.10 Schematic energy level diagram of H_3^+ and the lowest asymptotic limits of H_3. The $^2E'$ state is the electronic ground state of H_3. (Reprinted from *Adv. Gas Phase Ion Chem.* **4**, M. Larsson, "Merged-beams studies of electron–molecular ion interactions in ion storage rings," pp. 179–211, Copyright (2001), with permission from Elsevier.)

2001, 2002a, 2004, Zajfman, Schwalm, & Wolf 2003, Lammich *et al.* 2003a, Wolf *et al.* 2004). The energetics of the dissociative recombination of H_3^+ is:

$$\begin{aligned} H_3^+ + e^- &\rightarrow H(1s) + H(1s) + H(1s) + 4.76 \text{ eV}, \\ H_3^+ + e^- &\rightarrow H(1s) + H_2(v) + E_{\text{KER}}, \end{aligned} \quad (7.4)$$

where $E_{\text{KER}} \leq 9.23$ eV. Dalitz (1953) developed a method of analyzing data on τ-mesons decaying into three π-mesons, and this method was applied to chemical reaction dynamics in general by Strauss and Houston (1990)[1] and to the photo-induced breakup of H_3 in particular by Müller *et al.* (1999). It was applied to the three-body breakup in Reaction (7.4) by Strasser *et al.* (2001, 2002a). If ε_1, ε_2, and ε_3 are used to represent the kinetic energy of each of the three H atoms, and the individual momenta are given by $\mathbf{p}_1$, $\mathbf{p}_2$, and $\mathbf{p}_3$, the following relations hold:

$$\begin{aligned} \varepsilon_1 + \varepsilon_2 + \varepsilon_3 &= 4.76 \text{ eV}, \\ \mathbf{p}_1 + \mathbf{p}_2 + \mathbf{p}_3 &= 0. \end{aligned} \quad (7.5)$$

---

[1] Strauss and Houston (1990) do not refer to the paper by Dalitz, so it is unclear to what extent they were aware of it.

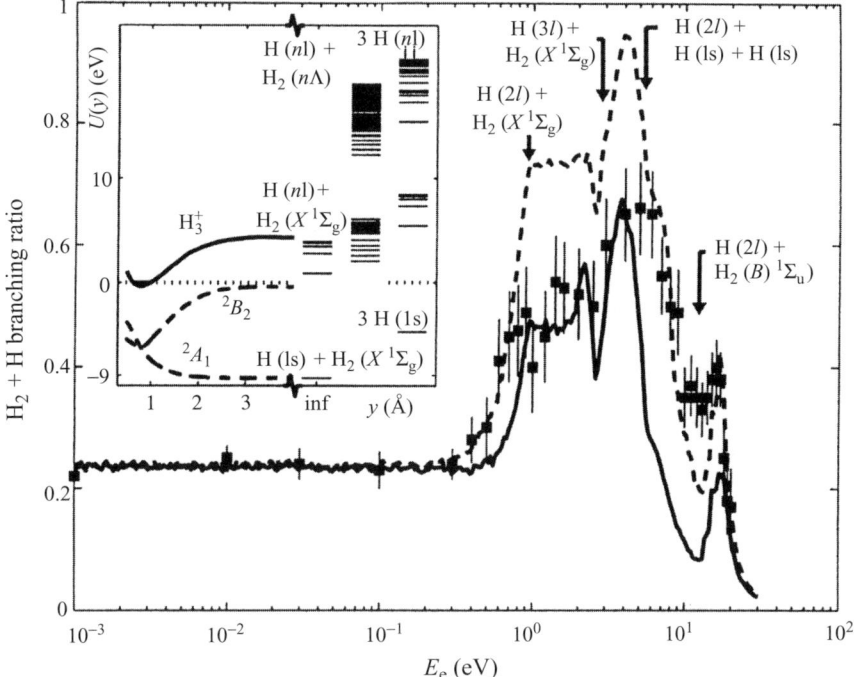

Figure 7.11 Branching ratios for the $H_3 + e^- \to H + H_2$ channel as a function of electron energy. Filled squares are the experimental data from CRYRING (Datz et al. 1995b), corrected for the toroidal effect, and the predictions by the statistical model; the dashed curve shows the results when all final states are included, whereas the solid line the results when only states with $l = 0$ are included. The arrows show the opening of channels. (Reprinted figure by permission from D. Strasser, J. Levin, H. B. Pedersen, et al., "Branching ratios in the dissociative recombination of polyatomic ions: The H_3^+ case," Phys. Rev. A **65**, pp. 010702-R1–R4, (2002). Copyright (2002) by the American Physical Society.)

Figure 7.12 shows an equilateral triangle for which the sum PL + PM + PN of the perpendiculars to the sides from any point P equals the altitude of the triangle. The perpendiculars (PL, PM, PN) are proportional to the individual H atom kinetic energies ($\varepsilon_1, \varepsilon_2, \varepsilon_3$). The Cartesian coordinates (x,y) of the point P can be expressed in the kinetic energies as

$$x = (\varepsilon_1 - \varepsilon_3)/\sqrt{3}E_{\text{KER}} \qquad y = (2\varepsilon_3 - \varepsilon_1 - \varepsilon_2)/3E_{\text{KER}}, \tag{7.6}$$

where the altitude of the triangle is chosen to be 1 unit. Momentum conservation implies that events are confined to the interior or periphery of the circle. If H_3^+ were to break up instantaneously after electron capture, with no modification of the geometric shape, each H atom would carry one third of the kinetic energy and $|\mathbf{p}_1| = |\mathbf{p}_2| = |\mathbf{p}_3|$; such an event is represented by a point at the origin in Fig. 7.12. A breakup in linear geometry leaves one H atom with no kinetic energy while the

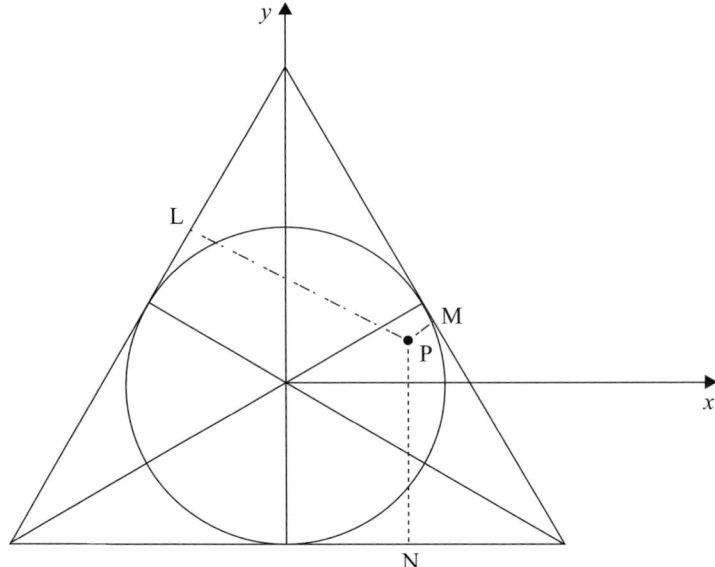

Figure 7.12 Dalitz triangle for a three-body decay such as $H_3^+ + e^- \rightarrow H + H + H$. The perpendiculars (PL, PM, PN) are proportional to the individual H-atom kinetic energies. (Reproduced with permission from Larsson and Thomas 2001.)

two others share the available energy equally. This gives events at the periphery at 30°, 150°, and 270°.

The TSR experiment measured only the positions of arrival on the imaging detector, and hence the projection of the fragment momentum triangle, but by using a Monte-Carlo reconstruction method, Strasser *et al.* (2001, 2002a) were able to obtain reconstructed Dalitz plots. They found a clustering of events at the positions that correspond to a linear breakup (30°, 150°, and 270°) for both H_3^+ and D_3^+. The studies also revealed the high rotational temperature of the precursor ions, as discussed in Section 7.2.9. Figure 7.13 shows a simulated Dalitz plot with all geometries present and potential energy surfaces calculated by Kokoouline and Greene (2003b). In the TSR experiment (Strasser *et al.* 2001), events were clustered at the crossing between the grey lines and the circle's periphery (the Dalitz plot is rotated +60° as compared with the original one, so that this occurs at 90°, 210°, and 330°).

In a later study, Strasser *et al.* (2004) studied the breakup dynamics of D_2H^+ and H_2D^+. These isotopologs also decay to the three-body channel in a linear geometry, with a propensity for the D atom to be in the middle, i.e., in the H_2D^+ case, H–D–H was observed, and in the D_2H^+ case, D–H–D was not observed. It was also found

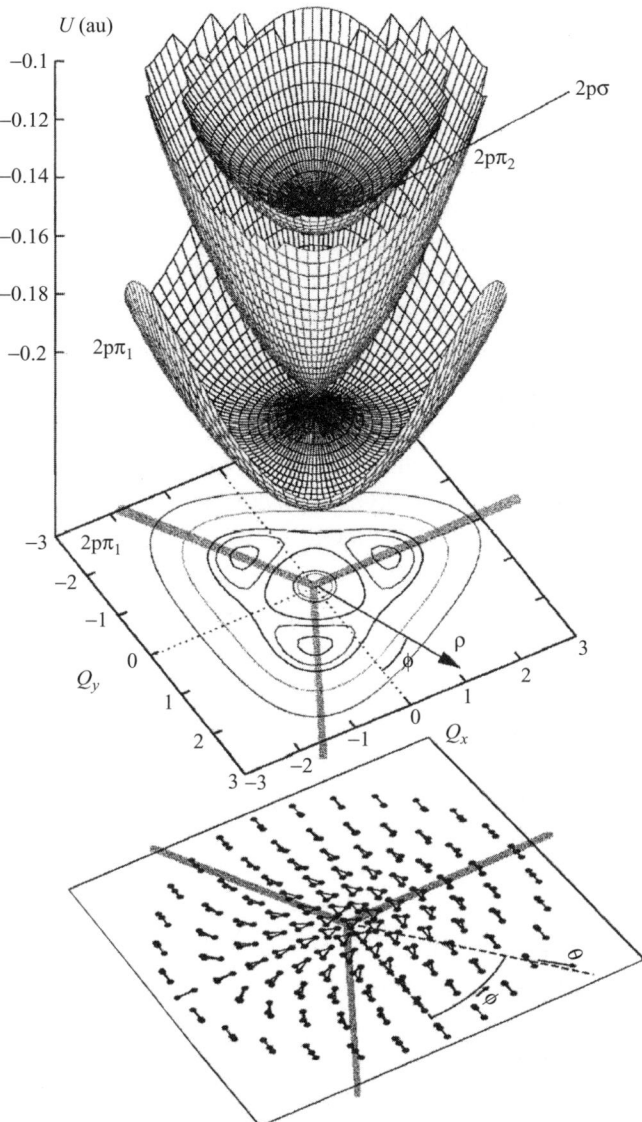

Figure 7.13 Electronic potential energy surfaces as a function of the bending coordinates Q_x, Q_y calculated by Kokoouline and Greene (2003b); the breathing coordinate (sum of internuclear distances) is set to the H_3^+ equilibrium geometry. The grey lines in the upper sheet show the routes towards dissociation in linear geometry along the $(2p\pi_1)$ electronic ground state of H_3. The relation between locations in the upper (Q_x, Q_y) plot and in the lower plot using the hyperangles θ and ϕ is only indicative. (Reproduced with permission from Wolf *et al.* 2004.)

that rotational relaxation occurred effectively, as would be anticipated for these symmetry-breaking molecules.

Müller and Cosby (1999), Müller et al. (1999), and Helm et al. (2003 and references cited therein) studied the breakup of H_3 molecules prepared by laser excitation to predissociating Rydberg states below the ionization limit. They found the Dalitz plots to be highly structured and to depend sensitively on the predissociating electronic state. In these experiments, they made use of the metastable $2pA'_2$ ($J = N = 0$) state as the initial state in the laser preparation of predissociating Rydberg states; thus, the experiments were truly single quantum state experiments. In contrast, the TSR experiments (Strasser et al. 2001, 2002a) concerned the recombination of ions populating many rotational levels. It is inconceivable that this broad rotational distribution washed out structures in the TSR Dalitz plots. Continetti's group (Laperle et al. 2004, 2005) has used a technique similar to that of Helm et al. (2003), but instead of studying the high-lying Rydberg states, they studied the three-body predissociation of low-lying ($n = 2, 3$) Rydberg states. By analyzing Dalitz plots in different kinetic energy release ranges, they found strikingly different, highly structured plots for different Rydberg states. Studies of D_3 also revealed considerable isotope effects.

The TSR group has also studied the two-body channels H_2 (v) + H and D_2 (v) + D (Strasser et al. 2001, 2002a) in the recombination of H_3^+ and D_3^+ respectively. They found broad vibrational distributions peaking at $v = 5$ and 10, respectively. This is much broader than predicted theoretically by Michels (1984), but he only included the direct mechanism and electron energies larger than 1 eV.

7.2.11 Theoretical studies

Theoretical work has been discussed briefly in previous sections; in this section we will focus on dissociative recombination with low-energy electrons. The theoretical work by Orel, Schneider, and Suzor-Weiner (2000), Schneider, Suzor-Weiner, and Orel (2000), and Schneider et al. (2000a) was clearly inspired by the success that Guberman (1994), and independently Sarpal, Tennyson, and Morgan (1994), had had in explaining dissociative recombination of HeH^+, a molecule for which the ground electronic state is crossed by *no* repulsive neutral state. Guberman showed that dissociative recombination of HeH^+ is driven by nonadiabatic coupling to repulsive HeH states lying outside the classical turning point of the HeH^+ electronic ground state. The question was whether the nuclear motion in H_3^+ could give rise to similar nonadiabatic couplings and drive dissociative recombination via predissociating Rydberg states.

Schneider and Orel (1999) started by considering the predissociation of the $2s(^2A_1')$, $3s(^2A_1')$, and $3p(^2E')$ Rydberg states to the $^2E'$ repulsive electronic ground state of H_3, for which the rates have been measured (Müller & Cosby 1996). Analytic nonadiabatic coupling elements between the $2s(^2A_1')$, $3s(^2A_1')$, and $3p(^2E')$ Rydberg states and the electronic ground state, $^2E'$, were calculated and used as the input in two-dimensional wave-packet calculations in order to obtain theoretical predissociation rates. Schneider and Orel (1999) obtained good agreement with the experimental results of Müller and Cosby (1996). This was an encouraging first step towards a calculation of the recombination cross section. The possible resemblance of the predissociation of Rydberg states below the ionization limit and dissociative recombination (above the ionization limit) was put forward in the early 1990s by Helm (1993).

Schneider, Suzor-Weiner, and Orel (2000) found that the cross section for *direct* dissociative recombination is very small by the application of wave-packet calculations, and that recombination is totally dominated by the *indirect* mechanism. This latter conclusion was arrived at by two-dimensional MQDT calculations in C_{2v} symmetry. Nevertheless, the calculated cross section was at least two orders of magnitude smaller than the one measured in CRYRING (Larsson *et al.* 1993a).

The final theoretical breakthrough came with the work of Kokoouline, Greene, and Esry (2001), in which they identified a new mechanism that had not been included in previous work: the Jahn–Teller distortion of the H_3^+ ion induced by the incoming electron. They showed that the Jahn–Teller coupling of electronic and nuclear motion is by far the most important mechanism for thermal energy electron recombination. It is interesting to note in passing that they were encouraged to include the Jahn–Teller effect by a previous study by Stephens and Greene (1995) of the photoabsorption spectrum of H_3 (Bordas, Lembo, & Helm 1991). Kokoouline, Greene, and Esry did not perform a fully quantitative calculation, but were able to establish upper and lower limits for the cross section. Their upper limit corresponded to a rate coefficient of 1.2×10^{-8} cm^3 s^{-1} at 300 K. In a "news and views" article in the same issue as the publication by Kokoouline, Greene, and Esry (2001), Suzor-Weiner and Schneider commented that concerning the effect of rotational excitations "Moreover, the effect should increase for rotationally hot H_3^+ target ions, perhaps explaining the larger value measured in storage ring experiments" (Suzor-Weiner & Schneider 2001, p. 872). At this point in time, theory definitely supported the FALP results of Smith and Španěl (1993a,b) and the new stationary afterglow results of Glosík *et al.* (2000), and it seemed that it could also make reasonable suggestions as to why the storage ring results were large. This position was further underlined by a publication in connection with the symposium "Dissociative Recombination of Molecular Ions with Electrons" in August 2001 (Greene, Kokoouline, & Esry 2003).

In order to perform a fully quantum mechanical treatment of dissociative recombination of H_3^+, Kokoouline and Greene (2003a,b) identified five problems that needed to be solved: (i) the vibrational degrees of freedom have three dimensions, not one as for diatomics; (ii) Jahn–Teller symmetry breaking effects had not earlier been included in treatment of dissociative recombination; (iii) a correct treatment of indirect recombination via intermediate Rydberg states was needed; (iv) MQDT had previously been applied only to the recombination of diatomic molecular ions, not triatomic ones; (v) rotational excitations needed to be taken into account. Kokoouline and Greene (2003a) also found a convention inconsistency that, when corrected, increased their calculated recombination rate by a factor π^2.

As a first step, Kokoouline, Greene, and Esry (2001) transformed the problem of three internuclear distances to one distance by making use of hyperspherical coordinates: $R^2 = \sqrt{3}(r_1^2 + r_2^2 + r_3^2)$, where R is the hyperspherical radius and r_1, r_2, and r_3 are the internuclear distances of each nucleus from the center of mass. By reducing the problem to one adiabatic parameter, the hyperspherical radius, the problem became amenable to MQDT, for which the adiabatic parameter is usually the internuclear distance in a diatomic molecule. The transformation to hyperspherical coordinates also yielded the qualitatively encouraging effect that the lowest vibrational level in the ion ground state, in hyperspherical coordinates, was crossed by adiabatic hyperspherical potentials of the neutral molecule. A full *ab initio* inclusion of the Jahn–Teller coupling was the next crucial element in the calculations. Kokoouline and Greene (2003a,b) managed to deal with all five problems described above, and obtained $\alpha(300\text{ K}) = (7.2 \pm 1.1) \times 10^{-8}$ cm^3 s^{-1}; the result is compared with the experimental results from CRYRING (McCall *et al.* 2003) and TSR (Kreckel *et al.* 2005b) in Fig. 7.9. For an electron temperature of 23 K, and a rotational temperature of 40 K (i.e., interstellar conditions), the theoretical value of $\alpha = 2.2 \times 10^{-7}$ cm^3 s^{-1} should be compared with the experimental value of $\alpha(T_e = 23\text{ K}, T_{\text{rot}} \approx 30\text{ K}) = 2.6 \times 10^{-7}$ cm^3 s^{-1} (McCall *et al.* 2003, 2004, 2005). One can note that the theoretical value for $\alpha(T_e = 300\text{ K}, T_{\text{rot}} = 300\text{ K})$ is only slightly larger than the theoretical value for $\alpha(T_e = 300\text{ K}, T_{\text{rot}} = 40\text{ K})$, which is very close to the experimental value 6.8×10^{-8} cm^3 s^{-1} (McCall *et al.* 2003, 2004). The calculations were performed for rotational temperatures up to 600 K, for which the cross section was found to be slightly below that of Jensen *et al.* (2001); in that experiment the rotational temperature was estimated to be in the range 1000–3000 K. It seems very plausible that the results obtained at CRYRING (Larsson *et al.* 1993a, Sundström *et al.* 1994b) and ASTRID (Jensen *et al.* 2001), giving values just above or at 10^{-7} cm^3 s^{-1}, were slightly too high because of rotational excitations. The FALP-MS result of Laubé *et al.* (1998b), for which the rotational temperature was likely to be 300 K, is in very good agreement with the theoretical result.

Three additional important points should be emphasized. The branching ratio calculated for the three-body channel by Kokoouline, Greene, and Esry (2001), 0.70

± 0.07, is in very good agreement with the CRYRING results (Datz et al. 1995b, McCall et al. 2004), the peak in the vibrational distribution of H_2 (v) was found to occur at $v = 5-6$ (Kokoouline, Greene, & Esry 2001), which agrees very well with the TSR experiment (Strasser et al. 2001), and the theoretical cross section for D_3^+ (Kokoouline & Greene 2003b) was found to be in excellent agreement with the results from CRYRING (Larsson et al. 1997, Le Padellec, Sheehan, & Mitchell 1998).

While the calculated and measured cross sections for D_3^+ agree very well, the agreement in the electron energy range 0.02–0.1 eV for H_3^+ is less satisfactory (Kokoouline & Greene 2005a, McCall et al. 2004, Wolf et al. 2004, Kreckel et al. 2005b, Wolf et al. 2006), with the experimental cross section being higher than the theoretical one. The influence of this disagreement on the deduced rate coefficients is, however, small.

Kokoouline and Greene (2005b) have also performed cross section calculations on the two C_{2v} isotopologs H_2D^+ and D_2H^+. They found that the adiabatic hyperspherical approach used for H_3^+ and D_3^+ (with D_{3h} symmetry) was not sufficiently accurate in the calculation of the vibrational energy levels and had to be improved. They obatined a very good agreement with the CRYRING results (Datz et al. 1995a, Larsson et al. 1996)[2] for electron energies below 10 meV, but also structures in the cross section above 10 meV that were not present in the experimental results. D_2H^+ posed more problems, since the theoretical result was a factor of 3–4 lower than the experimental one (Lammich et al. 2003b, 2005). This is, as yet, unexplained and should inspire further experimental and theoretical work.

7.2.12 Afterglow experiments since 2000, and discussion of the afterglow results

New afterglow experiments in Prague by Glosík's group were mentioned briefly in Sections 7.2.2 and 7.2.7 (Glosík et al. 2000, 2001a, Plašil et al. 2002, 2003, Poterya et al. 2002). The group has published further work (Pysanenko, Plašil, & Glosík 2004) on the recombination of H_3^+ using the AISA apparatus, and also results from flowing afterglow experiments (Glosík, Novotný, & Pysanentio 2003, Novotný et al. 2006) and cavity ring-down spectroscopy probing of recombining H_3^+ ($v = 0$) in an afterglow plasma (Macko et al. 2004a,b, Plašil et al. 2005). Thus, in Prague three different apparatuses have been used to study the recombination of H_3^+.

The first results from the AISA apparatus (Glosík et al. 2000, 2001a, Plašil et al. 2002, 2003, Poterya et al. 2002) came during a period when the measurement of a low rate could legitimately be claimed to agree with theory, and this was put

---

[2] The experimental data shown in Fig. 3 of Kokoouline and Greene (2005b) are correct, but the reference to Larsson et al. (1997) is incorrect since that paper concerns D_3^+.

forward repeatedly with reference to the work of Orel, Schneider, and Suzor-Weiner (2000), Schneider, Suzor-Weiner, and Orel (2000), Kokoouline, Greene, and Esry (2001), and Greene, Kokoouline, and Esry (2003). The publications in 2003 by Kokoouline and Greene (2003a,b) changed the situation radically, and it could no longer be claimed that results well below 10^{-8} cm^3 s^{-1} were supported by theory. This may have motivated the Prague group to try other methods of measuring the recombination rate of H_3^+.

The discharge vessel used in AISA is large in order to minimize diffusion, and it makes it possible to follow the afterglow plasma for a long time, of the order of 50 ms. A mixture of He/Ar/H$_2$ was used for the H_3^+ studies (He was the carrier gas, with 0.1–1.0% Ar and 0.1% H$_2$). In their first experiment on H_3^+, Glosík *et al.* (2000) found a dependence of the measured rate coefficient on the H$_2$ number densities (but not on the He or Ar densities) below $n[H_2] \approx 2 \times 10^{12}$ cm^{-3}. This was interpreted as being due to three-body collisions, in which the electron was first captured into an H$_3$ Rydberg state followed by stabilization of the capture by collision with an H$_2$ molecules. Measurements at $[H_2] = 2 \times 10^{11}$ cm^{-3} gave a recombination rate coefficient of about 1.3×10^{-8} cm^3 s^{-1} at 270 K. The excellent agreement with the FALP result of Smith and Španěl (1993a,b) was pointed out (Glosík *et al.* 2000), but this was an argument that was bound to backfire, as realized by Oka (2003a); the FALP experiment was performed at $[H_2] = 2 \times 10^{14}$ cm^{-3}. In the second series of experiments (Glosík *et al.* 2001a, Plašil *et al.* 2003), the H$_2$ number density was reduced to 5×10^{10} cm^{-3} and the recombination rate coefficient at interstellar conditions assumed to be $\ll 10^{-8}$ cm^3 s^{-1}. Other afterglow data by Amano (1990), Gougousi, Johnsen, and Golde (1995), Canosa *et al.* (1992), and Laubé *et al.* (1998b) were included in Fig. 4 of Glosík *et al.* (2001a) and Fig. 11 of Plašil *et al.* (2003), which shows the effective recombination coefficient as a function of $[H_2]$, and was explained as being due to three-body recombination. Interestingly, the data of Smith and Španěl (1993a,b), obtained at about the same H$_2$ number density as those of Gougousi, Johnsen, and Golde (1995), Canosa *et al.* (1992), and Laubé *et al.* (1998b), were not included in the plot, and the results of Smith and Španěl (1993a,b) are not discussed in papers subsequent to Glosík *et al.* (2000). As mentioned in Section 7.2.7, Glosík's group also studied the isotopolog D_3^+ (Plašil *et al.* 2002, 2003, Poterya *et al.* 2002); they measured the electron temperature dependence of the recombination rate (Pysanenko, Plašil, & Glosík 2004) and found the same strong pressure dependence at 130 K as at a higher temperature.

Glosík *et al.* (2003) used a flowing afterglow apparatus to study recombination of H_3^+ for $[H_2] = 2 \times 10^{13}$ cm^{-3} to 10^{16} cm^{-3} and found good agreement in that density range with the AISA results and hence also with other FALP experiments apart from that of Smith and Španěl (1993a,b).

7.2 The dissociative recombination of H_3^+

The third afterglow technique used in Prague (Macko et al. 2004a,b, Plašil et al. 2005) was based on the same principle as Amano's experiment (1988, 1990), but adapted to make use of the highly sensitive, cavity ring-down spectroscopy method. At a hydrogen density of $\sim 5 \times 10^{14}$ cm^{-3} and electron temperature of 330 K, they measured $\alpha(H_3^+ (v = 0)) = (1.6 \pm 0.6) \times 10^{-7}$ cm^3 s^{-1} (Macko et al. 2004a), essentially the same result as the "H_2 pressure saturated" AISA measurements (Glosík et al. 2001a, Plašil et al. 2002) and the FALP experiment (Glosík et al. 2003). A second paper (Plašil et al. 2005) reported results for a wider range of H_2 densities, and in the abstract of the paper a rate coefficient of $\alpha(H_3^+ (v=0), T_e = 330$ K$) = (0.8 \pm 0.3) \times 10^{-7}$ cm^3 s^{-1} is reported, but, curiously, it is not given or discussed in the paper itself; the scattered data points in Fig. 8 of that paper make it unclear how the data reduction was done. The ring-down spectroscopy experiments were done in a test chamber; hopefully Glosík and his group will succeed in using the cavity ring-down technique also at their AISA apparatus. Results from the three different afterglow experiments in Prague were summarized by Glosík et al. (2005).

Afterglow data from several experiments have been discussed in this section and in other sections of Chapter 7. If one adopts the view that the very good agreement between storage rings results (McCall et al. 2003, 2004, Kreckel et al. 2005b, Wolf et al. 2006) and theory (Kokoouline & Greene 2003a,b) points to a rate coefficient of $\alpha(T_e = T_{\rm rot} = 300$ K; $v_1 = v_2 = 0) = 7 \times 10^{-8}$ cm^3 s^{-1}, most afterglow results appear plausible and can be accommodated reasonably well in this framework. The early stationary afterglow results of Leu, Biondi, and Johnsen (1973b) and Macdonald, Biondi, and Johnsen (1984) are somewhat larger. Amano's spectroscopy result (Amano 1990) is also larger, but there are reasons to regard this result as an upper limit. The very first, preliminary, result from the Pittsburgh flowing afterglow apparatus (Gougousi, Golde, & Johnsen 1994) is identical to Amano's result, and the suggestion of three-body recombination put forward in the full paper (Gougousi, Johnsen, & Golde 1995) is no longer warranted. The result from the Rennes group (Laubé et al. 1998b) is in perfect agreement with the theoretical result, and two of the experiments from the Prague group also agree fairly well with the 7×10^{-8} cm^3 s^{-1}.

Two series of afterglow experiments stand out, however, as being clearly in disagreement with those using storage rings and with theory: those performed with the Birmingham FALP apparatus and those performed with the Prague AISA apparatus. The very low recombination coefficients in the range $10^{-11} - 10^{-10}$ cm^3 s^{-1} (Adams & Smith 1987, 1988b, 1989, Smith & Adams 1987, Smith, Adams, & Ferguson 1990) have been declared wrong by Smith (Smith & Španěl 1993b, p. 454): "In retrospect, and certainly in view of new measurements presented in this Letter, the latter very small value [10^{-11} cm^3 s^{-1}] of $\alpha(H_3^+)$ cannot be substantiated and must be considered wrong." The exact reasons why it must be considered wrong were not explained, and it has been up to others to speculate about them; Johnsen (2005) suggested that He$^+$ was present in the afterglow plasma. It is quite remarkable, and in hindsight difficult to understand, how the very low rate of 10^{-11} cm^3 s^{-1} could have such impact on chemical model calculations for the diffuse interstellar medium (e.g. van Dishoeck & Black (1986), Lepp & Dalgarno (1988)) when it is considered that the result was reported only briefly in conference proceedings, and that a full description of how the low rate was arrived at never appeared. A full description of how the low rate disappeared never appeared either.

Johnsen (2005), who participated in the very first mass selective afterglow experiment (Leu, Biondi, & Johnsen 1973b), and since then has both actively participated in and followed the different turns of the H_3^+ recombination rate, has reviewed the experimental work to determine this rate during the last three decades in general, and the afterglow experiments in particular. In general, but not always, afterglow experiments are performed in an H_2/He/Ar gas mixture. Penning ionization of argon by metastable helium leads to the following reactions with molecular hydrogen (Johnsen 2005):

$$Ar^+ + H_2 \rightarrow ArH^+ + H + 1.53 \text{ eV}, \tag{7.7}$$

$$ArH^+ + H_2 \rightarrow Ar + H_3^+ + 0.55 \text{ eV}. \tag{7.8}$$

The energy released refers to the production of ground-state products. The proton affinities of H_3^+ ($v_1 = v_2 = 0$) and Ar are 4.38 eV and 3.83 eV (Hunter & Lias 1998), respectively, but for H_3^+ ions in vibrational level $v_2 = 2$ or above, proton transfer to Ar occurs. In an afterglow plasma, this leaves H_3^+ ions in the ground vibrational level $v_1 = v_2 = 0$, and in the vibrationally excited levels $v_1 = 1$ and $v_2 = 1$. It should be noted that the Birmingham FALP experiments (Adams, Smith, & Alge 1984, Adams & Smith 1987, 1988b, 1989, Smith & Adams 1987) did not include argon. It has often been argued that the vibrationally excited levels below $3v_2$ in the afterglow plasma are removed by collisions, but none of the electron-density afterglow experiments have had the diagnostic tools to verify this. Laubé et al. (1998b) used both ArH$^+$ and KrH$^+$ to produce H_3^+, and since they found the rate coefficients to be independent of whether Ar or Kr was used, and since KrH$^+$ should produce H_3^+ only in its zeroth vibrational level, they inferred that their Ar data also concerned a population of only that level. This seems plausible considering the good agreement of their data with those of storage rings/theory. Johnsen (2005) has pointed out that the rate of H_2 collisional removal of vibrationally excited levels in H_3^+ is uncertain, and speculated that the slowly recombining component in Fig. 7.2 observed also by Smith & Španěl (1993a,b) and Gougousi, Johnsen, and Golde (1995), could be due to H_3^+ in the $1v_1$ level. This requires, then, that H_3^+ ($v_1 = 1$) recombines very slowly. While this does not seem very likely, it is something that could be tested by theory, and it has the advantage of explaining flowing afterglow data that are difficult to explain in other ways. Preliminary theoretical calculations by Kokoouline suggest that H_3^+ ($v_1 = 1$) recombines at a rate not very different from H_3^+ ($v_1 = 0$) (Kokoouline, private communication 2007).

The experimental results from the AISA apparatus (Glosík et al. 2000, 2001a, Plašil et al. 2002, 2003, Poterya et al. 2002) are difficult to understand, but their explanation of the onset of a saturation of the apparent rate coefficient somewhere between $[H_2] = 10^{12}$ cm^{-3} and 10^{13} cm^{-3} will not do. They conceive of a capture of an electron by H_3^+ into a long-lived Rydberg state, followed by collisional stabilization of the capture by H_2. First, if the capture takes place to a long-lived Rydberg state, the capture rate would be low, and a low apparent rate coefficient would result. Second, the collision rate at $[H_2] = 10^{13}$ cm^{-3} is only about 430 s^{-1} if the H_2 geometric cross section is assumed. Even if we assume that a Rydberg state has a cross section that is factor of 100 larger, a Rydberg state would live on the average 23 µs before a stabilizing collision occurs; this is an enormous time scale for autoionization and dissociative recombination processes. Some other explanation for the anomalous pressure dependence must be sought. Johnsen (2005) noted that a slow recombination rate for $v_1 = 1$ combined with a decreasing quenching efficiency of this level with decreasing H_2 concentration would explain Glosík et al.'s results. A theoretical calculation of the dissociative recombination cross section of H_3^+ ($v_1 = 1$) is highly desirable.

7.2.13 The rate coefficient's dependence on electron temperature

Kokouline and Greene (2003b) calculated $\alpha(H_3^+ (v_1 = v_2 = 0))$ for $T_e = T_{rot} = 0.1–600$ K, $T_{rot} = 600$ K (maximum rotational temperature in the calculations) and $T_e = 600–2000$ K, and $T_{rot} = 40$ K and $T_e = 0.1–2000$ K. Since the cross section measured at CRYRING concerned ions for which $T_{rot} \approx 30$ K, the relevant comparison is that of the experimental and theoretical rate coefficient for the low rotational temperature. The agreement was found to be very good, but with a divergence of results occurring below $T_e = 20$ K. This is related to the problem of obtaining a sufficient number of experimental data points at very low detuning energies to allow the cross section to be extrapolated to electron energies sampled at 10 K.

The theoretical data for $T_e = T_{rot} = 0.1–600$ K allow a comparison with afterglow data, some of which were obtained at variable temperature. Leu, Biondi, and Johnsen (1973b) measured the rate coefficient at 205 K, 300 K, and 450 K. If their 300 K value is scaled to 7×10^{-8} cm^3 s^{-1}, the result at 205 K is in good agreement and that at 450 K in satisfactory agreement with theory. Macdonald, Biondi, and Johnsen (1984) used microwave heating to cover the 500 K $\leq T_e \leq$ 3000 K, in addition to a measurement at 240 K, and found a T_e^{-1} variation. The agreement with theory is good in the 500 K $\leq T_e \leq$ 600 K interval if rescaling is applied. Adams, Smith, and Alge (1984) performed measurements at 95 K and 300 K, but no difference was found at these temperatures. Amano (1990) measured the rate at 273 K, 205 K, and 110 K. Rescaling his result at 273 K gives perfect agreement with theory at 205 K, but a somewhat too large value at 110 K. Since collisional-radiative recombination is known to increase with decreasing electron temperature, its relative contribution to the total rate coefficient is larger at 110 K. Smith and Španěl (1993a) studied the recombination rate at 140 K, 210 K, and 300 K and made the conservative statement that ". . . all that can be said is that $\alpha(H_3^+ (0))$ appears to increase slowly with decreasing temperature T the increase being not inconsistent with a T^{-1} variation" (Smith & Španěl 1993a, p. 173; the label "0" in the quotation means an assumed mixture of H_3^+ in $v = 0$ and 1).

7.2.14 Summary and conclusions

Almost all possible aspects of the dissociative recombination of H_3^+ have been investigated experimentally at ion storage rings, with the bulk of the effort concentrated at CRYRING and the TSR, but with important contributions also from ASTRID and TARN II. The entire database collected at these storage rings, and described in Sections 7.2.3–7.2.10, is consistent, with the exception of the cross section for D_2H^+, which requires further study. Many of the properties of the dissociative recombination of H_3^+ derived from the ion storage rings have been compared with fully *ab initio* quantal calculations, and in all aspects found to agree with the storage ring results. There is still a slight disagreement between experiment and theory in the cross section in the electron energy range 0.02−0.1 eV, but this disagreement has a marginal effect on the quantity of main interest for applications, the thermal rate coefficient. Theory predicts a difference in recombination rate between the $J = 1, K = 0$ (*ortho*) and $J = 1, K = 1$ (*para*) levels of the ground vibrational level, and this remains to be tested by ion storage ring experiments. Preliminary tests at CRYRING and the TSR suggest that it should be possible.

No other experimental technique has been able to address nearly so many aspects of dissociative recombination as the ion storage rings. A disturbing circumstance is that the combined results from other techniques, described in detail in the previous sections, do not provide a totally coherent picture. However, much progress has been made towards the understanding of different experimental results, and at this point it seems likely there are three experiments that are at variance with the ion storage ring/theoretical results, and results from other techniques. These experiments are the FALP measurement by Smith and Španěl (1993a,b) the experiments by means

Table 7.1. *Experimental and theoretical rate coefficients for* $H_3^+(v_1 = v_2 = 0)$

$\alpha(H_3^+)$ [10^{-7} cm^3 s^{-1}]	T_e [K]	T_{rot} [K]	Comment	Reference
1.15	300	~2000	ISR; estimated T_{rot}	Sundström et al. 1994b
0.68	300	~30	ISR	McCall et al. 2003, 2004
0.68	300	<20	ISR	Kreckel et al. 2005b, Wolf et al. 2006
1.00	300	1000–3000	ISR; estimated T_{rot}	Jensen et al. 2001
1.80	273	273	IR	Amano 1990
0.78	300	300	FALP-MS	Laubé et al. 1998b
1.6	330	330	IR	Macko et al. 2004b
0.8	330	330	IR	Plašil et al. 2005
0.72	300	300	Theory	Kokoouline & Greene 2003b
0.62	300	40	Theory	Kokoouline & Greene 2003b

ISR: ion storage ring; IR: infrared absorption spectroscopy; FALP-MS: flowing afterglow/Langmuir probe with mass spectrometer.

of the AISA apparatus (Glosík et al. 2000, 2001a, Plašil et al. 2002, 2003 Poterya et al. 2002), and the single-pass merged-beam experiment by Hus et al. (1988). These experiments were discussed in previous sections.

Complete listings of all dissociative recombination experiments of H_3^+ have been given by Oka (2003a) and Plašil et al. (2002), and all these experiments, apart from the very first ones by means of the non-mass-selective stationary afterglow technique, which were most certainly distorted by the presence of H_5^+, are discussed in the preceding sections. Instead of giving the results of all these experiments, we list in Table 7.1 those that were performed with H_3^+ ions that with some confidence could be regarded as vibrationally relaxed to $v_1 = v_2 = 0$.

Just as it was very satisfactory that two independent storage ring experiments were carried out with rotationally cold ions (McCall et al. 2003, 2004, Kreckel et al. 2005b, Wolf et al. 2006), and the agreement found to be perfect, it is highly desirable that a second, independent, theoretical calculation should be carried out. Work in this direction was started by Tashiro and Kato (2002a,b, 2003), but has unfortunately been discontinued. The application of the cavity ring-down infrared absorption spectroscopy technique to H_3^+ ions in the AISA apparatus is also highly desirable.

The experimental rate coefficient for interstellar conditions $\alpha(T_e = 23$ K, $T_{rot} \approx 30$ K$) = 2.6 \times 10^{-7}$ cm^3 s^{-1} (McCall et al. 2003) is in very good agreement with the theoretical value of $\alpha(T_e = 23$ K, $T_{rot} = 40$ K$) = 2.2 \times 10^{-7}$ cm^3 s^{-1} (Kokoouline

& Greene 2003a,b). With this number firmly established, the destruction of H_3^+ in diffuse clouds is three orders of magnitude faster than in the model of van Dishoeck and Black (1986). Since H_3^+ is ubiquitous in diffuse clouds, a formation mechanism that can balance the fast destruction must be formulated. McCall *et al.* (2003) suggested that the cosmic-ray ionization rate is 40 times higher in the diffuse cloud towards ζ Persei than previously assumed. More details of how this elevated cosmic ionization rate was arrived at are given by McCall (2006). Dalgarno (2006) has discussed the cosmic ionization rate in some detail, and concluded that it may fall in the narrow range 10^{-16}–10^{-15} s^{-1}, with the lower rate applicable in the dense core of molecular cloud and the higher value valid for the intercloud medium.

8
Polyatomic ions

There has been much work devoted to the study of dissociative recombination of polyatomic molecular ions, and it is not possible to describe all this work at the same level of detail as for some of the diatomic molecules or H_3^+.

8.1 Dissociation dynamics in recombination of XH_2^+ ions (X = C, N, O, S, P)

The three-body breakup dynamics of H_3^+ recombining with electrons was discussed in Section 7.2.10. Similar studies have been performed for XH_2^+ (X = C, N, O, S, P) (Datz *et al.* 2000a,b, Datz 2001, Thomas *et al.* 2002, Thomas 2004, Thomas *et al.* 2005b, Zhaunerchyk *et al.* 2005, Hellberg *et al.* 2005). These studies had their origin in the finding that the three-body breakup for these ions typically has branching ratios in the 60–80% range (Vejby-Christensen *et al.* 1997, Jensen *et al.* 1999, Vikor *et al.* 1999, Larson *et al.* 2000, Rosén *et al.* 2000). The experimental procedure for studying the dynamics of the three-body breakup is described in Section 2.2.1. Here we will choose H_2O^+ as the case study, and then make comparison with the other systems.

Figure 8.1 shows the energetics for electron recombination of H_2O^+. The $X\,^1A_1$ ground state of H_2O has the dominant configuration $(1a_1)^2(2a_1)^2(1b_2)^2(3a_1)^2(1b_1)^2$. The electron in the highest occupied molecular orbital (HOMO) is bound with 12.6 eV, and removal of this leads to the formation of H_2O^+ in its $X\,^2B_1$ ground state. Dissociative recombination with a zero eV electron gives:

$$H_2O^+ + e^- \rightarrow OH\,(X\,^3\Sigma^-) + H\,(^2S) + 7.5 \text{ eV}, \qquad (8.1a)$$

$$H_2O^+ + e^- \rightarrow O\,(^3P) + H_2\,(X\,^1\Sigma^+) + 7.6 \text{ eV}, \qquad (8.1b)$$

$$H_2O^+ + e^- \rightarrow O\,(^3P) + H\,(^2S) + H\,(^2S) + 3.04 \text{ eV}, \qquad (8.1c)$$

$$H_2O^+ + e^- \rightarrow O\,(^1D) + H\,(^2S) + H\,(^2S) + 1.07 \text{ eV}, \qquad (8.1d)$$

where the energies in (8.1a) and (8.1b) are given for H_2O^+, OH, and H_2 in their zeroth vibrational levels and with negligible rotational energy.

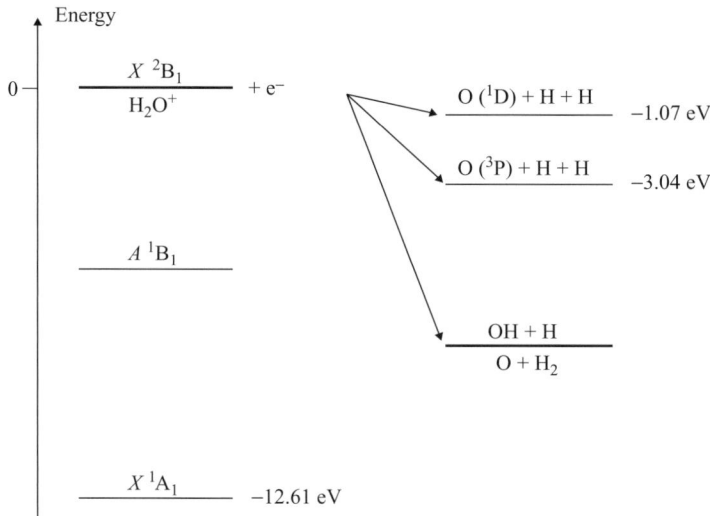

Figure 8.1 Energetics for H_2O^+ and H_2O. The zero energy is chosen to coincide with the zeroth vibrational level of H_2O^+. (Reproduced with permission from Larsson and Thomas 2001.)

The branching ratios for channel (8.1a), channel (8.1b), and channels (8.1c), (8.1d) have been measured (Rosén *et al.* 2000) and found to be 0.20, 0.09, and 0.71, respectively. This experiment was unable to distinguish between channels (8.1c) and (8.1d). The pertinent questions for the three-body breakup are:

- What is the branching ratio for (8.1c) and (8.1d)?
- How is the kinetic energy shared between the two H atoms?
- What is the H–O–H angle when the molecule dissociates?
- What is the time sequence of the breakup?

Datz *et al.* (2000a) did preliminary experiments at CRYRING using the particle imaging detector shown in Fig. 2.15, and these were followed by more extensive experiments and a more developed data reduction scheme (Datz *et al.* 2000b). Momentum conservation in the three-body breakup implies that the sum of the individual momenta is equal to zero:

$$\mathbf{u}_O + \mathbf{u}_{H1} + \mathbf{u}_{H2} = 0, \qquad (8.2)$$

where $\mathbf{u}_O$ is the momentum of the O atom, and $\mathbf{u}_{Hi}$ ($i = 1,2$) are the momenta of the H atoms in the center-of-mass frame. If the two bonds in the H_2O molecule are broken simultaneously, the absolute value of the momenta of the H atoms is equal:

$$|\mathbf{u}_{H1}| = |\mathbf{u}_{H2}| = p. \qquad (8.3)$$

8.1 Dissociation dynamics in recombination of XH_2^+ ions

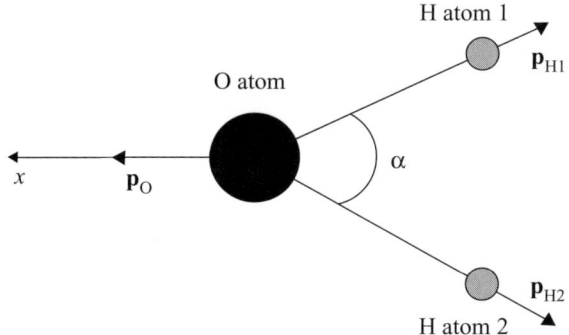

Figure 8.2 Momentum distribution in dissociative recombination of H_2O^+. The electron has been captured and in the highly excited H_2O molecule the two O–H bonds break simultaneously, which render the two H-atoms to carry the same amount of momentum. The angle α is the bond angle when the two H-atoms are released (redrawn with permission from Larsson and Thomas (2001)).

This is the situation shown in Figure 8.2. If E_{KER} is the total kinetic energy available to the three atoms, the kinetic energies of the individual atoms are related to E_{KER} through

$$E_{K,O} + E_{K,H1} + E_{K,H2} = E_{KER}, \qquad (8.4)$$

where $E_{K,X}$ (X = O, H1, H2) is the kinetic energy for each atom.

Let us consider some numbers based on the CRYRING experiment on H_2O^+. If a recombination event leads to the (8.1c) channel, about 3 eV of kinetic energy is available to the three atoms. If the H_2O^+ beam energy is 4.5 MeV, the detector is located 6 m from the interaction region, and the three-body breakup occurs parallel to the detector plane (i.e., perpendicular to the ion-beam axis), the two H atoms each move about 11 mm from the x-axis whereas the O atom will move only about 1 mm from the center-of-mass position. This explains why the foil technique described in Section 2.2.1 works so well. The O atoms are concentrated in the narrow zone on the detector where the differential stopping foil is located. Figure 8.3 shows a three-atom hit on the detector. If we assume that the center-of-mass coincides with the O atom, which is quite a good approximation, the kinetic energy release, E_{KER}, is directly related to $(R_1^2 + R_2^2)^{1/2}$, where R_1 and R_2 are defined in Fig. 8.3.

The timing system can be used to select events in which the atoms are approximately in the detector plane. This is done by imposing a time window on the events, so that atoms that arrive within less than 0.8 ns are registered. Figure 8.4 shows the number of events within a time window of 0.8 ns as a function of $(R_1^2 + R_2^2)^{1/2}$. The two peaks correspond to recombination into channels (8.1c) and (8.1d), respectively, and an analysis of the peak areas gives an $O(^3P)$ to $O(^1D)$ ratio of (3.5 ± 1) to 1. This answers the first of the four questions posed in the beginning of this section.

230   8 *Polyatomic ions*

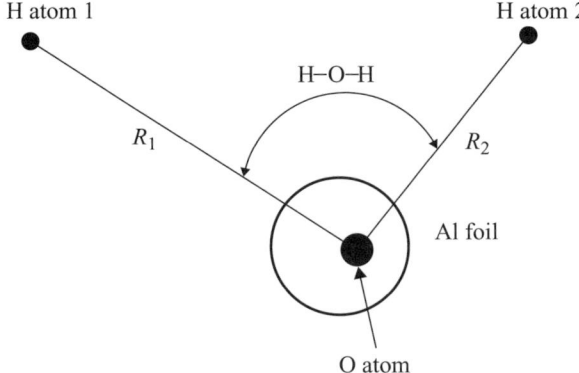

Figure 8.3 A three-particle hit on the imaging detector following dissociative recombination of H_2O^+. The diagram is simplified since the particles should be referred to the position of the center-of-mass. The large mass difference between the O- and H-atoms puts the center-of-mass close to the O-atom (redrawn with permission from Larsson and Thomas (2001)).

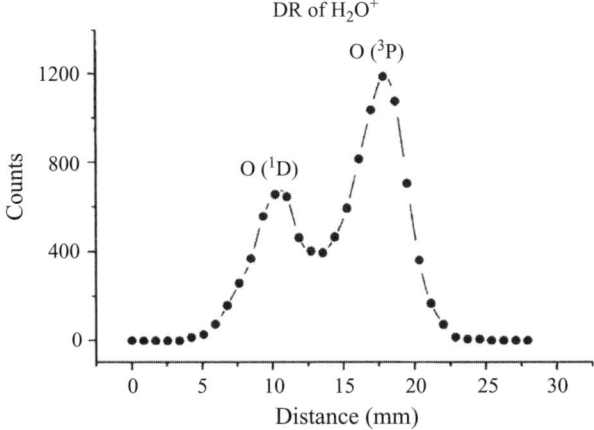

Figure 8.4 Number of events in dissociative recombination of H_2O^+ as a function of total recoil distance, $(R_1^2 + R_2^2)^{1/2}$, where R_1 and R_2 are defined in Fig. 8.3. (Reprinted figure by permission from S. Datz, R. Thomas, S. Rosén, *et al.*, "Dynamics of three-body breakup in dissociative recombination: H_2O^+," *Phys. Rev. Lett.* **85**, pp. 5555–5558, (2000). Copyright (2000) by the American Physical Society.)

The analytic representation of the breakup of a diatomic molecule cannot be extended to three-body breakup, and other methods must be used to extract physically meaningful information from the data. In their study of the photo-induced three-body breakup of H_3, Müller and Cosby (1999) performed a Monte-Carlo simulation of the detector response in order to obtain kinematic information.

8.1 Dissociation dynamics in recombination of XH_2^+ ions

A similar approach was used by Thomas *et al.* (2002) in the analysis of dissociative recombination of H_2O^+. We will describe this procedure here.

The three-body breakup of a triatomic molecule is governed by momentum conservation in the laboratory frame:

$$\sum_{i=1}^{3} m_i \mathbf{V}_i = M \mathbf{V}_0, \qquad (8.5)$$

where m_i are the fragment masses, M is the molecular mass, $\mathbf{V}_i$ are the fragment velocity vectors, and $\mathbf{V}_0$ is the molecular velocity vector. In the center-of-mass frame this becomes

$$\sum_{i=1}^{3} \mathbf{u}_i = \mathbf{0}, \qquad (8.6)$$

where $\mathbf{u}_i = m_i \mathbf{v}_i$ and $\mathbf{v}_i = \mathbf{V}_i - \mathbf{V}_0$. The three vectors $\mathbf{u}_i$ are expressed in terms of nine components, which can be reduced to six by means of Eq. (8.6), thus defining a plane in space-fixed reference frame (x, y, z). This means that it is always possible to find a coordinate system (x', y', z') for which the three vectors lie in the (x', y') plane. In this plane the three momentum vectors are labeled $\mathbf{u}'_i$. Thomas *et al.* (2002) introduced two independent parameters into the Monte-Carlo simulation. The first parameter χ is the angle between two of the momentum vectors in the (x', y', z') frame. In the case of the XH_2^+ ions, the angle was chosen to be the angle between the momentum vectors of the two H atoms. The second parameter, ρ, is defined as

$$\rho = \frac{\mathbf{v}_2^2}{\mathbf{v}_1^2} \quad (0 \le \rho \le 1), \qquad (8.7)$$

where $\mathbf{v}_1$ and $\mathbf{v}_2$ are the velocities of the two H atoms, with $\mathbf{v}_1 > \mathbf{v}_2$. Since $m_1 = m_2 = m_H$, $\mathbf{u}_1/\mathbf{u}_2 = \mathbf{v}_1/\mathbf{v}_2$. The fragment momentum vectors in the (x', y', z') coordinate system can be expressed in terms of $\mathbf{u}_1$, ρ, and χ by

$$\begin{aligned}
\mathbf{u}'_1 &= (|\mathbf{u}_1|, 0, 0), \\
\mathbf{u}'_2 &= (|\mathbf{u}_1| \rho^{1/2} \cos \chi, |\mathbf{u}_1| \rho^{1/2} \sin \chi, 0), \\
\mathbf{u}'_3 &= (-|\mathbf{u}_1|(1 + \cos \chi), -|\mathbf{u}_1| \rho^{1/2} \sin \chi, 0),
\end{aligned} \qquad (8.8)$$

where the absolute value of $\mathbf{u}_1$ can be expressed in terms of E_{KER} and the parameters ρ and χ as

$$|\mathbf{u}_1| = \left[\frac{2 m_1 E_{KER}}{(1 + \rho) + (m_1/m_3)(1 + \rho + 2 \rho^{1/2} \cos \chi)} \right]. \qquad (8.9)$$

The coordinate system (x', y', z'), which is the molecule-fixed frame, is related to the space-fixed frame (x, y, z) by the Euler angles, as illustrated in Fig. 8.5.

The space-fixed frame is oriented such that the ion beam propagates along the x-axis and the imaging detector lies in the (y, z) plane. The $\mathbf{u}'_1$ vector is oriented along

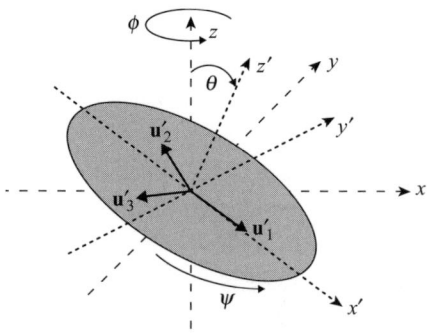

Figure 8.5 Euler angles θ, ϕ, and ψ relating the space-fixed (x, y, z) and molecule-fixed (x', y', z') frames. (Reprinted figure by permission from R. Thomas, S. Rosén, F. Hellberg, *et al.*, "Investigating the three-body fragmentation dynamics of water via dissociative recombination and theoretical modeling calculations," *Phys. Rev. A* **66**, pp. 032715-1–16, (2002). Copyright (2002) by the American Physical Society.)

the x'-axis. The Euler angles θ and ϕ are the polar and azimuthal angles in spherical polar coordinates with respect to the space-fixed frame, while ψ is the azimuthal angle in the molecule-fixed frame. Thus, ϕ and ψ run from 0 to 2π, and θ runs from π to $-\pi$. The volume element in spherical coordinates is $dV \equiv r^2\, dr\, d(\cos\theta)\, d\phi$, which implies that in the Monte-Carlo simulation, ϕ and ψ are random between 0 and 2π, and $\cos\theta$ is random between 1 and -1.

The second question, i.e., how the kinetic energy is shared between the hydrogen atoms, can now be addressed. The parameter ρ represents how the kinetic energy is shared, and Fig. 8.6 shows a comparison of the experimental data and three different Monte Carlo simulations with the parameter ρ set to 1.0 (equal sharing of kinetic energy for all events), to 0.1 (one H atom getting 90% of the kinetic energy and the other one 10%), and with ρ being a random number between 0 and 1 (denoted $\rho = c$ in Fig. 8.6). For those events that predominantly dissociate parallel to the detector it is possible to directly extract the energy sharing information from these experimental data. This is done by using the data for which the total recoil distance is large; for the channel (8.1c) this is straightforward, whereas for channel (8.1d) the data are overlapped by those from the (8.1c) channel. Thus, we consider only the (8.1d) channel. Figure 8.6 shows that the experimental data are very well reproduced by the Monte-Carlo simulation with a random ρ between 0 and 1. The answer to the second question is thus that the kinetic energy is randomly shared between the two hydrogen atoms.

The third question concerns the H−O−H angle when the molecule breaks up, or, to be more precise, the angle H−center-of-mass−H. In Fig. 8.3 the center-of-mass is assumed to coincide with the O atom, which is a good approximation. The analysis in Thomas *et al.* (2002), however, uses the center-of-mass rather than the

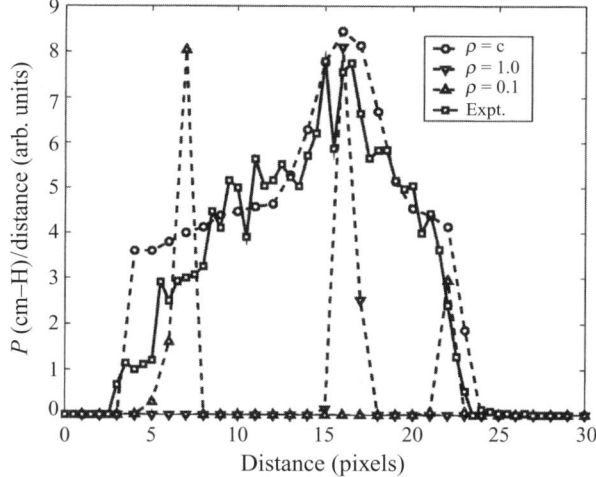

Figure 8.6 Energy distribution for the two hydrogen atoms in dissociative recombination of H_2O^+. The x-axis is the total recoil distance $(R_1^2 + R_2^2)^{1/2}$ and the y-axis gives the number of events divided by $(R_1^2 + R_2^2)^{1/2}$. Experimental data ($\square$, solid line) compared with Monte Carlo simulated data (dotted lines) for ρ = random (o), $\rho = 0.1$ ($\triangle$), and $\rho = 1.0$ (∇). The experimental data have been selected to include mainly those events that involve dissociation parallel to the detector plane. (Reprinted figure by permission from R. Thomas, S. Rosén, F. Hellberg, et al., "Investigating the three-body fragmentation dynamics of water via dissociative recombination and theoretical modeling calculations," Phys. Rev. A **66**, pp. 032715-1–16, (2002). Copyright (2002) by the American Physical Society.)

O atom. Once again selecting the events with large total recoil distances and by means of Monte-Carlo simulations, the distribution of the angle χ can be retrieved, as shown in Fig. 8.7. The bond angle in the H_2O^+ ion is about 109° (Zajfman et al. 1991). This means that there is a large change towards either smaller or larger angles before the H_2O molecule breaks up.

The time sequence of the breakup is related to whether it is concerted or sequential. In their review of photo-induced three-body decay, Maul and Gericke (1997) made a further classification of the concerted mechanism into synchronous and asynchronous concerted. Table 8.1 summarizes Maul and Gericke's classification.

It is clear that dissociative recombination of H_2O^+ is not a synchronous concerted process, or else the experimental data in Fig. 8.6 would agree with the $\rho = 1.0$ simulation. It is more difficult to distinguish between sequential and asynchronous concerted breakup. Provided only a small amount of energy went into internal excitations of OH in Reaction (8.1a), the first H atom would carry the main part of the kinetic energy, and the data would resemble the $\rho = 0.1$ simulation in Fig. 8.6. This is clearly not the case. The problem, however, is that we do not know whether the first H atom release leads to internal excitation of OH, which would reduce the kinetic energy of the H atom released first. It is interesting to compare photodissociation of

Table 8.1. *Classification of three-body decay mechanisms as given by Maul and Gericke (1997). Δt designates the time difference between the two bond cleavages, and $\tau_{\rm rot}$ the mean rotational period of the intermediate complex.*

Time difference	Three-body decay process
$\Delta t/\tau_{\rm rot} > 1$	Sequential
$\Delta t/\tau_{\rm rot} < 1$	Concerted
$\Delta t/\tau_{\rm rot} = 0$	Synchronous concerted
$0 < \Delta t/\tau_{\rm rot} < 1$	Asynchronous concerted

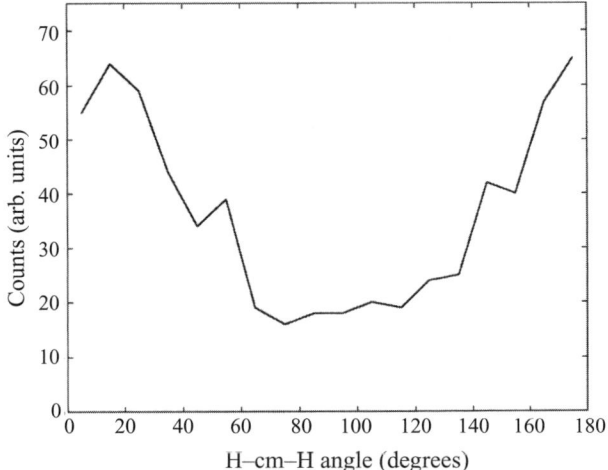

Figure 8.7 Measured H−center-of-mass−H angles, binned in 10° steps, for fragments from events that have dissociated parallel to the detector. (Reprinted figure by permission from R. Thomas, S. Rosén, F. Hellberg, *et al.*, "Investigating the three-body fragmentation dynamics of water via dissociative recombination and theoretical modeling calculations," *Phys. Rev. A* **66**, pp. 032715−1−16, (2002). Copyright (2002) by the American Physical Society.)

H_2O with dissociative recombination of H_2O. Slanger and Black (1982) photodissociated water using the Lyman-α line at 121.6 nm and measured the branching ratio of the three-body channel (8.1c) to be 12%. Mordaunt, Ashfold, and Dixon (1994), also using the 121.6 nm line, found the three-body breakup to be synchronous concerted. Photoexcitation at 121.6 nm, i.e., 10.2 eV, of water in its ground state brings the molecule to only 2.4 eV from its ionization limit (see Fig. 8.1). This energy difference has a large effect on the three-body breakup: from 12% to more than 60%. Mordaunt, Ashfold, and Dixon (1994) found that channel (8.1a) gives vibrationally fairly cold OH, but with a highly inverted rotational state distribution. The OH radical can be formed either in its ground state $X\,^2\Pi$, or in the excited

$A\,^2\Sigma^+$ state. The latter state predissociates even in its lowest vibrational level for sufficiently high rotational levels, and Mordaunt, Ashfold, and Dixon (1994), using wave-packet calculations, found that the three-body channel gained in importance when the wavelength was decreased. Dixon et al. (1999) and Harich et al. (2000) extended the work by Mordaunt, Ashfold, and Dixon (1994) with time-of-flight detection of photodissociated H atoms and quasiclassical trajectory calculations. Figure 8.8 shows a qualitative view of Lyman-α photodissociation of water. The initial excitation of H_2O at 10.2 eV (121.6 nm) is to the third singlet electronic state $\tilde{B}\,^1A_1$, which correlates adiabatically with H + OH($A\,^2\Sigma^+$). The dominant channel, however, is H + OH ($X\,^2\Pi$), which implies that the dissociation does not occur adiabatically. Nonadiabatic crossings from the $\tilde{B}\,^1A_1$ state to the $\tilde{A}\,^1B_1$ and $\tilde{X}\,^1A_1$ states occur, and it is these crossings through conical intersections that result in a high torque acting on OH and, as a consequence, the formation of OH ($X\,^2\Pi$) in very highly excited rotational levels. The dominant product is OH ($X\,^2\Pi$, $v = 0$), but OH ($X\,^2\Pi$, $v = 1-9$) and OH ($A\,^2\Sigma^+$, $v = 0-3$) are also formed (Harich et al. 2000). The three-body channel was also measured by Harich et al. (2000) and found to be 21%, i.e., slightly higher than the 12% measured by Slanger and Black (1982).

Thomas et al. (2002) performed quasiclassical trajectory calculations of a type similar to that used to model the photodissociation of H_2O (Dixon et al. 1999, Harich et al. 2000), but adapted to handle dissociative recombination occurring 12.61 eV above the water ground state rather than the 10.2 eV in photodissociation. One striking feature going from 10.2 eV to 12.61 eV is the increase in electronic states; the number increases from about 25 to 50. The calculations clearly showed that three-body breakup is significantly enhanced when the initial conditions are changed from Lyman-α excitation to those prevailing for dissociative recombination. The calculations showed that dominant products from the $2\,^1A'\,(\tilde{B}\,^1A_1)$, $2\,^1A''\,(^1A_2)$, and $2\,^3A'\,(^3A_2)$ states were three-body, although only if the OH ($A\,^2\Sigma^+$) predissociation was included in the calculation for the $2\,^1A'\,(\tilde{B}\,^1A_1)$ state. The calculations did not allow a detailed comparison with the experimental data, but showed that three-body fragmentation could occur with relative ease. It was not possible to include all electronic states in the calculations. Noting that the more surfaces that are involved, the greater the tendency for a statistical distribution, Thomas et al. (2002) calculated that the O (^3P) channel (8.1c) dominated the O (^1D) channel by a ratio of 3.0:1, which was obtained by taking the ratio of the electronic degeneracy factors, 9/5, and multiplying it by the translational energy factor $(3.04/1.07)^{\frac{1}{2}}$ (the kinetic energy release in reactions (8.1c) and (8.1d)); this is very close to the experimental value of (3.5 ± 0.5):1. The good agreement supports a statistical distribution, although it cannot be ruled out that it may be fortuitous.

The two-body channel OH + H in dissociative recombination of H_2O^+ has been studied by Sonnenfroh, Caledonia, and Lurie (1993) using a crossed-beam

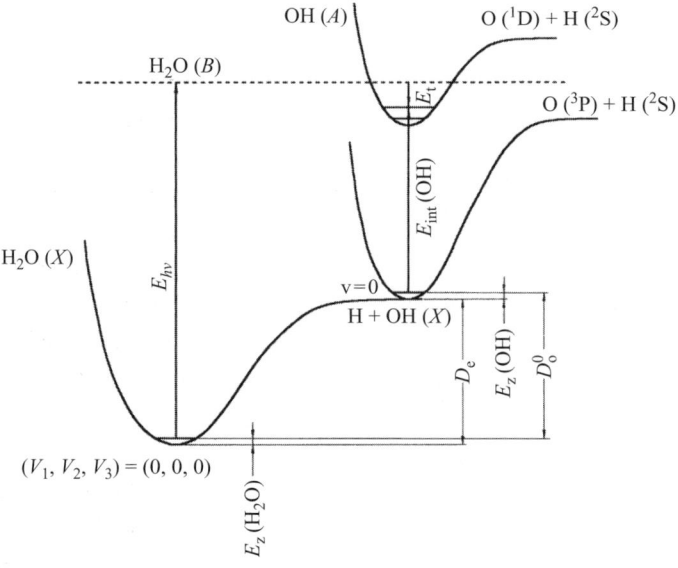

Figure 8.8 Energy diagram for Lyman-α photodissociation of H_2O. (Reused with permission from S. A. Harich, D. W. H. Hwang, X. Yang, et al. 2000, *Journal of Chemical Physics* **113**, 10073, (2000). Copyright 2000, American Institute of Physics.)

apparatus. They measured the branching ratio for the formation of OH in its $A\,^2\Sigma^+$ state to be $0.076^{+0.08}_{-0.04}$, with a vibrational distribution of 1.00:0.87 for $v=0$:$v=1$. They also measured a rotational temperature of 4000 K, which is qualitatively consistent with the photodissociation work of Harich et al. (2000).

Water's heavier homolog dihydrogen sulfide, H_2S, has an ionization energy of 10.47 eV, which is close to the Lyman-α energy 10.2 eV. In contrast to H_2O, Cook et al. (2001) found negligible production of SH in its ground electronic state. On the other hand, Liu et al. (1999) found almost exclusively the molecular ground-state channel H + SH ($X\,^2\Pi$) when photodissociating H_2S at 157.6 nm. The excitation energy at 157.6 nm is approximately twice the dissociation energy for the loss of a single hydrogen atom, as is that for H_2O at 121.6 nm. In both cases passage from the $\tilde{B}\,^1A_1$ to the $\tilde{X}\,^1A_1$ potential energy surfaces via conical intersection plays the dominant role in the dissociation to yield H + SH/OH ($X\,^2\Pi$). But for H_2S this product channel has become negligible at 121.6 nm, even though the excitation is to the same $\tilde{B}\,^1A_1$ state, and is replaced by the two-body product H + SH ($A\,^2\Sigma^+$) and the three-body products H + H + S (^3P). The cause of this change is purely energetic. The increase in energy leads to much faster radial acceleration on the $\tilde{B}\,^1A_1$ state surface, such that by the time the H–S–H angle is linear, which is the requirement for the conical intersection, the chemical bonds are stretched

8.1 *Dissociation dynamics in recombination of XH_2^+ ions* 237

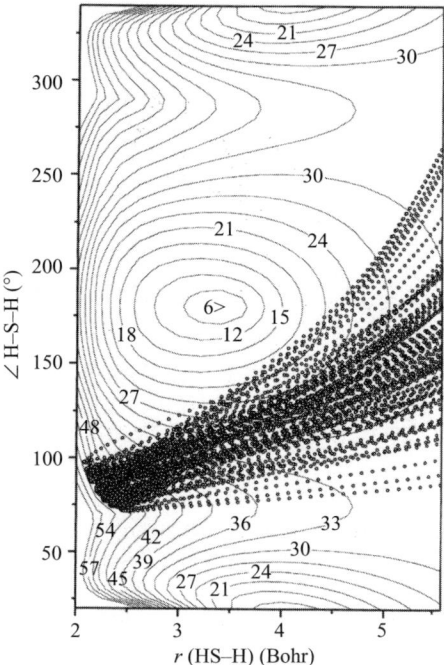

Figure 8.9 Two-dimensional representation of the $2^1A'$ ($\tilde{B}^1A_1$) potential energy surface of H_2S calculated by Cook *et al.* (2001). The H−S−H conical intersection with $1^1A'$ ($\tilde{X}^1A_1$) is clearly evident at H−S−H = 180°, with the corresponding minimum associated with the S−H−H conical intersection evident at small bond angles. Superimposed on the contours are the results of classical trajectory calculations performed with this surface, showing the likely evolution of Lyman-α excited molecules towards asymptotic products. The trajectories pass through the linear H−S−H configuration "outside" of the conical intersection with the ground-state surface, thus precluding the possibility of forming ground-state molecular products. (Reused with permission from P. A. Cook, S. R. Langford, R. N. Dixon, and M. N. R. Ashford, *Journal of Chemical Physics* **114**, 1672, (2001), Copyright 2001, American Institute of Physics.)

well beyond their values for the conical intersections, thereby inhibiting the surface crossing (see Fig. 8.9).

The three-body breakup of H_2S has two decay channels in Lyman-α photodissociation and in dissociative recombination of SH_2^+:

$$SH_2(\text{excited}) \rightarrow S(^3P) + H(^2S) + H(^2S), \quad (8.10a)$$

$$SH_2(\text{excited}) \rightarrow S(^1D) + H(^2S) + H(^2S). \quad (8.10b)$$

Cook *et al.* (2001) found that channel (8.10b) dominates at Lyman-α photodissociation at 10.2 eV. Hellberg *et al.* (2005) studied dissociative recombination[1] of SD_2^+,

---

[1] For technical reasons, SD_2^+ was chosen instead of SH_2^+ in order to minimize H atoms missing the detector.

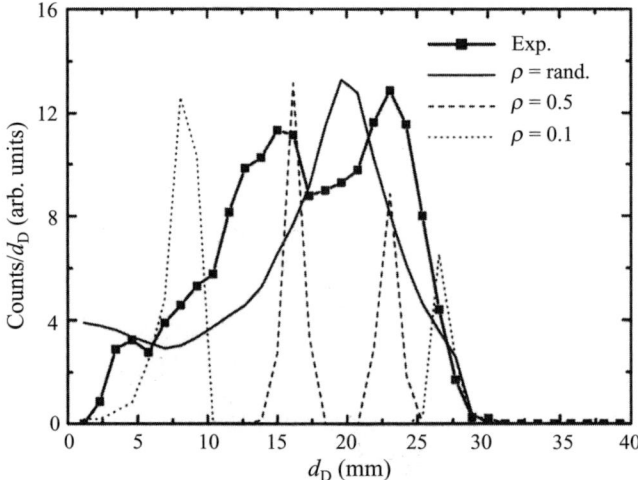

Figure 8.10 Energy distribution for the two deuterium atoms in dissociative recombination of SD_2^+. The axes are the same as in Fig. 8.6. (Reused with permission from F. Hellberg, V. Zhaunerchyk, A. Ehlerding, et al., *Journal of Chemical Physics* **122**, 224314, (2005). Copyright 2005, American Institute of Physics.)

and found the branching ratio for the S (^{3}P) channel to be 0.6 ± 0.1 and that for the S (^{1}D) channel to be 0.4 ± 0.1. This is marginally within the predictions of the statistical model discussed above for H_2O^+.

Figure 8.10 shows the distribution of the deuterium atoms in dissociative recombination of SD_2^+ in the same way that Fig. 8.6 shows the H atoms' energy distribution following recombination of H_2O^+. It is immediately obvious that the random distribution does not work as well in this case as it did for water. In addition to an asynchronous concerted three-body breakup, it is also conceivable that the breakup occurs through a two-step process via the predissociating SD ($A\,^2\Sigma^+$) state. Figure 8.11 shows the potential curves involved in the predissociation of the $A\,^2\Sigma^+$ state (Wheeler, Orr-Ewing, & Ashfold 1997). The sequential breakup

$$SD_2^+ + e^- \to SD(A\,^2\Sigma^+) + D \to S(^3P) + D + D \quad (8.11)$$

has two limiting cases. In one limit the $A\,^2\Sigma^+$ state is formed in its lowest rovibrational level and the first D atom carries about 2.5 eV in kinetic energy. When the $A\,^2\Sigma^+$ state predissociates, only 0.25 eV remains to be shared between the heavy sulfur atom and the second D atom. Momentum conservation leads to the D atom taking the lion's share of this energy, and the ratio of kinetic energies for the D atoms becomes $\rho = 0.1$. It is obvious from Fig. 8.10 that this gives a poor representation of the experimental data. In the other limit the $A\,^2\Sigma^+$ state is formed

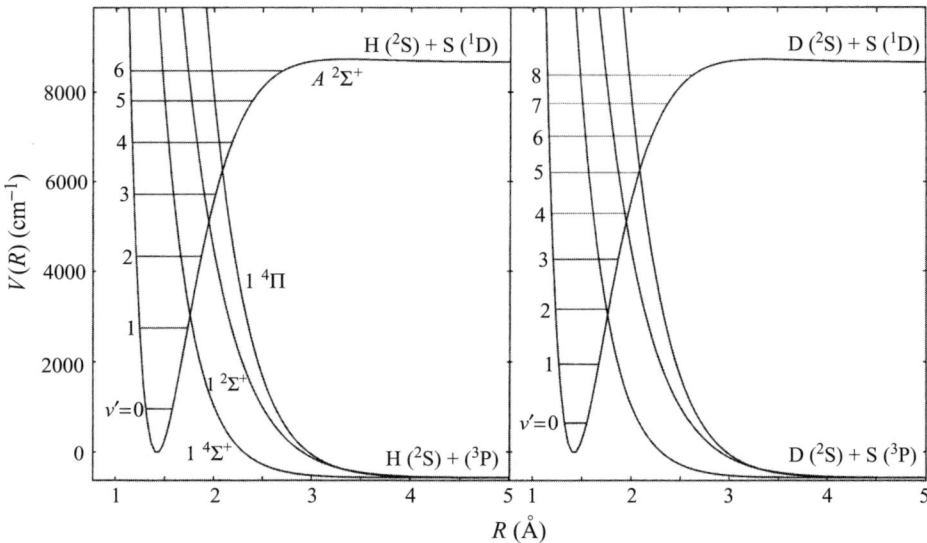

Figure 8.11 The potential curves for the $A\ ^2\Sigma^+$ state of SH and SD, and the repulsive electronic states causing predissociation of the $A\ ^2\Sigma^+$ state. (Reused with permission from M. D. Wheeler, A. J. Orr-Ewing, M. N. R. Ashfold, *Journal of Chemical Physics* **107**, 7591, (1997). Copyright 1997, American Institute of Physics.)

close to the dissociation limit in $v = 8$, with 1.15 eV of internal energy. This leads to a ratio of $\rho = 1.15/1.6 = 0.7$. The case $\rho = 0.5$ is shown in Fig. 8.10, and this gives a better representation of the experimental data. The best representation of the experimental data is obtained by a Monte-Carlo simulation for which 60% of the events are random, and 40% of the events have $\rho = 0.3-0.5$. Note that purely random ρ does not give adequate agreement with the data, as shown in Fig. 8.10. Thus, the dissociative recombination of SH_2^+ is best described as a combination of an asynchronous concerted (60%) and a sequential (40%) breakup.

The dissociative recombination of XH_2^+ for X = C, N, and P cannot be compared with results from photodissociation experiments, since the precursor for such experiments would be a free radical rather than a stable molecule, which drastically increases the experimental difficulties (of course, theory does not have this problem, see for example Kroes *et al.* (1997)). Figures 8.12 and 8.13 show the energy distributions for the two H atoms released in dissociative recombination of CH_2^+ and NH_2^+ (Thomas *et al.* 2005b). The experimental data for NH_2^+ are well reproduced in terms of the peak position for $\rho = 1.0$, i.e., the synchronous concerted case, but not in terms of the width. Just as for SD_2^+, the data are best reproduced for both CH_2^+ and NH_2^+ by a mixture of a random and fixed ρ, as can be seen in Fig. 8.13.

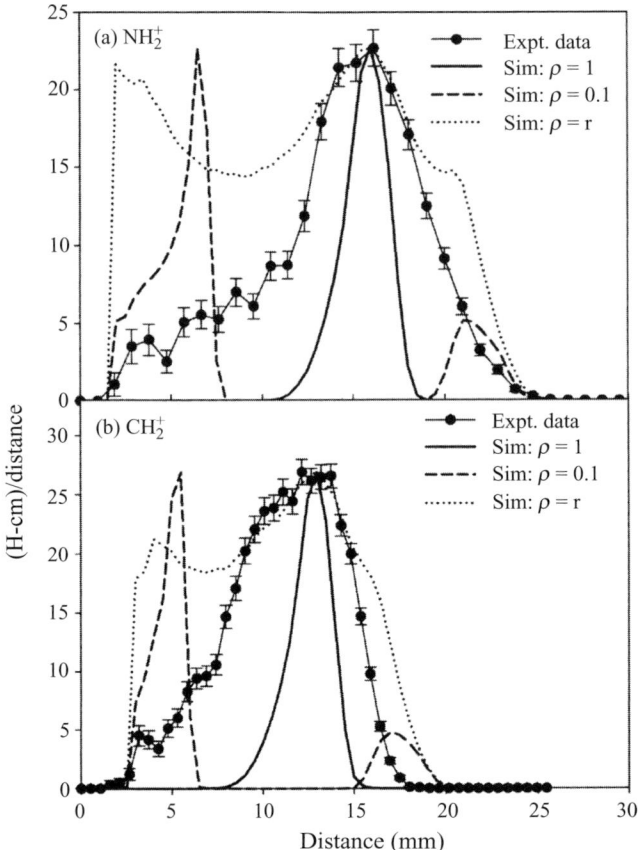

Figure 8.12 Energy distributions for the two H-atoms for dissociative recombination events which preferentially dissociate parallel to the detector. Experimental data for NH_2^+ are given in the upper panel (a) and for CH_2^+ in the lower panel (b). Experimental data for each case (−•−) are plotted and compared with three different simulated cases for the ρ parameter (defined in Eq. (8.7)). The three cases are $\rho = 1.0$ (solid line), $\rho = 0.1$ (dashed line), and $\rho = r$ (random; dotted line). None of the simulations gives a good representation of the experimental data. (Reprinted figure by permission of R. D. Thomas, F. Hellberg, A. Neau, et al., "Three-body fragmentation dynamics of amidogen and methylene radicals via dissociative recombination," Phys. Rev. A **71**, pp. 032711–1–16, (2005). Copyright (2005) by the American Physical Society.)

Table 8.2 summarizes the results for H_2O^+ (Thomas et al. 2002), CH_2^+, and NH_2^+ (Thomas et al. 2005b). For each ion, column 2 lists the initial state data, i.e., the ionization energy IE, the bond angle Θ_e, and the character of the state. Columns 3 and 4, respectively, list the available fragmentation channels with their kinetic energies assuming ground-state fragments and dissociative recombination with 0 eV electrons. Column 5 gives the results for the dissociative recombination

8.1 Dissociation dynamics in recombination of XH_2^+ ions

Table 8.2. *Table of dissociative recombination results for CH_2^+, NH_2^+, and H_2O^+.*

Ion	Initial state data	Chemical branching			Three-body competition			
		Neutral channels	Ground-state energies	Branching fraction	Three-body channels	Branching fraction expt. model[g]	Ground state angular distribution	H-energy distribution
CH_2^+	IE(CH_2):10.396 eV[a]	C + H_2	6.80 eV	0.12[d]			[e]	[e]
	$\Theta_e := 138°$[b]	CH + H	5.80 eV	0.25	C (^3P),2.45 eV	0.51[e] 0.72 (0.64)		
	State: $^2A_1$	C + H + H	2.45 eV	0.63			⊔	⋀
	Radical species				C (^1D),1.19 eV	0.49 0.28 (0.36)		
NH_2^+	IE(NH_2):11.163 eV[c]	N + H_2	8.24 eV	0.04[e]			[e]	[e]
	$\Theta_e := 150°$[b]	NH + H	6.97 eV	0.39	N (^4S),3.72 eV	0.53[e] 0.36 (0.20)		
	State: $^3B_1$	N + H + H	3.72 eV	0.57	N (^2D),1.34 eV	0.45 0.54 (0.50)	⊔	⋀
	Closed shell species				N (^2P),0.14 eV	0.02 0.10 (0.30)		
OH_2^+	IE(H_2O):12.296 eV[c]	O + H_2	6.97 eV	0.09[f]			[h]	[h]
	$\Theta_e := 108°$[b]	OH + H	6.96 eV	0.20	O (^3P),3.04 eV	0.78[h] 0.75 (0.64)		
	State: $^2B_1$	O + H + H	3.04 eV	0.71			⋈	⋀
	Open shell species				O (^1D),1.07 eV	0.22 0.25 (0.36)		

[a]Data taken from the NIST webbook: http://webbook.nist.gov/
[b]Values for Θ_e taken from the data reported by Graber *et al.* (1997).
[c]Data for IE (NH_2) taken from Dyke (quoted as private communication by Thomas *et al.* (2005b)).
[d]Branching data from the dissociative recombination of CH_2^+ taken from Larson *et al.* (1998).
[e]Data reported by Thomas *et al.* (2005b).
[f]Branching data from the dissociative recombination of H_2O^+ taken from Rosén *et al.* (2000).
[g]Values obtained from models used by Thomas *et al.* (2002) and, in brackets, Hellberg *et al.* (2003).
[h]Data reported by Thomas *et al.* (2002).

reaction as measured at CRYRING. Columns 6–10 list those data associated with the investigations into the competition between the available three-body fragmentation channels. Column 6 lists the available channels, giving the excitation state of the heavy atom as well as the kinetic energy associated with that channel. Column 7 lists the obtained experimental branching ratios while column 8 shows the predictions obtained from models used by Thomas *et al.* (2002) and Hellberg *et al.* (2003).

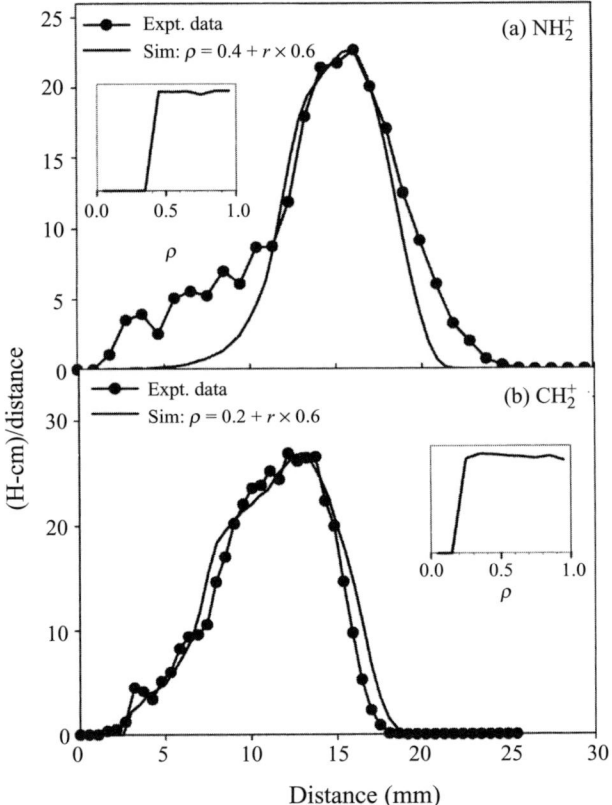

Figure 8.13 Energy distributions for the two H-atoms for dissociative recombination events which preferentially dissociate parallel to detector. Experimental data for NH_2^+ are given in the upper panel (a) and for CH_2^+ in the lower panel (b). These data are compared with a "best fit" simulation for the ρ parameter (solid line), $\rho = 0.4 + r \times 0.6$ and $\rho = 0.2 + r \times 0.6$ for NH_2^+ and CH_2^+, respectively, where r is a random number from a flat distribution of values between 0 and 1. The inset graph in each panel plots the distribution of ρ used in that particular simulation. (Reprinted figure by permission of R. D. Thomas, F. Hellberg, A. Neau, et al., "Three-body fragmentation dynamics of amidogen and methylene radicals via dissociative recombination," *Phys. Rev. A* **71**, pp. 032711-1–16, (2005). Copyright (2005) by the American Physical Society.)

Finally, columns 9 and 10 show a graphical representation of the angular distribution and kinetic-energy distribution between the H atoms, respectively, for selected reactions leading to the production of the ground-state heavy fragments only. For the plots in column 9, open linear geometry is to the left and small-angle geometry to the right.

There are some striking differences between on the one hand H_2O^+ and on the other hand CH_2^+ and NH_2^+. It is clear from column 7 in Table 8.2 that the atomic

8.1 Dissociation dynamics in recombination of XH_2^+ ions

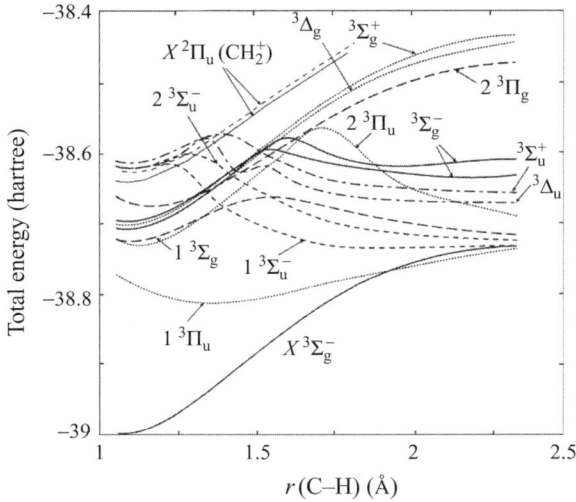

Figure 8.14 Potential energy curves for symmetric three-body dissociation of the linear CH_2 to the $C + 2H$ end products calculated by multiconfiguration SCF (MCSCF) linear response method. The $1\,^3\Sigma_u^-$ state correlates with the $3\,^3A_2$ of the bent CH_2 molecule. (Reprinted from *Chem. Phys.* **280**, B. Minaev and M. Larsson, "MCSCF linear response study of the three-body dissociative recombination $CH_2^+ + e \to C + 2H$," pp. 15–30, Copyright (1995), with permission from Elsevier.)

branching ratios for C and N are very close to being equal, in sharp contrast to the situation for O, where the ^3P state dominates. We discussed earlier that the results for the atomic branching ratio in the three-body breakup $H + H + O$ (^1D or ^3P) were in very good agreement with a statistical prediction, but this is clearly not the case for CH_2^+ and NH_2^+, where the statistical model fails. It is also clear from Figs. 8.12 and 8.13 that the random distribution of kinetic energy to the two H atoms, which worked so well for H_2O^+, fails for CH_2^+ and NH_2^+. It is possible that there is a connection between the atomic branching ratios and the H atoms energy sharing in that CH_2^+ and NH_2^+ recombine into the three-body channels in a fashion dictated more by dynamics, with a stronger component of a two-step sequential dissociation process. There is not much theoretical work addressing the three-body breakup for CH_2^+ and NH_2^+. Minaev and Larsson (2002) calculated ten excited singlet and triplet states of CH_2 in order to identify the states that could cause a three-body breakup. They concluded that in linear geometry the $1\,^3\Sigma_u^-$ state is the only plausible candidate for the three-body dissociative recombination process. Figure 8.14 shows the potential curves for CH_2^+ and CH_2 in linear geometry. The $1\,^3\Sigma_u^-$ state has a favorable crossing with the ion ground state and a very small barrier going from Rydberg character at shorter internuclear distances to valence state character at larger internuclear distance. It correlates with the $3\,^3A_2$ state of the bent CH_2 molecule. The electronic transition moment from the ground state of CH_2 to the $1\,^3\Sigma_u^-$ state is very small, which implies that this state does not play a role in photodissociation of CH_2.

The comparison of H_2O^+ and SH_2^+, and the tendency of NH_2^+ to dissociate by the synchronous concerted mechanism make it tempting to guess that PH_2^+ would show even clearer signs of simultaneous bond breaking. However, this is not the case. Zhaunerchyk et al. (2005) studied PD_2^+ in CRYRING and found that the best representation of the energy sharing between the D atoms was obtained by having random ρ, just as was the case for H_2O^+. This is not the first time that intuition failed in dissociative recombination.

The dynamics of three-body breakup in dissociative recombination was uncharted territory prior to the work on the XH_2^+ ion (X = H, C, N, O, S, P) described in Section 7.2.10 and in this section. After only about 5–6 years of storage ring experiments combined with imaging techniques, the level of detail now challenges that obtained in photodissociation studies (see, e.g., Einfeld et al. (2004), Roth, Maul, & Gericke (2004)).

8.2 Astrophysical molecular ions
8.2.1 HCO^+ and HCS^+

As far as interstellar molecular ions are concerned, the HCO^+ ion is second only to H_3^+ in importance. It has an interesting history in that its spectrum was first observed by means of radio astronomy by Buhl and Snyder (1970). They observed a line at 89.190 GHz, but were unable to identify the carrier and therefore labeled it X-ogen. Simultaneously, Klemperer (1970) proposed that the line originated from the $J = 1-0$ transition in HCO^+. Klemperer referred to the work of Buhl and Snyder (1970), but the reference given in the note is incorrect.[2] The discovery of the 89.190 GHz line was announced at the IAU General Assembly in Brighton in August 1970, where Klemperer heard about it. His paper to *Nature* was submitted a few weeks before that of Buhl and Snyder, and hence the interpretation was published before the discovery. The 89.190 GHz line was the first observed rotational spectrum of a molecular ion (Oka 2003b), and it was several years before the assignment to $H^{12}CO^+$ was confirmed. Electronic structure calculations by Wahlgren et al. (1973), Bruna (1975), and Kraemer and Diercksen (1976) provided support of the identification, and the observation of a line at 86.754 GHz (Snyder et al. 1975) was concurrent with a rotational emission line in $H^{13}CO^+$. The observation of a

---

[2] Klemperer refers to *Nature* **227**, p. 867 (1970), which is incorrect. In fact, the paper by Buhl and Snyder (1970) was published *after* Klemperer's paper. The incorrect reference has made its way into the literature and has more citations than the correct one.

Table 8.3. *Experimental rate coefficients for HCO^+*

$\alpha(HCO^+)$ $(10^{-7}\ cm^3\ s^{-1})$	T_e (K)	Comment	Reference
2.0 ± 0.3	300	Stationary afterglow	Leu, Biondi, & Johnsen 1973c
1.1	300	FALP	Adams, Smith, & Alge 1984
2.4 ± 0.4	300	SA, temperature dep.	Ganguli et al. 1988
3.1	273	Infrared absorption	Amano 1990
2.4	300	FALP	Canosa et al. 1991b
2.2	300	FALP	Rowe et al. 1992
1.5	300	FALP	Smith & Španěl 1993a
1.9	295	FALP	Gougousi, Golde, & Johnsen 1997
1.7	300	Merged beams	Le Padellec et al. 1997a[a]
2.0 ± 0.4	300	FALP-MS	Laubé et al. 1998b
1.4	300	FALP	Poterya et al. 2005
1.7	300	ISR, CRYRING	W. Geppert, private communication

FALP: flowing afterglow/Langmuir probe; FALP-MS: flowing afterglow/Langmuir probe with moving mass spectrometer; ISR: ion storage ring
[a] In the review by Florescu-Mitchell and Mitchell (2006) this result is changed to $0.7 \times 10^{-7}\ cm^3\ s^{-1}$, based on work by Sheehan (2000).

laboratory microwave spectrum of HCO^+ (Woods et al. 1975) provided the definitive evidence that Klemperer's conjecture was correct.

Even before the laboratory microwave spectrum of HCO^+ had been observed, the first measurement of the destruction of HCO^+ by electrons was performed by Leu, Biondi, and Johnsen (1973b) using the stationary afterglow technique. This measurement has been followed by many others, but they have never given rise to the controversies surrounding the recombination of H_3^+. Table 8.3 summarizes the results for HCO^+ obtained so far by means of different experimental techniques. There is some spread in the data, but no serious discrepancies. The most puzzling result is not shown in Table 8.3; it is the peculiar temperature dependence below 300 K measured by Poterya et al. (2005). They found a slightly decreasing rate coefficient with reducing temperature, in conflict with the cross section data (Le Padellec et al. 1997a, Geppert et al. 2006b) and measurements of the rate at different temperatures (Leu et al. 1973c, Adams, Smith, and Alge 1984, Smith and Španěl 1993a, and Amano 1990). Poterya et al. (2005) speculated that the recombination mechanism might differ from the direct and indirect mechanisms.

The branching ratios for dissociative recombination of DCO^+ have been measured (Geppert et al. 2004d) and found to be totally dominated by the D + CO channel, which was measured to be 88%. This experiment did not determine the state of excitation of the CO product, but optical emission studies of recombining HCO^+ in a flowing afterglow plasma have shown that the $a\ ^3\Pi$ state is formed

(Adams & Babcock 1994a,b, Butler, Babcock, and Adams 1997, Johnsen et al. 2000), and that it had a yield of 30% when a mixture of HCO$^+$ and HOC$^+$ was used (Johnsen et al. 2000). Tomashevsky, Herbst, and Kraemer (1998) calculated a vibrational state distribution for the $a\,^3\Pi$ state in good agreement with the experimental results of Butler, Babcock, and Adams (1997). Gougousi, Johnsen, and Golde (1997b) searched for the channel leading to OH formation using laser-induced fluorescence, but found very small amounts of OH.

The controversies about HCO$^+$ have concerned theory rather than experiment. Kraemer and Hazi (1985) used SCF/CI to calculate potential energy curves for HCO$^+$ and HCO in a linear geometry and with the CO distance fixed to 2.10 a_0. In a later work (Kraemer & Hazi 1989), they used complete active space SCF (CASSCF) and arrived at the same conclusion as in their earlier work. In linear geometry, the $X\,^1\Sigma^+$ electronic ground state of HCO$^+$ has the leading configuration ... $4\sigma^2 5\sigma^2 1\pi^4$. Kraemer and Hazi (1989) considered three different excitations of the bound electron, namely $5\sigma \to 6\sigma$, $5\sigma \to 1\pi$, and $1\pi \to 2\pi$, which gave six electronic states in HCO when letting the free electron enter valence orbitals. When the $5\sigma \to 6\sigma$ excitation is followed by the free electron entering the 5σ orbital, the $X\,^2\Sigma^+$ state is formed, which dissociates into H (^{2}S) + CO ($X\,^1\Sigma^+$). Kraemer and Hazi (1989) found the $X\,^2\Sigma^+$ ground-state potential of HCO to be below that of the ground state of HCO$^+$ at all C–H distances, and they were unable to identify any resonance state that could cause dissociative recombination of HCO$^+$ $X\,^1\Sigma^+$ ($v = 0$). They concluded that recombination with thermal-energy electrons could only occur through an unspecified indirect mechanism.

Talbi and collaborators arrived at a different conclusion in a series of papers stretching over decade (Talbi, Pauzat, & Ellinger 1988, Talbi et al. 1989, Talbi & Ellinger 1993, Le Padellec et al. 1997a). They, too, used *ab initio* methods to calculate adiabatic potential curves of HCO, but also included Rydberg states, and constructed a quasidiabatic curve dissociating to H (^{2}S) + CO ($X\,^1\Sigma^+$) and crossing the ion ground state near its minimum. The favorable crossing made it possible to invoke the direct mechanism, and Talbi and Ellinger (1993) and Le Padellec et al. (1997a) gave a theoretical rate coefficient of 2.3×10^{-7} cm^3 s^{-1}.

Bates (1992b) sided with Kraemer and Hazi and proposed that recombination of HCO$^+$ occurs through a multistep indirect mechanism, and although he was not able to calculate a quantitative value for the rate coefficient, he felt convinced that he had identified the correct mechanism.

Larson et al. (2005) performed extensive multireference configuration interaction (MRCI) calculations coupled with quantum defect analysis but were unable to find a diabatic state crossing the ion ground state. Thus, they concluded, just like Kraemer and Hazi (1989) and Bates (1992b), that the initial capture must take place to a Rydberg state. They also assessed the influence of the Renner–Teller effect, with the idea that it might influence the recombination just like the Jahn–Teller effect was found to do for H_3^+ (Kokoouline, Greene, & Esry 2001). In a preliminary report, Larson *et al.* (2005) assumed that the interaction of vibrational and electronic angular momentum (Renner–Teller effect) was too small to have any influence on the recombination cross section. On the other hand, they obtained a cross section that was small compared with that measured by Le Padellec *et al.* (1997a). Further work (Mikhailov *et al.* 2006) has shown that the conclusion concerning the Renner–Teller effect was premature, and more detailed calculations including this effect have brought theory into good agreement with experiment. The dissociative recombination of HCO^+ is now better understood.

Replacing oxygen in HCO^+ by sulfur gives the heavier homolog HCS^+. When this was discovered in the interstellar medium by Thaddeus, Guélin, and Linke (1981) it became the fifth interstellar ion after CH^+, HCO^+, N_2H^+, and CO^+. Millar (1983) addressed the two orders of magnitude larger HCS^+/CS ratio as compared with the HCO^+/CO ratio and proposed that it could be caused by a slow dissociative recombination of HCS^+. In a joint flowing afterglow experiment with the Birmingham group, Millar *et al.* (1985) measured the slowest rate coefficient of any heavy polyatomic ion, finding a preliminary value of $\sim 5 \times 10^{-8}$ cm^3 s^{-1} for dissociative recombination of HCS^+ at 300 K. This seemed to explain the HCS^+/CS abundance ratio. Talbi *et al.* (1989) did not find a favorable crossing for HCS^+, thus providing further arguments for a slow recombination. Abouelaziz *et al.* (1992) found a much larger rate coefficient, between 5.8×10^{-7} cm^3 s^{-1} and 7.9×10^{-7} cm^3 s^{-1}, when using the FALP apparatus in Rennes, which has a movable mass spectrometer. Ion impurities could have been a problem in the Millar *et al.* (1985) experiment, which in fact was only briefly described; a longer forthcoming paper was mentioned, but was never published. The FALP-MS result of Abouelaziz *et al.* (1992) is in very good agreement with the $9.7 \times 10^{-7}(T/300)^{-0.57}$ cm^3 s^{-1} rate coefficient measured in CRYRING (Montaigne *et al.* 2005). In contrast to HCO^+, the H atom loss channel is minor (19%) whereas the HC + S channel dominates (81%). Nevertheless, model calculations by Montaigne *et al.* (2005) show that the new data from CRYRING make it even more difficult to explain the high HCS^+/CS abundance.

8.2.2 N_2H^+

The N_2H^+ ion has two exothermic recombination channels:

$$N_2H^+ + e^- \rightarrow N_2 + H + 8.47 \text{ eV}, \tag{8.12a}$$

$$N_2H^+ + e^- \rightarrow NH + N + 2.25 \text{ eV}. \tag{8.12b}$$

Adams, Herd, and Smith (1990) noted in passing that H atoms in their flowing afterglow plasma were detected from recombination of N_2H^+ at the level of one

H atom per recombination, which implies that the $N_2 + H$ channel has essentially unity branching ratio. In a later study, Adams et al. (1991) added NO to the plasma while searching for laser-induced fluorescence from the OH radical. The absence of such fluorescence was interpreted as the absence of NH, which would react with NO to form $OH + N_2$. The branching ratio for the H atom abstraction channel (8.12a) measured by Adams et al. (1991) is very similar to the corresponding channel for HCO^+ (Geppert et al. 2004d). Experiments on N_2H^+ at CRYRING seemed to indicate that this was wrong (Geppert et al. 2004g), but new flowing afterglow and storage ring experiments now agree that Reaction (8.12b) is minor, with a branching ratio of about 0.1 or less, as was discussed at the seventh conference on dissociative recombination in Ameland, The Netherlands in 2007 (N. G. Adams, private communication, W. D. Geppert, private communication).

The first measurement of a recombination rate coefficient was done by Mul and McGowan (1979b) using the merged-beam technique. They measured the cross section down to 0.006 eV and extrapolated further to 0.001 eV; the cross section was then used to calculate the rate coefficient. No significant difference was found between N_2H^+ and N_2D^+, and the rate coefficient at 300 K was determined to be about 7.5×10^{-7} cm^3 s^{-1}, which should be divided by a factor of 2 to give 3.75×10^{-7} cm^3 s^{-1} (see Section 2.1.2). Smith and Adams (1984) and Adams, Smith, and Alge (1984) used the flowing afterglow technique and found a lower rate coefficient, 1.7×10^{-7} cm^3 s^{-1}, and argued that the Mul and McGowan rate was inflated by the presence of vibrationally excited ions. (At that time, of course, they were not aware of the factor of 2 problem.) Amano (1990) used the infrared absorption technique and obtained 7.0×10^{-7} cm^3 s^{-1} at 273 K, very close to the published merged-beam result (Mul & McGowan 1979b). An experiment at CRYRING gave $(1.0 \pm 0.2) \times 10^{-7}$ cm^3 s^{-1} (Geppert et al. 2004g), in quite good agreement with the flowing afterglow result, but clearly smaller than the results of Mul and McGowan (1979b) and Amano (1990). Both Smith (Smith & Španěl 1993a) and Adams (Poterya et al. 2005) revisited N_2H^+ with their flowing afterglow techniques, Smith and Španěl finding a 50% higher rate coefficient at 2.4×10^{-7} cm^3 s^{-1} and Poterya et al. (2005) a value only slightly higher (2.4×10^{-7} cm^3 s^{-1}). There are no theoretical calculations on this system for comparison. Poterya et al. (2005) also found the rate coefficient to be almost temperature independent in the range $T = 300-500$ K, similar to the behavior of the HCO^+ rate coefficient between 100 K and 300 K. Below 300 K, however, a normal temperature behavior was observed for N_2H^+.

Geppert et al. (2004g) also measured the product branching ratios and found that the $NH + N$ channel is much larger than previously believed, 64%, and consequently that the $N_2 + H$ channel has a branching ratio of only 36%. In a later experiment it was shown that this result was flawed due to contamination of the ion beam, and a

8.2 Astrophysical molecular ions

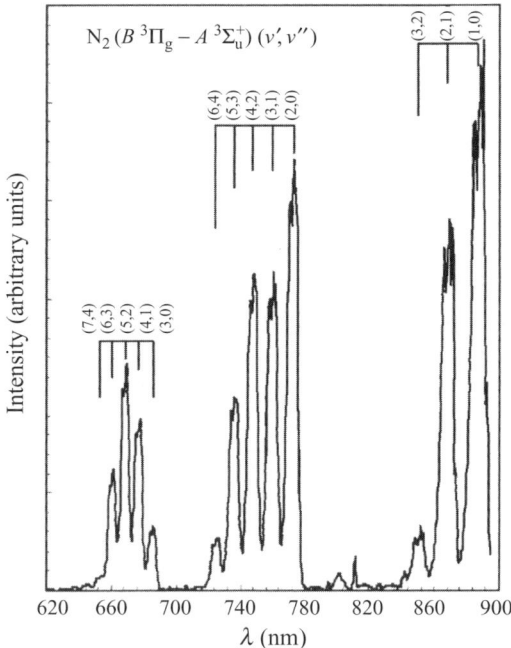

Figure 8.15 $N_2(B\,^3\Pi_g - A\,^3\Sigma_u^+)(v', v'')$ emission spectrum from $N_2H^+ + e^-$ in a He/Ar flowing afterglow. Data have been corrected for the spectral response of the detection system. Rosati, Johnsen, and Golde (2004) derived a $B\,^3\Pi_g$ state yield of 0.19±0.08. (Reused with permission from R. E. Rosati, R. Johnsen, and M. F. Golde, *Journal of Chemical Physics* **120**, 8025, (2004). Copyright 2004, American Institute of Physics.)

reanalysis gave a branching ratio of less than 10% for the NH + H channel (W. D. Geppert, 2007, private communication). The result is essentially in agreement with the flowing afterglow results of Adams, Herd, and Smith (1990) and Adams *et al.* (1991) who found H atom formation for every recombination event. The problem is that they did not study this reaction, and it is unclear whether NH + NO → OH + N_2 is an adequate diagnostic of NH. The reaction is not in question (Yamasaki *et al.* 1991), but others have failed to observe OH in NH + NO reactions (Harrison, Whyte, & Phillips 1986). The new technique developed by Adams and Babcock (2005) has been used to measure the branching ratios and the preliminary results suggest an NH + H branching ratio that is no larger than 5% (N. G. Adams, 2007, private communication), in agreement with the storage ring and earlier afterglow results. It has been shown that N_2 is also formed in electronically excited states (Adams & Babcock 1994a, Rosati, Johnsen, & Golde 2004, Poterya *et al.* 2004). Figure 8.15 shows an emission spectrum of $N_2\,B\,^3\Pi_g - A\,^3\Sigma_u^+$ recorded in a flowing afterglow plasma containing N_2H^+ (Rosati, Johnsen, & Golde 2004).

8.2.3 HCNH$^+$, HCN$^+$, HNC$^+$

The HCNH$^+$ ion has received considerable attention because of its astrophysical importance. HCNH$^+$ was first detected in the interstellar medium by Ziurys and Turner (1986), and it is an important ion in Titan's ionosphere (Petrie 2001). Adams and Smith (1988c) used HCN as source gas and H$_3^+$ as proton donator in their flowing afterglow apparatus and measured the recombination rate coefficient to be 3.5×10^{-7} cm^3 s^{-1} at 300 K. This value was confirmed much later in a storage ring experiment by Semaniak et al. (2001), who obtained $2.8 \times 10^{-7}(T/300)^{-0.65}$ cm^3 s^{-1}. Thus, the rate of destruction of HCNH$^+$ in interstellar clouds is not an issue of controversy. The more interesting question concerns the product branching ratios. The following channels are open in recombination with thermal-energy electrons

$$\text{HCNH}^+ + e^- \rightarrow \text{HCN}\left(\tilde{X}\,^1\Sigma^+\right) + \text{H} \qquad \Delta E = 5.9\,\text{eV}, \qquad (8.13\text{a})$$
$$\text{HCNH}^+ + e^- \rightarrow \text{HCN}\left(\tilde{X}\,^1\Sigma^+\right) + \text{H} \qquad \Delta E = 5.3\,\text{eV}, \qquad (8.13\text{b})$$
$$\text{HCNH}^+ + e^- \rightarrow \text{CN}\left(X\,^2\Sigma^+\right) + \text{H}_2 \qquad \Delta E = 5.1\,\text{eV}, \qquad (8.13\text{c})$$
$$\text{HCNH}^+ + e^- \rightarrow \text{CN}\left(X\,^2\Sigma^+\right) + \text{H} + \text{H} \qquad \Delta E = 0.6\,\text{eV}. \qquad (8.13\text{d})$$

The average HCN/HNC abundance ratio in dark cloud cores is 0.69 (Hirota et al. 1998), but it can also show drastic changes depending on the source. In the molecular cloud OMC-1 the HCN/HNC abundance ratio was found to be much larger than unity (Schilke et al. 1992). Since dissociative recombination of HCNH$^+$ is assumed to be a major formation route of interstellar HCN and HNC, the product branching ratios are important. Herbst (1978) developed a statistical theory of polyatomic ion–electron dissociative recombination building on the statistical phase-space theory of chemical reactions developed by Pechukas and Light (1965) and Light (1967). The basic assumption is that the decomposition of the collision complex, in our case a superexcited neutral state above the ionization limit, is determined by the phase space available to each product under conservation of angular momentum and energy. HCNH$^+$ was treated in Herbst's pioneering paper and in Table 8.4 these predictions are compared with the branching ratios derived from the ion storage ring experiment of Semaniak et al. (2001).

The agreement is quite good, and it is amusing to note that it is best for the three-body breakup, the channel which Herbst (1978) assumed would be somewhat underestimated by the statistical model. One must bear in mind, of course, that the three-body channel in this particular case is rather small in comparison with the class of polyatomic ions discussed in Section 8.1. The statistical model predicted essentially the same branching ratio for the HCN and HNC channels, and this was also the outcome of theoretical calculations based on *ab initio* electronic structure calculations (Talbi & Ellinger 1998, Shiba et al. 1998). What is quite surprising with these two papers is that they predict crossing (Talbi & Ellinger 1998) and

Table 8.4. *Products of* $HCNH^+ + e^-$ *(0 eV)*

Channel	Herbst 1978[a]	Semaniak et al. 2001[b]
HCN + H	0.23	
HNC + H	0.21	0.675 ± 0.016[c]
CN + H$_2$	0.19	0 ± 0.017
CN + H + H	0.36	0.325 ± 0.032

[a] 100 K and 2 Å in Table 2 of Herbst (1978).
[b] The experiment could not distinguish between the HCN and HNC channels.
[c] The combined H atom contribution is 1.325 (= 0.675 + 2 × 0.325), which is much higher than the 0.63 measured by Adams et al. (1991).

no crossing (Shiba et al. 1998) of the relevant neutral potential energy curves with the ion ground state. This means that whereas Talbi and Ellinger (1998) considered the direct mechanism to be dominant, Shiba et al. invoked Rydberg states and the indirect mechanism to explain the recombination. Jursic (1999) also used *ab initio* calculations to speculate that an uneven HNC/HCN ratio could be obtained by dissociative recombination of HCNH$^+$. None of these three papers went beyond calculations of potential curves. Figure 8.16 shows the electronic structure of HCNH$^+$ and HCNH, including the 1 $^2\Sigma^+$ and 2 $^2\Sigma^+$ states, which are believed to play the major role in the dissociative recombination of HCNH$^+$. Semaniak et al. (2001) used *ab initio* potential curves to identify the 2 $^2\Sigma^+$ as responsible for the three-body decay to CN + H + H.

Tachikawa (1999) was the first to attempt *ab initio* dynamics calculations and identified two dissociation mechanisms; one via a short-lived complex and the other via a long-lived one. His prediction of a channel HC + HN was not confirmed by experiment (Semaniak et al. 2001). Taketsugu et al. (2004), building on the work by Shiba et al. (1998), performed *ab initio* direct trajectory calculations, which include *ab initio* molecular orbital calculations at each point of the trajectory. Figure 8.17 shows snapshots of dissociating HCNH along one trajectory. Taketsugu et al. (2004) found essentially equal production of HCN and HNC. Talbi has also continued her work on HCNH$^+$ (Hickman et al. 2005), and with an approach very different from that of Taketsugu et al. (2004). Hickman et al. (2005) considered the direct mechanism and calculated rates for dissociative recombination through the C−H bond (i.e., formation of HNC + H) and the N−H bond (i.e., formation of HCN + H). They obtained almost identical rate coefficients for the two channels, and a total rate coefficient of 2.4×10^{-7} cm^3 s^{-1}. Channels other than those giving HCN and HNC were not included; the result was nevertheless in very good agreement with the experimental ones. The outstanding questions concerning HCNH$^+$ are whether it is possible to measure the HCN/HNC formation ratio, and what the outcome

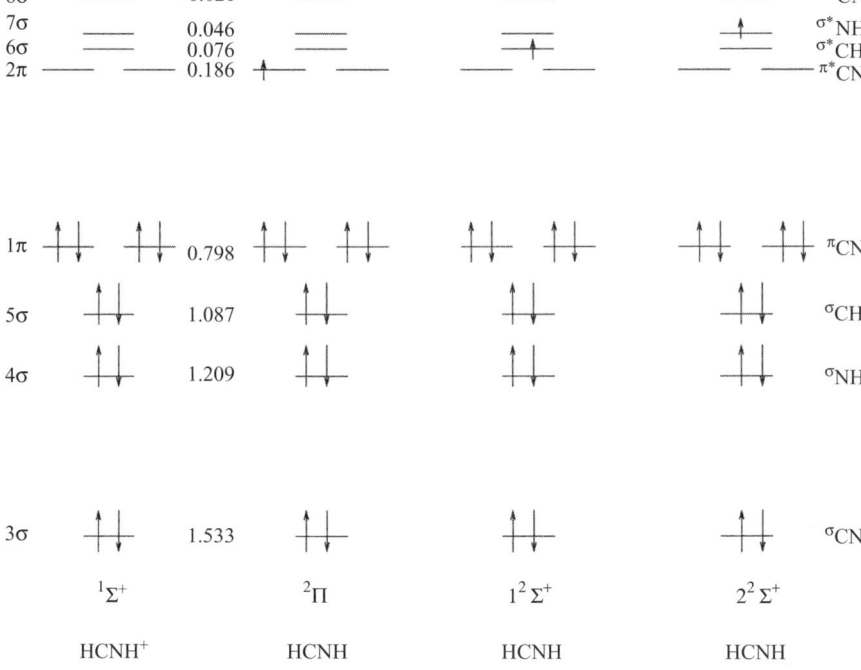

Figure 8.16 Molecular orbital energy levels scheme for the ground state of HCNH$^+$ and the main configurations for the most important state of HCNH. The ion molecular orbital energies are given in hartrees. (Reproduced by permission of the AAS from J. Semaniak, B. F. Minaev, A. M. Derkatch, *et al.*, "Dissociative recombination of HCNH$^+$: Absolute cross-sections and branching ratios," *Astrophys. J. Suppl. Ser.* **135**, pp. 275–283, (2001).)

will be of a fully *ab initio* calculation of the type that has been carried out for H$_3^+$ (Kokoouline & Greene 2003a,b). Such calculations have been started by Ngassam, Orel, and Suzor-Weiner (2005). Electron scattering results on this system have been published (Ngassam & Orel 2007).

HCN$^+$ and its isomer HNC$^+$ have received less attention than HCNH$^+$, and may be more important in Titan's atmosphere (Banaszkiewicz *et al.* 2000, Petrie 2001) than in the interstellar medium. Figure 8.18 shows energy level diagrams for HCN$^+$ and HNC$^+$ and their locations with respect to HCN and HNC. There is only one experiment addressing dissociative recombination of HCN$^+$ and HNC$^+$, a single-pass merged-beam experiment by Sheehan *et al.* (1999), and the results are shown in Fig. 8.19. The production of HNC$^+$ by means of CO$_2$ addition to the ion source, which was operated with the source gases N$_2$ (90%) and C$_2$H$_4$ (10%), was inferred rather than measured. It is clear that the experiments were performed with an undetermined vibrational state distribution, although arguments were later put forward that the HNC$^+$ ions might have been in their zeroth vibrational level

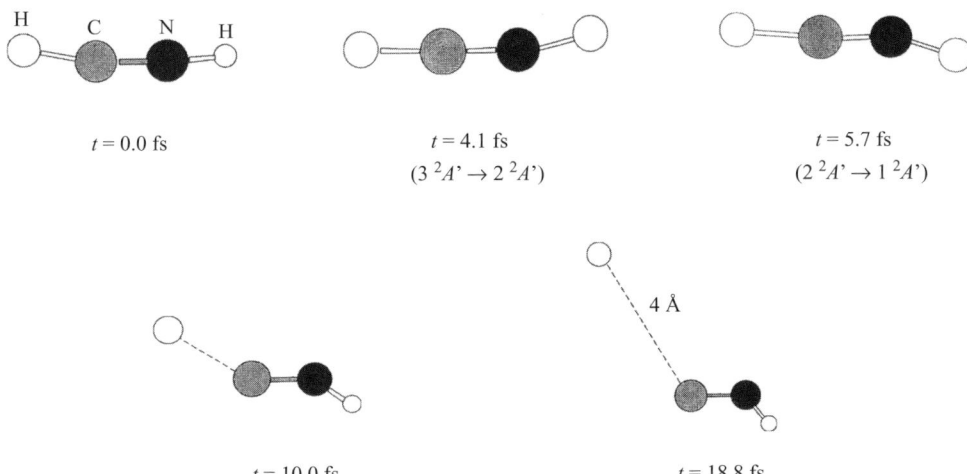

Figure 8.17 Snapshots of molecular structure along one trajectory. The $3\,^2A'$, $2\,^2A'$, and $1\,^2A'$ states in C_s symmetry corresponds to the $2\,^2\Sigma^+$, $1\,^2\Sigma^+$, and $1\,^1\Pi$ states in $C_{\infty v}$ (linear configuration), for which the molecular orbital structure is given in Fig. 8.16. (Reproduced by permission of the AAS from T. Taketsugu, A. Tajima, K. Ishii, and T. Hirano, 2004, "Ab initio direct trajectory simulation with nonadiabatic transitions of the dissociative recombination reaction HCNH$^+$ + $e^- \rightarrow$ HNC/HCN + H," *Astrophys. J.* **608**, pp. 323–329, (2004).)

(see below). Figure 8.19 shows that the cross section was higher when a mixed HCN$^+$/HNC$^+$ beam was used, which implies that the cross section for recombination of HCN$^+$ is higher than that for HNC$^+$. There is a clear change of slope in the cross section below 10 meV. The question is whether this is caused by the energy spread in the electron beam, as shown in Fig. 2.7. Sheehan *et al.* (1999) argued that this is not the case, but the argument is not totally convincing since an independent measurement of the electron temperature was not performed. Thus, it remains an open question whether or not the cross section levels off below 10 meV.

Talbi, Le Padellec, and Mitchell (2000) calculated potential energy surfaces for HCN$^+$ ($X\,^2\Pi$ and $A\,^2\Sigma$), HNC ($X\,^2\Sigma$), and for relevant electronic states of HCN in order to identify the recombination mechanisms. Figure 8.20 shows the HCN and HNC quasidiabatic potential curves of Σ and Π symmetry for the H–CN and H–NC reaction coordinates, respectively. The $A\,^2\Sigma$ state is only 0.4 eV above the ground state of HCN$^+$, and Sheehan *et al.* (1999) assumed that the $A\,^2\Sigma$ state was also populated in their ion-beam experiment. Talbi, Le Padellec, and Mitchell (2000) concluded, based on the potentials in Fig. 8.20(a),(b) that the ground state of HCN$^+$ is favorably crossed by the $2D\,^3\Sigma$, which would drive dissociative recombination via the direct mechanism, whereas HNC$^+$ in its lowest vibrational levels would only recombine through the indirect mechanism. The $v = 2$ level in HNC$^+$ is close to

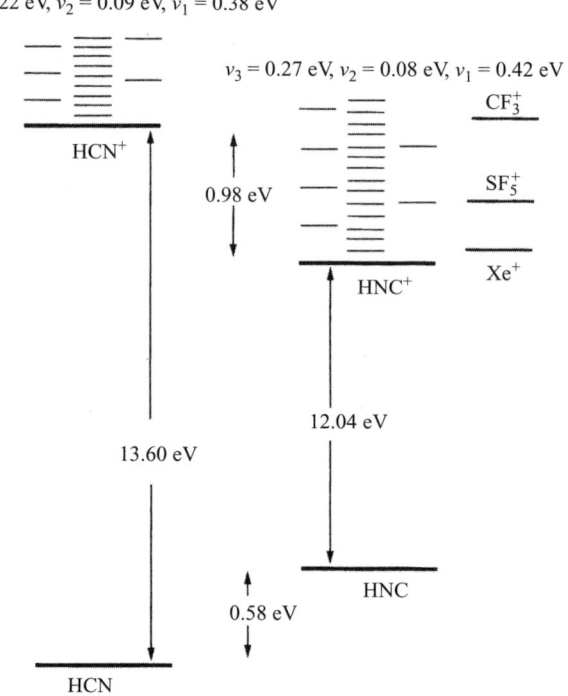

Figure 8.18 Energy level diagrams for HCN$^+$ and HNC$^+$. In an ion flow tube one can use monitors such as CF$_4$ (to produce CF$_3^+$ + HF + CN), Xe (to form Xe$^+$), and SF$_6$ (to produce SF$_5^+$ + HF + CN), and the energy thresholds of these monitors are shown. (Reused with permission from A. Hansel, M. Glantschnig, Ch. Scheiring, W. Lindinger, and E. E. Ferguson, *Journal of Chemical Physics* **109**, 1743, (1998). Copyright 1998, American Institute of Physics.)

the crossing between the $2D\ ^3\Sigma$ state and the electronic ground state of HNC$^+$, and one can assume that the direct mechanism is operative for $v \geq 2$. Electron scattering calculations are needed in order to uncover the recombination mechanism. Such calculations have been started (Orel 2005).

Sheehan *et al.* (1999) measured the cross section using a mixed HCN$^+$/HNC$^+$ beam and a pure HNC$^+$ beam, and the published rate coefficients based on these experiments are

$$\alpha(\text{HCN}^+/\text{HNC}^+) = 3.9 \times 10^{-7}(T/300)^{-0.96} \text{ cm}^3 \text{ s}^{-1}, \quad (8.14)$$

$$\alpha(\text{HCN}^+) = 1.82 \times 10^{-7}(T/300)^{-0.96} \text{ cm}^3 \text{ s}^{-1}. \quad (8.15)$$

Talbi, Le Padellec, and Mitchell (2000) used a set of ion–molecule reaction rates to model the conditions in the ion source used in the single-pass experiment (Sheehan *et al.* 1999), and came to the conclusion that the HNC$^+$ was 96.2% pure, with a

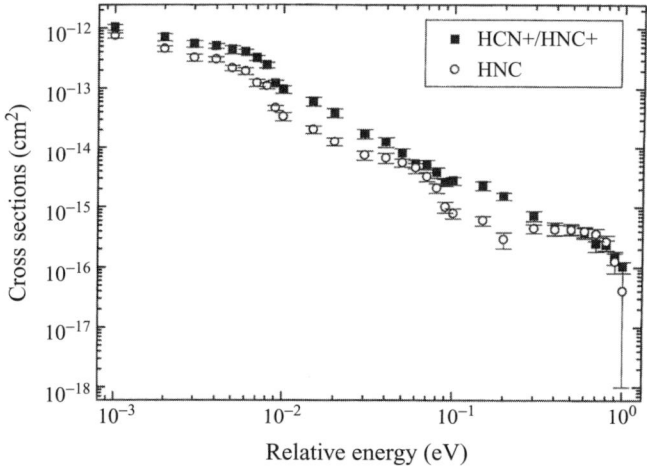

Figure 8.19 Absolute dissociative recombination effective cross sections, $<v_r \sigma>/v_d$, for the isomeric mixture HCN$^+$/HNC$^+$ (filled squares) and HNC (open circles) as a function of detuning energy between 1 meV and 1 eV. Statistical error bars are shown. (Reproduced with permission from C. Sheehan, A. Le Padellec, W. N. Lennard, D. Talbi, and J. B. A. Mitchell, "Merged beam measurement of the dissociative recombination of HCN$^+$ and HNC$^+$," *J. Phys. B* **32**, pp. 3347–3360 (1999), IOP Publishing Limited.)

3.8% HCN$^+$ impurity. They further inferred, based on the results of Hansel *et al.* (1998), that the HNC$^+$ beam was vibrationally relaxed. This gave

$$\alpha(\text{HCN}^+) = 25 \times 10^{-7}(T/300)^{-1.0} \text{ cm}^3 \text{ s}^{-1}, \tag{8.16}$$

$$\alpha(\text{HNC}^+) = 0.9 \times 10^{-7}(T/300)^{-1.0} \text{ cm}^3 \text{ s}^{-1}. \tag{8.17}$$

It is risky to draw such firm conclusions based on ion source modeling, since it is usually difficult to control and characterize ion source conditions to any degree of accuracy (see, for example, the discussion in Section 7.2.4). Mitchell realized this, and in the review by Florescu-Mitchell and Mitchell (2006), the rate coefficient (8.14) is given for mixed HCN$^+$/HNC$^+$ and (8.17) is given as a lower limit for the rate coefficient of recombination of HNC$^+$ (see Table 8.8 in the review).

Johnsen and coworkers (Rosati *et al.* 2007) have used the Pittsburgh flowing afterglow apparatus to measure the rate coefficient for dissociative recombination of HNC$^+$, and the absolute yields of formation of CN ($B\ ^2\Sigma^+$) and CN ($A\ ^2\Pi$). The rate coefficient for HNC$^+$ was measured to be 2×10^{-7} cm^3 s^{-1}, which is in good agreement with that of Sheehan *et al.* (1999). The yield of CN ($B\ ^2\Sigma^+$) formation was found to be 22% and that of CN ($A\ ^2\Pi$) formation 14%. The rotational temperature of the CN($B\ ^2\Sigma^+$) was estimated to be approximately 2500 K, and both the $B\ ^2\Sigma^+$ and $A\ ^2\Pi$ states were found to contain substantial vibrational excitations.

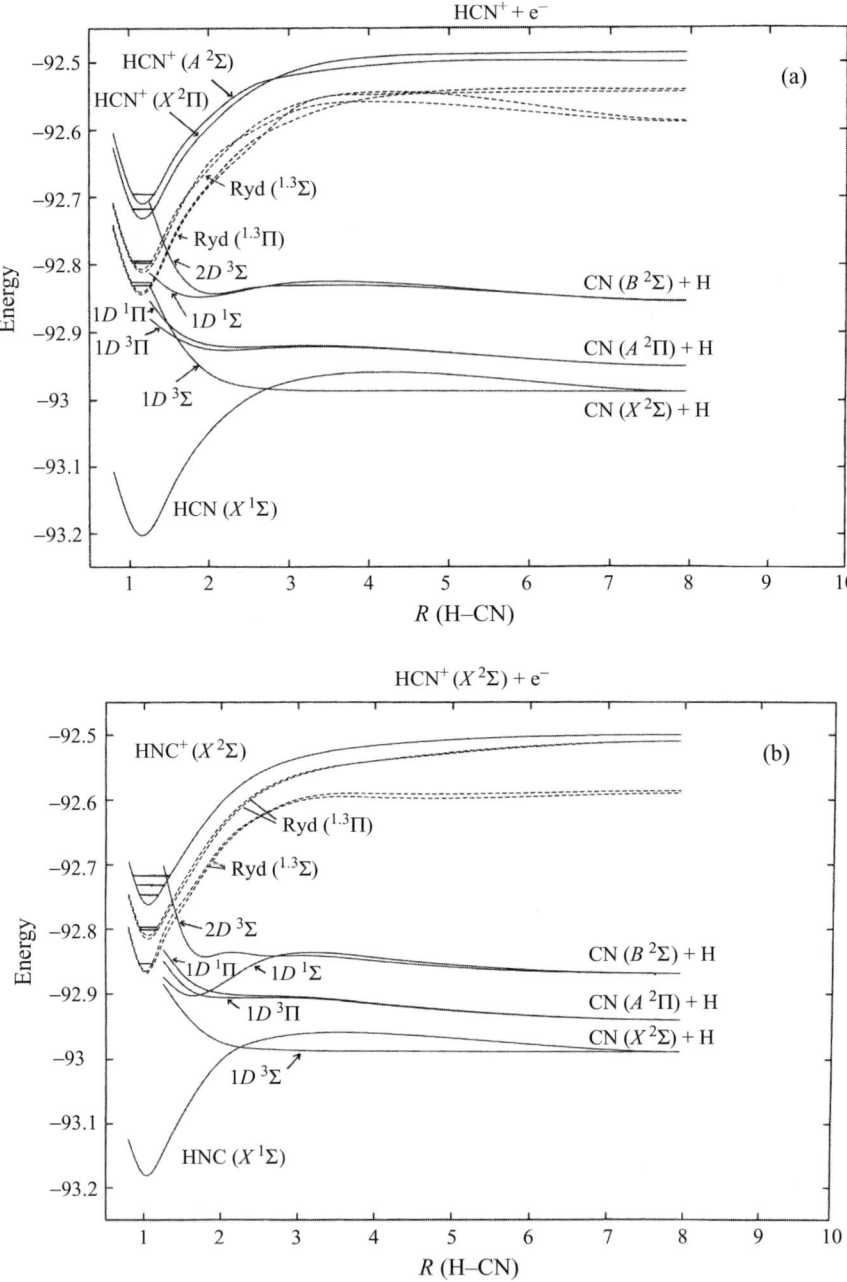

Figure 8.20(a),(b) HCN and HNC quasidiabatic potential energy surfaces of Σ and Π symmetry for the H–CN and H–NC dissociative reaction coordinates, respectively (in Fig. 8.20(b) the x-axis should be H–NC). Lowest Rydberg states of both singlet and triplet multiplicities are shown as dashed curves. Energies are in atomic units (hartree) and distances in ångströms. (Reproduced with permission from D. Talbi, A. Le Padellec, and J. B. A. Mitchell, "Quantum chemical calculations for the dissociative recombination of HCN^+ and HNC^+," *J. Phys. B* **33**, pp. 3631–3646 (2000), IOP Publishing Limited.)

This is in disagreement with the classical impulse model of Bates (1993b), which predicts that population of $v = 0$ and 1 should exceed 90%.

8.2.4 H_3O^+

It is not surprising that the hydronium ion, H_3O^+, is one of the best studied with respect to dissociative recombination. It is an interstellar ion and the recombination end products OH and H_2O are interstellar molecules, water of course having the extra attribute of being a prerequisite for life as we know it. It was discovered in the interstellar medium by Wooten *et al.* (1986), and the chemistries of the hydronium ion, water, and the hydroxyl radical in dark clouds were discussed shortly afterwards by Sternberg, Dalgarno, and Lepp (1987). Being the principal ion in flames, the loss rate of H_3O^+ by electron recombination has been measured many times. The first flame study was possibly that of Wilson (1931) who reported a rate coefficient of recombination of 8.5×10^{-7} cm^3 s^{-1} at 2000 K. King (1957) performed a recombination measurement in methane–air flames and obtained a rate coefficient of 2.5×10^{-7} cm^3 s^{-1}. Wilson and Evans (1967) used a shock-tube technique and a microwave system to measure the electron density to obtain the rate of recombination behind the shock waves. Their results are the oldest in Fig. 8.21, which shows results from a selection of measurements; the figure caption also gives the results of, and references to, work not displayed in the figure. A compilation of results from the many flame studies and other early studies can be found in Mul *et al.* (1983), and a review of flame studies has been given by Butler and Hayhurst (1996). The agreement between the result from CRYRING (Neau *et al.* 2000), the ion trap results (Heppner *et al.* 1976), and those from the flowing afterglow experiments (Herd, Adams, & Smith 1990, Gougousi, Golde, & Johnsen 1997) is very good. This is a very satisfactory situation, since there is no doubt that these are the most reliable results.

While the situation concerning the rate coefficient is very satisfactory, the measurements and calculations of product branching ratios are more controversial. This makes H_3O^+ an interesting case study for the experimental and theoretical methods that have been used to derive $H_3O^+ + e^-$ end products. Table 8.4 shows the results of those experimental and theoretical studies for which complete branching ratios have been published, and results from two flowing afterglow experiments for which the complete branching ratios was not obtained but instead the information on the

---

[3] The results of Mul *et al.* (1983) pose some problems. The original result is about 6×10^{-7} cm^{3}s^{-1} at 300 K. It was obtained during the period when the merged-beam apparatus gave a factor of two too large cross sections, and in Mitchell's review from 1990 the result is given as $3.15 \times 10^{-7}(T/300)^{-0.5}$ cm^{3}s^{-1}. This is what Neau *et al.* (2000) used to prepare Fig. 7 in their paper, reproduced here as Fig. 8.21. In the review by Florescu-Mitchell and Mitchell (2006), however, the rate coefficient from Mul *et al.* (1983) is given as $7.6 \times 10^{-7}(T/300)^{-0.83}$ cm^{3}s^{-1}, identical with the result of Neau *et al.* (2000).

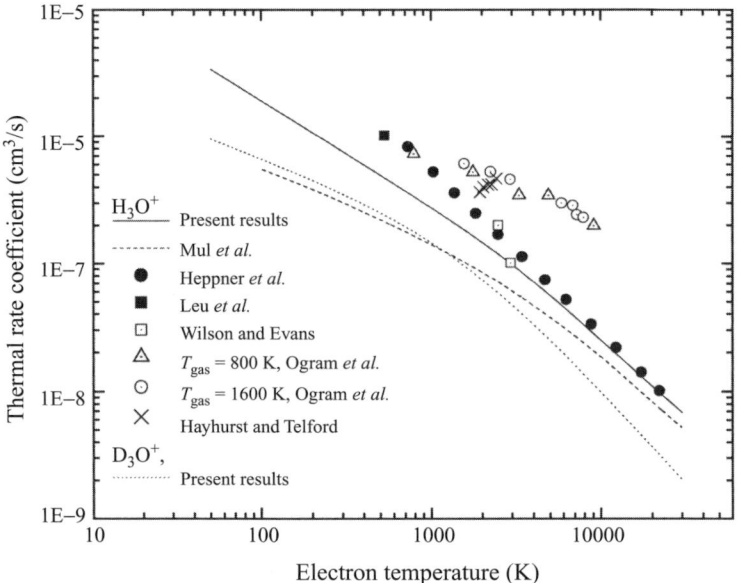

Figure 8.21 Thermal rate coefficients for H_3O^+ and D_3O^+ as a function of electron temperature. "Present results": ion storage ring, Neau et al. (2000); "Mul et al.": single-pass merged beams, Mul et al. (1983)[3]; "Heppner et al.": ion trap technique, Heppner et al. (1976); "Leu et al.": stationary afterglow, Leu, Biondi, and Johnsen (1973a); "Wilson and Evans": shock tube technique, Wilson and Evans (1967); "Ogram et al.": shock tube technique, Ogram, Chang, and Hobson (1980); "Hayhurst and Telford": flame study including mass spectrometer, Hayhurst and Telford (1974). In addition to the results displayed in this figure, the following results have been obtained: $(8.0\pm1.5)\times10^{-7}$ cm^{3}s^{-1} at 295 K, flowing afterglow (Gougousi, Golde, & Johnsen 1997a); $(3.6\pm1.5)\times T^{-2.1\pm0.7}$ cm^{3}s^{-1} in the range 1820 to 2400 K (relative variation), flame study (Butler & Hayhurst 1996); 1.0×10^{-6} cm^{3}s^{-1} at 300 K, flowing afterglow (Herd, Adams & Smith 1990); $(0.0132\pm0.0004)\times T^{-1.37\pm0.05}$ cm^{3}s^{-1}, flame study (Guo & Goodings 2000). (Reused with permission from A. Neau, A. Al-Khalili, S. Rosén, et al., *Journal of Chemical Physics* **113**, 1762, (2000). Copyright 2000, American Institute of Physics.)

OH radicals was more complete than in the other experiments (Herd, Adams, & Smith 1990, Gougousi, Johnsen, & Golde 1997a). Other results will be discussed in the text below. The branching ratios for H_3O^+ are also given in Table 17 in the review by Florescu-Mitchell and Mitchell (2006). Unfortunately, the results from CRYRING have been mixed up in that table, and the subsequent discussion on p. 361 in the review is based on these erroneous numbers. The branching ratio for the $H_2O + H$ channel was given as 0.67 by Florescu-Mitchell and Mitchell (2006), while the correct value is 0.18 (Neau et al. 2000). It is claimed by Florescu-Mitchell and Mitchell (2006)[4] that the production of OH measured in CRYRING is half that

---

[4] Errata have been distributed to part of the dissociative recombination community by Mitchell.

measured in ASTRID and in the afterglow experiments, but it follows from Table 8.5 that the agreement between the CRYRING result (Neau *et al.* 2000) and that of the flowing afterglow experiments (Herd, Adams, & Smith 1990, Gougousi, Johnsen, & Golde 1997a) is good.

Herbst (1978) made the first attempt to derive the branching ratios for electron recombination with H_3O^+ using the statistical phase-space method discussed above. It later became clear that this theory performed less well for H_3O^+ as compared with $HCNH^+$. Bates (1986, 1987a) used a more intuitive approach and argued that the channel involving the least rearrangement of valence bonds would be the most important; for H_3O^+ he concluded that this would be $H_2O + H$. In fact, Herbst had earlier also used an intuitive approach (Green & Herbst 1979) in order to obtain branching ratios in a much less computationally demanding way than the phase-space theory. Green and Herbst (1979) proposed that the rate at which potential energy is converted to kinetic energy, which is the very essence of dissociative recombination, would be inversely proportional to the mass. Thus, a hydrogen atom should be more easily ejected than heavy atoms or molecular fragments. Although they did not consider H_3O^+, the argument would apply to the $H_2O + H$, $OH + H_2$, and $OH + H + H$ channels. The approach of Bates (1986, 1987a) is quite different. In Bates's view a polyatomic ion was regarded as containing a single ionized atom having the valency of the isoelectronic neutral atom and a number of neutral atoms having their normal valency. The molecular ion is kept together by localized valence bonds, and when dissociative recombination occurs one of the valence bonds of the ionized atom is broken, and the molecule is broken into two parts. Bates and Herbst (1988) argued that charge dispersal would not invalidate this, basing their argument on Bates's work on bond energies in polyatomic ions (Bates 1987b), in which he conceived of polyatomic ions as being held together by localized valence bonds which are constant in strength from one to another. The valence-bond approach had an impact on the chemical modeling of interstellar clouds (Millar *et al.* 1988, Herbst 1988, 1989). Around 1988 the situation was such that one approach (Bates 1986, 1987a) predicted 100% water from dissociative recombination H_3O^+, whereas the other approaches (Herbst 1978, Green & Herbst 1979) predicted also the formation of OH.

The FALP experiment of Herd, Adams, and Smith (1990) in which they used laser-induced fluorescence to probe OH radical production from dissociative recombination of H_3O^+ was of decisive importance. Preliminary results of OH formation had already been reported in the summer of 1988 (see Adams and Smith (1989)), and Bates, knowing about it,[5] was able to publish a revised theory (Bates 1989) which incorporated a favorable potential energy curve crossing that gives importance

---

[5] See acknowledgement in Bates (1989)

Table 8.5. Products of dissociative recombination of H_3O^+ with thermal electrons

Channel	Herbst[a]	Williams[b]	Andersen[c] Vejby-Christensen[d]	Neau[e]	Jensen[f]	Kayanuma[g]	Gougousi[h]	Herd[i]
$H_2O + H$	0.09	0.05	0.33 ± 0.08	0.18 ± 0.05	0.25 ± 0.01	0.10		
$OH + H + H$	0.08	0.29	0.48 ± 0.08	0.67 ± 0.06	0.60 ± 0.02	0.87	0.65 ± 0.08[j]	0.65 ± 0.15[j]
$OH + H_2$	0.83	0.36	0.18 ± 0.07	0.11 ± 0.05	0.14 ± 0.01	0.015		
$O + H_2 + H$		0.30	0.01 ± 0.04	0.04 ± 0.06	0.013 ± 0.005	0.015		

[a] Statistical phase-space theory; $\lambda_{ROT} = 1$, 100 K, and 2 Å in Table 2 in Herbst (1978)
[b] Flowing afterglow and selected ion flow tube (SIFT); Williams et al. (1996)
[c] Ion storage ring ASTRID; Andersen et al. (1996a)
[d] Ion storage ring ASTRID; Vejby-Christensen et al. (1997)
[e] Ion storage ring CRYRING; Neau et al. (2000)
[f] Ion storage ring ASTRID; Jensen et al. (2000)
[g] *Ab initio* direct trajectory simulations; Kayanuma, Taketsugu, and Ishii (2006)
[h] Flowing afterglow; Gougousi, Johnsen, and Golde (1997a)
[i] Flowing afterglow; Herd, Adams, and Smith (1990)
[j] The sum of $OH + H + H$ and $OH + H_2$

to the OH + H$_2$ channel. Herd, Adams, and Smith (1990) measured the combined branching ratios of OH + H + H and OH + H$_2$, labeled f_{OH}, to be 0.65, and they found that branching into the zeroth vibrational level the $X\,^2\Pi$ state was 0.46. Adams *et al.* (1991) measured f_H, with contributions from OH + H + H, O + H$_2$ + H, and H$_2$O + H, to be in the range 0.80–1.16 depending on the precursor ion; $f_H = 0.93$ is quoted by Adams (1992). Bates (1991b) used the data from Herd, Adams, and Smith (1990) and Adams *et al.* (1991) to infer, under the assumption that the O + H$_2$ + H is negligible, branching ratios of approximately 1/3 for each of the first three channels in Table 8.5.

Galloway and Herbst (1991) developed the statistical phase-space theory to include the vibrational angular momentum in polyatomic product channels, processes leading directly to three-body breakup, the indirect mechanism involving Rydberg states, and the assumption that channels having unfavorable Franck–Condon factors can be neglected. They calculated f_{OH} to be 0.77, indeed in very good agreement with the afterglow result (Herd, Adams, & Smith 1990). They also predicted small values for the three-body channels, of the order of 0.01 and no larger than 0.1 (Galloway & Herbst 1991).

In 1996 two papers appeared which both presented results for the complete neutral product branching ratios (Williams *et al.* 1996, Andersen *et al.* 1996a); these results are shown in Table 8.5. Andersen *et al.* obtained their results from an experiment at ASTRID using the grid technique described in Section 2.2.1. Williams *et al.* applied a combination of techniques in a way that had never been attempted before nor has it been tried since. The detection of O atoms was accomplished by the reaction

$$O(^3P) + GeH_4 \rightarrow OH + GeH_3, \tag{8.18}$$

followed by detection of OH by laser-induced fluorescence. The problem with this reaction, as compared with the scheme to detect H atoms (see Fig. 2.21), is that the rate coefficient is much smaller for converting O atoms to OH. There is also a rapid secondary reaction in which OH is lost in collisions with GeH$_4$. An advantage in the recombination of H$_3$O$^+$ is that oxygen is only formed in its ground state, whereas for other ions, such as H$_2$O$^+$ (see Section 8.1), O (^{1}D) is also formed. Reactions of metastable oxygen with GeH$_4$ can lead to products other than OH. The results given in Table 8.5 are surprising, in particular the large branching ratio for the O + H$_2$ + H channel, which was previously believed to be negligible, and measured to be negligible by Andersen *et al.* (1996a). When the branching ratios in dissociative recombination of polyatomic ions were reviewed (Andersen, Heber, & Zajfman 2000) at the International Astronomical Union symposium on astrochemistry "From Molecular Clouds to Planetary Systems" in Sogwipo, Cheju, Korea, in August 1999, the following question was asked by B. Rowe (Andersen,

Heber, & Zajfman 2000, p. 271): "In the experiment by Williams, Adams, *et al.* (1996, *MNRAS*, **282**, 413) how do they know that H_3O^+ is vibrationally cold? To my knowledge, it is very difficult to ensure that protonated ions are vibrationally cold in a flowing afterglow experiment." The answer from D. Zajfman was: "This could be a central reason for the difference between our result and Williams, Adams, *et al.* result." Williams *et al.* (1996) claimed that the pressure in the flow tube was sufficient to ensure vibrationally cold H_3O^+.

Ion storage rings have been used to obtain more results. Vejby-Christensen *et al.* (1997) gave a more detailed account of the ASTRID experiment, and results from an improved experiment at ASTRID were published by Jensen *et al.* (2000). Those results essentially agree within experimental error with the results from CRYRING (Neau *et al.* 2000). Results for isotopologs have been published by Jensen *et al.* (2000) and Neau *et al.* (2000).

Gougousi, Johnsen, and Golde (1997a) used the Pittsburgh flowing afterglow apparatus in a detailed study of OH formation. They found the branching ratio for OH ($X\,^2\Pi$, $v = 0$) to be 0.48 ± 0.07, that of OH ($X\,^2\Pi$, $v = 1$) to be 0.12 ± 0.02, and that of OH ($X\,^2\Pi$, $v \geq 1$) to be 0.17 ± 0.04. The yield of OH ($A\,^2\Sigma^+$) was found to be very small. The agreement between the two flowing afterglow experiments by Herd, Adams, and Smith (1990) and Gougousi, Johnsen, and Golde (1997a) concerning f_{OH} is excellent, and the agreement with the storage ring results is clearly satisfactory.

Dissociative recombination of H_3O^+ has also been studied by means of *ab initio* calculations since the late 1990s, with the specific aim of obtaining the product branching ratios. Ketvirtis and Simons (1999) calculated potential energy surfaces associated with dissociative recombination of H_3O^+, and used these surfaces in discussions of the ASTRID results (Andersen *et al.* 1996a, Vejby-Christensen *et al.* 1997). Curiously, even though the paper of Ketvirtis and Simons is almost like a mini-review of dissociative recombination, with an extensive reference list, they do not discuss the results of Williams *et al.* (1996), which is included in their reference list. Regarding the $O + H_2 + H$ channel they just concluded that (Ketvirtis and Simons 1999, p. 6560): "As the ASTRID experimental studies did not result in the observation of $H_2 + O + H$ in significant quantities (branching ratio 0.01 ± 0.04) this reaction path was not considered in the present computational investigation."

Figure 8.22 shows the dissociation pathways calculated by Ketvirtis and Simons (1999). Based on the calculated potential energy surfaces and the experimentally determined branching ratios (Andersen *et al.* 1996a, Vejby-Christensen *et al.* 1997), Ketvirtis and Simons assumed that there should be sufficient excess energy in H_2O ($X\,^1A_1$) to cause further fragmentation, something which would enhance the $OH + H + H$ channel.

Tachikawa (2000) performed *ab initio* direct dynamics calculations and found that the dominating decay was $H_2O + H$, which is expected in view of the approximations used by Tachikawa. His focus was on the electron capture rather than on a complete description of the dissociation dynamics; for example, dynamics calculations on Rydberg states were not included. Kayanuma, Taketsugu, and Ishii (2006) used the same *ab initio* direct trajectory calculations, with surface hopping invoked in regions where two potential surfaces are close, that were used for HCNH$^+$ (Taketsugu *et al.* 2004). This is so far the most ambitious effort to calculate product branching ratios for recombining H_3O^+, and it follows from Table 8.5 that they predict the same trend as the ion storage rings, although the theoretical branching ratio for the $OH + H + H$ channel is larger than the experimental ones.

In summary, the situation concerning dissociative recombination of H_3O^+ is satisfactory. There is consensus on the rate coefficient, and theory seems to support the branching ratios measured in ion storage rings. On top of the wish list for future work is a flowing afterglow measurement of the complete product branching ratios.

8.2.5 CO_2^+

In the 1970 review by Bardsley and Biondi, CO_2^+ is one of only two covalently bonded polyatomic ions for which a rate coefficient is listed (the other one is H_3O^+). The motivation for the stationary afterglow measurement (Weller & Biondi 1967) was the emerging exploration of the Martian ionosphere and the desire to construct a theoretical model for it. Gutcheck and Zipf (1973) performed one of the very early experiments in which emission spectroscopy was used to measure the neutral recombination reaction products (see Section 2.3.1). By recording the fourth positive system ($A\ ^1\Pi - X\ ^1\Sigma^+$) of CO, they were able to infer the rate coefficient for $CO_2^+ + e^- \rightarrow CO\,(A\ ^1\Pi) + O$ and relate it to the total dissociative recombination rate coefficient; it was found to be about 5% of the total rate. Another early study of product formation was that of Wauchop and Broida (1972), who studied the CO ($a\ ^3\Pi - X\ ^1\Sigma^+$) emission spectrum in a He + CO_2 flowing afterglow. No rate coefficients were measured, but Wauchop and Broida (1972) measured the CO ($a\ ^3\Pi$) branching ratio to be 0.55. CO_2^+ was also one of the ions studied in the first flowing afterglow experiment (Mahdavi, Hasted, & Nakshbandi 1971). Naturally the ion was also studied in the FALP apparatuses in Birmingham and Pittsburgh (Geoghegan, Adams, & Smith 1991, Gougousi, Golde, & Johnsen 1997), and the results are in excellent agreement. Ion storage rings were later used to obtain cross sections and hence temperature-dependent rate coefficients (Seiersen *et al.* 2003c, Viggiano *et al.* 2005). Figure 8.23 shows the measured $<v_r\sigma>$ as a function of detuning energy in ASTRID and CRYRING. The experimentally

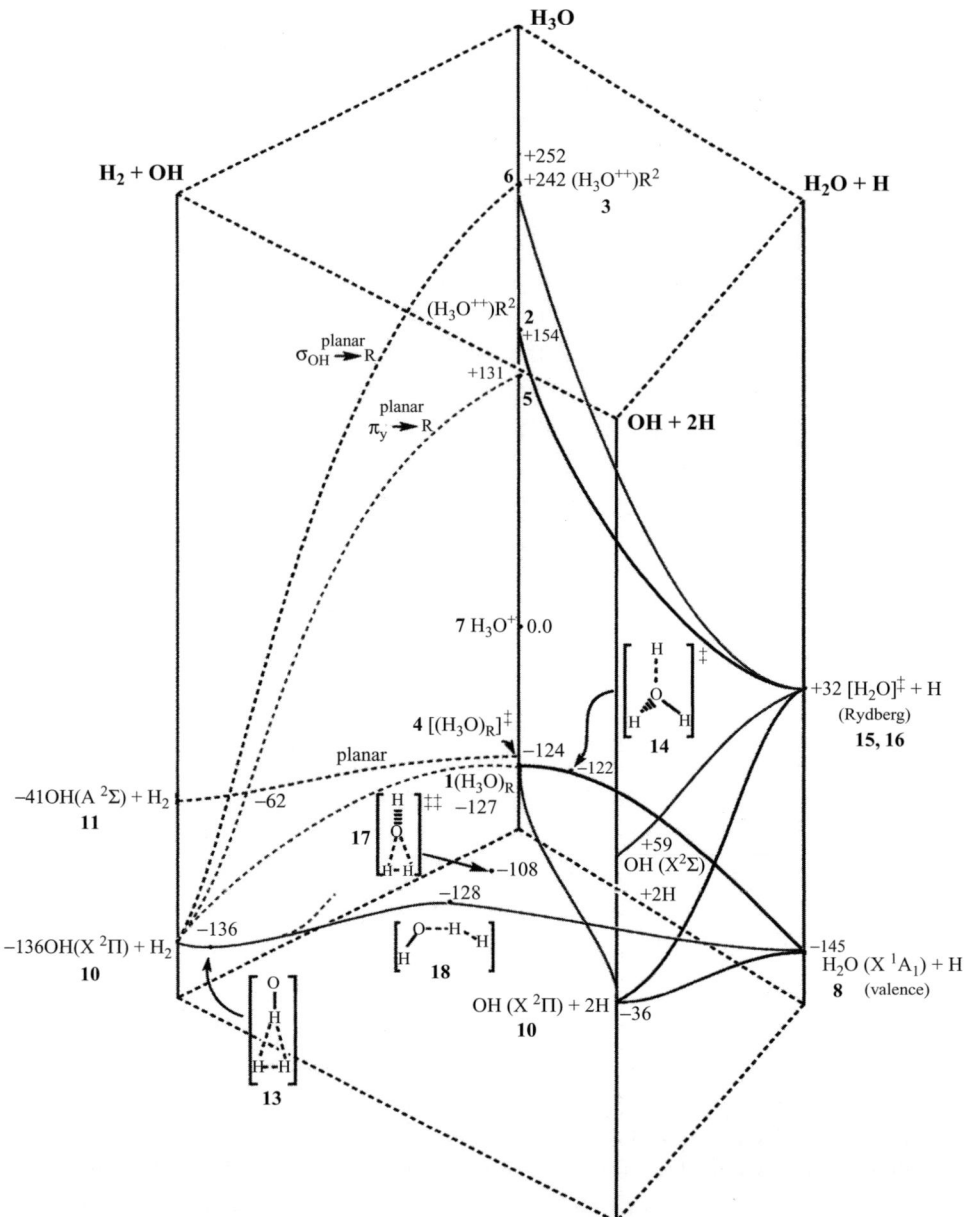

Figure 8.22 *Ab initio* calculated dissociation pathways for $H_3O^+ + e^-$ by Ketvirtis and Simons (1999). The electronic ground state of H_3O^+ is labeled **7**. The remaining species are as follows: **1** represents the ground state of $(H_3O)_R$, which is a Rydberg state with pyramidal C_{3v} structure and the intial product when the electron is captured by H_3O^+; **2** and **3** are electronically excited states of H_3O and proposed to be associated with structures in the cross section near 2–10 eV (see Vejby-Christensen *et al.* 1997, Neau *et al.* 2000); **4** is the planar (D_{3h}) version of molecule **1**, and the low inversion barrier between **1** and **4** makes the latter

8.2 Astrophysical molecular ions

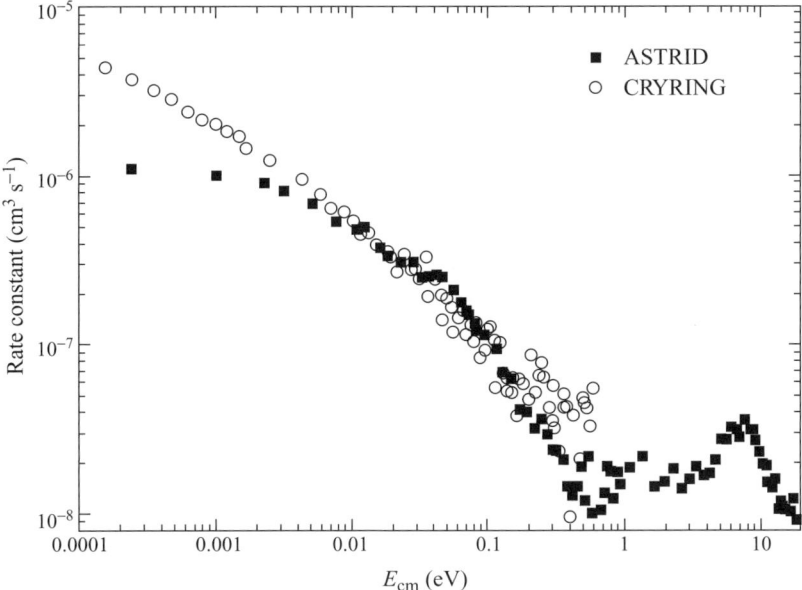

Figure 8.23 Measured ($v_r\sigma$((labeled Rate constant) as a function of the center-of-mass energy (detuning energy). Circles are data from CRYRING (Viggiano *et al.* 2005) and squares are data from ASTRID (Seiersen *et al.* 2003). The divergence of the data points below 10 meV is due to the difference in electron beam temperature. (Reused with permission from A. A. Viggiano, A. Ehlerding, F. Hellberg, *et al.*, *Journal of Chemical Physics* **122**, 226101, (2005). Copyright 2005, American Institute of Physics.)

measured rate coefficients around 300 K are given in Table 8.6. The consistency is very good. It is worth noting that Seiersen *et al.* (2003c) also measured the dissociative recombination rate coefficient for CO_2^{2+} and found it to be similar to that of CO_2^+. This is an important result for the modeling of the CO_2^{2+} density in the atmosphere of Mars (Witasse *et al.* 2002, 2003).

Figure 8.22 (*cont.*) important for the OH+H_2 channel; **5** and **6** represent electronically excited states of **4**; **8** is the ground state of the water molecule ($X\,^1A_1$); **9** is atomic hydrogen (the label is not shown); **10** and **11** represent the OH radical in the ground ($X\,^2\Pi$) and lowest electronically excited ($A\,^2\Sigma^+$) states, respectively; **12** is molecular hydrogen (the label is not shown); **13** is a loose van der Waals complex between OH($X\,^2\Pi$) (species **10**) and H_2 (species **12**); **14** is the transition state for dissociation of (H_3O)$_R$(species **1**) into $H_2O(X\,^1A_1)$ + H; **15** and **16** are H-atom ejection channels where H_2O is formed in Rydberg states; **17** is a second-order saddle point on the ground state H_3O surface and is of little significance in the context of dissociative recombination; **18** is the transition state along the hydrogen abstraction reaction OH+H→H_2O+H. (Reprinted with permission from A. E. Ketvirtis and J. Simons, "Dissociative recombination of H_3O^+," *J. Phys. Chem. A* **103**, pp. 6552–6563 (1999). Copyright (1999) American Chemical Society.)

Table 8.6. *Experimental rate coefficients for* CO_2^+

$\alpha(CO_2^+)$ $(10^{-7} \text{ cm}^3 \text{ s}^{-1})$	T_e (K)	Comment	Reference
3.8 ± 0.5	300	Stationary afterglow	Weller & Biondi 1967
3.4 ± 1.2	300	Flowing afterglow	Mahdavi, Hasted, & Nakshbandi 1971
4.0 ± 0.4	300	Stationary afterglow	Gutcheck and Zipf 1973
3.1 ± 0.6	300	FALP	Geoghean, Adams, & Smith 1991
3.5 ± 0.5	295	FALP	Gougousi, Golde, & Johnsen 1997
6.5 ± 1.9	300	ISR, ASTRID	Seiersen *et al.* 2003c
4.2 ± 0.9	300	ISR, CRYRING	Viggiano *et al.* 2005

FALP: flowing afterglow/Langmuir probe, ISR: ion storage ring.

The pioneering work of Gutcheck and Zipf (1973) was followed by other studies of the recombination reaction products, since they could be a possible source of the fourth positive system of CO observed in the atmospheres of Mars (Barth *et al.* 1969) and Venus (Durrance, Barth, & Stewart 1980). Vallée *et al.* (1986) found a direct correlation between the vibrational excitation in CO_2^+ and the CO $(A\,^1\Pi)$ vibrational distribution, and more detailed studies of $A\,^1\Pi$ state rovibrational distributions were performed by Tsuji *et al.* (1995). Skrzypowski *et al.* (1998) measured the formation yield of CO $(a\,^3\Pi)$ and agreed with Wauchop and Broida (1972) that it is quite large, although their result, 0.29 ± 0.10, was lower than the early result of 0.55 (Wauchop & Broida 1972). Tsuji *et al.* (1998) detected emission found from other triplets states in CO and estimated the relative formation yields of these states. More accurate measurements of the triplet state formation yields were performed by Rosati, Johnsen, and Golde (2003): $f(a'\,^3\Sigma^+) = 0.053 \pm 0.024$, $f(d\,^3\Delta_i) = 0.057 \pm 0.024$, and $f(e\,^3\Sigma^-) \sim 9 \times 10^{-4}$.

Seiersen *et al.* (2003c) studied the exothermic channels:

$$CO_2^+ + e^- \rightarrow CO_2 + 13.8 \text{ eV}, \quad (8.19a)$$

$$CO_2^+ + e^- \rightarrow C + O_2 + 2.3 \text{ eV}, \quad (8.19b)$$

$$CO_2^+ + e^- \rightarrow CO + O + 8.3 \text{ eV}. \quad (8.19c)$$

Somewhat surprisingly they found branching ratios of 0.04 ± 0.03 and 0.09 ± 0.03 for (8.19a) and (8.19b), respectively. In particular channel (8.19b), which is a rearrangement reaction, was predicted to be a large source of C atoms in the Martian thermosphere (Fox 2005). The branching ratios was measured again in CRYRING (Viggiano *et al.* 2005), resulting in a unity branching ratio for the CO + O channel (8.19c). A possible explanation for the unexpected results from ASTRID (Seiersen *et al.* 2003c) could be an incorrect background subtraction. Provided that the CRYRING result is correct, which means that CO_2^+ always recombines by a simple bond breaking to CO + O, there is no need to reanalyze the ionospheric models of Mars and Venus.

8.2.6 Other astrophysical polyatomic ions

We will end this section by listing other molecular ions of astrophysical interest (defined in a loose sense) for which studies of their recombination properties have been made (Table 8.7). No detailed results are given, since these can be found in the review by Florescu-Mitchell and Mitchell (2006), and in the web-based database URL: http://mol.physto.se/DRdatabase. Hydrocarbon ions are discussed in Section 8.3, but those with less than three C atoms are listed in Table 8.7.

8.3 Cluster ions

Cluster ions, because of their importance in the D region of the ionosphere (Wayne 2000), were among the early ions for which recombination with electrons was studied, and results for the dimers of the atmospheric molecular ions, $N_2 \cdot N_2^+$, $O_2 \cdot O_2^+$ and $NO \cdot NO^+$, and the water cluster ion $H_3O^+ \cdot H_2O$ are listed in the review by Bardsley and Biondi (1970). These results all derive from stationary afterglow measurements in Biondi's laboratory (Kasner & Biondi 1965, 1968, Weller & Biondi 1968). The rate coefficients are all larger than 10^{-6} cm^3 s^{-1} at 300 K, and in all cases larger than the rate coefficients for the corresponding monomer ions. The electron-temperature dependence of the recombination coefficient was measured for $N_2 \cdot N_2^+$ (Whitaker, Biondi, & Johnsen 1981b), $O_2 \cdot O_2^+$ (Dulaney, Biondi, & Johnsen, 1988), $CO^+ \cdot (CO)_n$ (Whitaker, Biondi, & Johnsen 1981a), $NH_4^+ \cdot (NH_3)_n$ (Huang, Biondi, & Johnsen 1976), and $H_3O^+ \cdot (H_2O)_n$ (Leu, Biondi, & Johnsen 1973b, Huang et al. 1978) using the stationary afterglow technique. Very weak temperature dependences were found for the $NH_4^+ \cdot (NH_3)_n$ and $H_3O^+ \cdot (H_2O)_n$ ion, but in his analysis of microwave afterglow measurements, Johnsen (1987) suspected that inelastic collisions might prevent an increase of the electron temperature (see Section 2.3.1). When Johnsen (1993b) and Skrzypkowsk and Johnsen (1997) applied the radio-frequency heating/radio-frequency probe stationary afterglow technique to $H_3O^+ \cdot (H_2O)_n$ and $NH_4^+ \cdot (NH_3)_{n=2,3}$, they found temperature dependences of $T^{-0.5}$ and $T^{-0.65}$, respectively.

Even if the near-zero temperature dependence turned out to be incorrect, and the temperature dependence measured by Whitaker, Biondi, and Johnsen (1981a,b) for $N_2 \cdot N_2^+$ and $CO^+ \cdot CO$ was slightly underestimated (Johnsen 1987), the high dissociative recombination rate for cluster ions is firmly established. In addition to those cluster ions already mentioned, studies using the FALP technique had been done on $(CH_3OH)_2 \cdot H^+$, $(CH_3OH)_3 \cdot H^+$, $(C_2H_5OH)_2 \cdot H^+$, and $(C_2H_5OH)_3 \cdot H^+$ by

Table 8.7. *Experimental studies of polyatomic molecular ions of astrophysical importance and not inlcuded in Sections 8.2.1–8.2.5*

Ion	Method	Information	Reference
SO_2^+	ISR, CRYRING	Cross section, rate coefficient, branching ratios	Geppert et al. 2004c
OCS^+	ISR, CRYRING	Cross section, rate coefficient, branching ratios	Montaigne et al. 2005
N_2O^+	ISR, CRYRING	Cross section, rate coefficient, branching ratios	Hamberg et al. 2005
	FALP	Rate coefficient	Gougousi, Golde, & Johnsen 1997
O_3^+	ISR, CRYRING	Cross section, rate coefficient, branching ratios	Zhaunerchyk et al. 2007
$C_n^+, 3 \leq n \leq 6$	ISR, ASTRID	Branching ratios	Heber et al. 2006
O_2H^+	Flowing afterglow	f_H and f_{OH}	Adams et al. 1991
H_2O^+	Flowing afterglow	Branching ratios	Rowe et al. 1988
	ISR, ASTRID	Branching ratios	Vejby-Christensen et al. 1997
H_2O^+/HDO^+	ISR, ASTRID	Cross section, rate coefficient, branching ratios	Jensen et al. 1999
	ISR, CRYRING	Cross section, branching ratios	Rosén et al. 2000
CH_2^+	Merged beams	Cross section, rate coefficient	Mul et al. 1981
	ISR, CRYRING	Cross section, rate coefficient, branching ratios	Larson et al. 1998
	Merged beams	Cross section, rate coefficient	Sheehan & St.-Maurice 2004a
C_2H^+	Merged beams	Cross section, rate coefficient	Mul & McGowan 1980
	ISR, CRYRING	Cross section, rate coefficient, branching ratios	Ehlerding et al. 2004
NH_2^+	ISR, CRYRING	Branching ratios	Vikor et al. 1999
	ISR, CRYRING	Branching ratios	Thomas et al. 2005b
CH_3^+	ISR, ASTRID	Relative cross section, branching ratios	Vejby-Christensen et al. 1997
	Merged beams	Cross section, rate coefficient	Mul et al. 1981
	Merged beams	Cross section, rate coefficient	Sheehan & St.-Maurice 2004a
N_2OH^+	FALP	Rate coefficient	Adams & Smith 1988c
	Flowing afterglow	f_H and f_{OH}	Adams et al. 1991
	FALP	Rate coefficient, f_{OH}	Herd, Adams, & Smith 1990
	FALP	Rate coefficient	Gougousi, Golde, & Johnsen 1997
	Flowing afterglow	$f_{OH,v}$	Gougousi, Johnsen, & Golde 1997b

Ion	Method	Information	Reference
N_2OD^+	ISR, CRYRING	Cross section, rate coefficient, branching ratios	Geppert et al. 2004f
HCO_2^+	FALP	Rate coefficient	Adams & Smith 1988c
	Flowing afterglow	f_H and f_{OH}	Adams et al. 1991
	FALP	Rate coefficient	Gougousi, Golde, & Johnsen 1997
	Flowing afterglow	$f_{OH,v}$	Gougousi, Johnsen, & Golde 1997b
$DOCO^+$	ISR, CRYRING	Cross section, rate coefficient, branching ratios	Geppert et al. 2004d
HSO_2^+	FALP	Rate coefficient	Geoghegan, Adams, & Smith 1991
$OCSH^+$	Flowing afterglow	f_H and f_{OH}	Adams et al. 1991
H_3S^+	FALP	Rate coefficient	Adams & Smith 1988c
	FALP	Rate coefficient	Abouelaziz et al. 1992
	Flowing afterglow	f_H	Adams et al. 1991
$C_2H_2^+$	Merged beams	Cross section, rate coefficient	Mul & McGowan 1980
	ISR, CRYRING	Branching ratios	Derkatch et al. 1999
H_5^+	Stationary afterglow	Rate coefficient	Leu, Biondi, and Johnsen 1973b
	Stationary afterlow	Rate coefficient	Macdonald, Biondi, & Johnsen 1984
D_5^+	ISR, CRYRING	Cross section, branching ratios	Andersson et al. 2003[a]
NH_4^+	Stationary afterglow	Rate coefficient	Huang, Biondi, & Johnsen 1976
	Ion trap	Cross section, rate coefficient	DuBois, Jeffries, & Dunn 1978
	FALP	Rate coefficient	Alge, Adams, & Smith 1983
	Flowing afterglow	f_H	Adams et al. 1991
NH_4^+/ND_4^+	ISR, CRYRING	Relative cross section, branching ratios	Vikor et al. 1999
CH_4^+	ISR, CRYRING	Cross section, rate coefficient, branching ratios	Öjekull et al. 2004
$CH_2OH^+/$	Merged beams	Cross section, rate coefficient	Mul et al. 1981
CD_2OD^+	Merged beams	Cross section, rate coefficient	Sheehan & St.-Maurice 2004a
$CD_2OD_2^+$	ISR, CRYRING	Cross section, rate coefficient, branching ratio	Hamberg et al. 2007

(cont.)

Table 8.7. (cont.)

Ion	Method	Information	Reference
$C_2H_3^+$	Merged beams	Cross section, rate coefficient	Mul & McGowan 1980
	ISR, CRYRING	Cross section, rate coefficient, branching ratios	Kalhori et al. 2002
$DCCCN^+$	ISR, CRYRING	Cross section, rate coefficient, branching ratio	Geppert et al. 2004b
$C_2H_4^+$	ISR, CRYRING	Cross section, rate coefficient, branching ratios	Ehlerding et al. 2004
CH_5^+	Merged beams	Cross section	Mul et al. 1981
	Merged beams	Cross section, rate coefficient	Sheehan & St.-Maurice 2004a
	FALP	Rate coefficient	Adams, Smith, & Alge 1984
	Flowing afterglow	f_H	Adams et al. 1991
	FALP	Rate coefficient	Smith & Španěl 1993a
	FALP-MS	Rate coefficient	Lehfaoui et al. 1997
	FALP	Rate coefficient	Gougousi, Golde, & Johnsen 1997
	ISR, CRYRING	Cross section, rate coefficient, branching ratios	Semaniak et al. 1998
	VT-FALP	Rate coefficient	McLain et al. 2004
$DCCCND^+$	ISR, CRYRING	Cross section, rate coefficient, branching ratio	Geppert et al. 2004b
$C_2H_5^+$	FALP	Rate coefficient	Adams & Smith 1988c
	FALP-MS	Rate coefficient	Lehfaoui et al. 1997
	FALP	Rate coefficient	Gougousi, Golde, & Johnsen 1997
	VT-FALP	Rate coefficient	McLain et al. 2004
$C_2D_5^+$	ISR, CRYRING	Cross section, rate coefficient, branching ratios	Geppert et al. 2004a
$C_3H_4^+$	CRYRING	Branching ratios	Geppert et al. 2004e
$CH_3OH_2^+$ /	FALP	Rate coefficient	Adams & Smith 1988
$CD_3OD_2^+$	ISR, CRYRING	Cross section, rate coefficient, branching ratios	Geppert et al. 2006a,b
$CH_3SH_2^+$	FALP	Rate coefficient	Adams & Smith 1988c
$CH_3NH_3^+$	FALP	Rate coefficient	Adams & Smith 1988c
$C_2H_5OH_2^+$	FALP	Rate coefficient	Adams & Smith 1988c

[a]The cross section is not reported in the abstract.

8.3 Cluster ions

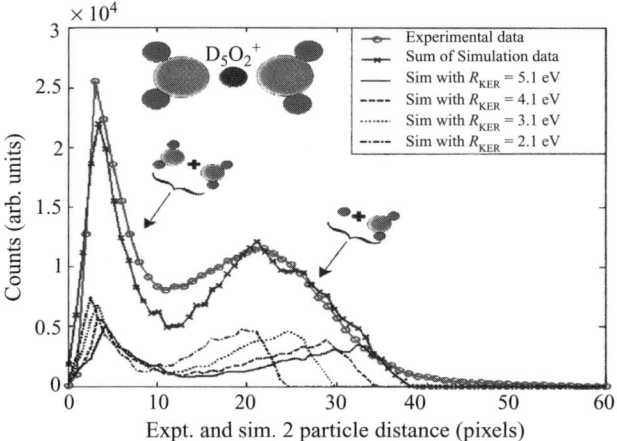

Figure 8.24 Two-particle distances D+D_2O, D_2+D_2O and D_2O+D_2O from the dissociative recombination of $D_3O^+\cdot(D_2O)$. (Reproduced with permission from Thomas *et al.* 2003.)

Adams and Smith (1988c), on $(HCOH)_2\cdot H^+$ and $(CH_3COH)_2\cdot H^+$ by Glosík *et al.* (2001b), and on $(CH_3COCH_3)_2\cdot H^+$ by Glosík and Plašil (2000).

Någård *et al.* (2002) used CRYRING to measure the $D_3O^+\cdot(D_2O)$ dissociative recombination cross section, thermal rate coefficient as a function of electron energy and temperature, and the product branching ratios. They found that $D_3O^+\cdot(D_2O)$ recombines at about 75% of the rate of $H_3O^+\cdot(H_2O)$ (Johnsen 1993b), which appears to be a reasonable isotope effect. The temperature was found to have a $T^{-0.7\pm0.1}$ dependence (Någård *et al.* 2002), somewhat larger than the $T^{-0.5}$ measured by Johnsen (1993b). However, one must bear in mind that whereas the CRYRING experiment was strictly mass selective, this was not the case in the afterglow plasma experiment by Johnsen (1993b). He found that the best conditions were obtained for $H_3O^+\cdot(H_2O)_{n=3}$, where it was possible to make the $n=3$ cluster ions ten times more abundant than the $n=2$ and 4 cluster ions. Thus, the $T^{-0.5}$ dependence is dominated by $H_3O^+\cdot(H_2O)_{n=3}$.

Någård *et al.* (2002) measured the branching ratio for the water channel $2D_2O$ + D to be 0.94 ± 0.04. Recombination to this channel puts 5.1 eV of energy at the disposal of the reaction products. The complete dominance of the three-body channel makes the imaging technique suitable for detailed studies of the dissociation dynamics (Thomas *et al.* 2003, 2005a, Thomas 2004). Figure 8.24 shows all two-particle distances, D + D_2O, D_2 + D_2O and D_2O + D_2O, together with the results from Monte-Carlo simulations. The best agreement with experimental data was obtained when a flat distribution of up to 3–4 eV going into internal excitation of the D_2O molecules was assumed. In other words, Thomas *et al.* (2003, 2005a)

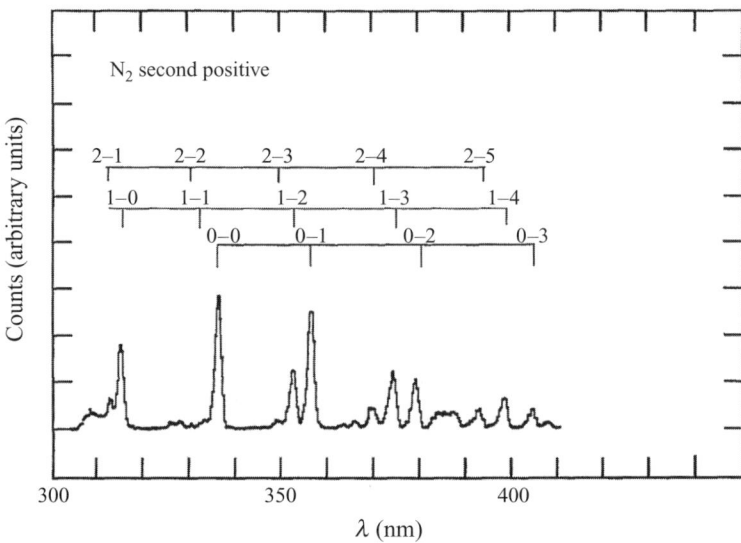

Figure 8.25 Emission spectrum of the $C\,^3\Pi_u - B\,^3\Pi_g$ transition in N_2 observed in the afterglow between 100 and 550 μs after termination of ionization. Initial and final vibrational levels are given as $v'-v''$. (Reused with permission from Y. S. Cao and R. Johnsen, *Journal of Chemical Physics* **95**, 7356, (1991). Copyright 1991, American Institute of Physics.)

and Thomas (2004) did not find a fixed energy partition between the D atom and the D_2O molecules. The lack of a $2D + OD + D_2O$ channel (Någård *et al.* 2002) suggests that the D_2O molecules are not formed with sufficient energy to allow secondary fragmentation to occur.

The large recombination rate for cluster ions interested Bates (1991a, 1992a), who invented the term superdissociative recombination to describe the encounter of a free electron with a cluster ion. Cluster ions have typical bond energies in the range $0.2 \leq D_0 \leq 1.0$ eV, falling between van der Waals clusters (0.01–0.05 eV) and normal valence bonds (2–5 eV) (Bowers 1989). Thus, Bates argued, the energy released in the recombination of $N_2 \cdot N_2^+$ should be enough to excite one of the molecules to a Rydberg state. The search by Cao and Johnsen (1991) for emission from N_2 in the recombination of $N_2 \cdot N_2^+$ was successful, as shown in Fig. 8.25, where band features of the $C\,^3\Pi_u - B\,^3\Pi_g$ transition in N_2 are easily identified. Cao and Johnsen (1991) noted that an emission spectrum of the $C\,^3\Pi_u - B\,^3\Pi_g$ transition had been observed almost 20 years earlier in a pulse radiolysis experiment (Sauer & Mulac 1972) and ascribed to dissociative recombination of $N_2 \cdot N_2^+$, although the ion composition was not determined in that experiment. In subsequent work, Johnsen (1993a) also observed emission spectra from the CO molecule following dissociative recombination of $CO \cdot CO^+$.

8.3 Cluster ions

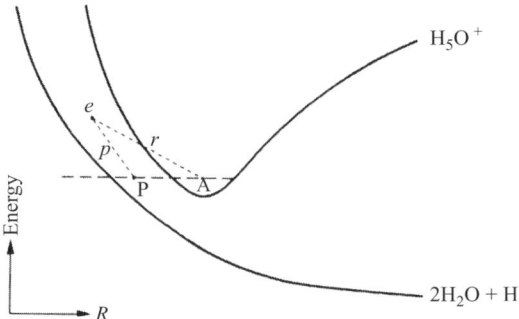

Figure 8.26 Single-electron transition in $H_5O_2^+$ dissociative recombination. The potential energy curves of ionic (note the typo H_5O^+) and neutral systems are schematic. The horizontal broken line marks the initial total energy. A is the equilibrium separation of the ion, and as a first approximation Bates used a wavefunction for the free electron assuming that the proton was at A. The wavefunction of the H(1s) atom was centered at P. (Reproduced with permission from D. R. Bates, "Single-electron transitions and cluster ion super-dissociative recombination," *J. Phys. B* **25**, pp. 3067–3073 (1992), IOP Publishing Limited.)

The basic idea put forward by Bates (1991a, 1992a) was that a cluster ion such as $H_3O^+\cdot(H_2O)$ recombines by means of a single-electron transition, where the formula for the water cluster ion is written to emphasize its proton-bridge bond structure:

$$H_2O\cdot H^+ \cdot H_2O + e^- \to 2H_2O + H\,(1s). \tag{8.20}$$

In this recombination reaction the free electron enters directly into the 1s orbital of atomic hydrogen. Figure 8.26 shows schematically how Bates (1992a) envisaged the process. This simple picture is in agreement with the measured branching ratio, but is more difficult to reconcile with the broad distribution of internal energy in the water molecules observed by Thomas *et al.* (2003).

Petrignani *et al.* (2005a) performed a detailed study of $NO\cdot NO^+$ using CRYRING. This study provided compelling evidence that the superdissociative recombination mechanism does not give the full picture. This is obvious from inspection of Table 8.8, which shows the product branching ratios.

A fragment imaging system was used to study the three-body channel (Petrignani *et al.* 2005a):

$$NO\cdot NO^+ + e^- \to NO + O\,(^3P) + N\,(^4S) + 2.17\ \text{eV}, \tag{8.21a}$$

$$NO\cdot NO^+ + e^- \to NO + O\,(^1D) + N\,(^4S) + 0.21\ \text{eV}, \tag{8.21b}$$

$$NO\cdot NO^+ + e^- \to NO + O\,(^3P) + N\,(^2D) - 0.21\ \text{eV}. \tag{8.21c}$$

Table 8.8. *Products of NO·NO$^+$ + e$^-$ (0 eV) with a branching ratio larger than 0.05*

Channel	Petrignani et al. 2005a[a]	Bates et al. 1991b[b]
NO + NO	0.23 ± 0.02	1.00
NO + O + N	0.69 ± 0.01	0.00

[a]CRYRING
[b]These branching ratios are only given implicitly by Bates.

where the energies apply for NO·NO$^+$ and NO in their zeroth vibrational levels and 0 eV electrons. One can first note that the dominant channel in the recombination of the monomer ion NO$^+$, O (^{3}P) + N (^{2}D), with a branching ratio of 0.80 (Vejby-Christensen et al. 1998) is closed for the dimer (Reaction (8.21c)), since storage in CRYRING removes vibrational excitations in NO·NO$^+$. One may speculate that the three-body channel would increase even further if the electron energy is increased to 0.21 eV or larger, but owing to the low cross section at this energy this could not be tested. Petrignani et al. (2005a) found that NO ($v = 0$) is formed in 45% of channel (8.21a) events, and that another 24% lead to NO ($v = 1$ and 2). The picture that emerges is one in which NO is largely a passive spectator while its charged partner NO$^+$ undergoes dissociative recombination. This picture is somewhat modified by the fact that when NO received less kinetic energy, the linear momentum correlation between N and O increased, but is still far away from Bates's single-electron mechanism.

Several attempts have been made to explain the large rate coefficients measured for cluster ions. Leu, Biondi, and Johnsen (1973a) proposed that the initial electron capture could be associated with excitations of internal modes in the complex, something that might slow down autoionization and ensure stabilization by dissociation. Smirnov (1977) developed a simple model building on the indirect process and the large number of vibrational modes in a cluster ion. Bates (1992a) levelled criticism at this while advocating the single-electron transition superdissociative recombination mechanism. Ion storage rings have two advantages in addressing this issue as compared with afterglow techniques. The cross section can be measured, and it can be measured for mass selective ions. Petrignani et al. (2005a) found that the cross section for NO·NO$^+$ at 0.2 eV was comparable to that for NO$^+$, but that it increased more steeply when the electron energy was reduced. Figure 8.27 shows the cross sections for NO·NO$^+$ and NO$^+$. Whereas the cross section for NO$^+$ is approximately proportional to E_c^{-1}, the cross section for the dimer ion scales with $E_c^{-1.4}$. This causes a significant change in the rate coefficient. Integrated according to Eq. (2.37), Petrignani et al. (2005a) obtained α(NO·NO$^+$, $T = 300$ K) = 1.5×10^{-6} cm^3 s^{-1}, in very good agreement with the stationary afterglow result

8.3 Cluster ions

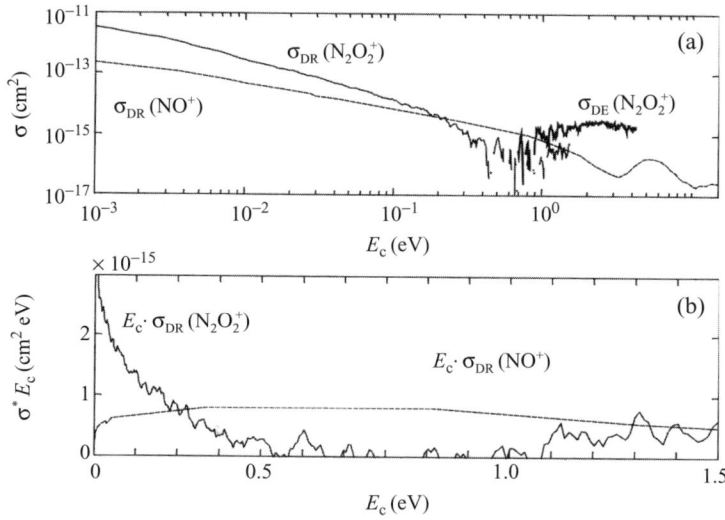

Figure 8.27 (a) The cross sections for dissociative recombination and dissociative excitation (see Section 9.1) of NO·NO$^+$ as a function of interaction (center-of-mass) energy measured in CRYRING by Petrignani *et al.* (2005) and the dissociative recombination of NO$^+$ measured in ASTRID by Vejby-Christensen *et al.* (1998). (b) The reduced cross section, $E_c \sigma_{DR}$; note that a linear scale is used for both axes. (Reused with permission from A. Petrignani, P. U. Andersson, J. B. C. Pettersson, *et al.*, *Journal of Chemical Physics* **123**, 194306, (2005). Copyright 2005, American Institute of Physics.)

of 1.4×10^{-6} cm^3 s^{-1} (Weller & Biondi 1968), and about 4 times higher than the monomer rate (Vejby-Christensen *et al.* 1998). A similar trend can be noted for D$_3$O$^+$·(D$_2$O) (Någård *et al.* 2002) and D$_3$O$^+$ (Neau *et al.* 2000) (see Fig. 3 in Någård *et al.*). The trend, however, is not so clear for NH$_4^+$·(NH$_3$)$_n$. Table 8.9 shows temperature-dependent rate coefficients measured in CRYRING (Öjekull *et al.* 2004, 2006). The temperature dependence is somewhat stronger for the dimer and trimer ions, reflecting a slightly steeper cross section at low electron energy. Otherwise it is difficult to deduce a clear trend from the data in Table 8.9. The rates for the monomer, NH$_4^+$, and dimer, NH$_4^+$·NH$_3$, are almost the same, whereas for the D atom substituted isotopolog there is a large jump when going from the monomer to the dimer, but almost no effect when going from the dimer to the trimer.

Prior to the ion storage ring experiments on cluster ions, the afterglow experiment suggested that the dissociative recombination rate for H$_3$O$^+$·(H$_2$O)$_n$ was size dependent (Leu, Biondi, & Johnsen 1973a, Huang *et al.* 1978, Johnsen 1993b) whereas that for NH$_4^+$·(NH$_3$)$_n$ (Huang, Biondi & Johnsen 1976, Skrzykpowski & Johnsen 1997) was not. (Glosík *et al.* (1999) studied only NH$_4^+$·(NH$_3$)$_{n=2}$.) Notwithstanding the experimental difficulties in adjusting conditions of temperature and pressure in the afterglow to enhance each member of a family of ions, the results for

Table 8.9. *Rate coefficients (in $cm^3\ s^{-1}$) for $NH_4^+ \cdot (NH_3)_n$ and $ND_4^+ \cdot (ND_3)_n$*

n	$NH_4^+ \cdot (NH_3)_n$	$ND_4^+ (ND_3)_n$	Reference
0	(9.43 ± 0.03) $\times 10^{-7}(T/300)^{-(0.605 \pm 0.001)}$	(5.65 ± 0.03) $\times 10^{-7}(T/300)^{-(0.687 \pm 0.001)}$	Öjekull et al. 2004
1	(1.07 ± 0.04) $\times 10^{-6}(T/300)^{-(0.724 \pm 0.003)}$	(2.31 ± 0.09) $\times 10^{-7}(T/300)^{-(0.777 \pm 0.003)}$	Öjekull et al. 2006
$2^{a,b}$	(3.34 ± 0.02) $\times 10^{-6}(T/300)^{-(0.831 \pm 0.003)}$	(1.58 ± 0.09) $\times 10^{-7}(T/300)^{-(0.762 \pm 0.003)}$	Öjekull et al. 2006

[a] Skrzypkowski and Johnsen (1997) obtained $(4.8 \pm 0.5) \times 10^{-6}(T/300)^{-0.65}$ $cm^3\ s^{-1}$ using radio-frequency heated afterglow.
[b] Glosík et al. (1999) obtained 1.4×10^{-6} $cm^3\ s^{-1}$ at 600 K.

the water cluster ions have at least been obtained in three different experiments (Leu, Biondi, & Johnsen 1973a, Huang *et al.* 1978, Johnsen 1993b), and the rate increase in the step going from monomer to dimer has been confirmed by mass-selective ion storage ring experiments (Neau *et al.* 2000, Någård *et al.* 2002). Bates had some considerable difficulties in reconciling the size dependence of $H_3O^+ \cdot (H_2O)_n$ and the absence (as believed at the time) of size dependence of $NH_4^+ \cdot (NH_3)_n$ with his single-electron transition hypothesis (Bates 1991a, 1992a). In his first paper on the subject (Bates 1991a) he pointed out that there was no reason to expect the ions to exhibit pronounced differences as a function of size. In the second paper (Bates, 1992a), however, he conjectured that near-equality between the ionization potentials of H and O would make the electron cloud mobile, thus enabling the proton bond to move in the $H_3O^+ \cdot (H_2O)_n$ cluster in such a way that the incoming free electron would always easily find the positive charge (Bates 1992a, 1994). Bates argued that a similar mobility would not occur in $NH_4^+ \cdot (NH_3)_n$. The CRYRING results (Öjekull *et al.* 2004, 2006) show that the situation is more complex.

Zhaunerchyk *et al.* (2004) studied the effect of replacing the hydronium ion in $D_3O^+ \cdot (D_2O)$ with Na. The recombination end products were not affected, with Na + D_2O being the completely dominant channel. The rate coefficient, however, dropped from 1.4×10^{-6} $cm^3\ s^{-1}$ to 2.3×10^{-7} $cm^3\ s^{-1}$. Measurements on $Na^+ \cdot O_2$ and $Na^+ \cdot CO_2$, performed by Keller and Beyer (1971) using a drift tube, gave a rate that was 20 times higher. It would be valuable to repeat these measurements in a storage ring. Preliminary results from a study of $Li \cdot H_2^+$ (Thomas *et al.* 2005a) showed that the Li + H_2 channel is the largest but is not as dominant as the Na + D_2O channel is in the dissociative recombination of $Na^+ \cdot (D_2O)$.

In summary, it is fair to say that, despite decades of efforts, there are still questions surrounding the dissociative recombination of cluster ions. It is unclear why the rate coefficients are so high, and the reasons proposed so far (Smirnov 1977,

Bates 1991b, 1992b, 1994) do not seem credible. The work on NO·NO$^+$ + e$^-$ by Petrignani *et al.* (2005a) represents the most detailed study of any cluster ion. The suggestion that NO acts mainly as a spectator to NO$^+$ recombination has found support from a CRYRING experiment on H$_2$O·NO$^+$, where the branching ratio for NO + OH + H was measured to be 0.80 (Österdahl 2006). Theoretical calculations would be challenging but are the only option if an explanation for the high rate coefficients is to be found.

8.4 Hydrocarbon ions

Hydrocarbon ions are not only of interest in astrophysics. In thermonuclear fusion reactors with highly exposed carbon-based divertor plate segments of the vacuum chamber, C$_x$H$_y$ hydrocarbons are released into the plasma. These hydrocarbons are ionized and dissociated, and a broad range of hydrocarbon ions are formed (Janev & Reiter 2002, 2004). The modeling of the edge region in fusion devices is complicated and requires as input large amounts of molecular data. The International Atomic Energy Agency (IAEA) initiated a coordinated research project (CRP) during 2001–4 with the aim of producing data for a range of molecular processes of significance in fusion edge plasmas (Clark 2006). A new CRP along the same lines started in 2005 (Humbert 2006). Another application concerns plasma-assisted combustion. An engine that takes in air from its surrounding in order to burn fuel is called an air-breathing engine. This is in contrast to a rocket which carries its own oxidizer and can operate in space. Examples of air-breathing engines include the scramjet, the ramjet, and the turbojet. Flying aircraft at hypersonic speed (>Mach 5) with air-breathing engines is very difficult, particularly if hydrocarbon fuels are used. A possible remedy is to use a plasma injector to speed up combustion and improve performance. Plasma modeling requires molecular data input (Viggiano *et al.* 2005), in particular the end products of dissociative recombination of hydrocarbon ions. Finally, hydrocarbon ions play an important role in the atmospheres of Jupiter and Titan (Fox 1996).

This section is divided into a part dealing with cross sections and rate coefficients, and a part about product branching ratios. Work on C$_{x\leq 2}$H$_y^+$ ions is listed in Table 8.7.

8.4.1 Cross sections and rate coefficients

The bulk of data on recombination rate coefficients for hydrocarbon ions of type C$_{x\geq 3}$H$_y^+$ comes from the FALP-MS apparatus in Rennes. The approach is to study many different hydrocarbons and look for trends, as described in the review by Mitchell and Rebrion-Rowe (1997) and Hassouna *et al.* (2003). A complicating factor in studies of hydrocarbon ions is the presence of different isomeric forms.

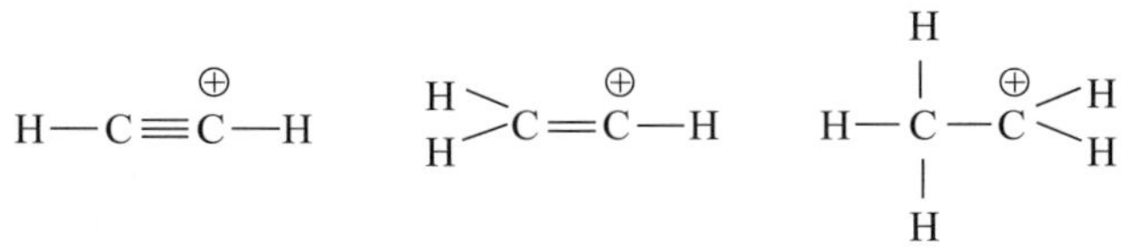

Figure 8.28 The ions $C_2H_2^+$, $C_2H_3^+$, and $C_2H_5^+$. (Reproduced with permission from J. B. A. Mitchell and C. Rebrion-Rowe, "The recombination of electrons with complex molecular ions," *Int. Rev. Phys. Chem.* **16**, pp. 201–213 (1997). Copyright (1997) Taylor & Francis Ltd., http://www.informaworld.com)

The first measurement of a rate coefficient for a $C_{x\geq 3}H_y{}^+$ ion seems to have been the flame study by Graham and Goodings (1984). Abouelaziz *et al.* (1993) used the FALP-MS technique to measure the recombination rate coefficient for $C_3H_3^+$, $C_5H_3^+$, $C_6H_6^+$, $C_7H_5^+$, and $C_{10}H_8^+$ and found it to be in the range $(3-10) \times 10^{-7}$ cm^3 s^{-1}, with essentially no variation as a function of ion size. Similar results were obtained for the protonated alkanes C_nH_{2n} ($n = 1-8$) (Lehfaoui *et al.* 1997), and $C_4H_5^+$, $C_4H_{11}^+$, $C_5H_9^+$, $C_6H_4^+$, $C_6H_5^+$, and $C_8H_7^+$ (Rebrion-Rowe *et al.* 1998). Replacing H atoms with the CH$_3$ group in cyclic benzene ring compounds did not give rise to a measurable effect (Rebrion-Rowe *et al.* 2000a), however, one must also keep in mind that these were difficult measurements because of the presence of different ions in the afterglow plasma.

Some correlation between electronic structure and the inclination for recombination has been found. Mitchell and Rebrion-Rowe (1997) noted a correlation for the ions shown in Fig. 8.28. Adding H atoms to $C_2H_2^+$ reduces the C≡C triple bond to double and single bonds, which increases the electron mobility and makes electron capture easier. The measured rate coefficients at 300 K are: $C_2H_2^+$ 2.7×10^{-7} cm^3 s^{-1} (Mul & McGowan 1980, Mitchell 1990a); $C_2H_3^+$ 4.5×10^{-7} cm^3 s^{-1} (Mul & McGowan 1980, Mitchell 1990a), 5.0×10^{-7} cm^3 s^{-1} (Kalhori *et al.* 2002); $C_2H_5^+$ 7.4×10^{-7} cm^3 s^{-1} (Adams & Smith 1988), 6.0×10^{-7} cm^3 s^{-1} (Lehfaoui *et al.* 1997), 9.0×10^{-7} cm^3 s^{-1} (Gougousi *et al.* 1997), and 12×10^{-7} cm^3 s^{-1} (McLain *et al.* 2004). The correlation corroborates Bates's idea (Bates 1992a) about the electron mobility facilitating electron capture.

McLain *et al.* (2004) used the variable temperature FALP apparatus in Georgia to investigate the temperature dependence of the rate coefficients for CH_5^+, $C_2H_5^+$, and $C_6H_7^+$. For CH_5^+ they found a $T^{-0.7}$ dependence below an electron temperature of 300 K, not so different from the beam results of $T^{-0.5}$ (Mul *et al.* 1981) and $T^{-0.52}$ (Semaniak *et al.* 1998). McLain *et al.* (2004) seem to have been unaware of any of the beam results, since they are not discussed nor referenced in their paper. The exponent for the temperature dependence of $C_2H_5^+$ was measured to be -0.8, which is very close to the storage ring result for $C_2D_5^+$ of -0.79 (Geppert *et al.* 2004a). It is from this perspective that the absence of temperature dependence for linear $C_3H_3^+$

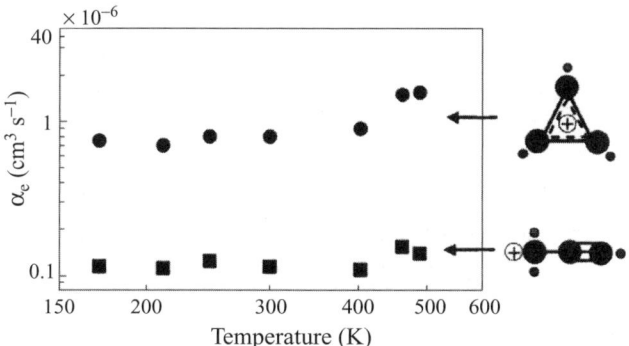

Figure 8.29 The thermal rate coefficient as a function of electron temperature for dissociative recombination of propargyl and cyclic $C_3H_3^+$ isomers. (Reprinted with permission from J. L. McLain, V. Poterya, C. D. Molek, D. M. Jackson, L. M. Babcock, and N. G. Adams, "$C_3H_3^+$ isomers: Temperature dependencies of production in the H_3^+ reaction with allene and loss by dissociative recombination with electrons," *J. Phys. Chem. A* **109**, pp. 5119–5123 (2005). Copyright (2005) American Chemical Society.)

(propargyl) and cyclic $C_3H_3^+$ should be seen (McLain *et al.* 2005). Figure 8.29 shows the rate coefficient measured by McLain *et al.* (2005) for linear and cyclic $C_3H_3^+$ at different temperatures between 172 K and 489 K. It is difficult to find data for comparison. The flame study by Graham and Goodings (1984) was performed at 2000 K, whereas the FALP-MS study by Abouelaziz *et al.* (1993) was done at room temperature. It seems risky to combine these different data to suggest that they indicate a T^{-1} dependence (Abouelaziz *et al.* 1993). We noted in Section 8.2.1 that VT-FALP in the apparatus in Georgia produced a temperature-independent rate coefficient for HCO$^+$ (Poterya *et al.* 2005), but also that this was in clear conflict with other results. Whether or not the data for $C_3H_3^+$ by McLain *et al.* (2005) are flawed has still to be determined. A very interesting point, that is most likely correct, is the factor of 7 larger rate coefficient for cyclic $C_3H_3^+$ as compared with propargyl (McLain *et al.* 2005). Talbi (2003) has shown by quantum chemical calculations that the ground state of c-$C_3H_3^+$ is crossed by the lowest $^2A_1$ state of c-C_3H_3 near its minimum energy, thus lending theoretical support to a large rate coefficient.

The only merged-beam experiments on $C_{x\geq 3}H_y^+$ ions measuring cross sections are those performed at CRYRING. Ehlerding *et al.* (2003) measured the cross section for the recombination of $C_3H_7^+$ and deduced a rate coefficient of 1.9×10^{-6} cm^3 s^{-1}, a factor of 2 larger than the FALP-MS result (Lehfaoui *et al.* 1997). There are reasons to believe that the CRYRING result is inflated. In this particular experiment, because of problems with current measurement in CRYRING, a two-step procedure was used to establish the number of circulating ions in the ring.

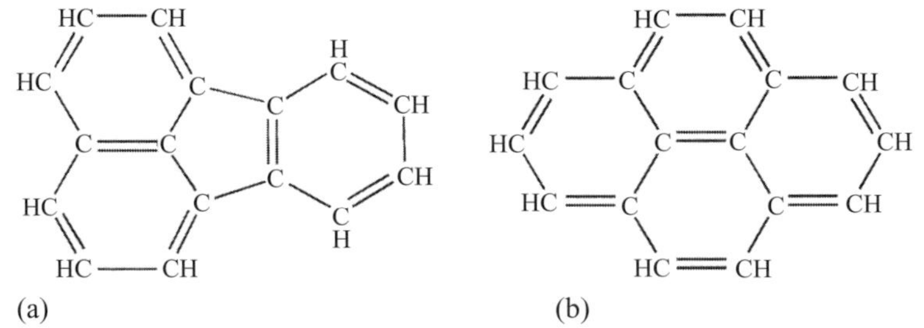

Figure 8.30 Structure of $C_{16}H_{10}$: fluoranthene (left) and pyrene (right).

It is likely that this procedure was less reliable than anticipated at the time. Cross section measurements for $C_3D_7^+$ and $C_4D_9^+$ (Larsson et al. 2005) resulted in lower rate coefficients, 2.3×10^{-7} cm^3 s^{-1} and 5.8×10^{-7} cm^3 s^{-1}, respectively, and these results are consistent with those from the FALP-MA apparatus (Lehfaoui et al. 1997), if one takes into account a plausible isotope effect. Geppert et al. (2004e) found a large recombination rate coefficient for $C_3H_4^+$, $\alpha(300 \text{ K}) = 2.95 \times 10^{-6}$ cm^3 s^{-1}, but there is nothing in this experiment to suggest that there is something wrong. There are no other measurements with which to compare.

The group in Rennes has studied PAH ions. In an early publication (Rowe & Rebrion-Rowe 1996) it was already noted that PAH cations do not recombine exceedingly fast (typically in the range $10^{-6} - 10^{-7}$ cm^3 s^{-1}) and there is no correlation between their complexity and their rate of recombination. But the experimental difficulties in determining rate coefficients for PAH cations were also acknowledged (Rebrion-Rowe et al. 2000a), and in a later study (Hassouna et al. 2003) a slight increase of the rate coefficient with size was observed. The main question in an experiment in which the electron density is measured as a function of distance from ion production in a flow tube is whether the electrons disappear only by recombination with the ion under study, or whether there are other reasons for their disappearance. Lack of knowledge of the chemical and physical properties of heavier PAH cations introduced uncertainties in how the data should be interpreted.

The FLAPI technique (see Section 2.4) was introduced to overcome some of the problems of measuring recombination rates for PAH cations by means of FALP-MS. Figure 8.30 shows the geometric structure of $C_{16}H_{10}$ in its isomeric forms, fluoranthene and pyrene. A FALP-MS measurement (Rebrion-Rowe et al. 2003) of the recombination coefficient of the fluoranthene cation gave $(2.5 \pm 1.5) \times 10^{-6}$ cm^3 s^{-1}. A FLAPI measurement on the pyrene cation gave the recombination coefficient as $(4.1 \pm 1.2) \times 10^{-6}$ cm^3 s^{-1} (Novotný et al. 2005c). The error bars in these measurements are too large to allow any conclusion as to whether there is an isomeric effect. Novotný et al. (2005c) also measured the recombination rate for $C_{14}H_{10}^+$ (anthracene cation) and found it to be smaller, $(2.4 \pm 0.8) \times 10^{-6}$ cm^3 s^{-1}, than for $C_{16}H_{10}^+$. The difference is too small and the error bars are too large to infer that there is a size effect. It is likely, however, that the FLAPI technique will allow these questions (isomeric and size effects) to be addressed in the future.

In conclusion, the dissociative recombination rate coefficients of hydrocarbon ions do not vary much, and even though different experimental techniques do not always give the same result for a given hydrocarbon ion, there is sufficient consistency in the data to allow reliable modeling in plasma chemistry.

8.4.2 Branching ratios

The end products in dissociative recombination of hydrocarbon ions are difficult to measure. In fact, most data have been obtained since 2000 by means of ion storage rings, and very few results have been obtained with afterglow methods. The interest in branching ratios for astrochemical modeling has been discussed earlier in this chapter. The main motivation from the point of view of applications for the study of the product branching ratios of hydrocarbon ion recombination, apart from astrophysics, concerns thermonuclear fusion reactors and air-breathing engines. In the modeling of plasma-assisted combustion it was earlier assumed that the termination step for neutralizing the plasma was dissociative recombination of hydrocarbon ions leading to the release of an H atom. The number of free radicals formed in the termination step was found to be the important parameter for improving performance. By 1997, the branching ratio measurements at ion storage rings (Larsson & Thomas 2001) had made it clear that dissociative recombination of polyatomic molecular ions leads to considerably more fragmentation than had been anticipated, and the consequences of this for the modeling of interstellar molecular clouds were investigated (Herbst & Lee 1997). The fragmentation pattern for astrophysically important polyatomic ions made it most relevant to ask whether hydrocarbon ions would follow the same trend, in particular since one of them already had been shown to do so (Vejby-Christensen *et al.* 1997). Systematic studies were initiated at both ASTRID and CRYRING. At ASTRID the focus was on studies of end products in the recombination of $C_{x \geq 3}H_y^+$ ions (Mitchell *et al.* 2003, Angelova *et al.* 2004a,b) with emphasis on the carbon–carbon bond, while the experiments at Stockholm were more focused on the complete breakup pattern for smaller systems (see Table 8.7); a few measurements have also been performed on larger ions (Ehlerding *et al.* 2003, Larsson *et al.* 2005).

Mitchell *et al.* (2003) found that the number of H atoms in a hydrocarbon ion plays a role in terms of how the C−C bonds are broken. Dissociative recombination of $C_4H_9^+$ was found to give C_4 and $C_3 + C$ (the location of the H atoms could not be determined because of insufficient resolution), whereas $C_4H_5^+$ recombined to give C_4 and $C_2 + C_2$. In a systematic study of $C_4H_n^+$ ($n = 1-9$), the $C_2 + C_2$ channel was

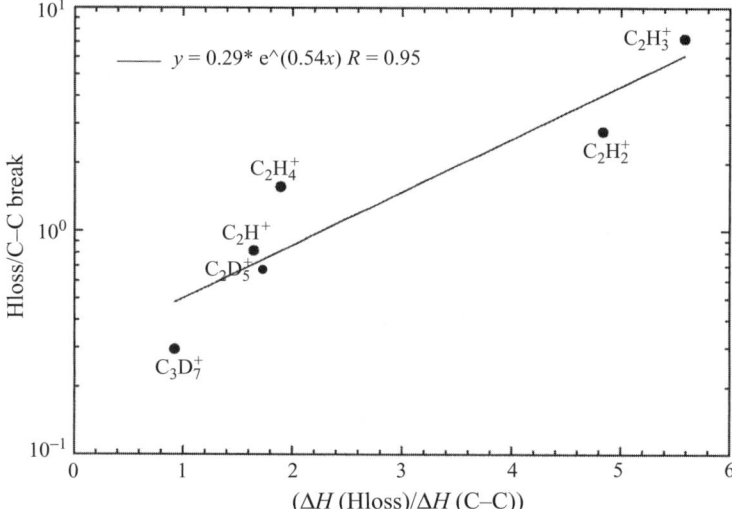

Figure 8.31 The ratio of the branching ratio for single H-atom loss to that of C−C cleavage vs. the exothermicities for the two processes. The exothermicity for C−C cleavage is the one involving no rearrangement or H-atom loss. The branching ratio for the C−C cleavage, however, is the one obtained by summation of all C−C bond breaking channels. (Reproduced with permission from A. A. Viggiano, A., Ehlerding, S. T. Arnold, and M. Larsson, "Dissociative recombination of hydrocarbon ions," *J. Phys.: Conf. Ser.* **4**, pp. 191−197 (2005), IOP Publishing Limited.)

found to peak (branching ratio 0.447) for $n = 5$. The presence of different isomeric forms in the stored ion beam in ASTRID could not be excluded, and it is not clear to what extent the results reflect mixed isomers in the beam or rearrangements following the electron capture.

Viggiano *et al.* (2005b) considered the complete product distributions for the entire $C_2H_n^+$ series ($n = 1–5$). The exothermicity of a dissociative recombination reaction is sometimes expressed differently depending on whether a physicist's or a chemist's language is used. For example,

$$C_2H^+ + e^- \rightarrow C_2 + H + 6.60 \text{ eV}, \tag{8.22a}$$

$$C_2H^+ + e^- \rightarrow C_2 + H, \Delta H = -6.60 \text{ eV}, \tag{8.22b}$$

where (8.22a) represents the physicist's language and (8.22b) the chemist's, where ΔH is the enthalpy change in the reaction, which has a negative value for an exothermic reaction. Viggiano *et al.* (2005b) searched for correlations between the enthalpy change for single H atom loss and C−C cleavage, and the branching ratios for these channels. Figure 8.31 shows the ratio of the enthalpy changes as a function of the ratio of branching ratios. The enthalpy change for the C−C bond breaking was taken as that involving no rearrangement or H atom loss, whereas the branching ratio was obtained by summation of all C–C bond breaking channels. There is an exponential relationship with a high correlation coefficient (0.95) over a wide range of ratios. But not all ions follow this correlation, $C_3H_4^+$ has a much higher single H atom loss channel than would be anticipated from its enthalpy change (Geppert *et al.* 2004e). The number of free radicals produced by dissociative recombination of $C_2H_y^+$ ions ($n = 1-5$) falls in the narrow range 1.9–2.3. This is a surprisingly narrow range and there is no obvious explanation.

8.5 Other polyatomic ions

Most studies of polyatomic molecular ions have been covered in the previous subsections. Strikingly few metal-containing polyatomic molecular ions have been studied with respect to recombination. Gooding and coworkers applied mass spectrometric sampling to an atmospheric pressure flame system to investigate the ionization of barium, yttrium, magnesium, lanthanum, and scandium, and the recombination of these metal ions bound to molecules and radicals from the flame (Goodings, Patterson, & Hayhurst 1995, Chen & Goodings 1998, Chen *et al.* 1999, Chen & Goodings 1999, Guo & Goodings 2002). These metals are found as naturally occurring trace additives in hydrocarbon and biomass fuels, and it is therefore of interest to understand the ion chemistry of these metals.

The widespread use of fluorocarbon plasmas in the semiconductor processing industry inspired experiments on CF_2^+ and CF_3^+ (see Chapter 6 for CF^+) by means of ion storage rings (Ehlerding *et al.* 2006) and FALP-MS (Angelova *et al.* 2004). The recombination rate coefficient was found to be similar for CF_2^+ and CF_3^+. Single F atom release was the dominant decay channel for both ions (0.71 and 0.80, respectively).

8.6 Electron capture dissociation

Mass spectrometry has an enormous range of analytical applications. A very important development in mass spectrometry was the invention of techniques for softly ionizing biological macromolecules so that their masses could be determined. The inventors of these techniques were awarded the Nobel Prize in chemistry in 2002 (Fenn 2003, Tanaka 2003). Once a biological macromolecule is available in the gas phase, the key to obtaining structural information about it is to break it down into smaller components. The combination of an electrospray ion source and Fourier transform ion cyclotron resonance (FT-ICR) mass spectrometry is a powerful tool for biomolecular analysis, and the fragmentation of ions can be performed by any of several "slow-heating" methods: infrared multiphoton dissociation (IRMPD), collision-induced dissociation (CID), and blackbody infrared radiative dissociation

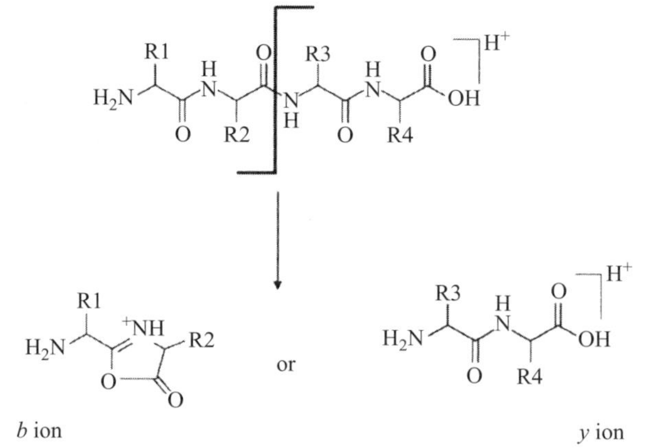

Figure 8.32 Cleavage of the peptide amide bond. The peptide bond is a covalent bond between the oxygen-bearing carbon of one amino-acid and the amino nitrogen of a second amino-acid (C_O-N). In the formation of a peptide bond between two amino-acids, a water molecule is eliminated. In the terminology of mass spectrometry, the cleavage of the peptide bond shown in the figure leads to the formation of an N-terminal (containing the NH_2 amino group) b ion and a C-terminal (containing the $COOH$ carboxyl group) y ion. (Reproduced with permission from H. J. Cooper, K. Håkansson, and A. G. Marshall, "The role of electron capture dissociation in biomolecular analysis," *Mass Spectrom. Rev.* **24**, pp. 201–222 (2005). Copyright (2005) John Wiley & Sons, Inc.)

(BIRD). "Slow-heating" means that the ions are heated (vibrationally excited) to a higher Boltzmann temperature and the energy is randomized so that fragmentation occurs along the lowest energy fragmentation pathway (Cooper, Håkansson, & Marshall 2005). For peptides, which are compounds composed of two or more amino acids, cleavage occurs at the peptide amide bond, as shown in Fig. 8.32.

In 1998 Zubarev, Kelleher, and McLafferty discovered, serendipitously as they point out, that low-energy electrons could fragment proteins (peptides with more than about 50 amino acids) in a way which is different from the "slow-heating" methods. Instead of causing a rupture of the weak peptide bond, they found that electrons with energy < 0.2 eV can break the stronger bond next to the peptide bond, the $N-C_\alpha$ bond between the alpha carbon and the nitrogen. Zubarev, Kelleher, and McLafferty (1998) labeled the process electron capture dissociation (ECD) while recognizing the similarity with dissociative recombination. They concluded that the process, in contrast to the ergodic "slow-heating" processes, is nonergodic. The ECD process provides advantages in the structural analysis of biologial macromolecules. Zubarev *et al.* (2002) pointed out that a fragmentation technique in tandem mass spectrometry (MS/MS) must combine a number of seemingly contradictory features: many bonds must be cleaved in order to establish the molecular structure uniquely; the fragmentation pattern should be simple; fragmentation

8.6 Electron capture dissociation

should be faster than intramolecular rearrangements, so that the fragments reflect the original structure; important labile bonds should remain intact so that the information on their location is preserved; and the fragment ion intensities should be reproducible and characteristic of the molecular composition and structure. They noted that ECD comes close to fulfilling these criteria. The topic of structural analysis of biological molecules falls outside the scope of this book, and the interested reader is referred to the reviews by Zubarev (2003) and Cooper, Håkansson, and Marshall (2005). What is of concern here is the ECD mechanism.

Zubarev et al. (2002) gave an overview of ECD and put it in perspective with other, related, processes, such as dissociative recombination. In the first ECD paper, Zubarev, Kelleher, and McLafferty (1998) envisaged a mechanism in which the electron attached to the protonated site of the peptide or protein, causing this site to become highly excited, followed by cleavage of the the N–C_α bond:

$$C_\alpha\text{–}C_O\overset{\overset{+\text{OH}}{\|}}{\text{–}}\text{NH–}C'_\alpha\text{–} + e^- \rightarrow C_\alpha\text{–}C_O\overset{\overset{\text{OH}}{|}}{\text{–}}\text{NH–}C'_\alpha\text{–} \rightarrow C_\alpha\text{–}C_O\overset{\overset{\text{OH}}{|}}{=}\text{NH} + {}^\bullet C'_\alpha\text{–}. \quad (8.23)$$

This mechanism worked less well to explain the discovery that electrons can effectively cleave S–S disulfide bonds in multiply-protonated proteins (Zubarev et al. 1999), the reason being that a proton is unlikely to be attached directly to the disulfide bond. On the other hand, the disulfide bridge has a high hydrogen affinity, which inspired the proposal of a mechanism in which a hydrogen is released by electron capture and then, instead of escaping the protein, is collisionally de-excited until it has an energy favorable for attacking the disulfide bond and causing S–S fragmentation. It was also assumed that the same "hot H atom" mechanism could cause rupture of the N–C_α bond. Various theoretical methods have been used in attempts to describe the mechanisms outlined above (Turuček et al. 2000, Sawicka et al. 2003, Uggerud 2004).

Al-Khalili et al. (2004) used CRYRING to study dissociative recombination of $CD_3COHNHCH_3^+$ (protonated N-methylacetamide) and $CH_3SSHCH_3^+$ (protonated dimethyl disulfide). These ions are sufficiently small to allow studies of the type described in this chapter for other polyatomic ions, while at the same time containing amide and amine bonds (protonated N-methylacetamide), and disulfide bonds (protonated dimethyl disulfide). They found that the branching ratio for H atom release from $CD_3COHNHCH_3^+ + e^-$ was 0.817 ± 0.011, which would support the "hot H atom" mechanism. Recombination of $CH_3SSHCH_3^+$ was dominated by the channel for which CSS + C was formed (neglecting the hydrogen atoms); the branching ratio for this channel was measured to be 0.625 ± 0.09. A small system like protonated dimethyl disulfide apparently behaves differently from a protein in terms of S−S fragmentation. Studies of electron interaction with heavier ions are better performed in electrostatic storage rings (see Section 2.2.2), and Tanabe and coworkers have used the ring shown in Fig. 2.16 to study singly charged peptides (Tanabe *et al.* 2003, 2005) and DNA anions (Tanabe *et al.* 2004, 2005). The main effort in understanding the ECD mechanism(s) has, however, been done using theoretical methods (e.g. Syrstad & Turuček (2005)).

Conferences dedicated to ECD and related topics have been organized annually since 2003.

9
Related processes

Dissociative recombination is part of the broader field of electron–molecule scattering, which dates back to the famous Franck–Hertz experiment in 1914 (Franck & Hertz 1914). There are also similarities between dissociative recombination and photodissociation, and being a reactive process, recombination can also be considered as branch of chemical reaction dynamics. It is not possible to give a comprehensive presentation of all these topics, or this book would take an encyclopedic format. Instead we will focus on the topics that are most closely related to dissociative recombination.

A broad overview of the entire field of atomic collisions, including electron–ion recombination, is given in McDaniel (1989) and McDaniel, Mitchell, and Rudd (1993). The theory of electron–atom and electron–molecule collisions has been covered in Khare (2002). Since the recombination of atomic ions is not covered in the present book, the reader is referred to Dunn *et al.* (1984), Graham *et al.* (1992), McDaniel, Mitchell, and Rudd (1993), Hahn (1997), and Phaneuf *et al.* (1999) for reviews of this topic. Edited volumes by Christophorou (1984a,b), Märk and Dunn (1985), Ehrhardt and Morgan (1994), and Becker (1998) cover well the development in electron–molecule collisions during the 1980s and 1990s. Review articles covering dissociative recombination are given in Chapter 1. Other aspects of the interaction of electrons and molecules have been reviewed: cross section measurements (Trajmar, Register, & Chutjian 1983, Filippelli *et al.* 1994, Trajmar & McConkey 1994, Crompton 1994, Itikawa 1994, Tanaka & Sueko 2001, Brunger & Buckman 2002, Buckman, Panajotovic, & Jelisavcic 2004), associative ionization, i.e., the reverse process to dissociative recombination (Weiner, Masnou-Seeuws, & Giusti-Weiner 1989), theoretical developments (Huo & Gianturco 1995, Winstead & McKoy 1996, 2000, Rescigno & McCurdy 1998), electron-induced rotational excitation (Itikawa & Mason 2005), electron interaction with excited molecules (Christophorou & Olthoff 2001), low-energy electron attachment (Smith & Španěl 1994, Dunning 1995, Chutjian, Garscadden, & Wadehra 1996, Illenberg 2000,

Hotop et al. 2003), and threshold phenomena (Hotop et al. 2003). The interaction of positrons with atoms and molecules has been reviewed by Surko, Gribakin, and Buckman (2005). A very useful guide to bibliographies and review articles in atomic and molecular collision physics has been compiled by McDaniel and Manskey (1994), and very good sourcebooks of activities in this area are the proceedings of the biennial International Conference on the Physics of Electronic and Atomic Collisions (ICPEAC), since 1999 renamed International Conference on Photonic, Electronic and Atomic Collisions (ICPEAC).

9.1 Dissociative excitation and ionization of molecular ions

Dissociative excitation of molecular ions was discussed briefly in Chapter 2 in connection with ion sources (Section 2.1.3). The process was used in order to determine the state of excitation of the ions extracted from an ion source. Dissociative excitation has received much less attention than dissociative recombination, and there have been very few investigations of dissociative ionization. Cross sections for excitations between bound molecular ion states are also rare (Crandall et al. 1974). Dissociative excitation and ionization have been reviewed by Dunn and Djurić (1998), and their review also contains a listing of all results up to 1998. Dissociative excitation and ionization are, in contrast to recombination, studied exclusively by beam methods. Nearly 50% of all experimental studies concern hydrogen molecular ions, i.e., H_2^+ and H_3^+ and their isotopologs. This section is not intended to be a comprehensive review; instead we will focus on a few examples and give a tabulated update of experiments performed since the review of Dunn and Djurić (1998).

It is natural that H_2^+ being the simplest molecule, received the interest from theorists with respect to dissociative excitation (Saltpeter 1950, Kerner 1953, Ivash 1958). The first experimental studies were performed in Dunn's laboratory in JILA, Boulder, Colorado, in the mid-1960s by means of crossed electron and ion beams (Dunn, Van der Zyl, & Zare 1965, Dunn & Van der Zyl 1967). The results were in very good agreement with theoretical work by Peek (1967). One of the incentives for this work was the need for cross sections in order to understand high-energy plasma devices such as those being developed for controlled thermonuclear fusion.

The dissociative processes discussed in this section are of the following types:

$$XY^+(B) + e^- \rightarrow XY^+(D) + e^- \rightarrow X + Y^+ + e^- + \varepsilon_{DE}, \tag{9.1}$$
$$XY^+(B) + e^- \rightarrow XY^{**}(D) \rightarrow XY^+(D) + e^- \rightarrow X + Y^+ + e^- + \varepsilon_{RDE}, \tag{9.2}$$
$$XY^+(B) + e^- \rightarrow XY^{++}(D) + 2e^- \rightarrow X^+ + Y^+ + 2e^- + \varepsilon_{DI}, \tag{9.3}$$

where ε_{DE}, ε_{RDE}, and ε_{DI} are the kinetic energy releases in dissociative excitation, resonant dissociative excitation, and dissociative ionization, respectively. Figure 9.1 illustrates these processes with potential energy curves for a diatomic molecule.

9.1 Dissociative excitation and ionization of molecular ions

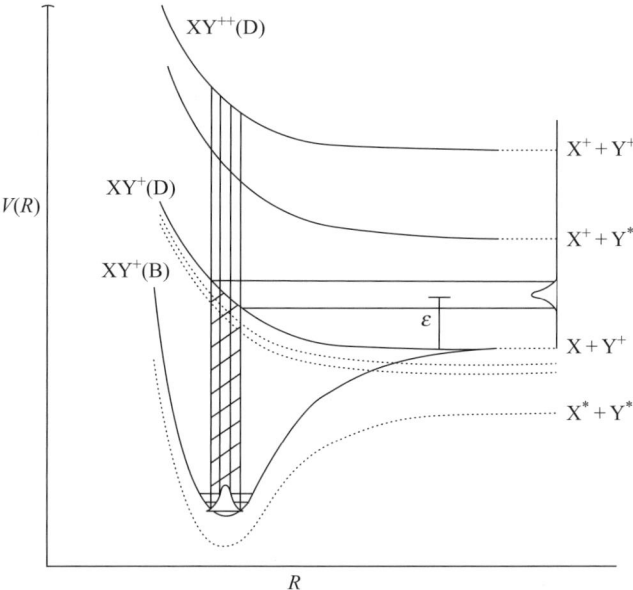

Figure 9.1 Schematic diagram of dissociative excitation, resonant dissociative excitation, and dissociative ionization. Dissociative excitation involves an electronic transition from the bound ion state $XY^+(B)$ to the dissociative state $XY^+(D)$ induced by an electron. Resonant dissociative excitation involves electron capture into a compound state $XY^{**}(D)$ followed by autoionization back to the ion state and a free electron. Dissociative ionization leads to the formation of a doubly charged molecular ion $XY^{++}(D)$, which dissociates to X^+ and Y^+. (Reproduced with permission from Dunn, G.H. & Djuric, N. 1998, "Electron impact dissociative excitation and ionization of molecular ions," in *Novel Aspects of Electron-Molecule Collisions*, ed. K.H. Becker, World Scientific, Singapore, pp. 241–281. Copyright (1998) World Scientific.)

The first experiments at JILA (Dunn, Van der Zyl, & Zare 1965, Dunn & Van der Zyl 1967) were followed by experiments in other laboratories, such as those in Culham (Dance *et al.* 1967) and Newcastle (Peart & Dolder 1971) in England. These crossed-beam apparatuses were quite similar in ion optics design, with magnetic sector fields to obtain mass-selected ion beams and electrostatically focused electron beams crossing perpendicular to the ion beams. The crossed-beam technique has also been used at the Oak Ridge National Laboratory in Tennessee (Schulz *et al.* 1986, Bannister *et al.* 2003, Bahati *et al.* 2005) and Louvain-la-Neuve, Belgium (Bahati *et al.* 2001b). Although the merged-beam technique does not offer any advantage for studies of excitation and ionization as compared with the crossed-beam technique, it has been used in both the single-pass configuration (Mitchell & Hus 1985) and ion storage rings (Forck *et al.* 1993a, Andersen *et al.* 1997a, Le Padellec *et al.* 1998).

The H_2^+ ion remains the best studied with respect to dissociative excitation. The measurement of the vibrational state distribution in H_2^+ ions produced by electron impact of H_2 (von Busch & Dunn 1972) made it possible to compare results from different experiments, provided that de-population of higher vibrational levels was avoided. Peart and Dolder (1971) measured the proton production in electron impact on H_2^+ ions having a von Busch and Dunn vibrational distribution. Some criticism of this experiment was leveled by Rundel (1972), but satisfactory counterarguments were given by Dolder and Peart (1972). In their second experiment on H_2^+, Peart and Dolder (1972a) were able for the first time to take proper account of the fact that H^+ can also be produced by dissociative ionization by implementing a coincidence technique. In a third experiment, Peart and Dolder (1972b) accessed the low-energy regime, down to 3.45 eV, by making use of the inclined-beam technique. The data sets taken between 1965 and 1972 were compared with theoretical calculations (Peek 1974) and the overall agreement was found to be very good. In particular the excellent agreement between the JILA (Dunn & Van der Zyl 1967) and Newcastle experiments (Peart & Dolder 1971, 1972a) received the following comment by Dunn and Djurić (1998, p. 250): "One might consider these sets of experiments and their results as paradigms in the area of electron impact dissociation of molecular ions." Yousif and Mitchell (1995) applied the merged-beam technique to access collisions down to 10 meV, and their results at higher energies joined smoothly with the crossed- and inclined-beam data. The consistency of data from five different experiments and theory makes it difficult to understand why the more recent set of data, from a crossed-beam experiment in Louvain-la-Neuve by Abdellahi El Ghazaly et al. (2004), agrees with previous results only at energies higher than 80 eV. Also the ion storage ring experiment by Andersen et al. (1997b) gave data in some disagreement with the majority of results. It is possible that the H_2^+ ions in the storage ring experiment had a von Busch and Dunn vibrational distribution when extracted from the ion source, but that the highest vibrational levels were depleted by dissociative recombination after a few seconds of storage, as noticed by van der Zande et al. (1996). Abdellahi El Ghazaly et al. (2004) measured the vibrational distribution in the H_2^+ ions used in their experiment and found a cut-off around $v = 12$. It is unclear whether this is sufficient to explain the deviation from older experiments.

Hydrocarbon ions occur in the edge plasmas of tokamaks (Janev & Reiter 2002, 2004), in the interstellar medium, and in the atmospheres of the giant planets and their satellites, such as Titan. In order to allow a modeling of such plasmas, the JILA group embarked on an ambitious program to study dissociative excitation of CD_n^+ ($n = 1-5$) ions (Djurić et al. 1997, 1998). They measured the production of light charged fragments, D^+ and D_2^+, which required a redesign of the crossed-beam machine at JILA, as described by Djurić et al. (1997) and Dunn and Djurić (1998).

9.1 Dissociative excitation and ionization of molecular ions

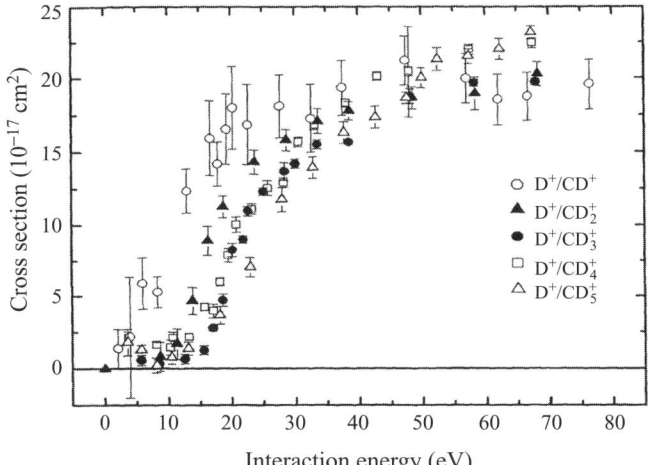

Figure 9.2 Absolute cross sections for D^+ fragment ion production from CD_n^+ ($n = 1-5$). (Reused with permission from Nada Djurić, in *Recent Experimental Studies of Electron-Impact Excitation of Atomic and Molecular Ions*, T. Kato (ed), Conference Proceeding 771, 162 (2005). Copyright 2005, American Institute of Physics.)

The most remarkable outcome of these studies was that the cross section for D^+ formation for all ions ($n = 1-5$) reached a plateau value of 2×10^{-16} cm^2 when the electron energy was increased past the threshold level, as shown in Fig. 9.2. The most immediate reaction to such an effect would be to suspect some systematic error in the experiment, but this was carefully checked for and there is no doubt that the effect is real. No explanation for this surprising result has been put forward. In subsequent work (Popović *et al.* 2001), the C_2H^+ and $C_2H_2^+$ ions were studied, and for these ions also essentially the same dissociative excitation cross section for H^+ production was measured ($(1.3-1.7) \times 10^{-16}$ cm^2). The propensity for reaching the same cross section independent of the number of hydrogen (deuterium) atoms present in the molecule was also shown by molecular systems in which carbon was replaced by oxygen or nitrogen (Djurić *et al.* 2000), with a plateau value of $(1.3-2) \times 10^{-16}$ cm^2. Typically in all these studies, the cross section for H_2^+ (D_2^+) production was about an order of magnitude smaller than for H^+(D^+) production.

Resonant dissociative excitation was recognized as an important process in low-temperature plasmas during the 1990s. Consider electron impact dissociation of HeH$^+$:

$$HeH^+ + e^- \rightarrow He^+ + H + e^-, \quad (9.4a)$$

$$HeH^+ + e^- \rightarrow He + H^+ + e^-. \quad (9.4b)$$

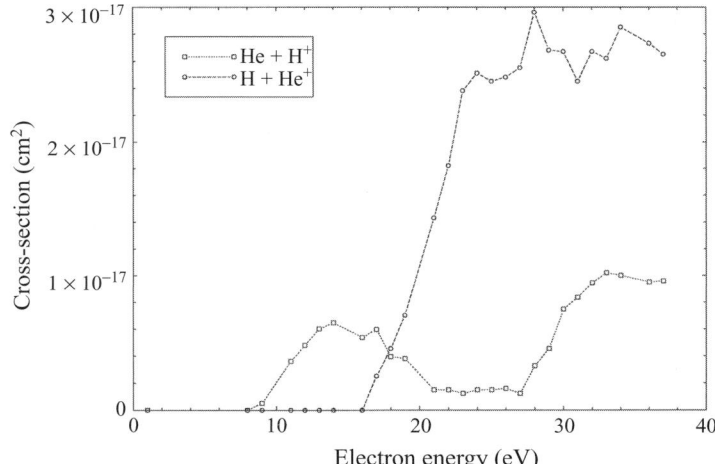

Figure 9.3 Cross section for dissociative excitation of ^{4}HeH$^+$ as a function of electron energy. (Reprinted figure by permission from C. Strömholm, J. Semaniak, S. Rosén, et al., "Dissociative recombination and dissociative excitation of ^{4}HeH$^+$: Absolute cross sections and mechanisms," *Phys. Rev. A* **54**, pp. 3086–3094, (1996). Copyright (1996) by the American Physical Society.)

The $X\,^1\Sigma^+$ electronic ground state of HeH$^+$ dissociaties to He (^{1}S) + H$^+$. Dissociative excitation takes place by a direct transition from the ground state to the $a\,^3\Sigma^+$ state dissociating to He$^+$ (^{2}S) + H ($n = 1$). In CRYRING, the production of H atoms was measured instead of the production of He$^+$ ions, and the dissociative excitation cross section for (9.4a) was as expected (Strömholm *et al.* 1996), as shown in Fig. 9.3. It was expected that it would require close to 30 eV to reach channel (9.4b), however, Strömholm *et al.* (1996) found the onset of production of He to require about 10 eV. The explanation is that a third decay route is accessed:

$$\text{HeH}^+ + e^- \rightarrow \text{HeH}^{**} \rightarrow \text{He}\left(^1\text{S}\right) + \text{H}^+ + e^-, \qquad (9.4c)$$

where the doubly excited HeH state is the same as that involved in dissociative recombination, hence the name resonant (or resonance-enhanced) dissociative excitation. Orel and Kulander (1996) did an *ab initio* study of Process (9.4c) by considering the electron capture into the resonant state, the evolution of the molecule on the resonant potential energy curve until the electron autoionizes, and the distribution of vibrational levels, bound or free, after the electron escapes. The electron scattering parameters came from complex Kohn variational calculations, and the dissociation dynamics was treated by the wave-packet method. They found very good agreement with the experimental results for the threshold and peak positions, but their calculated cross section was a factor of 2 larger than the experiment.

Table 9.1. *Absolute cross section data for dissociative excitation of molecular ions since 1999*

Ion	Products	Method	Reference
H_2^+	$(H^+) + H$	Crossed beams	Abdellahi El Ghazaly et al. 2004
N_2^+	$(N^+) + N$	Crossed beams	Bahati et al. 2001c
CO_2^+	$(C^+) + $ other	Crossed beams	Bahati et al. 2001b
	$(O^+) + $ other		
H_2O^+	$(OH) + H^+$	Ion storage ring	Jensen et al. 1999
	$(H_2) + O^+$	ASTRID	
	$(O) + H_2^+$		
	$(H) + OH^+$		
HDO^+	$(OH) + D^+$	Ion storage ring	Jensen et al. 1999
	$(OD) + H^+$	ASTRID	
D_nO^+	$(D^+) + $ other	Crossed beams	Djurić et al. 2000
	$(D_2^+) + $ other		
H_3O^+	$(H^+) + $ other	Crossed beams	Bahati et al. 2001a
HD_2O^+	$(H^+) + $ other	Crossed beams	Bahati et al. 2001a
	$(D^+) + $ other		
D_3O^+	$(D^+) + $ other	Crossed beams	Bahati et al. 2001a
ND_n^+	$(D^+) + $ other	Crossed beams	Djurić et al. 2000
	$(D_2^+) + $ other		
CH^+	$(C^+) + H$	Crossed beams	Bannister et al. 2003
$C_2H_n^+$	$(H^+) + $ other	Crossed beams	Popović et al. 2001
$D^{13}CO^+$	$(^{13}CO^+) + D$	Crossed beams	Bahati et al. 2005

However, when the factor of 2 error in the cross section formula was corrected, very good agreement with the experimental results was obtained (Orel 2000b).

An example of resonant dissociative recombination was also found for D_3^+ (Le Padellec et al. 1998). The peak found around 8 eV originates from the same electron capture that gives the imprints on the cross sections for recombination and ion-pair formation discussed in Sections 7.2.3 and 7.2.4.

For a long time there was only one direct measurement of dissociative ionization, the reaction being $H_2^+ + e^- \rightarrow H^+ + H^+ + 2e^-$ (Peart & Dolder 1973b). In a few cases an indirect approach has been applied in order to infer information about dissociative ionization cross sections, as described by Dunn and Djurić (1998). Since the early 2000s, however, cross sections for dissociative ionization have been measured directly by means of crossed-beam techniques in Louvain-la-Neuve and Oak Ridge, as reviewed by Bannister (2005).

Table 9.1 lists the dissociative excitation measurements performed after the Dunn and Djurić (1998) review. A complete listing of data until 2004 has also been given by Djurić (2005).

9.2 Ion-pair production

Ion-pair formation is a process for which the initial step is the same as in dissociative recombination, but instead of dissociation to neutral products, ion-pair formation leads, as the name suggests, to positive and negative ions:

$$XY^+ (B) + e^- \rightarrow XY^{**} (D) \rightarrow X^- + Y^+ + \varepsilon_{RIP}, \tag{9.5}$$

where ε_{RIP} is the kinetic energy release in the resonant ion-pair formation processes ("resonant" is sometimes added to emphasize that the process involves a neutral resonant state). It is the third stabilization channel following dielectronic capture into a doubly excited compound state XY^{**} (D).

There are not many measurements of electron impact induced ion-pair formation reported in the literature. The crossed- and merged-beam studies of $H_3^+ + e^- \rightarrow H^- + H_2^+$ (Peart, Foster, & Dolder 1979, Yousif, Van der Donk, & Mitchell 1993, Kalhori et al. 2004) are discussed in Section 7.2.4. The pioneering experiment was done by Peart and Dolder (1975) using H_2^+ as the target, and was part of their series of measurements on this ion (Peart & Dolder 1971, 1972a,b, 1973b, 1974a, 1975). Until the late 1990s these remained the only experiments, and just as for dissociative excitation, ion-pair formation cannot be studied by afterglow techniques.

In the late 1990s, the ion storage ring CRYRING was used to study ion-pair formation. In the first experiment the following process was studied (Zong et al. 1999):

$$HD^+ (v = 0) + e^- \rightarrow H^+ + D^-. \tag{9.6}$$

In the experiment by Peart and Dolder (1975), H_2^+ populated all vibrational levels, so one would anticipate a quite different outcome for the CRYRING experiment. This was indeed the case, as shown in Fig. 9.4.

The cross section in Fig. 9.4 shows two distinct features; a sharp threshold at 1.92 eV, and 14 peaks in the electron energy range 2–10 eV. The sharp threshold is a direct consequence of the energetics and the fact that only the ground vibrational level was populated in the experiment. The dissociation energy of HD^+ is 2.6677 eV and the electron affinity of atomic hydrogen is 0.7546 eV, which gives a threshold of 1.913 eV, which is very close to the observed one. The peak structure was unexpected and more difficult to explain. A tentative explanation was put forward by Zong et al. (1999), and this was followed by quantitative calculations by Larson et al. (2000). The potential curves of HD^+ and HD are discussed in Chapter 4; the $X\,^2\Sigma_g^+$ ground state of HD^+ is crossed by the resonant state $^1\Sigma_g^+ (2p\sigma_u)^2$, which correlates diabatically with the ion-pair limit at infinite internuclear separation. On its way to large separation, the $^1\Sigma_g^+ (2p\sigma_u)^2$ state crosses an infinite number of Rydberg states of type $^1\Sigma_g^+(1s\sigma_g n s\sigma_g)$ and $^1\Sigma_g^+ (1s\sigma_g (n+1)d\sigma_g)$ approaching the H (n) + D limit asymptotically. In a diabatic representation, the resonant state crosses many of these Rydberg states twice, first at about 5 a_0 and then at a very large

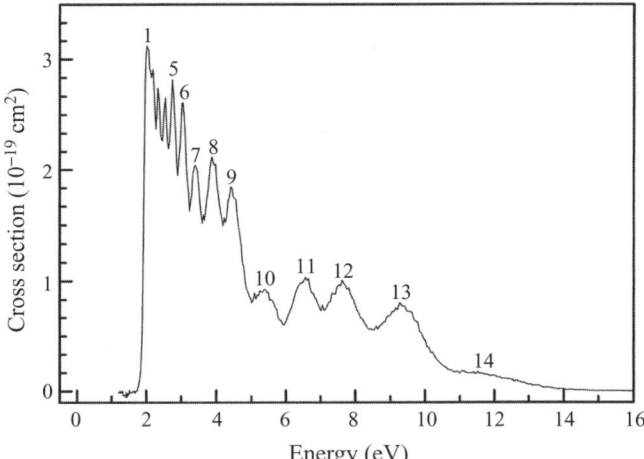

Figure 9.4 The absolute cross section for the process HD^+ $(v = 0) + e^- \to H^+ + D^-$ measured in CRYRING by recording the production of D^- as a function of electron energy. (Reprinted figure by permission from W. Zong, G.H. Dunn, N. Djurić, *et al.*, "Resonant ion pair formation in electron collisions with ground state molecular ions," *Phys. Rev. Lett.* **83**, pp. 951–954, (1999). Copyright (1999) by the American Physical Society.)

distance, in the case of $n = 3$ at 35.6 a_0. Thus, the molecule has different routes for reaching the ion-pair limit; it can either dissociate along the $^1\Sigma_g^+$ $(2p\sigma_u)^2$ state potential, or it can make a transition to one of the Rydberg states induced by the electronic part of the Hamiltonian (diabatic coupling), propagate along the Rydberg potential until the second crossing, and then make a transition back to the ion-pair state. With two different routes to dissociation, quantum mechanical interferences could therefore be anticipated. Quantitative calculations based on a time-dependent Landau–Zener–Stückelberg model gave very satisfactory agreement with the experimental peaks (Larson *et al.* 2000). When the outgoing flux was treated incoherently, the cross section was still quite well reproduced, but the interference pattern giving rise to the 14 peaks was absent. The conclusions that there was quantum mechanical interference was reinforced by a wave-packet calculation that gave very good agreement with experiment (Larson & Orel 2001). A natural question to pose was whether the $H^- + D^+$ channel would show a similar behavior, and to this end the experiment in CRYRING was repeated, this time by monitoring the production of H^- (Neau *et al.* 2002). The cross section was found to be identical to the one for $H^+ + D^-$, and hence the total cross section for ion-pair formation is twice that of the individual channels. This is a beautiful example of a molecular system displaying what could be labeled chemical "double slits." Another striking example is the photodissociation of water at 121.6 nm, manifested in oscillations in the OH ($X\,^2\Pi$) rotational distribution (Dixon *et al.* 1999).

The cross section for ion-pair formation in electron collisions with HD$^+$ is smaller than the dissociative recombination cross section at 2 eV. By a careful measurement requiring a long time for data taking, Lange *et al.* (1999) at the TSR found a small decrease in the recombination cross section at 1.9 eV as a result of the opening of the ion-pair channel.

There are only two other studies of Reaction (9.5), both performed at CRYRING. The cross section for ion-pair formation in NO$^+$ + e$^-$ collisions gave a sharp peak at 12.44 eV, and a structured cross section in the electron energy range 8–18 eV (Le Padellec *et al.* 2001b). The ion-pair limit can also be accessed by the process NO + $h\nu$ → N$^+$ + O$^-$ (Erman *et al.* 1995), and a comparison of the electron- and photo-induced ion-pair formation processes revealed effects derived from both different Franck–Condon regions (internuclear separations 1.063 Å and 1.151 Å for NO$^+$ and NO, respectively) and selection rules.

The HF$^+$ ion is unusual in the respect that the H$^+$ + F$^-$ limit almost coincides with the H ($n = 2$) + F (^{2}P$_{3/2}$) limit. This results in competition between ion-pair formation and dissociative recombination for which the ion-pair channel is much stronger than in other systems, accounting for about 25% of the total "recombination" at low energies. For OH$^+$ no measurable ion-pair signal has been found and an upper limit of 10^{-21} cm^2 has been determined (Larson *et al.* 2000). The cross section for ion-pair formation for HeH$^+$ has been calculated theoretically (Larson & Orel 1999) and is sufficiently large to be measurable; however, no such experiment has yet been performed.

9.3 Electron impact detachment of negative ions

The study of negative molecular ions is a topic in its own right, and this section will only consider aspects related to electron impact phenomena. Very good introductions, including historical perspectives, to negative ions are provided by Massey's book (Massey 1950, 1976) and a chapter in Branscomb's book *Atomic and Molecular Processes* (Branscomb 1962); the modern reviews by Andersen, Andersen, and Hvelplund (1997) and Pegg (2004) cover both atomic and molecular ions, and photo- and electron-impact detachment, whereas Andersen (2004) covers only atomic ions. It is now well established that electron-impact detachment cross sections for atomic ions are well described by a classical reaction zone model (Andersen *et al.* 1995), with a smooth cross section rising from a threshold that is larger than the electron affinity, and with no resonance structures.

The process we are concerned with in this section is electron-impact detachment of negative molecular ions:

$$\begin{align} AB^- + e^- &\to A + B + 2e^- \\ AB^- + e^- &\to AB + 2e^-. \end{align} \quad (9.7)$$

9.3 Electron impact detachment of negative ions

This process is technically difficult to study since negative ions are fragile systems and easily destroyed by collisions with residual gas molecules even under ultrahigh vacuum conditions. Studies of Process (9.7) have been confined to ion storage rings, and the first ion for which an electron-impact detachment cross section was measured was C_2^- (Andersen *et al.* 1996b). The technique for measuring an electron-impact detachment cross section in a storage ring is very similar to the technique for measuring dissociative recombination. Even the grid technique described in Section 2.2.1 can be used advantageously to determine the branching ratio for the two channels in Reaction (9.7). An important difference as compared with dissociative recombination experiments is that there is a threshold for detachment which depends on the electron affinity, and with an electron energy onset for electron detachment larger than the electron affinity. Andersen *et al.* (1996b) found a structure in the C_2 + 2e$^-$ channel at 10 eV, which they interpreted as the formation of a short-lived C_2^{2-} ion:

$$C_2^- + e^- \rightarrow C_2^{2-} \rightarrow C_2 + 2e^-. \tag{9.8}$$

In a subsequent study of both C_2^- and B_2^-, Pedersen *et al.* (1998) found no resonances in the detachment cross section for B_2^-, but a very clear peak in the B$^-$ + B^0 channel, i.e., the channel corresponding to resonant dissociative excitation: $B_2^- + e^- \rightarrow B_2^{2-} \rightarrow B^- + B^0 + e^-$. A second ion for which a detachment resonance was found (Pedersen *et al.* 1999) is BN$^-$. The ASTRID team then increased the complexity of the molecular target further (Andersen *et al.* 2001a), and found NO_2^- to have two detachment resonances, of which one was identified as being due to the ground state of NO_2^{2-}. Figure 9.5 shows the results for NO_2^-. From the widths of the resonances, 3–4 eV, they estimated the lifetime of the dianion to be about $(2-3) \times 10^{-16}$ s. It is not a general trend, however, that polyatomic negative ions have detachment resonances; in an experiment involving O_3^-, NO_3^-, and SO_2^- (Seiersen *et al.* 2003b), only NO_3^- showed evidence of a short-lived dianion state.

Also at CRYRING a program on atomic and molecular negative ions was started by Hanstorp's group from Göteborg, with CN$^-$ and C_4^- being the first molecular systems investigated. The statistical uncertainty of the data for CN$^-$ was too large to allow a definitive conclusion of a detachment resonance (Le Padellec *et al.* 2001a, Le Padellec 2003). Simultaneous measurements at ASTRID, however, gave data of higher quality and these showed clear evidence of a detachment resonance (Andersen *et al.* 2001b). This is not too surprising considering the large electron affinity of CN, 3.86 eV. The importance of high-statistics data in the threshold region is well illustrated by the studies of C_4^- in CRYRING, in which indications of a near-threshold resonance was found (Le Padellec *et al.* 2001c). In a subsequent experiment, this resonance was measured with much better statistics, which made it possible to determine the width to be 1.4 eV (Fritioff *et al.* 2004). The

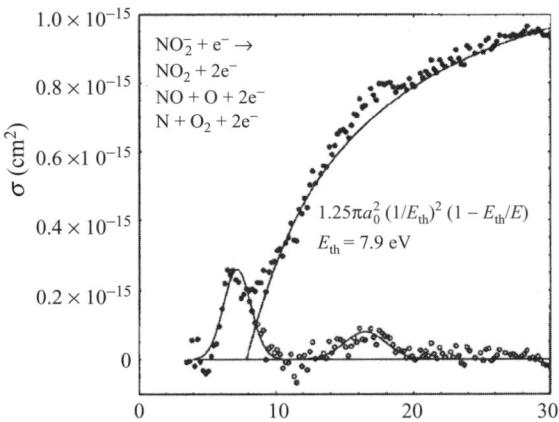

Figure 9.5 Electron-impact detachment cross section for NO_2^-. The smooth curve is a fit to the data in the region of no resonances. The nonresonant cross section is well reproduced by the classical reaction zone model of Andersen et al. (1995), which describes how a reaction takes place once the particles are inside a certain reaction radius. The open circles result from the subtraction of the measured data points by the smooth curve. The grid technique was used to determine that the branching ratio for the nondissociative channel (NO_2) was 0.75 and the dissociative channels ($NO + O$ and $N + O_2$) was 0.25 over the entire electron energy range 10–30 eV. (Reused with permission from L.H. Andersen, R. Bilodeau, M.J. Jensen, S.B. Nielsen, C.P. Sfvan, and K. Seiersen, *Journal of Chemical Physics* **114**, 147, (2001). Copyright 2001, American Institute of Physics.)

threshold region is usually analyzed by means of a classical over-the-barrier model developed for this particular application by Andersen et al. (1995). It allows the direct detachment cross section to be subtracted in order to analyze the resonance. Figure 9.6 shows the peak at 8.8 eV measured for C_4^- (Fritioff et al. 2004). It is narrower than the corresponding peak for C_2^-, which means that C_4^{2-} is somewhat longer lived than C_2^{2-}.

Small doubly charged negative ions, dianions, are stable in the liquid phase, in which polarization interaction with the solvent can stabilize the bonding of a second electron. This interaction is absent in the gas phase, making small dianions unstable. The question then arises of what happens if solvent water molecules are added one by one to a negative molecular ion. To answer this question, $OH^- \cdot (H_2O)_n$ cluster ions were studied in ASTRID with n ranging from 0 to 4 (Svendsen et al. 2004). Detachment of OH^- was found to be uneventful: it was basically the same as the smooth curve in Fig. 9.5, with a threshold at 4.5 eV. Addition of water molecules had the effect of shifting the threshold to higher energies, with the largest effect occurring when the first water molecule was added. For $OH^- \cdot (H_2O)_4$ the threshold was found to be 12.1 eV. The branching ratios were measured for $n \geq 1$, and in all cases the detachment was found to be dominated by detachment plus

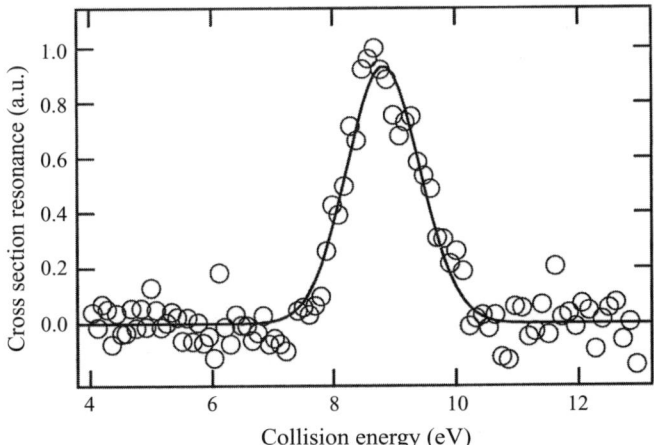

Figure 9.6 Cross section for $C_4^- + e^- \to C_4^{2-} \to C_4 + 2e$ measured in CRYRING. (Reproduced with permission from K. Fritioff, J. Sandström, P. Andersson, et al., "Observation of an excited C_4^{2-} ion," J. Phys. B **37**, pp. 2241–2246 (2004), IOP Publishing Limited.

complete dissociation into OH and $n \times H_2O$ with a branching ratio of about 0.80. No resonances were found for $n \leq 2$ clusters, but for the largest clusters resonances were found at about 15 eV, indicating temporary formation of $OH^{2-} \cdot (H_2O)_{n=3,4}$. Svendsen et al. (2004) showed that these dianions decayed to $OH^- \cdot (H_2O)_{n-1} + H_2O + e^-$, which is a different decay route than that for the negative molecular ions, where ejection of two electrons dominates. In experiments at ASTRID and ELISA, Svendsen et al. (2005b) showed how the resonance at threshold for $NO_2^- \cdot (H_2O)_n$ ($n = 0-2$) could be "tuned" by the presence of water molecules, in effect lowering the position of the short-lived dianions with respect to the anion. Hydration is known to have this effect on bound state dianions (Wang et al. 2001), but Svendsen et al. (2005b) were the first to show that it is also the case for unbound dianion states.

The picture emerging from the studies of electron-impact detachment of negative molecular ions is as follows. At long range, the interaction is mainly electrostatic repulsion, and the potential energy increases as the electron approaches the target ion. If the electron energy is too low, the electron will scatter off the ion before it has a chance to experience the attraction caused by the short-range polarization interaction. The combined long-range repulsive and short-range attractive forces give rise to a repulsive Coulomb barrier, which the incident electron will experience when its kinetic energy is sufficiently high (Dreuw & Cederbaum 2000). If the electron penetrates the barrier and is trapped, a doubly charged negative ion is formed. The barrier has to be sufficiently narrow for the electron to tunnel through, which means that a dianion in a short-lived excited state is formed. Examples of this are C_2^{2-} and C_4^{2-}, where the larger C_4^{2-} lives a little longer than C_2^{2-}. In the same way the addition of water molecules to OH$^-$ reduces the Coulomb barrier, making it possible for short-lived dianions to form when $n \geq 3$. Diner *et al.* (2004) used a trapped ion-beam technique (Dahan *et al.* 1998) to study electron-impact detachment of C_n^- and found, surprisingly, that the cross section increased as a function of cluster size. Since the binding energy increases as a function of cluster size, the cross section would be expected to decrease. In contrast, the cross section for electron-impact detachment of Al_n^- showed a normal behavior (Diner *et al.* 2004).

The observation of long-lived C_7^{2-} and other carbon cluster dianions in the gas phase (Shauer, Williams, & Compton 1990), where long-lived in this context means that they could be observed in a mass spectrometer and hence must have a lifetime of about 10 μs, has stimulated interest in multiply charged anions (MCAs), as evident from reviews by Dreuw and Cederbaum (2002) and Mathur (2004). The first electron-impact experiment using dianions as the target was performed in the electrostatic storage ring ELISA by Andersen's group (El Ghazaly *et al.* 2004). They showed that the thresholds for neutralization of $Pt(CN)_4^{2-}$ and $Pt(CN)_6^{2-}$ are 17.2 and 18.7 eV, respectively, and they also found evidence for resonances arising from short-lived trianions.

We list in Table 9.2 all electron-impact experiments on negative molecular ions known to us. Several experiments have also been carried out at ASTRID and CRYRING with atomic ions, but this work is not included here.

9.4 Electron–molecule scattering; dissociative attachment

Up to this point we have covered electron–molecular ion scattering comprehensively. The main topic of this book, dissociative recombination, is by far the largest subfield of electron–molecular ion scattering, with a much smaller volume of work having been done in the areas described in Sections 9.1–9.3. The situation is radically different, of course, when we consider electron scattering on neutral molecules. Even if we try to be selective and confine this section to the process which most closely resembles dissociative recombination, namely dissociative attachment of neutral molecules by electrons, we still have to be selective. Before discussing this process in more detail, it is worth putting electron collisions with neutral molecules into a slightly broader perspective.

An electron colliding with a molecule can induce the following processes without breaking or ionizing the molecule:

$$AB\,(\eta, v, J) + e^-\,(E_1) \rightarrow AB\,(\eta', v', J') + e^-\,(E_2), \quad (9.9)$$

where η is the electronic state, v the vibrational level, and J the rotational level of the molecule AB, and the prime on the right hand side of the arrow indicates an excitation or a de-excitation. The energies of the incident and scattered electrons

9.4 Electron–molecule scattering; dissociative attachment

Table 9.2. *Electron-impact detachment of negative molecular and cluster ions*

Ion	Threshold	Resonances	Storage ring	Reference
C_2^-	~7 eV	10 eV	ASTRID	Andersen et al. 1996b
				Pedersen et al. 1998
B_2^-	~3.5 eV	~5 eV	ASTRID	Pedersen et al. 1998
				Pedersen et al. 1999
O_2^-	4.5 eV	No	ASTRID	Pedersen et al. 1999
BN^-	~5.9 eV	5.6 eV	ASTRID	Pedersen et al. 1999
CN^-	8.5^a eV	10 eV	ASTRID	Andersen et al. 2001b
CN^-	7 eV	?	CRYRING	Le Padellec et al. 2001a
BO^-	5.2 eV	No		Andersen et al. 2001b
F_2^-	10 eV	No	ASTRID	Pedersen et al. 2001
Cl_2^-	4.8 eV	15 eV	CRYRING	Collins et al. 2005
LiH_2^-	5.5 eV	No	TSR	Lammich et al. 2007
NO_2^-	7.9^a eV	7.2, 16.5 eV	ASTRID	Andersen et al. 2001a
SO_2^-	7.8 eV	No	ASTRID	Seiersen et al. 2003b
O_3^-	7.3 eV	No	ASTRID	Seiersen et al. 2003b
NCO^-	9.1 eV	9.3 eV	ASTRID	Svendsen, El Ghazaly, &
				Andersen (2005a)
NCS^-	8.9 eV	8.4, 19.0 eV	ASTRID	Svendsen, El Ghazaly, &
				Andersen (2005a)
NO_3^-	10.5 eV	18.6 eV	ASTRID	Seiersen et al. 2003b
C_4^-	7.3 eV	8.8 eV	CRYRING	Le Padellec et al. 2001c
				Fritioff et al. 2004
C_n^- ($1 \leq n \leq 9$)	Not measured	Not measured	Ion beam trap	Diner et al. 2004
C_n^- ($2 \leq n \leq 11$)	~8 eV for C_8^-	Maybe	Ion beam trap	Eritt et al. 2006
Al_n^- ($2 \leq n \leq 5$)	Not measured	Not measured	Ion beam trap	Diner et al. 2004
Al_n^- ($2 \leq n \leq 10$)	Not measured	Not measured	Ion beam trap	Eritt et al. 2006
$OH^-(H_2O)_0$	3.7 eV	No	ASTRID	Pedersen et al. 1999
	4.5 eV			Svendsen et al. 2004
$OH^-(H_2O)_1$	7.9 eV	No	ASTRID	Svendsen et al. 2004
$OH^-(H_2O)_2$	10.1 eV	No	ASTRID	Svendsen et al. 2004
$OH^-(H_2O)_3$	11.4 eV	15.6 eV	ASTRID	Svendsen et al. 2004
$OH^-(H_2O)_4$	12.1 eV	15.2 eV	ASTRID	Svendsen et al. 2004
$NO_2^-(H_2O)_0$	9.0 eV^a	7.2, 18.8 eV^b	ASTRID, ELISA	Svendsen et al. 2005b
$NO_2^-(H_2O)_1$	9.5 eV^a	6.4 eV	ASTRID, ELISA	Svendsen et al. 2005b
$NO_2^-(H_2O)_2$	Not given	5.6 eV	ASTRID, ELISA	Svendsen et al. 2005b
$Pt(CN)_4^{2-}$	17.2 eV^a	17.0 eV	ELISA	El Ghazaly et al. 2004
$Pt(CN)_6^{2-}$	18.7^a eV	15.3, 18.1, 20.1 eV	ELISA	El Ghazaly et al. 2004
$C_6H_4CO^-$	8.4 eV	No	ELISA	El Ghazaly et al. 2005

^{a}This is the classical threshold in the absence of resonances.
^{b}Probably a more accurate value than the 16.5 eV found in the earlier work (Andersen et al. 2001b)

are E_1 and E_2, respectively. If the quantum numbers are the same on both sides, the electron is elastically scattered; a change in the rotational quantum number gives a rotational excitation; a change in the vibrational quantum number gives a vibrational excitation; and a change in the electronic state gives an electronic excitation (or de-excitation in all those cases in which $E_1 < E_2$). Two types of experiments are used to study the processes contained in (9.9): beam techniques and swarm techniques (Brunger & Buckman 2002). Beam techniques will be discussed later in this section. Swarm techniques measure the macroscopic properties of an ensemble ("swarm") of electrons drifting, as the result of an applied electric field, through a chamber filled with gas (e.g. Crompton (1994)).

If the electron energy is sufficiently high, dissociation or ionization or both can occur:

$$AB + e^- \to A + B + e^-, \tag{9.10}$$

$$AB + e^- \to AB^+ + 2e^-, \tag{9.11}$$

$$AB + e^- \to A + B^+ + 2e^-. \tag{9.12}$$

Even higher electron energies can cause double ionization, etc. Process (9.10) has been discussed by Zipf (1984), and, more speculatively, by Mitchell and Rowe (2000), and Processes (9.11) and (9.12) have been reviewed by Märk (1984). Process (9.12) can be studied by the (e^-, $2e^-$) technique, as described by Leung (1998), or by measuring the production of ions (Märk 1992). Then there is finally the possibility of electron attachment:

$$AB\,(v) + e^-(E_1) \to AB^{-*} \to AB(v') + e^-(E_2), \tag{9.13a}$$

$$AB\,(v) + e^-(E_1) \to A + B^-, \tag{9.13b}$$

$$AB\,(v) + e^-(E_1) \to AB^-. \tag{9.13c}$$

The resonance AB^{-*} is usually formed by the electron entering an unoccupied molecular orbital. The complex AB^{-*} is called the shape resonance, the compound state, or the temporary negative ion state (Schulz 1973), and usually has a lifetime in the range 10^{-10}–10^{-15} s. In addition, some systems, for example H_2O, exhibit Feschbach resonances (see Chapter 3; Haxton, Rescigno, and McCurdy (2005)). The compound state has three decay modes, which are labeled scattering (9.13a), dissociative attachment (9.13b), and nondissociative attachment (9.13c). If the final vibrational level v' lies in the continuum, AB dissociates to A + B; if v' represents a discrete vibrational level, the electron attachment leads to vibrational excitation. Electron attachment can resonantly enhance the cross section for vibrational excitation. It is a process with broad applications in, for example, negative ion source development, plasma-enhanced processing, combustion, and aeronomy. The reader

9.4 Electron–molecule scattering; dissociative attachment

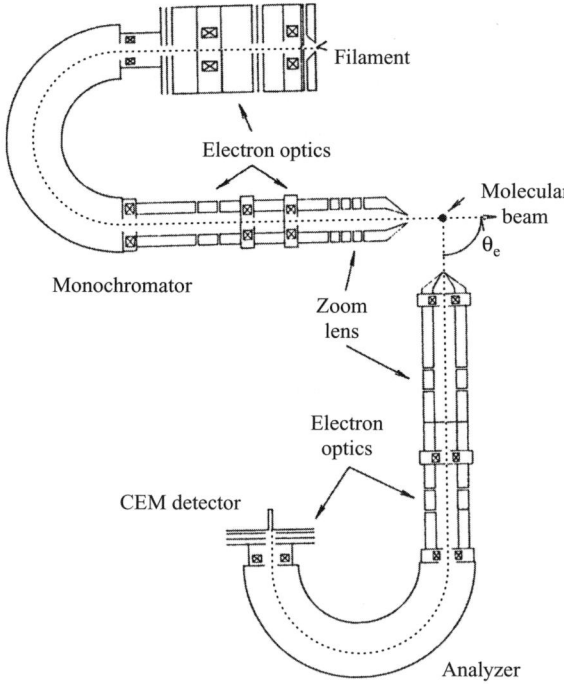

Figure 9.7 Crossed molecular-beam electron-beam apparatus. The molecular beam is perpendicular to the paper plane and is formed by effusive flow through a multi-channel capillary array with a total active area 1 mm². The electron optics were specially designed to transport low-energy electrons (0–5 eV) with high efficiency, and the electrons are detected by a channeltron electron multiplier (CME). Crossed-beam spectrometers can reach a resolution of 10 meV. (Reproduced with permission from the *Australian Journal of Physics* **43**: 665–682 (M. J. Brunger, S. J. Buckman, & D. S. Newman). Copyright CSIRO 1990. Published by CSIRO PUBLISHING, Melbourne Australia – http://www.publish-csiri.au/journals)

is referred to the review by Chutjian, Garscadden, and Wadehra (1996) for more details of the applications. More recent reviews, focusing on the fundamentals, are those of Hotop *et al.* (2003), and, in a shorter format, by Hotop, Ruf, and Fabrikant (2004).

Process (9.9) has been studied extensively with stable molecules as targets. The review by Brunger and Buckman (2002), gives complete coverage for diatomic molecules. Some material on polyatomics is given in the shorter review by Buckman, Panajotovic, and Jelisavcic (2004). In the review of Brunger and Buckman (2002), the cross section is only enhanced by negative ion resonances (9.13). Figure 9.7 shows an electron-beam–molecular-beam apparatus; the molecular beam intersects the electron beam at an angle of 90°. In that respect it resembles the crossed-beam experiments discussed in Section 9.1, but there are large differences. The molecular beam has a sub-eV energy and much higher density

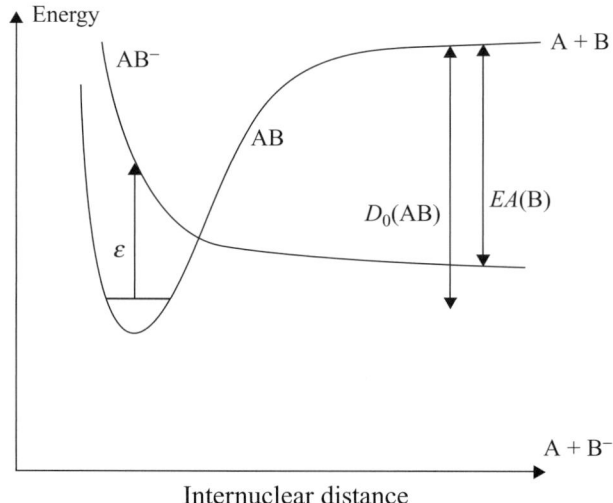

Figure 9.8 Potential energy curves showing the dissociative attachment of an electron by a molecule AB to form $A + B^-$. The molecule resides in the rovibrational level (v, J) and collides with an electron with kinetic energy ε. The electron is captured and a compound state AB^{-*} is formed. Immediately after the capture, the anion starts to dissociate. As long as it moves on the potential energy curve to the left of the crossing point with the neutral state, autodetachment can occur, resulting in (ro)vibrational excitation. When the system has passed the crossing point, the capture is stabilized and the anion continues to dissociate to $A + B^-$. The resemblance with dissociative recombination is obvious. The energy difference between the dissociation energy of AB, $D_0(AB)$, and the electron affinity of B, $EA(B)$, gives the threshold energy for dissociative attachment.

than the keV ion beams. The scattered electron is detected rather than the fragment ions; alternatively, light from the electron–molecule interaction region can be detected (Trajmar & McConkey 1994). Extracting the differential and total cross sections from a measurement of the scattered electron intensity is nontrivial (Brunger & Buckman 2002) and is beyond the scope of this book. The field of electron–molecule scattering is completely dominated by molecules that are accessible from gas tubes or liquids, and the difficult area of electron scattering on free radicals is limited to theoretical work (e.g. Lee *et al.* (2002)). The third type of beam crossing is that involving an ion beam and a molecular beam, and is used for the study of ion–molecule reactions (e.g. Herman (2001)). Crossed molecular beams are used to study chemical reaction dynamics (Casavecchia 2001), and crossed ion beams to study charge transfer between ions (Braeuning & Salzborn 2005).

The process of dissociative attachment and resonant vibrational excitation ((9.13a), (9.13b)) can be illustrated by potential energy curves in much the same way as in dissociative recombination. Figure 9.8 shows schematically how the process works.

9.4 Electron–molecule scattering; dissociative attachment

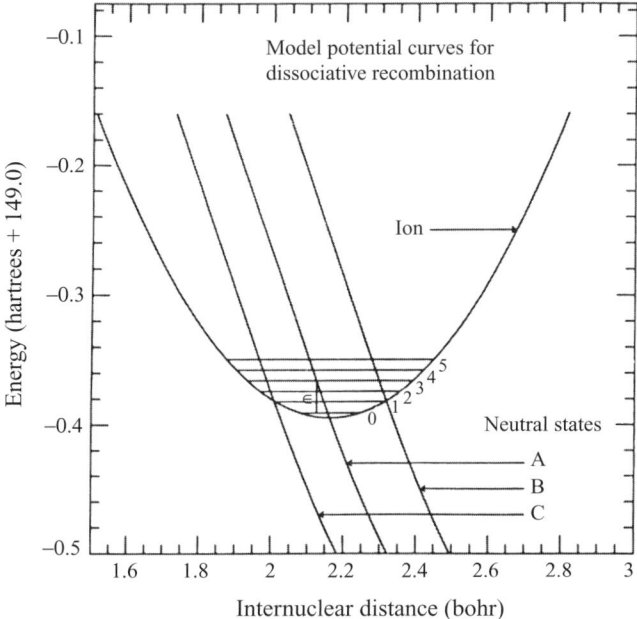

Figure 9.9 Potential energy curves for three different cases of dissociative recombination. (From Guberman (1986); reprinted by permission of the American Institute of Aeronautics and Astronautics, Inc.)

Electron attachment has been studied by means of a range of experimental techniques, including those using beams, swarms, and flowing afterglows. These techniques are discussed in some detail by Chutjian, Garscadden, and Wadehra (1996), and the application of the flowing afterglow technique to electron attachment has been reviewed by Smith & Španěl (1994). The potential energy curves in Fig. 9.8 illustrate a situation resembling dissociative recombination in which the resonant state crosses the ion state at the right turning point of a vibrationally excited level, as shown in Fig. 9.9. It is case B in Fig. 9.9 that gives rise to the high-energy peak in H_3^+ discussed in Section 7.2.3. However, as shown in Fig. 9.8, dissociative attachment would occur only in the Franck–Condon zone where the vibrational overlap is large, and the cross section has a threshold determined by $D_0(AB) - EA(B)$. An example is dissociative attachment of H_2, which has a threshold at 3.75 eV (Schulz & Asundi 1967).

The situation shown in Fig. 9.10 is more complex. The compound state XY^{-*} crosses the neutral state XY as in Fig. 9.8, but now there is also a negative ion state XY^- which has stable vibrational levels. At electron energies close to 0 eV, the electron can be captured by XY ($v = 0$) into the negative ion state XY^-, which cannot dissociate to $X + Y^-$ for energetic reasons. Thus, stabilization of the capture by dissociation is not possible. In small molecules the capture can only

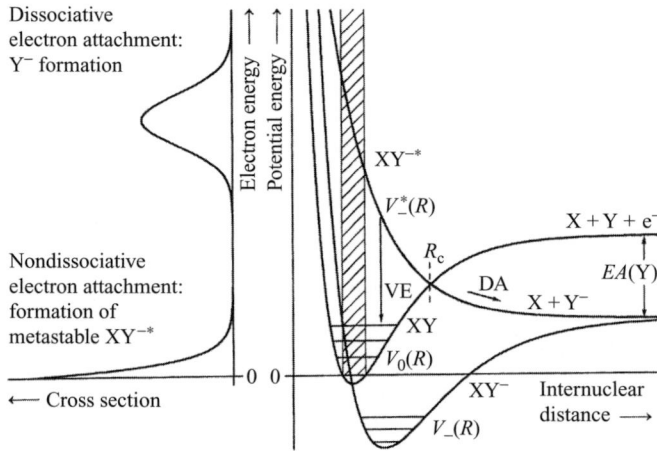

Figure 9.10 Potential energy curve diagram for dissociative attachment (DA), nondissociative attachment, and vibrational excitation (VE). The potential curve of the compound state XY^{-*} is labeled $V_-^*(R)$, the potential curve of the neutral molecule XY is labeled $V_0(R)$, and the potential curve of the stable negative ion XY^- is labeled $V_-(R)$. The shaded area is the Franck–Condon region for the primary electron capture process by the molecule XY in its lowest vibrational level, $v = 0$. The energy dependences of the cross sections are shown in the left part of the diagram. (Reprinted from *Advances in Atomic, Molecular & Optical Physics*, Vol. **49**, 2003, (ISBN 0120038498), Bederson (ed.), pp. 85–216, Hotop et al.: "Resonances and threshold phenomena in low-energy electron collisions with molecules and clusters." Copyright (2003), with permission from Elsevier.)

be stabilized by collisions, and this is only possible in high-density media or in clusters. Larger molecules, however, can redistribute the energy by intramolecular vibrational redistribution (IVR) so that autodetachment becomes less favorable and a metastable anion XY^- is formed.

Experiments aimed at probing the cross section at very low electron energies are technically demanding. The merged-beam technique, which avoids the problem of using electron beams at very low energies in the laboratory frame by a transformation to the center-of-mass frame, is not really an option for neutral molecules. Instead one must deal with the problem of handling electrons at very low energies in the laboratory frame. The reviews of, for example, Chutjian, Garscadden, and Wadehra (1996), Brunger and Buckman (2002), and Hotop et al. (2003) cover this in some detail. There are two basic problems: how to create electrons at meV and even sub-meV energy, and how to prevent these electrons from being accelerated by stray electric fields. The second problem is avoided by very careful shielding. The first problem has been addressed by using Rydberg atoms (Finch & Dunning 2000), and the threshold photoelectron spectroscopy for attachment (TPSA) method,

9.4 Electron–molecule scattering; dissociative attachment

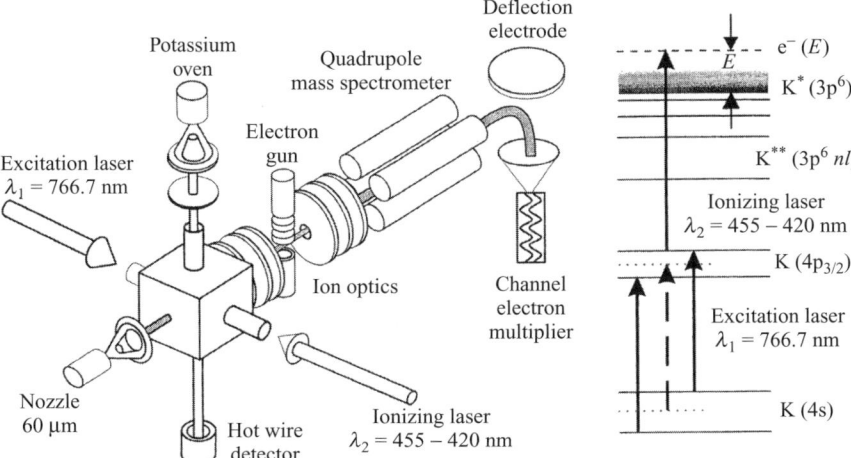

Figure 9.11 Laser photodetachment apparatus for very-low-energy electron attachment experiments. The photoelectrons are created by two-color laser photodetachment of potassium. The auxiliary electron gun is used for diagnostics of the target beam and of the residual gas by means of electron impact ionization. A supersonic beam serves as the target, the electron energy can be tuned continuously from zero to 200 meV, and the negative ions are imaged into the quadrupole mass spectrometer. The energy level diagram to the right shows how the photoelectrons are produced. (Reprinted from *Advances in Atomic, Molecular & Optical Physics*, Vol. **49**, 2003, (ISBN 0120038498), Bederson (ed.), pp. 85–216, Hotop *et al.*: "Resonances and threshold phenomena in low-energy electron collisions with molecules and clusters." Copyright (2003), with permission from Elsevier.)

developed by Ajello and Chutjian (1979). In this method low-energy electrons are created using threshold photoionization by vacuum ultraviolet radiation exposure of rare gas atoms (Ajello & Chutjian 1979). Figure 9.11 shows the TPSA apparatus developed and used by Hotop's group (Hotop *et al.* 2003), for which laser photodetachment is used in order to increase the resolution. Set-ups such as the one shown in Fig. 9.11 have opened up the possibility of studying resonances and threshold phenomena near the onset for vibrational excitation, where the channels (9.13a)–(9.13c) are coupled and can interfere. The reader is referred to the review by Hotop *et al.* (2003) for an in-depth discussion. Here we will illustrate the power of the TPSA method by one example.

SF_6 is one of the most commonly used insulating gases because, in addition to being electronegative ($EA = 1.05$ eV) and having a large cross section for low-energy electron attachment, it is nontoxic, not flammable, and relatively inexpensive. It lacks dipole and quadrupole moments and it has no threshold for electron attachment. SF_6 is therefore a molecule for which it is of interest to study the electron attachment cross section for $E_e \to 0$. Theoretically, the cross section should

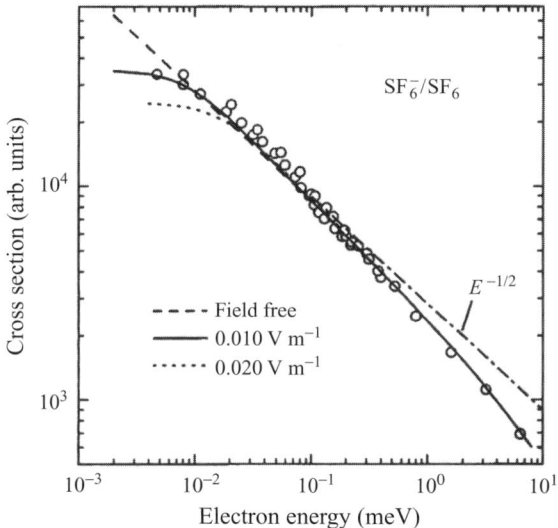

Figure 9.12 The relative cross section for formation of SF_6^- in collsions with free electrons measured by means of the apparatus shown in Fig. 9.11. The residual electric field was 0.01 V m^{-1} and the resolution 20 μeV at an electron energy of 20 μeV and 150 μeV at 6.5 meV. (Reprinted figure by permission from A. Schramm, J. M. Weber, J. Kreil, D. Klar, M.-W. Ruf, and H. Hotop, "Laser photoelectron attachment to molecules skimmed in a supersonic beam: Diagnostics of weak electric fields and attachment cross sections down to 20 μeV," *Phys. Rev. Lett.* **81**, pp. 778–781, (1998). Copyright (1998) by the American Physical Society.)

diverge as $\sigma(E_e) \propto E_e^{L-1/2}$ (Bethe 1935, Wigner 1948, Vogt & Wannier 1954), and for *s*-wave electrons ($L = 0$) this gives an $E_e^{-1/2}$ dependence. Chutjian and Alajajian (1985) used the TPSA method and achieved a resolution of 8 meV, which was sufficient to verify that their data were consistent with *s*-wave attachment. Hotop *et al.* (1995) measured the electron attachment cross section in the electron energy range 20 μeV–250 meV at sub-meV resolution using the apparatus shown in Fig. 9.11 and conclusively verified the $E_e^{-1/2}$ dependence. At the lowest energies, the measurements are very sensitive to stray electric fields (Schramm *et al.* 1998). Figure 9.12 shows the relative attachment cross sections for SF_6 and their sensitivity to stray electric fields as low as 0.01 V m^{-1}. A comparison with different theoretical results is given by Hotop *et al.* (2003).

9.5 Photodissociation and photoionization

The interaction of molecules with electromagnetic radiation is the origin of molecular spectroscopy, photochemistry, and radiation chemistry, topics which are too broad to be discussed here. Instead we will confine this section to those processes

9.5 Photodissociation and photoionization

induced by the absorption of a photon by a molecule that are closest to dissociative recombination. There are several different ways to approach the topic. We choose here to restrict ourselves to processes that lead to either dissociation or ionization, and we divide the topic according to the target.

The interaction of photons with positive molecular ions gives rise to the following processes:

$$AB^+ + h\nu \rightarrow A + B^+, \quad (9.14)$$

$$AB^+ + h\nu \rightarrow AB^{2+} + e^-. \quad (9.15)$$

If the ion AB^+ absorbs a single photon, light in the visible region is usually required to cause dissociation. Multiphoton dissociation can be induced by infrared photons, and this is widely exploited in Fourier transform ion cyclotron resonance mass spectrometry (FT-ICR MS) (Little et al. 1994) for biomolecular analysis. As described in Chapter 8, electron capture dissociation is a new tool in mass spectrometry (Cooper, Håkansson & Marshall 2005). Single photon dissociation is an entirely different matter. In the late 1970s the fast ion beam laser spectroscopy technique was developed to study molecular ions at rotational resolution (see, e.g., Cox et al. (1999) for a review). In many respects the technique shares features of the electron–molecular ion merged-beam technique, but with the electron beam replaced by a laser beam. The high spectral resolution one can obtain by using fast ions is explained in Section 2.1.1. A particularly sensitive method for detecting the absorption of a photon by a molecular ion is to excite it to a level at which it predissociates and then measure the production of fragment ions. Thus, in Process (9.14) the laser photon excites AB^+ to a rovibrational level that is unstable with respect to dissociation, and by tuning the laser or Doppler-tuning the ion beam, many such short-lived rovibrational levels can be probed. The photodissociation spectrum in such cases is highly structured, and analysis of such a spectrum provides information about the molecular structure and the predissociation dynamics. The process leading to dissociation is the same as the one giving rise to indirect dissociative recombination. The technique was used by Hechtfischer et al. (1998) to, among other things, measure the rotational temperature of CH^+ stored in the TSR. It is also possible to photon induce a direct transition to a repulsive electronic state, just like in dissociative excitation. The laser beam can be either collinear or perpendicular to the ion beam, and measurement of the kinetic energy of the charged fragment gives information about the repulsive state. The work by Broström et al. (1991) on ArN^+ is an example of photofragment kinetic energy spectroscopy.

Ionization of positive ions by the absorption of a single photon requires access to a light source capable of delivering photons with an energy of at least about 20 eV.

This cannot be done with table-top laser systems, but at third generation synchrotron light sources such as the Advance Light Source (ALS) at the Lawrence Berkeley National Laboratory and ASTRID in Aarhus (the ring can also be operated as a light source by storage of electrons) merged ion–photon beams have been used to study the photoionization of ions: studies of $CO^+ + h\nu \to CO^{2+} + e^-$ at ALS (Hinojosa *et al.* 2001) and ASTRID (Andersen *et al.* 2001c) are rare examples. Most studies of this type have so far been focused on atomic ions.

Much less photon energy is required to remove an electron from a molecular anion:

$$AB^- + h\nu \to AB + e^-, \quad (9.16a)$$

$$AB^- + h\nu \to A + B + e^-. \quad (9.16b)$$

Depending on what type of information one is interested in, crossed or collinear beam configurations can be used. Andersen (2004) and Pegg (2004) have reviewed the photodetachment of negative atomic ions. High-resolution photodetachment spectroscopy of molecular ions was developed by Lineberger's group at JILA, Boulder, and applied to C_2^- (Hefter *et al.* 1983). C_2^- has bound electronically excited states as first observed by Herzberg and Lagerqvist (1968). They rotationally analyzed the $B\,^2\Sigma_u^+ - X\,^2\Sigma_g^+$ transition and provided spectroscopic constants up to the $v' = 4$. Levels of the $C_2^-\,B\,^2\Sigma_u^+$ state with $v' \geq 5$ lie in the $C_2 + e^-$ continuum and this was exploited by Hefter *et al.* (1983) using a fast ion beam merged collinearly with a narrow bandwidth laser. The $B\,^2\Sigma_u^+ - X\,^2\Sigma_g^+$ transition was used to access vibrational levels above $C_2^-\,B\,^2\Sigma_u^+\,v' = 4$, which autodetach to $C_2 + e^-$. By measuring the production of neutral C_2 as a function of laser wavelength, Hefter *et al.* (1983) measured the positions and widths of the autodetachment resonances. Further high-resolution autodetachment work by the Lineberger group has been reviewed by Lykke *et al.* (1988).

Process (9.16a) can also be used to measure electron affinities and electronic state term energies by measuring the kinetic energy of the electron. The crossed-beam geometry is better in this case because the energy resolution needed to obtain a photoelectron spectrum requires a geometrically well-defined electron source. An example is the work on organic radicals by Lineberger's group (Wenthold & Lineberger 1999). The groups of Neumark at Berkeley and Lineberger at JILA have also very successfully applied negative ion photodetachment spectroscopy to probe reactive potential energy surfaces, which has led to the development of transition state spectroscopy in both the frequency domain and the time domain (Neumark 2005).

Dissociative detachment (9.16b) can proceed either directly or sequentially. Direct dissociative detachment involves a transition from the negative ion state

9.5 Photodissociation and photoionization

to the repulsive region of the neutral state. The photoelectron spectrum in such a case is structureless. The sequential process leads to a structured photoelectron spectrum that does not carry information on the subsequent dissociation process. By measuring the coincidence between the photoelectron and the neutral fragments A and B, the complete dissociative detachment process can be probed (Continetti 2000).

Photodissociation and photoionization of neutral molecules give:

$$AB + h\nu \to A + B, \tag{9.17}$$
$$AB + h\nu \to AB^+ + e^-. \tag{9.18}$$

Contained in these two simple formulas are the vast fields of photodissociation dynamics (9.17) and photoelectron spectroscopy (9.18). The book edited by Ashfold and Baggot (1987) and the monograph by Schinke (1993) give very good overviews of photodissociation, and a book on electron spectroscopy that is somewhere between a textbook and a monograph, is that of Ellis, Feher, and Wright (2005). The two volumes edited by Ng (2000a,b) give broad coverage of photoionization and photodetachment processes, and the volume edited by Sham (2002) covers their chemical applications. Wave-packet theory of photodissociation has been reviewed by Balint-Kurtis (2003).

Hatano (1999, 2002, 2003a,b) has reviewed the interaction of vacuum ultraviolet radiation (photon energies 10–50 eV) with molecules, and based his discussion to a large extent on the concept of superexcited states. This is a concept introduced by Platzman (1962a,b) in a theoretical analysis of the interaction of ionizing radiation with matter. A superexcited state is a state located above the first ionization limit. Figure 9.13 shows potential energy curves explaining the autoionization process and is taken from Nakamura's review article (Nakamura 1991). Figure 9.13(a) concerns autoionization of a superexcited state of the first kind, which is a two-valence-electron excited state, or alternatively an inner-shell-electron excited state. This superexcited state autoionizes as a result of electron correlation, or more specifically by electronic coupling of the superexcited state and the electronic continuum. This is a process that competes with direct dissociative recombination. The superexcited state of the second kind can autoionize as a result of the coupling between electronic and nuclear motion. This process competes with indirect dissociative recombination.

Nakamura (1984) also constructed a diagram based on the superexcited state and the connection between different processes. Figure 9.14 is based on his diagram. Dissociative recombination via the Rydberg states gives the indirect mechanism. As we have seen in Sections 9.1 and 9.2, the vibrational excitation exit channel leads to resonant dissociative excitation when the vibrational level lies in the nuclear

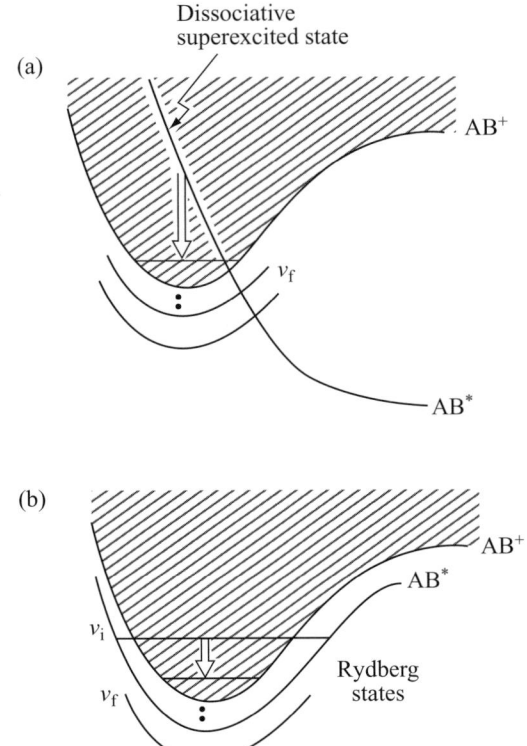

Figure 9.13 Schematic potential energy curves illustrating autoionization: a) autoionization of the dissociative superexcited state of the first kind by electronic coupling to the electron continuum; b) autoionization of the superexcited state of the second kind by coupling between the nuclear and electronic motion. (Reproduced with permission from H. Nakamura, "What are the basic mechanisms of electronic transitions in molecular dynamic processes?" *Int. Rev. Phys. Chem.* **10**, pp. 123–188 (1991). Copyright (1991) Taylor & Francis Ltd., http://www.informaworld.com)

continuum, and there is an exit channel that produces ion pairs as a result of both photon absorption and electron capture.

Synchrotron radiation is suitable to use for the study of superexcited states since it is continuously tunable in the photon energy range needed to excite a stable, neutral molecule to a superexcited state. As discussed in Section 9.2, a direct comparison between photon absorption and electron capture into a superexcited state is possible when the decay into ion-pair states is measured. However, there are only a few cases for which the ion-pair channel has been measured in electron–ion collisions. It is more difficult to compare the neutral channels, photodissociation and dissociative recombination, since, due to the geometry of synchrotron radiation experiments, it is difficult to detect neutral fragments. Optically emissive fragments are certainly

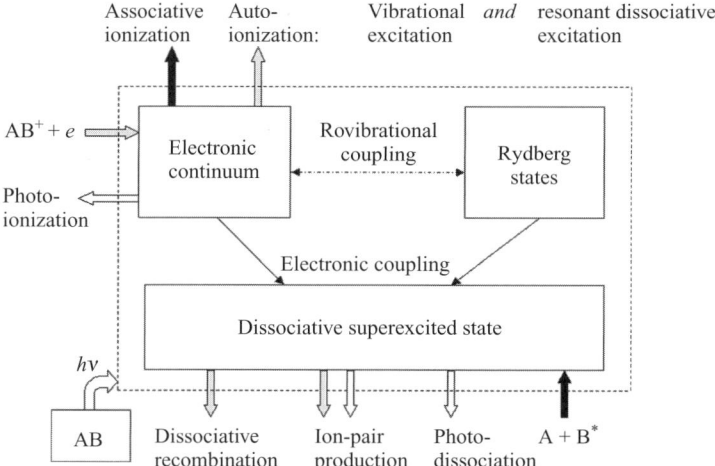

Figure 9.14 Different processes related to superexcited states. In the box within the dashed lines, vibrationally excited Rydberg states (superexcited states of the second kind), the dissociative superexcited state, and the electronic continuum are residing. There are three entry routes to the box: photoexcitation from the ground state of the neutral molecule AB, $AB^+ + e^-$ collision, and $A + B^*$ collision. The first entry has two exit routes, photoionization and photodissociation, the second one has two exit routes, dissociative recombination and vibrational excitation, and the third one exit route, associative ionization. The arrows are shaded accordingly. (Redrawn with permission from H. Nakamura, "Electronic transitions in atomic and molecular dynamic processes," *J. Phys. Chem. A* **88**, pp. 4812–4823 (1984). Copyright (1984) American Chemical Society.)

possible to detect, but these are usually not the dominant channels of superexcited states. The autoionization channels give plenty of opportunities for coincidence experiments between ions and electrons (Ullrich *et al.* 2003), but these cannot be compared directly with dissociative recombination.

A molecule can also dissociate as a result of the absorption of a photon with energy smaller than the ionization energy. In such cases, a superexcited state is not involved. In addition to the books cited earlier, Sato (2001, 2004) has reviewed more recent developments. Photodissociation is a subfield of chemical reaction dynamics and is basically half of a bimolecular reaction:

$$AB + C \rightarrow ABC^* \rightarrow A + BC, \qquad (9.19)$$
$$ABC + h\nu \rightarrow ABC^* \rightarrow AB + C. \qquad (9.20)$$

The first reaction is a bimolecular reaction; we refer the reader to Casavecchia (2001) for a review of chemical reaction dynamics. The second reaction, often called a "half-collision," can be viewed as a unimolecular reaction driven by light energy. The compound state ABC* is an electronically excited state of ABC, but not a superexcited state. The three-body decay of molecules induced by photons has been reviewed by Maul and Gericke (1997).

We refer the reader to Chapter 8 on polyatomic molecules for a discussion of the relation between dissociative recombination of XH_2^+ and photodissociation of XH_2.

10
Applications

The applications of dissociative recombination have been described in the previous chapters in connection with discussions about different molecular ions. Thus, this chapter is short and serves to direct the reader to some key references.

10.1 Molecular astrophysics

The development of astrochemistry is well covered by the proceedings of the International Astronomical Union: Andrew (1980), Vardya and Tarafdar (1987), Singh (1992), Van Dishoeck (1997), Minh and Van Dishoeck (2000), and Lis, Blake, and Herbst (2006). Astrochemistry was also a topic at the Fifth International Chemical Congress of Pacific Basin Societies (PACIFICHEM) in December 2005 (Kaiser *et al.* 2006). A good introduction to the field of molecular astrophysics is the book of the same name edited by T. W. Hartquist in honor of the sixtieth birthday of Alex Dalgarno (Hartquist 1990). This was followed almost 10 years later by another book in honor of Alex Dalgarno (Hartquist & Williams 1998). During 2006, the Nobel Symposium No 133 was dedicated to cosmic chemistry and molecular astrophysics (URL: www.nobel133.physto.se), and PNAS[1] published a special feature issue on interstellar chemistry (Klemperer 2006).

The role played by dissociative recombination in molecular astrophysics is particularly important to the chemistry of interstellar space. Eddington's 1926 Bakerian Lecture marks the starting point of the study of the interstellar medium (Eddington 1926), and the possibility of molecules in the interstellar medium was discussed even though at that time only atomic absorption lines had been observed. The first molecules observed in interstellar space were CH, CH^+, and CN in the late 1930s, as described in Herzberg's textbook on diatomic molecules (Herzberg 1950, p. 496). This page is famous because of the statement it contains about the rotational

---

[1] *Proceedings of the National Academy of Sciences of the United States of America.*

temperature of the CN molecules. Had Herzberg realized that the temperature of 2.3 K was due to the interaction of CN with the cosmic microwave background, he would most likely have been considered for a second Nobel Prize. A personal account of the development of interstellar chemistry has been given by Klemperer (1995), and the detailed development of the field, including the key references, has been given by Solomon (1973) and Van Dishoeck (1990).

Herbst and Klemperer (1973) and, independently, Watson (1973) proposed that molecules in interstellar space are formed by ion–molecule reactions. The number of observed molecules was at that time increasing rapidly owing to observations by radio astronomy, and when their article directed at a broader physics readership (Herbst & Klemperer 1976) was published, the number of observed interstellar molecules was 39. Molecular cloud chemistry starts with the slow cosmic-ray (CR) ionization of H_2:

$$H_2 + CR \rightarrow H_2^+ + e^- + CR'. \qquad (10.1)$$

The rapid ion–molecule reaction (see Eq. (7.1))

$$H_2 + H_2^+ \rightarrow H_3^+ + H \qquad (10.2)$$

was first pointed out as the source of interstellar H_3^+ by Martin, McDaniel, and Meeks (1961). The proton affinity of H_2 is relatively low, about 4.5 eV, which means that H_3^+ donates a proton to most atoms and molecules it collides with in a molecular cloud:

$$H_3^+ + X \rightarrow XH^+ + H_2. \qquad (10.3)$$

This reaction is the starting point for further ion–molecule reactions that build successively more complex molecules, for which dissociative recombination is the terminating step. There is an extensive literature on ion–molecule reactions with applications in the interstellar medium, in comets, and in planetary atmospheres (Huntress 1977, Anicich & Huntress 1986, Anicich 1993a,b, 2003). Figure 10.1 illustrates such a reaction network, in which the dissociative recombination steps can be found at the end of the branches. In this scheme, dissociative recombination is regarded as a process for molecule formation. When Bates and Spitzer (1951) discussed molecule formation in interstellar space, dissociative recombination of CH^+ was included as a loss channel. Bates and Spitzer saw no possibility of estimating how rapidly CH^+ is destroyed by electrons and labeled the rate coefficient as unknown but added that is was probably small, though conceivably it was large.

Dissociative recombination of molecular ions and its importance in astrochemistry has been discussed at every dissociative recombination conference since the first one in 1988 at Chateau Lake Louise, Alberta, Canada: Herbst (1989), Black

10.1 Molecular astrophysics

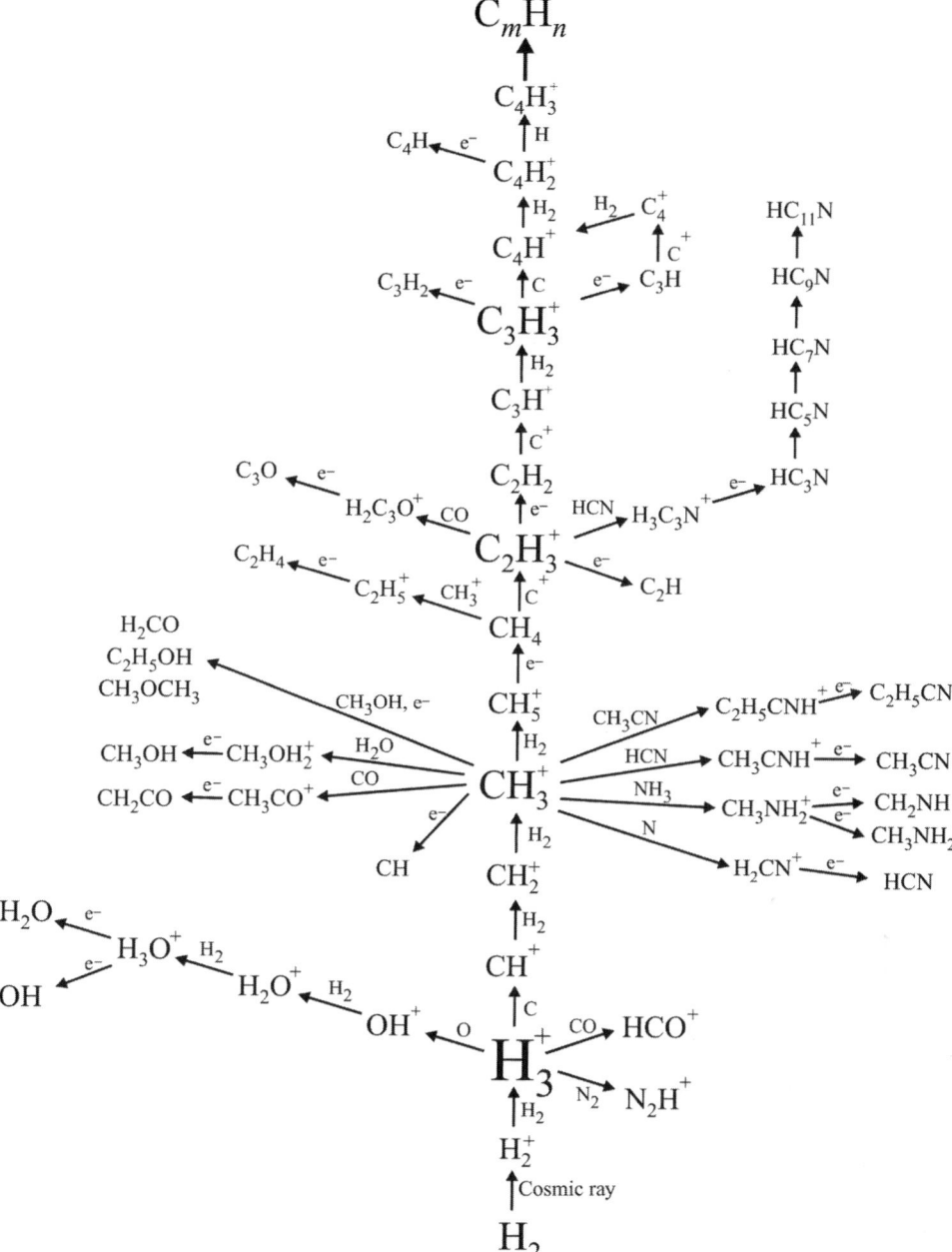

Figure 10.1 Network of ion–molecule chemistry (reproduced with permission from McCall (2001)).

and Van Dishoeck (1989), Turner (1989), Dalgarno (1993), Roueff and Pineau des Forêts (1993), Roueff, Le Bourlot, and Pineau des Forêts (1996), Dalgarno (2000), Oka (2000), Herbst (2003a), McCall and Oka (2003), Le Petit and Roueff (2003), Cravens (2003), Roueff (2005), Dalgarno (2005), Herbst (2005), and Geppert *et al.* (2005).

The most extensive database for astrochemistry is the one at the University of Manchester Institute of Science and Technology (UMIST), named UDFA (UMIST Database For Astrochemistry). The database is described by Le Teuft, Millar, and Markwick (2000), and can be found at URL: www.udfa.net. Another database has been developed at the Ohio State University by the group of Eric Herbst; URL: www.physics.ohio-state.edu/~eric/idex.html.

One aspect of the interstellar ion–molecule reactions that was realized early (Watson 1974) was that they would lead to isotope fractionation. Deuterium is extracted from its reservoir in molecular clouds, the HD molecule, and put into molecules instead of hydrogen. During the last 30 years, about 20 deuterated molecules have been observed in molecular clouds (Millar 2003). These molecules have an abundance ratio much higher than the cosmic D/H of 1.5×10^{-5} (Linsky 2003). The fractionation occurs because of the low temperature in clouds combined with the difference in zero-point energy between different isotopologs. The first fractionation step is (Le Petit & Roueff 2003, Herbst 2003b):

$$H_3^+ + HD \rightarrow H_2D^+ + H_2 + 232 \text{ K}. \tag{10.4}$$

This reaction is slightly exothermic, which means that in cold interstellar clouds the backward reaction is inefficient. This leads to a H_2D^+/H_3^+ ratio which is much larger than the cosmic D/H ratio. In a second step, H_2D^+ reacts with, for example, CO to form DCO^+:

$$H_2D^+ + CO \rightarrow DCO^+ + H_2. \tag{10.5}$$

Statistically, this reaction occurs in 1/3 of the collisions. The increased sensitivity of radio astronomy has led to the observation of multiply deuterated molecules (Roueff & Gerin 2003). The ND_3/NH_3 ratio of 10^{-4} is an enhancement of 10^{11} over the cosmic ratio of $(D/H)^3 \approx 10^{-15}$ (Millar 2003).

Le Bourlot *et al.* (1993) discovered multiple solutions to the gas-phase chemical rate equations of dark molecular clouds. Two different steady state solutions were found and labeled the high- and the low-ionization phase, respectively. A subsequent "News and views" appeared in *Nature* under the title "Chaos on interstellar clouds" (Lepp 1993). It was also found that the dissociative recombination rate coefficient of H_3^+ played a crucial role whether bistable solutions existed or not (Le Bourlot *et al.* 1993, Roueff & Pineau des Forêts 1993). A unique steady state occurred when $\alpha(H_3^+) \leq 2.5 \times 10^{-8}(T/300)^{-0.5}$ cm^3 s^{-1} (Le Bourlot *et al.* 1993), whereas bistable solutions existed for a larger rate coefficient. For a sufficiently large rate coefficient, the steady state returned (Pineau des Forêts & Roueff 2000). It is unclear whether bistability exists in reality. Boger and Sternberg (2006) have shown that the bistability is damped by inclusion of gas-grain neutralization.

It has become clear that gas-phase reactions alone cannot explain all molecular abundances (Millar 2006, Herbst & Cuppen 2006). One such example is methanol, for which experiments have shown that dissociative recombination of $CH_3OH_2^+$ leads to formation of methanol in only about 5% of the reactions (Geppert et al. 2006a,b). It is possible that methanol is formed through grain surface reactions, and that such reactions could also lead to isotope fractionation (Nagaoka, Watanabe, & Kouchi 2006).

10.2 Atmospheric physics and chemistry

The historical development of dissociative recombination and its close relation with atmospheric science is described in Chapter 1, and Section 5.2 is devoted to the atmospheric ions O_2^+, N_2^+, and NO^+. The role of dissociative recombination in atmospheric science has been a theme at every dissociative recombination conference since the beginning: Fox (1989), Yee, Abreau, and Colwell (1989), Fox (1993b, 1996, 2000), Ng (2003), Fox (2005), and Bultel and Chéron (2005). The research monograph by Wayne (2000) provides an overview of the chemistry of the Earth's atmosphere as well as the extraterrestrial atmospheres such as those of Venus, Mars, Jupiter, Saturn, and Titan. Space missions, which allow *in situ* measurements of planetary atmospheres, have contributed enormously to our knowledge of far distant atmospheres.

The Cassini–Huygens spacecraft passed through Titan's upper atmosphere in a first fly-by in October 2004. This and later fly-bys have generated a lot of new and partially unexplained data (Waite et al. 2005, Cravens et al. 2006, Vuitton, Yelle, & Anicich 2006). The Cassini–Huygens mission has shown that a substantial ionosphere exists on the nightside of Titan. It also seems clear that hydrocarbon chemistry cannot explain the analysis of the composition of Titan's upper atmosphere. The observation of ions at $m/z = 18, 30, 42, 54$, and 56 suggests that Titan's atmosphere contains many more nitrogen-bearing molecules than previously anticipated. Studies of dissociative recombination of nitrile ions ($C_xN_yH_z^+$) is required in order to understand the ion chemistry of Titan's upper atmosphere.

During the last 20 years there has been growing interest in the layer of metallic atoms in Earth's upper atmosphere that derive from meteoric ablation (Plane 2003). Iron is the most abundant meteoric metal in the upper atmosphere, but Na, K, and Ca are also present. Understanding the chemistry of these meteoric metals involves the dissociative recombination of metallic oxide ions such as FeO^+, which is an unexplored field.

10.3 Plasma physics and fusion research

Dissociative recombination in the context of plasma physics and fusion research has not been a recurrent theme at every recombination conference. Mitchell (1996) discussed the role of electron–molecular ion processes in ITER[2] edge plasmas, with emphasis on the H_2^+, H_3^+, and HeH^+ ions. Janev (2000) covered similar aspects but also brought the hydrocarbons into focus. This latter subject is well covered in the reviews by Janev and Reiter (2002, 2004). The IAEA reports by Clark (2006) and Humbert (2006) concern the need for atomic and molecular physics data in plasma modeling, in particular in fusion research.

A conference series was established in 1997 on the subject of atomic and molecular data and their applications. It is now an established conference series under the acronym ICAMDATA (International Conference on Atomic and Molecular Data and Their Applications), with the first meeting in Gaithersburg, Maryland in 1997, followed by meetings in Oxford, England (2000), Gatlinburg, Tennessee (2002), Toki, Japan (2004), and Meudon, France (2006).

---

[2] ITER used to be an acronym used for the International Thermonuclear Experimental Reactor. Now it is simply "The way" in Latin. The agreement to establish the international organization that will implement the ITER fusion energy project was signed by China, the European Union, India, Japan, the Republic of Korea, the Russian Federation, and the United States of America on November 21, 2006. Cadarache in the south of France will be the site of ITER.

References

Abdellahi El Ghazaly, M. O., Jureta, J., Urbain, X., & Defrance, P. 2004, "Total cross sections and kinetic energy release for the electron impact dissociation of H_2^+ and D_2^+," *J. Phys. B* **37**, pp. 2467–2483.

Abouelaziz, H., Gomet, J. C., Pasquerault, D., & Rowe, B. R. 1993, "Measurements of $C_3H_3^+$, $C_5H_3^+$, $C_6H_6^+$, $C_7H_5^+$, and $C_{10}H_8^+$ dissociative recombination rate coefficients," *J. Chem. Phys.* **99**, pp. 237–243.

Abouelaziz, H., Queffelec, J. L., Rebrion, C., Rowe, B. R., Gomet, J. C., & Canosa, A. 1992, "Dissociative recombination of HCS^+ and H_3S^+ ions studied in a flowing afterglow apparatus," *Chem. Phys. Lett.* **194**, pp. 263–267.

Abrahamsson, K., Andler, G., Bagge, L., *et al.* 1993, "CRYRING – a synchrotron cooler and storage ring," *Nucl. Instr. Methods Phys. Res. B* **79**, pp. 269–272.

Abreu, V. J., Solomon, S. C., Sharp, W. E., & Hays, P. B. 1988, "The dissociative recombination of O_2^+: The quantum yield of $O(^1S)$ and $O(^1D)$," *J. Geophys. Res.* **88**, pp. 4140–4144.

Adams, N. G. 1992, "Spectroscopic determination of the products of electron–ion recombination," *Adv. Gas Phase Ion. Chem.* **1**, pp. 271–310.

1993, "Flowing afterglow studies of electron–ion recombination using Langmuir probes and optical spectroscopy," in *Dissociative Recombination: Theory, Experiment, and Applications*, eds. B. R. Rowe, J. B. A. Mitchell, & A. Canosa, NATO ASI Series B: Physics Vol. 313, New York: Plenum Press, pp. 99–111.

1994, "Afterglow techniques with spectroscopic detection for determining the rate coefficients and products of dissociative electron–ion recombination," *Int. J. Mass Spectrom. Ion Proc.* **132**, pp. 1–27.

Adams, N. G. & Babcock, L. M. 1994a, "Optical emissions for the dissociative recombination of N_2H^+ and HCO^+," *J. Phys. Chem.* **98**, pp. 4564–4569.

1994b, "Vibrational excitation in the products of electron–ion recombination: A test of theory for ions of interstellar significance," *Astrophys. J.* **434**, pp. 184–187.

2005, "Molecular ion recombination in trapped and flowing plasmas: methods, recent results, new goals, open questions," *J. Phys.: Conf. Ser.* **4**, pp. 38–49.

Adams, N. G., & Smith, D. 1987, "Recent advances in the studies of reaction rates relevant to interstellar chemistry," in *Astrochemistry, IAU Symp. 120*, eds. M. S. Vardya & S. P. Tarafdar, Dordrecht: Reidel Publishing Company, pp. 1–18.

1988a, "Flowing afterglow and SIFT," in *Techniques for the Study of Ion Molecule Reactions*, eds. J. M. Farrar & J. W. H. Saunders, Techniques of Chemistry, Vol. 20, New York: Wiley Interscience, pp. 165–220.

1988b, "Laboratory studies of dissociative recombination and mutual neutralization and their relevance to interstellar chemistry," in *Rate Coefficients in Astrochemistry*, eds. T. J. Millar & D. A. Williams, Dordrecht: Kluwer, pp. 173–192.

1988c, "Measurements of the dissociative recombination coefficients for several polyatomic ion species at 300 K," *Chem. Phys. Lett.* **144**, pp. 11–14.

1989, "FALP studies of positive ion/electron recombination," in *Dissociative Recombination: Theory, Experiment and Applications*, eds. J. B. A. Mitchell & S. L. Guberman, Singapore: World Scientific, pp. 124–140.

Adams, N. G., Babcock, L. M., & McLain, J. L. 2003, "Electron–ion recombination," in *Encyclopedia of Mass Spectrometry*, Vol. 1: *Theory and Ion Chemistry*, ed. P. Armentrout, Amsterdam: Elsevier, pp. 542–555.

Adams, N. G., Herd, C. R., & Smith, D. 1989, "Development of the flowing afterglow/Langmuir probe technique for studying the neutral products of dissociative recombination using spectroscopic techniques: OH production in the $HCO_2^+ + e$ reaction," *J. Chem. Phys.* **91**, pp. 963–973.

1990, "Determination of the products of dissociative recombination reactions," in *The Physics of Electronic and Atomic Collisions: XVI International Conference (XVI ICPEAC, New York, NY)*, eds. A. Dalgarno, R. S. Freund, P. M. Koch, M. S. Lubell, & T. B. Lucatorto, AIP Conf. Proceedings Vol. 205, New York: American Institute of Physics, pp. 90–95.

Adams, N. G., Poterya, V., & Babcock, L. M. 2006, "Electron molecular ion recombination: Product excitation and fragmentation," *Mass Spectrom. Rev.* **25**, pp. 798–828.

Adams, N. G., Smith, D., & Alge, E. 1984, "Measurements of dissociative recombination rate coefficients of H_3^+, HCO^+, N_2H^+, and CH_5^+ at 95 K and 300 K using the FALP apparatus," *J. Chem. Phys.* **81**, pp. 1778–1784.

Adams, N. G., Herd, C. R., Geoghegan, M, et al. 1991, "Laser induced fluorescence and vacuum ultraviolet spectroscopic studies of H-atom production in dissociative recombination of some protonated ions," *J. Chem. Phys.* **94**, pp. 4852–4857.

Ajello, J. M., & Chutjian, A. 1979, "Line shapes for attachment of threshold electrons to SF_6 and $CFCl_3$: Threshold photoelectron (TPSA) studies of Xe, CO, and C_2H_2," *J. Chem. Phys.* **71**, pp. 1079–1087.

Al-Khalili, A., Danared, H., Larsson, M., et al. 1998, "Dissociative recombination of $^3HeH^+$: comparison of spectra obtained with 100, 10 and 1 meV temperature electron beams," *Hyperfine Interact.* **114**, pp. 281–287.

Al-Khalili, A., Thomas, R., Ehlerding, A., et al. 2004, "Dissociative recombination cross section and branching ratios of protonated dimethyl disulfide and N-methylacetamide," *J. Chem. Phys.* **121**, pp. 5700–5708.

Al-Khalili, A., Rosén, S., Danared, H., et al. 2003, "Absolute high-resolution rate coefficients for dissociative recombination of electrons with HD^+: Comparisons of results from three heavy-ion storage rings," *Phys. Rev. A* **68**, pp. 042702-1–14.

Alge, E., Adams, N. G., & Smith, D. 1983, "Measurements of the dissociative recombination coefficients of O_2^+, NO^+ and NH_4^+ in the temperature range 200–600 K," *J. Phys. B* **16**, pp. 1433–1444.

Amano, T. 1988. "Is the dissociative recombination of H_3^+ really slow? A new spectroscopic measurement of the rate constant," *Astrophys. J. Lett.* **329**, pp. L121–L124.

1990, "The dissociative recombination rate coefficient of H_3^+, HN_2^+, and HCO^+," *J. Chem. Phys.* **92**, pp. 6492–6501.

Amitay, Z., & Zajfman, D. 1997, "A new type of multiparticle three-dimensional imaging detector with subnanosecond time resolution," *Rev. Sci. Instrum.* **68**, pp. 1387–1392.

Amitay, Z., Baer, A., Dahan, M., *et al.* 1998, "Dissociative recombination of HD^+ in selected vibrational quantum states," *Science* **281**, pp. 75–78.

Amitay, Z., Baer, A., Dahan, M., *et al.* 1999, "Dissociative recombination of vibrationally excited HD^+: State-selective experimental investigation," *Phys. Rev. A* **60**, pp. 3769–3785.

Amitay, Z., Zajfman, D., Forck, P., *et al.* 1996a, "Dissociative recombination of cold OH^+: Evidence for indirect recombination through excited core Rydberg states," *Phys. Rev. A* **53**, pp. R644–R647.

Amitay, Z., Zajfman, D., Forck, P., *et al.* 1996b, "Dissociative recombination of CH^+: cross section and final states," *Phys. Rev. A* **54**, pp. 4032–4050.

Andersen, J. U., Hvelplund, P., Nielsen, S. B., *et al.* 2002, "The combination of an electrospray ion source and an electrostatic storage ring for lifetime and spectroscopy experiments on biomolecules," *Rev. Sci. Instr.* **73**, pp. 1284–1287.

Andersen, L. H. 1993, Electron–ion recombination at low energy, Thesis, Aarhus University, Aarhus.

Andersen, L. H., & Bolko, J. 1990, "Radiative recombination between fully stripped ions and free electrons," *Phys. Rev. A* **42**, pp. 1184–1191.

Andersen, L. H., Andersen, T., & Hvelplund, P. 1997, "Studies of negative ions in storage rings," *Adv. At. Mol. Opt. Phys.* **38**, pp. 155–191.

Andersen, L. H., Heber, O., & Zajfman, D. 2000, "Dissociative recombination of polyatomic ions: Branching ratios and isotopic effects," in *Astrochemistry: From Molecular Clouds to Planetary Systems, IAU Symp. 197*, eds. Y. C. Minh & E. F. Van Dishoeck, San Francisco: ASP, pp. 265–271.

2004, "Physics with electrostatic rings and traps," *J. Phys. B* **37**, pp. R57–R88.

Andersen, L. H., Bak, J., Boyé, S., *et al.* 2001b, "Resonant and nonresonant electron impact detachment of CN^- and BO^-," *J. Chem. Phys.* **115**, pp. 3566–3570.

Andersen, L. H., Bilodeau, R., Jensen, M. J., Nielsen, S. B., Sfvan, C. P., & Seiersen, K. 2001a, "Coulomb and centrifugal barrier bound dianion resonances of NO_2," *J. Chem. Phys.* **114**, pp. 147–151.

Andersen, L. H., Johnson, P. J., Kella, D., Pedersen, H. B., & Vejby-Christensen, L. 1997, "Dissociative-recombination and excitation measurements with H_2^+ and HD^+," *Phys. Rev. A* **55**, pp. 2799–2808.

Andersen, L. H., Heber, O., Kella, D., Pedersen, H. B., Vejby-Christensen, L., & Zajfman, D. 1996a, "Production of water molecules from dissociative recombination of H_3O^+ with electrons," *Phys. Rev. Lett.* **77**, pp. 4891–4894.

Andersen, L. H., Hvelplund, P., Kella, D., *et al.* 1996b, "Resonance structure in the electron-impact detachment cross section of C_2^- caused by the formation of C_2^{2-}," *J. Phys. B* **29**, pp. L643–L649.

Andersen, L. H., Mathur, D., Schmidt, H. T., & Vejby-Christensen, L. 1995, "Electron-impact detachment of D^-: near-threshold behavior and the nonexistence of D^{2-} resonances," *Phys. Rev. Lett.* **74**, pp. 892–895.

Andersen, T. 2004, "Atomic negative ions: structure, dynamics and collisions," *Phys. Rep.* **394**, pp. 157–313.

Andersen, T., Kjeldsen, H., Knudsen, H., & Folkmann, F. 2001c, "Absolute cross section for photoionization of CO^+ leading to longlived metastable CO^{2+}," *J. Phys. B* **34**, pp. L327–L332.

Andersson, P. U., Öjekull, J., Pettersson, J. B. C., et al. 2003, "Dissociative recombination of D_5^+: Cross-sections and branching ratios," in *23rd International Conference on Photonic, Electronic and Atomic Collisions (XXIII ICPEAC, Stockholm, Sweden), Abstracts of contributed papers*, Vol. II, eds. J. Anton et al., Stockholm: Universitetsservice US AB, p. Mo111.

Andrew, B. H. (ed.) 1980, *Interstellar Molecules, IAU Symp. 87*, Reidel Publishing Company, Dordrecht.

Angelova, G., LeGarrec, J. L., Rebrion-Rowe, C., Rowe, B. R., Novotny, O., & Mitchell, J. B. A. 2004, "The dissociative recombination of CF_3^+," *J. Phys. B* **37**, pp. 4135–4141.

Angelova, G., Novotny, O., Mitchell, J. B. A., et al. 2004a, "Branching ratios for the dissociative recombination of hydrocarbon ions. II. The cases of $C_4H_n^+$ ($n = 1$–9)," *Int. J. Mass Spectrom.* **232**, pp. 195–203.

Angelova, G., Novotny, O., Mitchell, J. B. A., et al. 2004b, "Branching ratios for the dissociative recombination of hydrocarbon ions. III. The cases of $C_3H_n^+$ ($n = 1$–9)," *Int. J. Mass Spectrom.* **232**, pp. 195–203.

Ångström, J. A. 1869, "Spectrum of the aurora borealis," *Phil. Mag.* **38**, pp. 246–247.

Anicich, V. G., 1993a, "Evaluated bimolecular ion-molecule gas phase kinetics of positive ions for use in modeling planetary atmospheres, cometary comae, and interstellar clouds," *J. Phys. Chem. Ref. Data* **22**, pp. 1469–1569.

1993b, "A survey of bimolecular ion–molecule gas phase kinetics of positive ions for use in modeling planetary atmospheres, cometary comae, and interstellar clouds: 1993 supplement," *Astrophys. J. Suppl. Ser.* **84**, pp. 215–315.

2003, *An Index of the Literature for Bimolecular Gas Phase Cation- Molecule Reaction Kinetics*, JPL Publication 03-19, Pasadena: Jet Propulsion Laboratory.

Anicich, V. G., & Huntress Jr., W. T. 1986, "A survey of bimolecular ion-molecule gas phase kinetics of positive ions for use in modeling planetary atmospheres, cometary comae, and interstellar clouds," *Astrophys. J. Suppl. Ser.* **62**, pp. 553–672.

Anisimov, A. I., Vinogradov, N. I., & Golant, V. E. 1964, "Measurement of the volume removal coefficient for electrons in plasma decay in oxygen," *Sov. Phys. Techn. Phys.* **8**, pp. 850–851.

Appleton, E. V. 1937, "The Bakerian Lecture – Regularities and irregularities in the ionosphere – I," *Proc. Roy. Soc. A* **162**, pp. 451–479.

1949, "The ionosphere," in *Les Prix Nobel en 1947*, ed. M. A. Holmberg, Stockholm: P. A. Nordstedt & Söner, pp. 101–107.

Appleton, E. V., & Barnett, M. A. F. 1925a, "Local reflection of wireless waves from upper atmosphere," *Nature* **115**, pp. 333–334.

1925b, "On some direct evidence of downward atmospheric reflection of electric rays," *Proc. Roy. Soc. A* **109**, pp. 621–641.

Appleton, E. V., & Weekes, K. 1939, "On lunar tides in the upper atmosphere," *Proc. Roy. Soc. A* **171**, pp. 171–187.

Ashfold, M. N. R., & Baggot, J. E. (eds.) 1987, *Molecular Photodissociation Dynamics*, Advances in Gas-Phase Photochemistry and Kinetics, London: The Royal Society of Chemistry.

Auerbach, D., Cacak, R., Caudano, R., et al. 1977, "Merged electron–ion beam experiments I. Methods and measurements of $(e-H_2^+)$ and $(e-H_3^+)$ dissociative recombination cross section," *J. Phys. B* **10**, pp. 3797–3820.

Azuma, T., Tanuma, H., & Shiromaru, H. 2004, "Present and future projects of TMU atomic physics group," *J. Phys.: Conf. Ser.* **2**, pp. 143–151.

Bahati, E. M., Jureta, J. J., Belić, D. S., Cherkani-Hassani, H., Abdellahi, M. O., & Defrance, P. 2001c, "Electron impact dissociation and ionization of N_2^+," *J. Phys. B* **34**, pp. 2963–2973.
Bahati, E. M., Jureta, J. J., Belić, D. S., Rachafi, S., & Defrance, P. 2001b, "Electron impact ionization and dissociation of CO_2^+ to C^+ and O^+," *J. Phys. B* **34**, pp. 1757–1767.
Bahati, E. M., Jureta, J. J., Cherkani-Hassani, H., & Defrance, P. 2001a, "Electron impact single ionization and dissociative excitation of H_3O^+, HD_2O^+ and D_3O^+," *J. Phys. B* **34**, pp. L333–L337.
Bahati, E. M., Thomas, R. D., Vane, C. R., & Bannister, M. E. 2005, "Electron impact dissociation of $D^{13}CO^+$ molecular ions to $^{13}CO^+$ ions," *J. Phys. B* **38**, pp. 1645–1655.
Baird, S., Chanel, M., Möhl, D., & Tranquille, D. 1990, "LEAR," *Part. Accel.* **26**, pp. 223–228.
Balint-Kurtis, G. B. 2003, "Wavepacket theory of photodissociation and reactive scattering," *Adv. Chem. Phys.* **128**, pp. 249–301.
Banaszkiewicz, M., Lara, L. M., Rodrigo, R., Lopéz-Moreno, J. J., & Molina-Cuberos, G. J. 2000, "A coupled model of Titan's atmosphere and ionosphere," *Icarus* **147**, pp. 386–404.
Bannister, M. E. 2005, "Experiments on electron-impact ionization of atomic and molecular ions," in *Atomic and Molecular Data and Their Applications*, eds. T. Kato, H. Funaba, & D. Kato, AIP Conf. Proceedings Vol. 771, New York: American Institute of Physics, pp. 172–179.
Bannister, M. E., Krause, H. F., Vane, C. R., et al. 2003, "Electron-impact dissociation of CH^+ ions: Measurement of C^+ fragment ions," *Phys. Rev. A* **68**, pp. 042714-1–6.
Bardsley, J. N. 1967, "The theory of dissociative recombination," in *Fifth International Conference on the Physics of Electronic and Atomic Collisions (V ICPEAC, Leningrad, USSR), Abstracts of Papers*, eds. I. P. Flaks & E. S. Solovyol, Leningrad: Nauka, pp. 338–340.
  1968a, "Configuration interaction in the continuum states of molecules," *J. Phys. B* **1**, pp. 349–364.
  1968b, "The theory of dissociative recombination," *J. Phys. B* **1**, pp. 365–380.
  1983, "Dissociative recombination of electrons with NO^+," *Planet. Space Sci.* **31**, pp. 667–671.
Bardsley, J. N., & Biondi, M. A. 1970, "Dissociative recombination," *Adv. At. Mol. Phys.* **6**, pp. 1–57.
Bardsley, J. N., & Junker, B. R. 1973, "Dissociative recombination of CH^+ ions," *Astrophys. J. Lett.* **183**, pp. L135–L137.
Barrios, A., Sheldon, J. W., Hardy, K. A., & Peterson, J. R. 1992, "Superthermal component in an effusive beam of metastable krypton: evidence of Kr_2^+ dissociative recombination," *Phys. Rev. Lett.* **69**, pp. 1348–1351.
Barth, C. A., Fastie, W. G., Hord, C. W., et al. 1969, "Mariner 6: Ultraviolet spectrum of Mars upper atmosphere," *Science* **165**, pp. 1004–1005.
Bates, D. R. 1950a, "Electron recombination in helium," *Phys. Rev.* **77**, pp. 718–719.
  1950b, "Dissociative recombination," *Phys. Rev.* **78**, pp. 492–493.
  1982, "Airglow and auroras," in *Applied Atomic Collisions*, Vol. 1: *Atmospheric Physics and Chemistry*, eds. H. S. W. Massey & D. R. Bates, New York: Academic Press, pp. 149–224.
  1986, "Products of dissociative recombination of polyatomic ions," *Astrophys. J.* **306**, pp. L45–L47.

1987a, "Interstellar cloud chemistry revisited," in *Recent Studies in Atomic and Molecular Processes*, ed. A. E. Kingston, New York: Plenum Press, pp. 1–27. This book is referenced in the literature under a plethora of names: "Modern Applications of Atomic and Molecular Processes," "Modern Applications of Atomic and Molecular Physics," "Recent Studies in Atomic and Molecular Physics." It is the proceedings of a conference held in honor of the 70th birthday of Professor David Bates, held November 17–18, 1986, at the Queen's University of Belfast, Belfast, Northern Ireland. It has ISBN 0306426870.

1987b, "Polyatomic ions: Bond energies, most stable isomeric form and low excited states," *Int. J. Mass Spectrom. Ion Proc.* **80**, pp. 1–16.

1989, "Dissociative recombination of polyatomic ions: Curve crossing," *Astrophys. J.* **344**, pp. 531–534.

1990, "Oxygen green and red line emission and O_2^+ dissociative recombination," *Planet. Space Sci.* **38**, pp. 889–902.

1991a, "Super dissociative recombination," *J. Phys. B* **24**, pp. 703–709.

1991b, "Dissociative recombination of polyatomic ions," *J. Phys. B* **24**, pp. 3267–3284.

1992a, "Single-electron transitions and cluster ion super-dissociative recombination," *J. Phys. B* **25**, pp. 3067–3073.

1992b, "Dissociative recombination when potential energy curves do not cross," *J. Phys. B* **25**, pp. 5479–5488.

1993a, "Prevalance of rapid dissociative recombination in absence of crossing of potentials," *Proc. R. Soc. Lond. A* **443**, pp. 257–264.

1993b, "Vibrational excitations of products of dissociative recombination," *Mon. Not. R. Astron. Soc.* **263**, pp. 369–374.

1994, "Dissociative recombination: crossing and tunneling modes," *Adv. At. Mol. Opt. Phys.* **34**, pp. 427–486.

Bates, D. R. & Dalgarno, A. 1962, "Electronic recombination," in *Atomic and Molecular Processes*, ed. D. R. Bates, New York: Academic Press, pp. 245–271.

Bates, D. R., & Herbst, E. 1988, "Dissociative recombination: Polyatomic positive ion reactions with electrons and negative ions," in *Rate Coefficients in Astrochemistry*, eds. T. J. Millar & D. A. Williams, Dordrecht: Kluwer, pp. 41–48.

Bates, D. R., & Massey, H. S. W. 1943a, "The negative ions of atomic and molecular oxygen," *Phil. Trans. R. Soc. Lond. A* **239**, pp. 269–304.

1943b, "The properties of neutral and ionized oxygen and their influence on the upper atmosphere," *Rep. Prog. Phys.* **9**, pp. 62–74.

1946, "The basic reactions in the upper atmosphere. I," *Proc. Roy. Soc. A* **187**, pp. 261–296.

1947, "The basic reactions in the upper atmosphere. II. The theory of recombination in the ionized layers," *Proc. Roy. Soc. A* **192**, pp. 1–16.

Bates, D. R., & Mitchell, J. B. A. 1991, "Rate coefficients for $N_2^+(v)$ dissociative recombination," *Planet. Space Sci.* **39**, pp. 1297–1300.

Bates, D. R., & Spitzer, L. 1951, "The density of molecules in interstellar space," *Astrophys. J.* **113**, pp. 441–463.

Bates, D. R., & Zipf, E. C. 1980, "The $O(^1S)$ quantum yield from O_2^+ dissociative recombination," *Planet. Space Sci.* **28**, pp. 1081–1086.

Bates, D. R., Guest, M. F., & Kendall, R. A. 1993, "Enigma of H_3^+ dissociative recombination," *Planet. Space Sci.* **41**, pp. 9–15.

Bates, D. R., Kingston, A. E., & McWhirter, R. W. P. 1962, "Recombination of electrons and ions I. Optically thin plasmas," *Proc. Roy. Soc. A* **267**, pp. 297–312.

Bauer, E., & Wu, T.-Y. 1956, "Cross sections of dissociative recombination," *Can. J. Phys.* **34**, pp. 1436–1447.

Beck, M. H., Jäckle, A., Worth, G. A., & Meyer, H.-D. 2000, "The multiconfiguration time-dependent Hartree (MCTDH) method: a highly efficient algorithm for propagating wave packets," *Phys. Rep.* **324**, pp. 1–105.

Becker, K. H. (ed.) 1998, *Novel Aspects of Electron-Molecule Collisions*, Singapore: World Scientific.

Berkner, K. H., Morgan, T. J., Pyle, R. V., & Stearns, J. W. 1971, "Dissociation cross sections for 410- to 1800-keV H_3^+ ions in collisions with H_2 and N_2 gases," in *The Physics of Electronic and Atomic Collisions: 7th International Conference (VII ICPEAC, Amsterdam, the Netherlands)*, eds. L. M. Brancomb *et al.*, Amsterdam: North-Holland Publishing Company, pp. 422–423.

Bernath, P., & Amano, T. 1982, "Detection of the infrared fundamental band of HeH^+," *Phys. Rev. Lett.* **48**, pp. 20–22.

Berry, R. S., & Leach, S. 1979, "Elementary attachment and detachment processes. II," *Adv. Electr. El. Phys.* **57**, pp. 1–144.

Bethe, H. A. 1935, "Theory of disintegration of nuclei by neutrons," *Phys. Rev.* **47**, pp. 747–759.

Bialecke, E. P., & Dougal, A. A. 1958, "Pressure and temperature variation of the electron–ion recombination coefficient in nitrogen," *J. Geophys. Res.* **63**, pp. 539–546.

Biondi, M. A. 1951, "Concerning the mechanism of electron-ion recombination. II," *Phys. Rev.* **83**, pp. 1078–1080.

1956, "High-speed, direct recording Fabry-Perot interferometer," *Rev. Sci. Instr.* **27**, pp. 36–39.

1963, "Studies of the mechanism of electron–ion recombination. I," *Phys. Rev.* **129**, pp. 1181–1188.

1964, "Electron-ion and ion-ion recombination," *Ann. Géophys.* **20**, pp. 34–46.

1973, "The effects of ion complexity on electron–ion recombination," *Comments At. Mol. Phys.* **4**, pp. 85–91.

1978, "Objections to the N_2^+ + e dissociative recombination coefficients inferred from analysis of Atmosphere Explorer measurements," *Geophys. Res. Lett.* **5**, pp. 661–664.

2003, "Dissociative recombination of electrons and ions: The early experiments," in *Dissociative Recombination of Molecular Ions with Electrons*, ed. S. L. Guberman, New York: Kluwer/Plenum Publishers, pp. 13–23.

Biondi, M. A., & Brown, S. C. 1949a, "Measurements of ambipolar diffusion in helium," *Phys. Rev.* **75**, pp. 1700–1705.

1949b, "Measurement of electron–ion recombination," *Phys. Rev.* **76**, pp. 1697–1700.

Biondi, M. A., & Holstein, T., 1951, "Concerning the mechanism of electron-ion recombination," *Phys. Rev.* **82**, pp. 962–963.

Birtwistle, D. T., & Herzenberg, A. 1971, "Vibrational excitation of N_2 resonance scattering of electrons," *J. Phys. B* **4**, pp. 53–70.

Bishop, D. M., & Cheung, L. M. 1979, "Theoretical investigation of HeH^+," *J. Mol. Spectrosc.* **75**, pp. 462–473.

Black, J. H. 1978, "Molecules in planetary nebulae," *Astrophys. J.* **222**, pp. 125–131.

Black, J. H., & Van Dishoeck, E. F. 1989, "Dissociative recombination in interstellar clouds," in *Dissociative Recombination: Theory, Experiment and Applications*, eds. J. B. A. Mitchell, & S. L. Guberman, Singapore: World Scientific, pp. 317–328.

Bluhme, H., Jensen, M. J., Brøndsted Nielsen, S., *et al.* 2004, "Electron scattering on stored mononucleotide anions," *Phys. Rev. A* **70**, pp. 020701-1–4.

Boger, G. I., & Sternberg, A. 2006, "Bistability in interstellar gas-phase chemistry," *Astrophys. J.* **645**, pp. 314–323.

Bohr, N. 1919, "On the model of a triatomic hydrogen molecule," *Meddel. från K. Vet.-Akad:s Nobelinstitut* **5**, pp. 1–16.

Bordas, M. C., Lembo, L. J., & Helm, H 1991, "Spectroscopy and multichannel quantum-defect theory analysis of the np Rydberg series of H_3," *Phys. Rev. A* **44**, pp. 1817–1827.

Bottcher, C. 1974, "Theory of dissociative recombination," *Proc. R. Soc. Lond. A* **340**, pp. 301–322.

1976, "Dissociative recombination of the hydrogen molecular ion," *J. Phys. B* **9**, pp. 2899–2921.

Bottcher, C., & Docken, K. 1974, "Autoionizing states of the hydrogen molecule," *J. Phys. B* **7**, pp. L5–L8.

Bowers, M. T. 1989, "Photodissociation dynamics of small cluster ions," in *Ion and Cluster Ion Spectroscopy and Structure*, ed. J. P. Maier, Amsterdam: Elsevier, pp. 241–273.

Boyé, S., Krogh, H., Nielsen, I. B., et al. 2003, "Vibrationally resolved photoabsorption spectroscopy of red fluorescent protein chromophore anions," *Phys. Rev. Lett.* **90**, pp. 118103-1–4.

Braeuning, H., & Salzborn, E. 2005, "Ion–ion collision processes: Experiment," in *Atomic and Molecular Data and their Applications*, eds. T. Kato, H. Funaba, & D. Kato, AIP Conf. Proceedings, Vol. 771, New York: American Institute of Physics, pp. 219–228.

Branscomb, L. M. 1962, "Photodetachment," in *Atomic and Molecular Processes*, ed. D. R. Bates, New York: Academic Press, pp. 100–140.

Broström, L., Larsson, M., Mannervik, S., & Sonnek, D. 1991, "The visible photoabsorption spectrum and potential curves of ArN^+," *J. Chem. Phys.* **94**, pp. 2734–2740.

Bruna, P. J. 1975, "Theoretical study of the properties of HCO^+ at equilibrium," *Astrophys. Lett.* **16**, pp. 107–113.

Brune, W. H., Schwab, J. J., & Anderson, J. G. 1983, "Laser magnetic resonance, resonance fluorescence, and resonance absorption studies of the reaction kinetics of $O + OH \rightarrow H + O_2$, $O + HO_2 \rightarrow OH + O_2$, $N + OH \rightarrow H + NO$, and $N + HO_2 \rightarrow$ products at 300 K between 1 and 5 torr," *J. Phys. Chem.* **87**, pp. 4503–4514.

Brunger, M. J., & Buckman, S. J. 2002, "Electron-molecule scattering cross sections. I. Experimental techniques and data for diatomic molecules," *Phys. Rep.* **357**, pp. 215–458.

Brunger, M. J., Buckman, S. J., & Newman, D. S. 1990, "Low energy electron scattering from H_2," *Aust. J. Phys.* **43**, pp. 665–682.

Buckman, S. J., Panajotovic, R., & Jelisavcic, M. 2004, "Low energy electron-molecule collision cross sections," *Phys. Scripta* **T110**, pp. 166–171.

Buhl, D., & Snyder, L. E. 1970, "Unidentified interstellar microwave line," *Nature* **228**, pp. 267–269.

Bultel, A., & Chéron, B. G. 2005, "Role of molecular ions in plasmas of atmospheric and energetic interest," *J. Phys.: Conf. Ser.* **4**, pp. 205–210.

Burdett, N. A., & Hayhurst, A. N. 1978, "Kinetics of gas phase electron–ion recombination by $NO^+ + e^- \rightarrow N + O$ from measurements in flames," *J. Chem. Soc. Faraday Trans. 1* **74**, pp. 53–62.

Burke, P. G., Hibbert, A., & Robb, W. D. 1971, "Electron scattering by complex atoms," *J. Phys. B* **4**, pp. 153–161.

Butler, C. J., & Hayhurst, A. N. 1996, "Kinetics of dissociative recombination of H_3O^+ ions with free electrons in premixed flames," *J. Chem. Soc. Faraday Trans.* **92**, pp. 707–714.

Butler, J. M., Babcock, L. M., & Adams, N. G. 1997, "Effects of deuteration on vibrational excitation in the products of the electron recombination of HCO^+ and N_2H^+," *Mol. Phys.* **91**, pp. 81–90.

Canosa, A., Gomet, J. C., Rowe, B. R., Mitchell, J. B. A., & Queffelec, J. L. 1992, "Further measurements of the H_3^+ ($v = 0,1,2$) dissociative recombination rate coefficient," *J. Chem. Phys.* **97**, pp. 1028–1037.

Canosa, A., Gomet, J. C., Rowe, B. R., & Queffelec, J. L. 1991a, "Flowing afterglow Langmuir probe measurement of N_2^+ ($v = 0$) dissociative recombination rate coefficient," *J. Chem. Phys.* **94**, pp. 7159–7163.

Canosa, A., Rowe, B. R., Mitchell, J. B. A., Gomet, J. C., & Rebrion, C. 1991b, "New measurements of the H_3^+ and HCO^+ dissociative recombination rate coefficient," *Astron. Astrophys.* **248**, pp. L19–L21.

Cao, Y. S., & Johnsen, R. 1991, "Recombination of N_4^+ ions with electrons," *J. Chem. Phys.* **95**, pp. 7356–7359.

Carata, L., Orel, A. E., & Suzor-Weiner, A. 1999, "Dissociative recombination of He_2^+ molecular ions," *Phys. Rev. A* **59**, pp. 2804–2812.

Carata, L., Orel, A. E., Raoult, M., Schneider, I. F., & Suzor-Weiner, A. 2000, "Core-excited resonances in the dissociative recombination of CH^+ and CD^+," *Phys. Rev. A* **62**, pp. 052711-1–10.

Carata, L., Schneider, I. F., Suzor-Weiner, A., Tennyson, J., & de Lange, C. A. 1997, "The role of Rydberg states in dissociative recombination, as revealed by ion storage ring experiments [and discussion]," *Phil. Trans. R. Soc. Lond. A* **355**, pp. 1677–1691.

Carney, G. D., & Porter, R. N. 1980, "*Ab initio* prediction of the rotation-vibration spectrum of H_3^+ and D_3^+," *Phys. Rev. Lett.* **45**, pp. 537–541.

Carrington, A., Kennedy, R. A., Softley, T. P., Fournier, P. G., & Richard, E. G. 1983, "Infrared bound to quasibound vibration–rotation spectrum of HeH^+ and its isotopes," *Chem. Phys.* **81**, pp. 251–261.

Casavecchia, P. 2001, "Chemical reaction dynamics with molecular beams," *Rep. Prog. Phys.* **63**, pp. 355–414.

Chang, J. S., Hobson, R. M., Ichikawa, Y., Kaneda, T., Maruyama, N., & Teii, S. 1989, "Dissociative recombination of Ne_2^+ at elevated electron and gas temperatures," *J. Phys. B*, **22**, pp. L665–L668.

Chapman, S. 1931, "The Bakerian Lecture – Some phenomena of the upper atmosphere," *Proc. Roy. Soc. A* **132**, pp. 353–374.

Chen, C. L., Leiby, C. C., & Goldstein, L. 1961, "Electron temperature dependence of the recombination coefficient in pure helium," *Phys. Rev.* **121**, pp. 1391–1400.

Chen, J. C. Y., & Mittleman, M. H. 1967, "The role of Rydberg states in dissociative recombination," in *Fifth International Conference on the Physics of Electronic and Atomic Collisions (V ICPEAC, Leningrad, USSR)*, Abstracts of papers, eds. I. P. Flaks, & E. S. Solovyol, Leningrad: Nauka, pp. 329–331.

Chen, Q., & Goodings, J. M. 1998, "Chemical kinetics of yttrium ionization in $H_2-O_2-N_2$ flames," *Int. J. Mass Spectrom.* **176**, pp. 1–12.

1999, "Chemical kinetics of lanthanum ionization in $H_2-O_2-N_2$ flames," *Int. J. Mass Spectrom.* **188**, pp. 213–224.

Chen, Q., Milburn, R. K., Hopkinson, A. C., Bohme, D. K., & Goodings, J. M. 1999, "Magnesium chemistry in the gas phase: calculated thermodynamic properties and experimental ion chemistry in $H_2-O_2-N_2$ flames," *Int. J. Mass Spectrom.* **184**, pp. 153–173.

Christoffersen, R. E., Hagstrom, S., & Prosser, F. 1964, "H_3^+ molecule ion. Its structure and energy," *J. Chem. Phys.* **40**, pp. 236–237.

Christophorou, L. G. (ed.) 1984a, *Electron–Molecule Interactions and Their Applications*, Vol. 1, New York: Academic Press.

1984b, *Electron–Molecule Interactions and Their Applications*, Vol. 2, New York: Academic Press.

Christophorou, L. G., & Olthoff, J. K. 2001, "Electron interactions with excited atoms and molecules," *Adv. At. Mol. Opt. Phys.* **44**, pp. 155–293.

Chutjian, A., & Alajajian, S. H. 1985, "s-wave threshold in electron attachment: Observations and cross sections in CCl_4 and SF_6 at ultralow electron energies," *Phys. Rev. A* **31**, pp. 2885–2892.

Chutjian, A., Garscadden, A., & Wadehra, J. M. 1996, "Electron attachment to molecules at low electron energies," *Phys. Rep.* **264**, pp. 393–470.

Clark, R. E. H. (ed.) 2006, *Summary Report of Final IAEA Research Co-ordination Meeting. Data For Molecular Processes in Edge Plasmas*, INDC International Nuclear Data Committee, INDC(NDS)-0491, Vienna: IAEA.

Cohen, J. S. 1976, "Multistate curve-crossing model for scalttering: associative ionization and excitation transfer in helium," *Phys. Rev. A* **13**, pp. 99–114.

Collins, C. B., & Robertson, W. W. 1965, "Comments on collisional–radiative recombination of He_2^+ into dissociative states," *J. Chem. Phys.* **43**, p. 4188.

Collins, L. A., & Schneider, B. I. 1983, "Linear algebraic approach to electronic excitation of atoms and molecules by electron impact," *Phys. Rev. A* **27**, pp. 101–111.

Collins, G. F., Pegg, D. J., Fritioff, K., et al. 2005, "Electron-impact fragmentation of Cl_2^-," *Phys. Rev. A* **72**, pp. 042708-1–7.

Compton, R. N., & Bardsley, J. N. 1984, "Dissociation of molecules by slow electrons," in *Electron–Molecule Collisions*, eds. I. Shimamura, & K. Takayanagi, New York: Plenum Press, pp. 275–349.

Connor, T. R., & Biondi, M. A. 1965, "Dissociative recombination in neon: spectral line-shape studies," *Phys. Rev.* **140**, pp. A778–A791.

Conroy, H. 1964, "Potential energy surfaces for the H_3^+ molecule-ion," *J. Chem. Phys.* **40**, pp. 603–604.

Continetti, R. E. 2000, "Dissociative photodetachment studies of transient molecules by coincidence techniques," in *Photoionization and Photodetachment*, Part II, ed C.-Y. Ng, Advanced Series in Physical Chemistry, Vol. 10B, Singapore: World Scientific, pp. 748–808.

Cook, P. A., Langford, S. R., Dixon, R. N., & Ashford, M. N. R. 2001, "An experimental and *ab initio* reinvestigation of Lyman-α photodissociation of H_2S and H_2D," *J. Chem. Phys.* **114**, pp. 1672–1684.

Cooper, H. J., Håkansson, K., & Marshall, A. G. 2005, "The role of electron capture dissociation in biomolecular analysis," *Mass Spectrom. Rev.* **24**, pp. 201–222.

Cooper, R., van Sonsbeek, R. J., & Bhave, R. N. 1993, "Pulse radiolysis of ion-electron recombination in gaseous argon," *J. Chem. Phys.* **98**, pp. 383–389.

Coulson, C. A. 1935, "The electronic structure of H_3^+," *Proc. Camb. Phil. Soc.* **31**, pp. 244–259.

1951, *Valence*, London: Oxford University Press.

Cox, S. G., Critchley, D. J., McNab, I. R., & Smith, F. E. 1999, "High-resolution spectroscopy of ion beams," *Meas. Sci. Technol.* **10**, pp. R101–R128.

Crandall, D. H., Kaupilla, W. E., Phaneuf, R. A., Taylor, P. O., & Dunn, G. H. 1974, "Absolute cross sections for electron impact excitation of N_2^+," *Phys. Rev. A* **9**, pp. 2545–2551.
Cravens, T. E. 2003, "Dissociative recombination in cometary ionospheres," in *Dissociative Recombination of Molecular Ions with Electrons*, ed. S. L. Guberman, New York: Kluwer/Plenum Publishers, pp. 385–400.
Cravens, T. E., Robertson, I. P., & Waite Jr., J. H. 2006, "Composition of Titan's ionosphere," *Geophys. Res. Lett.* **33**, pp. 07105-1–4.
Crofton, M. W., Altman, R. S., Haese, N. N., & Oka, T. 1989, "Infrared spectra of ^{4}HeH$^+$, ^{4}HeD$^+$, ^{3}HeH$^+$, and ^{3}HeD$^+$," *J. Chem. Phys.* **91**, pp. 5882–5886.
Crompton, R. W. 1994, "Benchmark measurements of cross sections for electron collisions: Electron swarm methods," *Adv. At. Mol. Opt. Phys.* **33**, pp. 97–148.
Cunningham, A. J., & Hobson, R. M. 1969, "Experimental measurement of dissociative recombination in vibrationally excited gases," *Phys. Rev.* **185**, pp. 98–100.
  1972a, "Dissociative recombination at elevated temperatures I. Experimental measurements in krypton afterglows," *J. Phys. B* **5**, pp. 1773–1783.
  1972b, "Dissociative recombination at elevated temperatures III. O_2^+ dominated afterglows," *J. Phys. B* **5**, pp. 2320–2327.
  1972c, "Dissociative recombination at elevated temperatures IV. N_2^+ dominated afterglows," *J. Phys. B* **5**, pp. 2328–2331.
Cunningham, A. J., O'Malley, T. F., & Hobson, R. M. 1981, "On the role of vibrational excitation in dissociative recombination," *J. Phys. B* **14**, pp. 773–782.
Čurík, R., & Greene, C. H. 2007a, "Indirect dissociative recombination of LiH$^+$ fueled by complex resonance manifolds," *Phys. Rev. Lett.* **98**, pp. 173201-1–4.
  2007b, "Vibrational excitation and dissociative recombination of LiH$^+$," *Mol. Phys.* **105**, pp. 1565–1574.
Dabrowski, I., & Herzberg, G. 1977, "The predicted infrared spectrum of HeH$^+$ and its possible astrophysical importance," *Trans. NY Acad. Sci.* **38**, pp. 14–25.
Dahan, M., Fishman, R., Heber, O., et al. 1998, "A new type of electrostatic ion trap for storage of fast ion beams," *Rev. Sci. Instr.* **69**, pp. 76–83.
Dalgarno, A. 1993, "Chemistry of supernova 1987A," in *Dissociative Recombination: Theory, Experiment, and Applications*, eds. B. R Rowe, J. B. A. Mitchell, & A. Canosa, NATO ASI Series B: Physics Vol. 313, New York: Plenum Press, pp. 243–248.
  1994, "Terrestrial and extraterrestrial H_3^+," *Adv. At. Mol. Opt. Phys.* **32**, pp. 57–68.
  2000, "Dissociative recombination in astrophysical environments," in *Dissociative Recombination: Theory, Experiment and Applications IV*, eds. M. Larsson, J. B. A. Mitchell, & I. F. Schneider, Singapore: World Scientific, pp. 1–12.
  2005, "Molecular processes in the early Universe," *J. Phys.: Conf. Ser.* **4**, pp. 10–16.
  2006, "The galactic cosmic ray ionization rate," *Proc. Natl. Acad. Sci. USA* **103**, pp. 12269–12273.
Dalitz, R. H. 1953, "On the analysis of τ-meson data," *Phil. Mag.* **44**, pp. 1068–1080.
Danared, H. 1993, "Fast electron cooling with a magnetically expanded electron beam," *Nucl. Instr. Meth. Phys. Res. A* **335**, pp. 397–401.
  1995, "Electron cooling for atomic physics," *Phys. Scripta* **T59**, pp. 121–125.
Danared, H., Andler, G., Bagge, L., et al. 1994, "Electron cooling with an ultracold electron beam," *Phys. Rev. Lett.* **72**, pp. 3775–3778.
Danared, H., Källberg, A., Liljeby, L., & Rensfelt, K. G. 1998, "The CRYRING super conducting electron cooler," in *Proceedings of the 6th Particle Acceleration Conference*, Bristol: Institute of Physics Publishing, pp. 1031–1033.

Dance, D. F., Harrison, M. F. A., Rundel, R. D., & Smith, A. C. H. 1967, "A measurement of the cross section for proton production in collisions between electrons and H_2^+ ions," *Proc. Phys. Soc.* **92**, pp. 577–588.

Danilov, A. D., & Ivanov-Kholodny, G. S. 1965, "Research on ion-molecule reactions and dissociative recombination in the upper atmosphere and in the laboratory," *Sov. Phys. Usp.* **8**, pp. 92–116.

Datz, S. 2001, "Dynamics of dissociative recombination of molecular ions: three-body breakup of triatomic di-hydrides," *J. Phys. Chem. A* **105**, pp. 2369–2373.

Datz, S., & Larsson, M. 1992, "Radiative lifetimes for all vibrational levels in the $X\,^1\Sigma^+$ state of HeH$^+$ and its relevance to dissociative recombination experiments in ion storage rings," *Phys. Scripta* **46**, pp. 343–347.

Datz, S., Larsson, M., Strömholm, C., *et al.* 1995a, "Dissociative recombination of H$_2$D$^+$: Cross sections, branching fractions, and isotope effects," *Phys. Rev. A* **52**, pp. 2901–2909.

Datz, S., Rosén, S., Al-Khalili, A., *et al.* 2000a, "Three-body breakup dynamics in dissociative recombination," in *Dissociative Recombination: Theory, Experiment and Applications IV*, eds. M. Larsson, J. B. A. Mitchell, & I. F. Schneider, Singapore: World Scientific, pp. 200–209.

Datz, S., Sundström, G., Biedermann, Ch., *et al.* 1995b, "Branching processes in the dissociative recombination of H_3^+," *Phys. Rev. Lett.* **74**, pp. 896–899.

Datz, S., Thomas, R., Rosén, S., *et al.* 2000b, "Dynamics of three-body breakup in dissociative recombination: H$_2$O$^+$," *Phys. Rev. Lett.* **85**, pp. 5555–5558.

Davidson, D. F., & Hobson, R. M. 1987, "The shock tube determination of the dissociative recombination rate of NO$^+$," *J. Phys. B* **20**, pp. 5753–5756.

Deloche, R., Monchicourt, P., Cheret, M., & Lambert, F. 1976, "High-pressure helium afterglow at room temperature," *Phys. Rev. A* **13**, pp. 1140–1176.

Dempster, A. J. 1916, "The ionization and dissociation of hydrogen molecules and the formation of H$_3$," *Phil. Mag.* **31**, pp. 438–443.

Derkatch, A. M., Al-Khalili, A., Vikor, L., *et al.* 1999, "Branching ratios in dissociative recombination of the C$_2$H$_2^+$ molecular ion," *J. Phys. B* **32**, pp. 3391–3398.

Derkits, C., Bardsley, J. N., and Wadehra 1979, "Dissociative recombination in e–D$_2^+$ collisions," *J. Phys. B* **12**, pp. L529–L531.

Dinelli, B. M., Miller, S., & Tennyson, J. 1992, "Bands of H_3^+ up to $4\nu_2$: rovibrational transitions from first principle calculations," *J. Mol. Spectrosc.* **153**, pp. 718–725; *ibid.* 1992, Erratum, **156**, p. 243.

Dinelli, B. M., Neale, L., Polyansky, O. L., & Tennyson, T. 1997, "New assignments for the infrared spectrum of H_3^+," *J. Mol. Spectrosc.* **181**, pp. 142–150.

Diner, A., Toker, Y., Strasser, D., *et al.* 2004, "Size-dependent electron-impact detachment of internally cold C_n^- and Al_n^-," *Phys. Rev. Lett.* **93**, pp. 063402-1–4.

Dittner, P. F., Datz, S., Miller, P. D., Pepmiller, P. L., & Fou, C. M. 1986, "Dielectronic recombination measurements of P^{4+}, S^{5+}, and Cl^{6+}," *Phys. Rev. A* **33**, pp. 124–130.

Dixon, R. N., Hwang, D. W., Yang, X. F., Harich, S., Lin, J. J., & Yang, X. 1999, "Chemical 'double slits': Dynamical interference of photodissociation pathways in water," *Science* **285**, pp. 1249–1253.

Djurić, N. 2005, "Recent experimental studies of electron-impact excitation of atomic and molecular ions," in *Atomic and Molecular Data and their Applications*, eds. T. Kato, H. Funaba, & D. Kato, AIP Conf. Proceedings Vol. 771, New York: American Institute of Physics, pp. 162–171.

Djurić, N., Chung, Y.-S., Wallbank, B., & Dunn, G. H. 1997, "Measurement of light fragments in dissociative excitation of molecular ions: CD$^+$," *Phys. Rev. A* **56**, pp. 2887–2892.

Djurić, N., Dunn, G. H., & Al-Khalili, A. 2001, "Resonant ion-pair formation and dissociative recombination in electron collisions with ground-state HF$^+$ ions," *Phys. Rev. A* **64**, pp. 022713-1–9.

Djurić, N., Neau, A., Rosén, S., Zong, W., & Dunn, G. H. 2000, "Light-ionic fragment production in dissociative electron–molecular-ion collisions: Detection of D$^+$ and D$_2^+$ from ND$_n^+$ ($n = 2-4$) and OD$_n^+$ ($n = 2,3$)," *Phys. Rev. A* **62**, pp. 032702-1–6.

Djurić, N., Zhou, S., Dunn, G. H., & Bannister, M. E. 1998, "Electron-impact dissociative excitation of CD$_n^+$ ($n = 2-5$): Detection of light fragment ions D$^+$ and D$_2^+$," *Phys. Rev. A* **58**, pp. 304–308.

Dolder K. T., & Peart, B. 1972, "Comments on Rundel's discussion of proton production by collisions between electrons and H$_2^+$," *J. Phys. B* **5**, pp. L129–L133.

1976, "Collisions between electrons and ions," *Rep. Prog. Phys.* **39**, pp. 693–749.

1986, "Electron–ion and ion–ion collisions with intersecting beams," *Adv. At. Mol. Phys.* **22**, pp. 197–241.

Domcke, W., & Estrada, H. 1988, "Friction and memory effects in the dynamics of short-lived negative ions," *J. Phys. B* **21**, pp. L205–L211.

Donahue, T. M., Parkinson, T., Zipf, E. C., Doering, J. P., Fastie, W. G., & Miller, R. E. 1968, "Excitation of the auroral green line by dissociative recombination of the oxygen molecular ion: analysis of two rocket experiments," *Planet. Space Sci.* **16**, pp. 737–747.

Dörner, R., Vogt, T., Mergel, V., et al. 1996, "Ratio of cross sections for single to double ionization of He by 85–400 eV photons," *Phys. Rev. Lett.* **76**, pp. 2654–2657.

Douglas, A. E., & Herzberg, G. 1941, "CH$^+$ in interstellar space and in the laboratory," *Astrophys. J.* **94**, p. 381.

Dreuw, A., & Cederbaum, L. S. 2000, "Nature of the repulsive Coulomb barrier in multiply charged negative ions," *Phys. Rev. A* **63**, pp. 012501-1–13.

2002, "Multiply charged anions in the gas phase," *Chem. Rev.* **102**, pp. 181–200.

Drossart, P., Maillard, J.-P., & Caldwell, J. 1989, "Detection of H$_3^+$ on Jupiter," *Nature* **340**, pp. 539–541.

Duane, W., & Wendt, G. L. 1917, "A reactivity modification of hydrogen produced by alpha-radiation," *Phys. Rev.* **10**, pp. 116–128.

Dubé, L., & Herzenberg, A. 1975, "Resonant electron–molecule scattering: The impulse approximation in N$_2$O," *Phys. Rev. A* **11**, pp. 1314–1325.

DuBois, R. D., Jeffries, J. B., & Dunn, G. H. 1978, "Dissociative recombination cross sections for NH$_4^+$ ions and electrons," *Phys. Rev. A* **17**, pp. 1314–1320.

Dubrovsky, G. V., & Ob'edkov, V. D. 1967, "Decay of molecular hydrogen ions through collisions with thermal electrons," *Sov. Astron. –AJ* **11**, pp. 305–307.

Dubrovsky, G. V., Ob'edkov, V. D., & Janev, R. K. 1967, "The deacy of two-atom molecules in collisions with electrons," in *Fifth International Conference on the Physics of Electronic and Atomic Collisions (V ICPEAC, Leningrad, USSR)*, Abstracts of papers, eds. I. P. Flaks & E. S. Solovyol, Leningrad: Nauka, pp. 342–345.

Dulaney, J. L., Biondi, M. A., & Johnsen, R. 1987, "Electron temperature dependence of the recombination of electrons with NO$^+$ ions," *Phys. Rev. A* **36**, pp. 1342–1350.

1988, "Electron temperature dependence of the recombination of electrons with O$_4^+$ ions," *Phys. Rev. A* **37**, pp. 2539–2542.

Dunn, G. H., & Djurić, N. 1998, "Electron impact dissociative excitation and ionization of molecular ions," in *Novel Aspects of Electron–Molecule Collisions*, ed. K. H. Becker, Singapore: World Scientific, pp. 241–281.

Dunn, G. H., & Van Zyl, B. 1967, "Electron impact dissociation of H$_2^+$," *Phys. Rev.* **154**, pp. 40–51.

Dunn, G. H., Van der Zyl, B., & Zare, R. N. 1965, "Dissociation of H_2^+ by electron impact," *Phys. Rev. Lett.* **15**, pp. 610–612.

Dunn, G. H., Belić, D. S., Djurić, N., & Mueller, D. W. 1984, "Dielectronic recombination," in *Atomic Physics 9*, eds. R. S. Van Dyck & E. Norval Fortson, Singapore: World Scientific, pp. 505–522.

Dunning, F. B. 1995, "Electron–molecule collisions at very low electron energies," *J. Phys. B* **28**, pp. 1645–1672.

Durrance, S. T., Barth, C. A., & Stewart, A. I. F. 1980, "Pioneer Venus observations of the Venus dayglow spectrum 1250–1430 Å," *Geophys. Res. Lett.* **7**, pp. 222–224.

Eddington, A. S. 1926, "Diffuse matter in interstellar space," *Proc. Roy. Soc. A* **111**, pp. 424–456.

Ehlerding, A., Arnold, S. T., Viggiano, A. A., et al. 2003, "Rates and products of the dissociative recombination of $C_3H_7^+$ in low-energy electron collisions," *J. Phys. Chem. A* **107**, pp. 2179–2184.

Ehlerding, A., Hellberg, F., Thomas, R., et al. 2004, "Dissociative recombination of C_2H^+ and $C_2H_4^+$: Absolute cross sections and product branching ratios," *Phys. Chem. Chem. Phys.* **6**, pp. 949–954.

Ehlerding, A., Viggiano, A. A., Hellberg, F., et al. 2006, "The dissociative recombination of fluorocarbon ions III: CF_2^+ and CF_3^+," *J. Phys. B* **39**, pp. 805–812.

Ehrhardt, H., & Morgan, L. A. (eds.) 1994, *Electron Collisions with Molecules, Clusters and Surfaces*, New York: Plenum Press.

Einfeld, T., Chichini, A., Maul, C., & Gericke, K.-H. 2004, "Photodissociation dynamics of phosgene: new observations by applying a three-dimensional imaging technique," *J. Chem. Phys.* **116**, pp. 2803–2810.

Eletskii, A. V., & Smirnov, B. M. 1982, "Dissociative recombination of electrons and molecular ions," *Sov. Phys. Usp.* **25**, pp. 13–30.

El Ghazaly, M. O. A., Svendsen, A., Bluhme, H., et al. 2004, "Electron scattering on centrosymmetric molecular dianions $Pt(CN)_4^{2-}$ and $Pt(CN)_6^{2-}$," *Phys. Rev. Lett.* **93**, pp. 203201-1–4.

El Ghazaly, M. O. A., Svendsen, A., Bluhme, H., et al. 2005, "Electron scattering on *p*-benzoquionone anions," *Chem. Phys. Lett.* **405**, pp. 278–281.

Ellis, A., Feher, M., & Wright, T. 2005, *Electronic and Photoelectron Spectroscopy: Fundamentals and Case Studies*, Cambridge: Cambridge University Press.

Eritt, M., Diner, A., Toker, Y., et al. 2006, "Size effects in the interaction between ionic clusters and low-energy electrons," *Phys. Scr.* **73**, pp. C32–C35.

Erman, P., Karawajczyk, A., Rachlew-Källne, E., & Strömholm, C. 1995, "Photoionization and photodissociation of nitric oxide in the range 9–35 eV," *J. Chem. Phys.* **102**, pp. 3064–3076.

Estrada, H., & Domcke, W. 1989, "Non-Markovian dynamics of electron–molecule collision complexes," *Phys. Rev. A* **40**, pp. 1262–1278.

Faire, A. C., & Champion, K. S. W. 1959, "Measurements of dissociative recombination and diffusion in nitrogen at low pressures," *Phys. Rev.* **113**, pp. 1–6.

Faire, A. C., Fundingland, O. T., Aden, A. L., & Champion, K. S. W. 1958, "Electron recombination coefficient in nitrogen at low pressures," *J. Appl. Phys.* **29**, pp. 928–930.

Fehér, M., Rohrbacher, A., & Maier, J. P. 1994, "Infrared laser kinetic spectroscopy of the H_3^+ ion," *Chem. Phys.* **185**, pp. 357–364.

Fehsenfeld, F. C., Schmeltekopf, A. L., Goldan, P. D., Schiff, H. I., & Ferguson, E. E. 1965, "Thermal energy ion–neutral reaction rates. I. Some reactions of helium ions," *J. Chem. Phys.* **44**, pp. 4087–4094.

Fenn, J. B. 2003, "Electrospray wings for molecular elephants," in *Le prix Nobel, The Nobel Prizes 2002*, ed. T. Frängsmyr, Stockholm: Edita Nordstedts Tryckeri AB, pp. 154–184.

Ferguson, E. E., Fehsenfeld, F. C., & Schmeltekopf, A. L. 1965, "Dissociative recombination in helium afterglow," *Phys. Rev.* **138**, pp. A381–A385.

Feynman, R. P. 1939, "Forces in molecules," *Phys. Rev.* **56**, pp. 340–343.

Filippelli, A. R., Lin, C. C., Anderson, L. W., & McConkey, J. W. 1994, "Principles and methods for measurement of electron impact excitation cross sections for atoms and molecules by optical techniques," *Adv. At. Mol. Opt. Phys.* **33**, pp. 1–62.

Finch, C. D., & Dunning, F. B. 2000, "Rydberg atoms: A nanoscale laboratory to examine the dynamics of electron attachment," *Comm. Mod. Phys.* **2**, pp. D89–D97.

Flannery, M. R. 1994, "Electron–ion and ion–ion recombination processes," *Adv. At. Mol. Opt. Phys.* **32**, pp. 117–147.

  1995, "Semiclassical-classical path hybrid theory of direct electron–ion recombination and a proposal for $e^- + H_3^+$ recombination," in *Atomic and Molecular Physics: Fourth US/Mexico Symposium*, eds. I. Alvarez, C. Cisneros, & T. J. Morgan, Singapore: World Scientific, pp. 329–341.

Florescu, A. I., Suzor-Weiner, A., Leininger, T., & Gadéa, F. X. 2004, "Non-adiabatic mechanisms in dissociative recombination," *Phys. Scripta* **T110**, pp. 172–177.

Florescu-Mitchell, A. I., & Mitchell, J. B. A. 2006, "Dissociative recombination," *Phys. Rep.* **430**, pp. 277–374.

Flower, D. R., & Roueff, E. 1979, "On the formation and destruction of HeH^+ in gaseous nebulae and the associated infrared emission line spectrum," *Astron. Astrophys.* **72**, pp. 361–366.

Forand, J. L., Mitchell, J. B. A., & McGowan, J. Wm. 1985, "Triatomic molecular dissociation: a method for measuring individual decay channel cross sections," *J. Phys. E* **18**, pp. 623–626.

Forck, P., Broude, C., Grieser, M., *et al.* 1994, "New resonances in the dissociative recombination of vibrationally cold CD^+," *Phys. Rev. Lett.* **72**, pp. 2002–2005.

Forck, P., Grieser, M., Habs, D., *et al.* 1993a, "Molecular physics in a storage ring: dissociative recombination and excitation of cold HD^+," *Nucl. Instr. Meth. B* **79**, pp. 273–275.

Forck, P., Grieser, M., Habs, D., *et al.* 1993b, "Dissociative recombination of HD^+ at the Test Storage Ring," *Phys. Rev. Lett.* **70**, pp. 426–429.

Fox, J. L. 1986, "The vibrational distribution of O_2^+ in the dayside ionosphere," *Planet. Space Sci.* **34**, pp. 1241–1252.

  1989, "Dissociative recombination in aeronomy," in *Dissociative Recombination: Theory, Experiment and Applications*, eds. J. B. A. Mitchell & S. L. Guberman, Singapore: World Scientific, pp. 264–285.

  1993a, "The production and escape of nitrogen atoms on Mars," *J. Geophys. Res.* **98**(E2), pp. 3287–3310.

  1993b, "Dissociative recombination in planetary ionospheres," in *Dissociative Recombination: Theory, Experiment, and Applications*, eds. B. R. Rowe, J. B. A. Mitchell, & A. Canosa, NATO ASI Series B: Physics Vol. 313, New York: Plenum Press, pp. 219–242.

  1996, "Hydrocarbon ions in the ionosphere of Titan and Jupiter," in *Dissociative Recombination: Theory, Experiment and Applications III*, eds. D. Zajfman, J. B. A. Mitchell, D. Schwalm, & B. Rowe, Singapore: World Scientific, pp. 40–46.

2000, "Applications of velocity distributions of atomic products in dissociative recombination to aeronomy," in *Dissociative Recombination: Theory, Experiment and Applications IV*, eds. M. Larsson, J. B. A. Mitchell, & I. F. Schneider, Singapore: World Scientific, pp. 25–30.

2005, "Effects of dissociative recombination on the composition of planetary atmospheres," *J. Phys.: Conf. Ser.* **4**, pp. 32–37.

Fox, J. L., & Hać, A. 1997, "Spectrum of hot O at the exabase of the terrestrial planets," *J. Geophys. Res.* **102** (A11), pp. 24005–24011.

Fox, J. N., & Hobson, R. M. 1966, "Temperature dependence of dissociative recombination coefficients in argon," *Phys. Rev. Lett.* **17**, pp. 161–163.

Franck, J., & Hertz, G. 1914, "Über zusammenstössen zwischen Elektronen und den Molekülen des Quecksiblerdampfes und die Ionisierungsspannung desselben," *Ver. d. D. Phys. Ges.* **16**, pp. 457–467.

Frederick, J. E., Rusch, D. W., Victor, G. A., Sharp, W. E., Hays, P. B., & Brinton, H. C. 1976, "The OI (λ5577 Å) Airglow: Observations and excitation mechanisms," *J. Geophys. Res.* **81**, pp. 3923–3930.

Frisch, P. 1972, "Abundances of interstellar CH and CH$^+$ radicals," *Astrophys. J.* **173**, pp. 301–316.

Fritioff, K., Sandström, J., Andersson, P., *et al.* 2004, "Observation of an excited C_4^2 ion," *J. Phys. B* **37**, pp. 2241–2246.

Frommhold, L., & Biondi, M. A. 1969, "Interferometric study of dissociative recombination radiation in neon and argon afterglows," *Phys. Rev.* **185**, pp. 244–252.

Frommhold, L., Biondi, M. A., & Mehr, F. J. 1968, "Electron-temperature dependence of electron–ion recombination in neon," *Phys. Rev. A* **165**, pp. 44–52.

Galli, D., & Palla, F. 1998, "The chemistry of the early Universe," *Astron. Astrophys.* **335**, pp. 403–420.

Galloway, E. T., & Herbst, E. 1991, "Can phase space theory reproduce experimental neutral product branching ratios for dissociative recombination reactions?," *Astrophys. J.* **376**, pp. 531–539.

Ganguli, B., Biondi, M. A., Johnsen, R., & Dulaney, J. L. 1988, "Electron-temperature dependence of the recombination of HCO$^+$ ions with electrons," *Phys. Rev. A* **37**, pp. 2543–2547.

Geballe, T. R., & Oka, T. 1996, "Detection of H_3^+ in interstellar space," *Nature* **384**, pp. 334–335.

2006, "A key molecular ion in the universe and in the laboratory," *Science* **312**, pp. 1610–1612.

Geballe, T. R., McCall, B. J., Hinkle, K. H., & Oka, T. 1999, "Detection of H_3^+ in the diffuse interstellar medium: The Galactic center and Cygnus OB2 number 12," *Astrophys. J.* **510**, pp. 251–257.

Geoghegan, M., Adams, N. G., & Smith, D. 1991, "Determination of the electron–ion dissociative recombination coefficients for several molecular ions at 300 K," *J. Phys. B* **24**, pp. 2589–2599.

Geppert, W., Ehlerding, A., Hellberg, F., *et al.* 2004a, "First observation of four-body breakup in electron recombination: $C_2D_5^+$," *Phys. Rev. Lett.* **93**, pp. 153201-1–4.

Geppert, W. D., Ehlerding, A., Hellberg, F., *et al.* 2004b, "Dissociative recombination of nitrile ions: DCCCN$^+$ and DCCCND$^+$," *Astrophys. J.* **613**, pp. 1302–1309.

Geppert, W. D., Hamberg, M., Thomas, R. D., *et al.* 2006a, "Dissociative recombination of protonated methanol," *Faraday Discuss.* **133**, pp. 177–190.

Geppert, W. D., Hellberg, F., Ehlerding, A., *et al.* 2004c, "Dissociative recombination of $S^{18}O_2^+$: Evidence for three-body breakup," *Astrophys. J.* **610**, pp. 1228–1233.

Geppert, W. D., Hellberg, F., Österdahl. F., et al. 2006b, "Dissociative recombination of $CD_3OD_2^+$," in *Astrochemistry: Recent Successes and Current Challenges, IAU Symp.* **231**, Cambridge: Cambridge University Press, pp. 117–124.

Geppert, W. D., Thomas, R., Ehlerding, A. et al. 2004d, "Extraordinary branching ratios in astrophysically important dissociative recombination reactions," *Faraday Discuss.* **127**, pp. 425–437.

Geppert, W. D., Thomas, R., Ehlerding, A., et al. 2004e, "Dissociative recombination of $C_3H_4^+$: preferential formation of the C_3H_3 radical," *Int. J. Mass Spectrom.* **237**, pp. 25–32.

Geppert, W. D., Thomas, R. D., Ehlerding, A. et al. 2005, "Dissociative recombination branching ratios and their influence on interstellar clouds," *Phys.: Conf. Ser.* **4**, pp. 26–31.

Geppert, W. D., Thomas, R., Hellberg, F., et al. 2004f, "Dissociative recombination of N_2OD^+," *Phys. Chem. Chem. Phys.* **6**, pp. 3415–3419.

Geppert, W. D., Thomas, R., Semaniak, J. et al. 2004g, "Dissociative recombination of N_2H^+: evidence for the fracture of the N−N bond," *Astrophys. J.* **609**, pp. 459–464.

Gillan, C. J., Tennyson, J., & Burke, P. G. 1995, "The UK molecular R-matrix scattering package: A computational perspective," in *Computational Methods for Electron Molecule Collisions*, eds. W. M. Huo & F. A. Gianturco, New York: Plenum Press, pp. 239–254.

Giusti, A. 1980, "A multichannel quantum defect approach to dissociative recombination," *J. Phys. B* **13**, pp. 3867–3894.

Giusti-Suzor, A. 1986, "Recent developments in the theory of dissociative recombination," in *Atomic Processes in Electron–Ion and Ion–Ion Collisions*, ed. F. Brouillard, New York: Plenum Press, pp. 223–237.

Giusti-Suzor, A., & Lefebvre-Brion, H. 1977, "The dissociative recombination of CH^+ ions," *Astrophys. J. Lett.* **214**, pp. L101–L103.

Giusti-Suzor, A., Bardsley, J. N., & Derkits, C. 1983, "Dissociative recombination in low-energy $e-H_2^+$ collisions," *Phys. Rev. A* **28**, pp. 682–691.

Glockler, G., & Fuller, D. L. 1933, "Helium hydride ion," *J. Chem. Phys.* **1**, pp. 886–887.

Glosík, J. 1992, "Dissociative electronic recombination – recent results," *Plasma Phys. Contr. Fusion* **34**, pp. 2091–2097.

Glosík, J., & Plašil, R. 2000, "The recombination rate coefficient of a protonated acetone dimer with electrons: indication of a temperature dependence," *J. Phys. B* **33**, pp. 4483–4494.

Glosík, J., Novotný, O., & Pysanenko, A. 2003, "The recombination of H_3^+ and H_5^+ ions with electrons in hydrogen plasma: dependence on temperature and on pressure of H_2," *Plasma Sources Sci. Techn.* **12**, pp. S117–S122.

Glosík, J., Bánó, G., Plašil, R., Luca, A., & Zakouřil, P. 1999, "Study of the electron ion recombination in high pressure flowing afterglow: recombination of $NH_4^+ \cdot (NH_3)_2$," *Int. J. Mass Spectrom.* **189**, pp. 103–113.

Glosík, J., Plašil, R., Poterya, V., Kudrna, P., & Tichý, M. 2000, "The recombination of H_3^+ ions with electrons: dependence of partial pressure of H_2," *Chem. Phys. Lett.* **331**, pp. 209–214.

Glosík, J., Plašil, R., Poterya, V., Kudrna, P., Tichý, M., & Pysanenko, A. 2001a, "Experimental study of recombination of H_3^+ ions with electrons relevant for interstellar and planetary plasmas," *J. Phys. B* **34**, pp. L485–L494.

Glosík, J., Plašil, R., Pysanenko, A., et al. 2005, "Recombination studies in a He-Ar-H_2 plasma," *J. Phys.: Conf. Ser.* **4**, pp. 104–110.

Glosík, J., Plašil, R., Zakouřil, P., & Poterya, V. 2001b, "Dissociative recombination of protonated dimer ions $H^+\cdot(HCOH)_2$ and $H^+\cdot(CH_3COH)_2$ with electrons at near thermal energies," *J. Phys. B* **34**, pp. 2781–2793.

Goldan, P. D., Schmeltekopf, A. L., Fehsenfeld, F. C., Schiff, H. I., & Ferguson, E. E. 1965, "Thermal energy ion-neutral reaction rates. I. Some reactions of ionospheric interest," *J. Chem. Phys.* **44**, pp. 4095–4103.

Golubkov, G. V., & Ivanov, G. K. 1990, "Interaction of auto-decayed state in molecule photodissociation processes," *Nuovo Cimento D* **12**, pp. 1–20.

Golubkov, M. G., Golubkov, G. V., & Ivanov, G. K. 1997, "Low-temperature dissociative recombination of electrons with H_2^+, HD^+ and D_2^+," *J. Phys. B* **30**, pp. 5511–5534.

Goodings, J. M., Patterson, P. M., & Hayhurst, A. N. 1995, "Mass-spectrometric study of $BaOH^+$ ions and free electrons from barium added to flames of H_2+O_2+Ar," *J. Chem. Soc. Faraday Trans.* **91**, pp. 2257–2267.

Gougousi, T., Golde, M. F., & Johnsen, R. 1994, "Electron–ion recombination measurements in flowing afterglow plasmas," *Bull. Am. Phys. Soc. Ser. 2* **39**, p. 1456.
  1997, "Electron–ion recombination rate coefficient measurements in a flowing afterglow plasma," *Chem. Phys. Lett.* **265**, pp. 399–403.

Gougousi, T., Johnsen, R., & Golde, M. F. 1995, "Recombination of H_3^+ and D_3^+ in a flowing afterglow plasma," *Int. J. Mass Spectrom. Ion Proc.* **149/159**, pp. 131–151.
  1997a, "Yield determination of OH ($v = 0,1$) radicals produced by the electron-ion recombination of H_3O^+ ions," *J. Chem. Phys.* **107**, pp. 2430–2439.
  1997b, "Yield determination of OH ($v = 0,1$) radicals produced by the electron-ion recombination of protonated molecules," *J. Chem. Phys.* **107**, pp. 2440–2443.

Graber, T., Kanter, E. P., Levin, J., Zajfman, D., Vager, Z., & Naaman, R. 1997, "Direct measurement of bending conformations in triatomic dihydride ions," *Phys. Rev. A* **56**, pp. 2600–2613.

Graham, S. M., & Goodings, J. M. 1984, "Metallic ions in hydrocarbon flames. II. Mechanism for the reduction of $C_3H_3^+$ by metals in relation to soot suppression," *Int. J. Mass Spectrom. Ion Proc.* **56**, pp. 205–222.

Graham, W. G., Fritsch, W., Hahn, Y., & Tanis, J. A. (eds.) 1992, *Recombination of Atomic Ions*, New York: Plenum Press.

Green, S., & Herbst, E. 1979, "Metastable isomers: A new class of interstellar molecules," *Astrophys. J.* **229**, pp. 121–131.

Greene, C. H., & Kokoouline, V. 2004, "Dissociative recombination of polyatomic molecules: A new mechanism," *Phys. Scripta* **T110**, pp. 178–182.
  2006, "Theoretical progress and challenges in H_3^+ dissociative recombination," *Phil. Trans. R. Soc. A* **364**, pp. 2965–2980.

Greene, C. H., Kokoouline, V., & Esry, B. D. 2003, "Importance of Jahn-Teller coupling in the dissociative recombination of H_3^+ by low energy electrons," in *Dissociative Recombination of Molecular Ions with Electrons*, ed. S. L. Guberman, New York: Kluwer/Plenum Publishers, pp. 221–233.

Gross, S. H., & Rasool, S. I. 1964, "The upper atmosphere of Jupiter," *Icarus* **3**, pp. 311–322.

Guberman, S. L. 1979, "Potential curves for dissociative recombination of O_2^+," *Int. J. Quant. Chem.: Quant. Chem. Symp.* **13**, pp. 531–540.
  1983a, "Potential energy curves for dissociative recombination," in *Physics of Ion–Ion and Electron–Ion Collisions*, eds. F. Brouillard, & J. W. McGowan, New York: Plenum Press, pp. 167–200.
  1983b, "The doubly excited autoionizing states of H_2," *J. Chem. Phys.* **78**, pp. 1404–1413.

1986a, "Windows in direct dissociative recombination cross sections," *Can. J. Phys.* **64**, pp. 1621–1625.
1986b, "Theoretical studies of dissociative recombination," in *Thermophysical Aspects of Re-Entry Flow*, eds. J. N. Moss & C. D. Scott, Progress in Astronautics and Aeronautics Vol. 103, New York: American Institute of Aeronautics and Astronautics, pp. 225–242.
1987, "The production of O(^{1}S) from dissociative recombination of O_2^+," *Nature* **327**, pp. 408–409.
1988, "The production of O(^{1}D) from dissociative recombination of O_2^+," *Planet. Space Sci.* **36**, pp. 47–53.
1989, "Ab initio studies of dissociative recombination," in *Dissociative Recombination: Theory, Experiment and Applications*, eds. J. B. A. Mitchell, & S. L. Guberman, Singapore: World Scientific, pp. 45–60.
1991, "Dissociative recombination of the ground state of N_2^+," *Geophys. Res. Lett.* **18**, pp. 1051–1054.
1993, "Electron–ion continuum-continuum mixing," in *Dissociative Recombination: Theory, Experiment, and Applications*, eds. B. R. Rowe, J. B. A. Mitchell & A. Canosa, NATO ASI Series B: Physics Vol. 313, New York: Plenum Press, pp. 47–57.
1994, "Dissociative recombination without a curve crossing," *Phys. Rev. A* **49**, pp. R4277–R4280.
1995a, "New mechanisms for dissociative recombination," in *The Physics of Electronic and Atomic Collisions: XIX International Conference (XIX ICPEAC, Whistler, Canada)*, eds. L. J. Dube, J. B. A. Mitchell, J. W. McConkey, & C. E. Brion, AIP Conf. Proceedings Vol. 360, New York: American Institute of Physics, pp. 307–316.
1995b, "The dissociative recombination of OH$^+$," *J. Chem. Phys.* **102**, pp. 1699–1704.
1997, "Mechanism for the green glow of the upper ionosphere," *Science* **278**, pp. 1276–1278.
2001, "Chemistry – Breaking up is hard to do without an electron," *Science* **294**, pp. 1474–1475.
(ed.) 2003a, *Dissociative Recombination of Molecular Ions with Electrons*, New York: Kluwer/Plenum Publishers.
2003b, "Dissociative recombination mechanisms," in *Dissociative Recombination of Molecular Ions with Electrons*, ed. S. L. Guberman, New York: Kluwer/Plenum Publishers, pp. 1–11.
2003c, "The dissociative recombination of N_2^+," in *Dissociative Recombination of Molecular Ions with Electrons*, ed. S. L. Guberman, New York: Kluwer/Plenum Publishers, pp. 187–196.
2004, "Product angular distributions in dissociative recombination," *J. Chem. Phys.* **120**, pp. 9509–9513.
2005, "Dissociative recombination angular distributions," *J. Phys: Conf. Ser.* **4**, pp. 58–65.
Guberman, S. L., & Giusti-Suzor, A. 1991, "The generation of O(^{1}S) from the dissociative recombination of O_2^+," *J. Chem. Phys.* **95**, pp. 2602–2613.
Guberman, S. L., & Goddard, W. A., III, 1975, "Nature of the excited states of He$_2$," *Phys. Rev. A* **12**, pp. 1203–1221.
Gunton, R. C. 1967, "Study of electrons, positive ions and negative ions in oxygen afterglows," *Bull. Am. Phys. Soc.* **12**, pp. 218–219.
Gunton, R. C., & Shaw, T. M. 1965, "Electron–ion recombination in nitric oxide in the temperature range 196 to 358 K," *Phys. Rev.* **140**, pp. A756–A763.

Guo, J., & Goodings, J. M. 2000, "Recombination coefficients for H_3O^+ ions with electrons e^- and with Cl^-, Br^- and I^- at flame temperatures 1820–2400 K," *Chem. Phys. Lett.* **329**, pp. 393–398.

2002, "Chemical kinetics of scandium ionization in $H_2-O_2-N_2$ flames," *Int. J. Mass Spectrom.* **214**, pp. 349–364.

Gutcheck, R. A., & Zipf, E. C. 1973, "Excitation of the CO fourth positive system by the dissociative recombination of CO_2^+ ions," *J. Geophys. Res.* **78**, pp. 5429–5436.

Habs, D., Bauman, W., Berger, J., et al. 1989, "First experiments with the Heidelberg test storage ring TSR," *Nucl. Instrum. Methods Phys. Res. B* **43**, pp. 390–410.

Habs, D., Kramp, J., Krause, P., Matl, K., Neumann, R., & Schwalm, D. 1988, "Ultracold ordered electron beam," *Phys. Scr.* **T22**, pp. 269–276.

Hackam, R. 1965, "Temperature dependence of electron–ion recombination and ion mobilities in nitrogen afterglows," *Planet. Space Sci.* **13**, pp. 667–674.

Hahn, Y. 1997, "Electron–recombination processes – an overview," *Rep. Prog. Phys.* **60**, pp. 691–759.

Hamberg, M., Geppert, W. D., Rosén, S., et al. 2005, "Branching ratios and absolute cross sections for dissociative recombination processes of N_2O^+," *Phys. Chem. Chem. Phys.* **7**, pp. 1664–1668.

Hamberg, M., Gepper, W. D., Thomas, R. D., et al. 2007, "Experimental determination of dissociative recombination reaction pathways and absolute reactions cross-sections of CH_2OH^+, CD_2OD^+ and $CD_2OD_2^+$," *Mol. Phys.* **105**, pp. 899–906.

Hansel, A., Glantschnig, M., Scheiring, C., Lindinger, W., & Ferguson, E. E. 1998, "Energy dependence of the isomerization of HCN^+ and HNC^+ via ion molecule reactions," *J. Chem. Phys.* **109**, pp. 1743–1747.

Hansen, K., Andersen, J. U., Hvelplund, P., Møller, S. P., Pedersen, U. V., & Petrunin, V. V. 2001, "Observation of a $1/t$ decay law for hot clusters and molecules in a storage ring," *Phys. Rev. Lett.* **87**, pp. 123401-1–4.

Harich, S. A., Hwang, D. W. H., Yang, X., et al. 2000, "Photodissociation of H_2O at 121.6 nm: A state-to-state dynamical picture," *J. Chem. Phys.* **113**, pp. 10073–10090.

Harrison, J. A., Whyte, A. R., & Phillips, L. F. 1986, "Kinetics of reactions of NH with NO and NO_2," *Chem. Phys. Lett.* **129**, pp. 346–352.

Hartquist, T. W. (ed.) 1990, *Molecular Astrophysics*, Cambridge: Cambridge University Press.

Hartquist, T. W., & Williams, D. A. (eds.) 1998, *The Molecular Astrophysics of Stars and Galaxies*, Oxford: Clarendon Press.

Hassouna, M., Le Garrec, J. L., Rebrion-Rowe, C., Travers, D., & Rowe, B. R. 2003, "Reactions of electrons with hydrocarbon cations: from linear alkanes to aromatic species," in *Dissociative Recombination of Molecular Ions with Electrons*, ed. S. L. Guberman, New York: Kluwer/Plenum Publishers, pp. 49–57.

Hasted, J. B. 1972, *Physics of Atomic Collisions*, New York: American Elsevier Publishing Company.

Hatano, Y. 1999, "Interaction of vacuum ultraviolet photons with molecules. Formation and dissociation dynamics of molecular superexcited states," *Phys. Rep.* **313**, pp. 109–169.

2002, "Spectroscopy and dynamics of molecular superexcited states," in *Chemical Applications of Synchrotron Radiation*, Part I: *Dynamics and VUV Spectroscopy*, Advanced Series in Physical Chemistry Vol. 12A, Singapore: World Scientific, pp. 55–111.

2003a, "Formation and dissociation dynamics of molecular superexcited states," *Bull. Chem. Soc. Jpn.* **76**, pp. 853–864.

2003b, "Spectroscopy and dynamics of molecular superexcited states. Aspects of primary processes of radiation chemistry," *Radiat. Phys. Chem.* **67**, pp. 187–198.

Haxton, D. J., McCurdy, C. W., & Rescigno, T. N., 2006, "Angular dependence of dissociative electron attachment to polyatomic molecules: Application to the $^2B_1$ metastable of the H_2O and H_2S anions," *Phys. Rev. A* **73**, pp. 062724-1–15.

Haxton, D. J., Rescigno, T. N., & McCurdy, C. W. 2005, "Topology of the adiabatic potential energy surfaces for the resonance states of the water anion," *Phys. Rev. A* **72**, pp. 022705-1–12.

Hayhurst, A. N., & Telford, N. R. 1974, "Kinetics of dissociative recombination of free electrons with hydronium ions in premixed flames," *J. Chem. Soc., Faraday Trans. 1* **70**, pp. 1999–2010.

Hays, P. B., & Sharp, W. E. 1973, "Twilight airglow 1. Photoelectrons and [OI] 5577-ångstrom radiation," *J. Geophys. Res.* **78**, pp. 1153–1166.

Hazi, A. U., & Taylor, H. S. 1970, "Stabilization method of calculating resonance energies: model problem," *Phys. Rev. A* **1**, pp. 1109–1120.

Hazi, A. U., Derkits, C., & Bardsley, J. N. 1983, "Theoretical study of the lowest $^1\Sigma_g^+$ doubly excited state of H_2," *Phys. Rev. A* **27**, pp. 1751–1759.

Heather, R., Jiang, X. P., Metiu, H., Bjorken, J. D., & Dunietz, I. 1988, "Time-dependent theory of Raman scattering for systems with several excited electronic states: Applications to a H_3^+ model system," *J. Chem. Phys.* **90**, pp. 6903–6915.

Heaviside, O. 1902, "Theory of the electric telegraph," in *Encyclopaedia Britannica*, tenth edition, Vol. 33, Encyclopaedia Britannica, Inc. pp. 213–218.

Heber, O., Seiersen, K., Bluhme, H., Svendsen, A., Andersen, L. H., & Maunoury, L. 2006, "Dissociative recombination of small carbon cluster ions," *Phys. Rev. A* **73**, pp. 022712-1–6.

Hechtfischer, U., Amitay, A., Forck, P., et al. 1998, "Near-threshold photodissociation of cold CH^+ in a storage ring," *Phys. Rev. Lett.* **80**, pp. 2809–2812.

Hefter, U., Mead, R. D., Schulz, P. A., & Lineberger, W. C. 1983, "Ultrahigh-resolution study of autodetachment study in C_2^-," *Phys. Rev. A* **28**, pp. 1429–1439.

Hellberg, F., Rosén, S., Thomas, R., et al. 2003, "Dissociative recombination of NO^+: Dynamics of the $X\,^1\Sigma^+$ and the $a\,^3\Sigma^+$ electronic states," *J. Chem. Phys.* **118**, pp. 6250–6259.

Hellberg, F., Zhaunerchyk, V., Ehlerding, A., et al. 2005, "Investigating the breakup dynamics of dihydrogen sulfide ions recombining with electrons," *J. Chem. Phys.* **122**, pp. 224314-1–9.

Heller, E. J. 1981, "The semiclassical way to molecular spectroscopy," *Acc. Chem. Res.* **14**, pp. 368–375.

Helm, H. 1993, "Predissociation of excited states of H_3," in *Dissociative Recombination: Theory, Experiment, and Applications*, eds. B. R. Rowe, J. B. A. Mitchell, & A. Canosa, NATO ASI Series B: Physics Vol. 313, New York: Plenum Press, pp. 145–153.

Helm, H., Galster, U., Mistrík, I., Müller, U., & Reichle, R. 2003, "Coupling of bound states to continuum states in neutral triatomic hydrogen," in *Dissociative Recombination of Molecular Ions with Electrons*, ed. S. L. Guberman, New York: Kluwer/Plenum Publishers, pp. 265–274.

Helm, H., Hazell, I., Walter, C. W., & Cosby, P. C. 1996, "On the branching in dissociative recombination of O_2^+," in *Dissociative Recombination: Theory, Experiment and*

Applications III, eds. D. Zajfman, J. B. A. Mitchell, D. Schwalm, & B. Rowe, Singapore: World Scientific, pp. 139–151.

Heppner, R. A., Walls, F. L., Armstrong, W. T., & Dunn, G. H. 1976, "Cross-section measurements for electron–H_3O^+ recombination," *Phys. Rev. A* **13**, pp. 1000–1011.

Herbst, E. 1978, "What are the products of polyatomic ion–electron dissociative recombination reactions?" *Astrophys. J.* **222**, pp. 508–516.

1988, "Dense interstellar cloud chemistry," in *Rate Coefficients in Astrochemistry*, eds. T. J. Millar & D. A. Williams, Dordrecht: Kluwer, pp. 239–262.

1989, "Dissociative recombination reactions in the chemistry of dense interstellar clouds," in *Dissociative Recombination: Theory, Experiment and Applications*, eds. J. B. A. Mitchell, & S. L. Guberman, Singapore: World Scientific, pp. 303–316.

2003a, "Dissociative recombination in interstellar clouds," in *Dissociative Recombination of Molecular Ions with Electrons*, ed. S. L. Guberman, New York: Kluwer/Plenum Publishers, pp. 351–363.

2003b, "Isotope fractionation by ion–molecule reactions," *Space Sci. Rev.* **106**, pp. 293–304.

2005, "Molecular ions in interstellar reaction networks," *J. Phys.: Conf. Ser.* **4**, pp. 17–25.

Herbst, E., & Cuppen, H. M. 2006, "Monte Carlo studies of surface chemistry and nonthermal desorption involving insterstellar grains," *Proc. Natl. Acad. Sci. USA* **103**, pp. 12257–12262.

Herbst, E., & Klemperer, W. 1973, "The formation and depletion of molecules in dense interstellar clouds," *Astrophys. J.* **185**, pp. 505–533.

1976, "The formation of interstellar molecules," *Physics Today* **29**(6), pp. 32–39.

Herbst, E., & Lee, H.-H. 1997, "New dissociative recombination product branching fractions and their effect on calculated interstellar molecular abundances," *Astrophys. J.* **485**, pp. 689–696.

Herbst, E., Miller, S., Oka, T., & Watson, J. K. G. (eds.) 2000, "Astronomy, physics and chemistry of H_3^+," *Phil. Trans. R. Soc. Lond. A* **358**, pp. 2359–2559.

Herd, C. R., Adams, N. G., & Smith, D. 1990, "OH production in the dissociative recombination of H_3O^+, HCO_2^+, and N_2OH^+: comparison with theory and interstellar implications," *Astrophys. J.* **349**, pp. 388–392.

Herman, Z. 2001, "The crossed-beam scattering method in studies of ion-molecule reaction dynamics," *Int. J. Mass Spectrom.* **212**, pp. 413–443.

Hernandez, N. 1971, "The signatures profiles of the $O(^1S)$ in the airglow," *Planet. Space Sci.* **19**, pp. 467–476.

Herzberg, G. 1950, *Molecular Spectra and Molecular Structure. I. Spectra of Diatomic Molecules*, New York: Van Nostrand.

1979, "A spectrum of triatomic hydrogen," *J. Chem. Phys.* **70**, pp. 4806–4807.

Herzberg, G., & Lagerqvist, A. 1968, "New spectrum associated with diatomic carbon," *Can. J. Phys.* **46**, pp. 2363–2373.

Herzenberg, A., 1968, "Oscillatory energy dependence of resonant electron–molecule scattering," *J. Phys. B* **1**, pp. 548–558.

Hickman, A. P. 1987, "Dissociative recombination of electrons with H_2^+," *J. Phys. B* **13**, pp. 3867–3894.

Hickman, A. P., Miles, R. D., Hayden, C., & Talbi, D. 2005, "Dissociative recombination of e + $HCNH^+$: Diabatic potential curves and dynamic calculations," *Astron. Astrophys.* **438**, pp. 31–37.

Hinojosa, G., Covington, A. M., Phaneuf, R. A., *et al.* 2002, "Formation of long-lived CO^{2+} via photoionization of CO^+." *Phys. Rev. A* **66**, pp. 032718-1-5.

Hirota, T., Yamamoto, S., Mikami, H., & Ohishi, M. 1998, "Abundances of HCN and HNC in dark cloud cores," *Astrophys. J.* **503**, pp. 717–728.

Hirschfelder, J. O. 1938, "The energy of the triatomic hydrogen molecule and ion, V," *J. Chem. Phys.* **6**, pp. 795–806.

Hogness, T. R., & Lunn, E. G. 1925, "The ionization of hydrogen by electron impact as interpreted by positive ray analysis," *Phys. Rev.* **26**, pp. 44–55.

Holt, E. M. 1959, "Electron loss processes in the oxygen afterglow," *Bull. Am. Phys. Soc.* **4**, pp. 112–113.

Holt, R. B., Richardson, J. M., Howland, B., & McClure, B. T. 1950, "Recombination spectrum and electron density measurements in neon afterglows," *Phys. Rev.* **77**, pp. 239–241.

Hotop, H., Ruf, M.-W., & Fabrikant, I. I. 2004, "Resonance and threshold phenomena in low-energy collisions with molecules and clusters," *Phys. Scr.* **T110**, pp. 22–31.

Hotop, H., Klar, D., Kreil, J., Ruf, M.-W., Schramm, A., & Weber, J. M. 1995, "Studies of low energy electron collisions at sub-meV resolution," in *The Physics of Electronic and Atomic Collisions: XIXth International Conference (XIX ICPEAC, Whistler, Canada)*, eds. L. J. Dube, J. B. A. Mitchell, J. W. McConkey, & C. E. Brion, AIP Conf. Proceedings Vol. 360, New York: American Institute of Physics, pp. 267–278.

Hotop, H., Ruf, M.-W., Allan, M., & Fabrikant, I. I. 2003, "Resonance and threshold phenomena in low-energy electron collisions with molecules and clusters," *Adv. At. Mol. Opt. Phys.* **49**, pp. 85–216.

Hu, X. K., Mitchell, J. B. A., & Lipson, R. H. 2000, "Resonance-enhanced multiphoton-ionization-photoelectron study of the dissociative recombination and associative ionization of Xe_2^+," *Phys. Rev. A* **62**, pp. 052712-1–8.

Huang, C.-M., Biondi, M. A., & Johnsen, R. 1975, "Variation of electron–NO^+-ion recombination coefficient with electron temperature," *Phys. Rev. A* **11**, pp. 901–905.

1976, "Recombination of of electrons with $NH_4^+ \cdot (NH_3)_n$-series ions," *Phys. Rev. A* **14**, pp. 984–989.

Huang, C.-M., Whitaker, M., Biondi, M. A., & Johnsen, R. 1978, "Electron-temperature dependence of recombination of electrons with $H_3O^+ \cdot (H_2O)_n$-series ions," *Phys. Rev. A* **18**, pp. 64–67.

Huestis, D. L. 1982, "Introduction and overview [to gas lasers]," *Pure Appl. Phys.* **43** (Appl. At. Collision Phys. **3**), pp. 1–34.

Hulthén, E. 1949, "The 1947 Nobel prize for physics," in *Les Prix Nobel en 1947*, ed. M. A. Holmberg, Stockholm: P. A. Nordstedt & Söner, pp. 20–22.

Humbert, D. (ed.) 2006, *Summary Report of First IAEA Research Co-ordination Meeting. Atomic and Molecular Data for Plasma Modelling*, INDC International Nuclear Data Committee, INDC(NDS)-0482, IAEA, Vienna.

Hunten, D. M. 1969, "The upper atmosphere of Jupiter," *J. Atmos. Sci.* **26**, pp. 826–834.

Hunter, E. P. L., & Lias, S. G. 1998, "Evaluated gas phase basicities and proton affinities of molecules: an update," *J. Phys. Chem. Ref. Data* **27**, pp. 413–656.

Huntress Jr., W. T. 1977, "Laboratory studies of bimolecular reactions of positive ions in interstellar clouds, in comets, and in planetary atmospheres of reducing composition," *Astrophys. J. Suppl. Ser.* **33**, pp. 495–514.

Huo, W. M., & Gianturco, F. A. (eds.) 1995, *Computational Methods for Electron Molecule Collisions*, New York: Plenum Press.

Hus, H., Yousif, F., Sen, A., & Mitchell, J. B. A. 1988, "Merged-beam studies of the dissociative recombination of H_3^+ ions with low internal energy," *Phys. Rev. A* **38**, pp. 658–663.

Hus, H., Yousif, F., Noren, C., Sen, A., & Mitchell, J. B. A. 1988, "Dissociative recombination of electrons with H_2^+ in low vibrational states," *Phys. Rev. Lett.* **60**, pp. 1006–1009.

Illenberg, E. 2000, "Electron capture processes by free and bound molecules," in *Photoionization and Photodetachment*, Part II, ed. C.-Y. Ng, Advanced Series in Physical Chemistry Vol. 10B, Singapore: World Scientific, pp. 1063–1160.

Itikawa, Y. 1994, "Electron collisions with N_2, O_2 and O: What we do and do not know," *Adv. At. Mol. Opt. Phys.* **33**, pp. 253–274.

Itikawa, Y., & Mason, N. 2005, "Rotational excitation of molecules by electron collisions," *Phys. Rep.* **414**, pp. 1–41.

Ivanov, G. K., & Golubkov, G. V. 1984, "Coupling of the processes of dissociative recombination and scattering of slow electrons by molecular ions," *Chem. Phys. Lett.* **107**, pp. 261–264.

  1985, "A simple version of the multichannel quantum defect analysis of inelastic atomic processes involving molecular Rydberg states," *J. Phys. B* **18**, pp. L383–L387.

Ivash, E. V. 1958, "Dissociation of the hydrogen molecule ion by electron impact," *Phys. Rev.* **112**, pp. 155–158.

Janev, R. K., 2000, "Role of dissociative recombination and related molecular processes in fusion edge plasmas," in *Dissociative Recombination: Theory, Experiment and Applications IV*, eds. M. Larsson, J. B. A. Mitchell, & I. F. Schneider, Singapore: World Scientific, pp. 40–47.

Janev, R. K., & Reiter, D. 2002, "Collision processes of CH_y and CH_y^+ hydrocarbons with plasma electron and protons," *Phys. Plasma* **9**, pp. 4071–4081.

  2004, "Collision processes of $C_{2,3}H_y$ and $C_{2,3}H_y^+$ hydrocarbons with plasma electrons and protons," *Phys. Plasma* **11**, pp. 780–829.

Jensen, M. J., Pedersen, U. V., & Andersen, L. H. 2000, "Stability of the ground state vinylidene anion H_2CC^-," *Phys. Rev. Lett.* **84**, pp. 1128–1131.

Jensen, M. J., Bilodeau, R. C., Heber, O., *et al.* 1999, "Dissociative recombination and excitation of H_2O^+ and HDO^+," *Phys. Rev. A* **60**, pp. 2970–2976.

Jensen, M. J., Bilodeau, R. C., Safvan, C. P., *et al.* 2000, "Dissociative recombination of the H_3O^+, HD_2O^+, and D_3O^+," *Astrophys. J.* **543**, pp. 764–774.

Jensen, M. J., Pedersen, H. B., Safvan, C. P., Seiersen, K., Urbain, X., & Andersen, L. H. 2001, "Dissociative recombination and excitation of H_3^+," *Phys. Rev. A* **63**, pp. 052701-1–5.

Jiang, L., Gutherie, J. A., Chaney, R. C., & Cunningham, A. J. 1989, "Dissociative recombination measurements at elevated temperatures in helium–neon mixtures," *J. Phys. B* **22**, pp. 3047–3054.

Jog, V. E., & Biondi, M. A. 1981, "Dissociative recombination of Hg_2^+ ions and electrons: dependence of the total rate coefficient and excited state production on electron temperature," *J. Phys. B* **14**, pp. 4719–4727.

Johnsen, R. 1986, "rf-probe method for measurements of electron densities in plasmas at high densities," *Rev. Sci. Instr.* **57**, pp. 428–432.

  1987, "Microwave afterglow measurements of dissociative recombination of molecular ions with electrons," *Int. J. Mass Spectrom. Ion Proc.* **81**, pp. 67–84.

  1989, "Recombination measurements in microwave plasma afterglows," in *Dissociative Recombination: Theory, Experiment, and Applications*, eds. J. B. A. Mitchell and S. L. Guberman, Singapore: World Scientific, pp. 141–150.

1993a, "Recombination of cluster ions," in *Dissociative Recombination: Theory, Experiment, and Applications*, eds. B. R. Rowe, J. B. A. Mitchell, & A. Canosa, NATO ASI Series B: Physics Vol. 313, New York: Plenum Press, pp. 135–143.

1993b, "Electron-temperature dependence of the recombination of $H_3O^+(H_2O)_n$ ions with electrons," *J. Chem. Phys.* **98**, pp. 5390–5395.

2005, "A critical review of H_3^+ recombination studies," *J. Phys.: Conf. Ser.* **4**, pp. 83–91.

Johnsen, R., & Mitchell, J. B. A. 1998, "Complex formation in electron–ion recombination of molecular ions," *Adv. Gas Phase Ion. Chem.* **3**, pp. 49–80.

Johnsen, R., Shuńko, E. V., Gougousi, T., & Golde, M. F. 1994, "Langmuir-probe measurements in flowing-afterglow plasmas," *Phys. Rev. E* **50**, pp. 3994–4004.

Johnsen, R., Skrzypkowski, M., Gougousi, T., & Golde, M. F. 2000, "Spectroscopic emissions from the recombination of N_2O^+, N_2OH^+/HN_2O^+, CO_2^+, CO_2H^+, HCO^+/COH^+, H_2O^+, NO_2^+, HNO^+, and LIF measurements of the H atom yield from H_3^+," in *Dissociative Recombination: Theory, Experiment and Applications IV*, eds. M. Larsson, J. B. A. Mitchell, & I. F. Schneider, Singapore: World Scientific, pp. 200–209.

Johnson, R. A., McClure, B. T., & Holt, R. B. 1950, "Electron removal in helium afterglows," *Phys. Rev.* **80**, pp. 376–379.

Joly, J. 1902, "Mr. Marconi's results in day and night wireless telegraphy," *Nature* **66**, p. 199.

Jursic, B. S. 1999, "Complete basis set ab initio study of potential energy surfaces of the dissociative recombination reaction $HCNH^+ + e^-$," *J. Mol. Struct. (Theochem)* **487**, pp. 211–220.

Kaiser, R. I., Bernath, P., Osamura, Y., Petrie, S., & Mebel, A. M. (eds.) 2006, *Astrochemistry – From Laboratory Studies to Astronomical Observations*, AIP Conf. Proceedings Vol. 855, New York: American Institute of Physics.

Kalhori, S., Thomas, R., Al-Khalili, A., *et al.* 2004, "Resonant ion-pair formation in electron collisions with rovibrationally cold H_3^+," *Phys. Rev. A* **69**, pp. 022713-1–11.

Kalhori, S., Viggiano, A. A., & Arnold, S. T. 2002, "Dissociative recombination of $C_2H_3^+$," *Astron. Astrophys.* **391**, pp. 1159–1165.

Kaplan, J. 1931, "The light of the night sky," *Phys. Rev.* **38**, pp. 1048–1051.

Kasner, W. H. 1967, "Study of the temperature dependence of electron–ion recombination in nitrogen," *Phys. Rev.* **164**, pp. 194–200.

1968, "Study of the pressure and temperature dependence of electron–ion recombination in neon," *Phys. Rev.* **167**, pp. 148–151.

Kasner, W. H., & Biondi, M. A. 1965, "Electron–ion recombination in nitrogen," *Phys. Rev.* **137**, pp. A317–A329.

1967, "Electron–ion recombination studies in oxygen," *Bull. Am. Phys. Soc.* **12**, p. 218.

1968, "Temperature dependence of electron–O_2^+-ion recombination coefficient," *Phys. Rev.* **174**, pp. 139–144.

Kasner, W. H., Rogers, W. A., & Biondi, M. A. 1961, "Electron–ion recombination coefficients in nitrogen and in oxygen," *Phys. Rev. Lett.* **7**, pp. 321–323.

Kayanuma, M., Taketsugu, T., & Ishii, K. 2006, "Ab initio surface hopping simulation on dissociative recombination of H_3O^+," *Chem. Phys. Lett.* **418**, pp. 511–518.

Kella, D., Johnson, P. J., Pedersen, H. B., Vejby-Christensen, L., & Andersen, L. H. 1996, "Branching ratios for dissociative recombination of $^{15}N^{14}N^+$," *Phys. Rev. Lett.* **77**, pp. 2432–2435.

Kella, D., Vejby-Christensen, L., Johnson, P. J., Pedersen, H. B., & Andersen, L. H. 1997, "The source of the green light emission determined from a heavy-ion storage ring experiment," *Science* **276**, pp. 1530–1533; Corrections and Clarifications, *ibid.* **277**, p. 167.

Keller, G. E., & Beyer, R. A. 1971, "Carbon dioxide and oxygen clustering to sodium ions," *J. Geophys. Res.* **76**, pp. 289–290.

Kennelly, A. E. 1902, "On the elevation of electrically-conducting strata of the earth's atmosphere," *Elect. World Eng.* **39**, p. 473.

Kenty, C. 1928, "The recombination of argon ions and electrons," *Phys. Rev.* **32**, pp. 624–635.

Kerner, E. H. 1953, "The dissociation of H_2^+ by electron impact," *Phys. Rev.* **92**, pp. 1441–1447.

Ketvirtis, A. E., & Simons, J. 1999, "Dissociative recombination of H_3O^+," *J. Phys. Chem. A* **103**, pp. 6552–6563.

Keyser, C. J., Froelich, H. R., Mitchell, J. B. A., & McGowan, J. W. 1979, "Beam-scanning system for determination of beam profiles and form factors in merged-beam experiments," *J. Phys. E* **12**, pp. 316–320.

Kharchenko, V., Dalgarno, A., & Fox, J. L. 2005, "Thermospheric distribution of fast O (^1D) atoms," *J. Geophys. Res.* **110**, pp. A12305-1–9.

Khare, S. P. 2002, *Introduction to the Theory of Collisions of Electrons with Atoms and Molecules*, New York: Kluwer Academic/Plenum Publishers.

Kilgus, G., Habs, D., Schwalm, D., Wolf, A., Badnell, N. R., & Müller, A. 1992, "High-resolution measurement of dielectronic recombination of lithiumlike Cu^{26+}," *Phys. Rev. A* **46**, pp. 5730–5740.

King, I. R. 1957, "Ion recombination rates in methane-air flames," *J. Chem. Phys.* **27**, pp. 817–818.

Kiyoshima, T., Sato, S., Pazyuk, E. A., Stolyarov, A. V., & Child, M. S. 2003, "Lifetime measurement and quantum-defect theory treatment of the $k\,^3\Pi_u^-$ state of hydrogen molecule," *J. Chem. Phys.* **118**, pp. 121–129.

Klemperer W. 1970, "Carrier of the interstellar 89.190 GHz line," *Nature* **227**, p. 1230.

  1995, "Some spectroscopic reminiscenses," *Annu. Rev. Phys. Chem.* **46**, pp. 1–26.

  2006, "Interstellar chemistry," *Proc. Natl. Acad. Sci. USA* **103**, pp. 12232–12234.

Kley, D., Lawrence, G. M., & Stone, E. J. 1977, "The yield of N(^2D) atoms in the dissociative recombination of NO^+," *J. Chem. Phys.* **66**, pp. 4157–4165.

Kokoouline, V., & Greene, C. H., 2003a, "Theory of dissociative recombination of D_{3h} triatomic ions applied to H_3^+," *Phys. Rev. Lett.* **90**, pp. 133201-1–4.

  2003b, "Unified theoretical treatment of dissociative recombination of D_{3h} triatomic ions: Applications to H_3^+ and D_3^+," *Phys. Rev. A* **68**, pp. 012703-1–23.

  2005a, "Theoretical study of the H_3^+ ion dissociative recombination process," *J. Phys.: Conf. Ser.* **4**, pp. 74–82.

  2005b, "Theoretical study of C_{2v} triatomic ions: Application to H_2D^+ and D_2H^+," *Phys. Rev. A* **72**, pp. 022712-1–12.

Kokoouline, V., Greene, C. H., & Esry, B. D. 2001, "Mechanism for the destruction of H_3^+ ions by electron impact," *Nature* **412**, pp. 891–894.

Korolov, I., Novotný, O., Plašil, R., *et al.* 2006, "Recombination of XeH^+ and KrH^+ with electrons in low temperature plasma," *Czech. J. Phys.* **56** (Suppl. B), pp. B854–B864.

Kraemer, W. P., & Diercksen, G. H. F. 1976, "Indentification of interstellar X-ogen as HCO^+," *Astrophys. J.* **205**, pp. L97–L100.

Kraemer, W. P., & Hazi, A. U. 1985, "Dissociative recombination of interstellar ions: electronic structure calculations for HCO^+," in *Molecular Astrophysics*, ed. H. F. Diercksen *et al.*, Dordrecht: Reidel, pp. 575–581.

1989, "Dissociative recombination of HCO$^+$: Complete Active Space (CAS) SCF electronic structure calculations," in *Dissociative Recombination: Theory, Experiment and Applications*, eds. J. B. A. Mitchell, & S. L. Guberman, Singapore: World Scientific, pp. 61–72.

Krause, J. L., Orel, A. E., Lengsfield III, B. H., & Kulander, K. C., 1992, "Wave packed studies of the predissociation of H$_3$," in *Time Dependent Quantum Molecular Dynamics: Experiments and Theory*, eds. J. Broeckhove & L. Lathouwers, New York: Plenum Press, pp. 131–142.

Krauss, M., & Julienne, P. S. 1973, "Dissociative recombination of $e + CH^+$ ($X\,^1\Sigma^+$)," *Astrophys. J. Lett.* pp. L139–L141.

Kreckel, H., Krohn, S., Lammich, L., et al. 2002, "Vibrational and rotational cooling of H$_3^+$," *Phys. Rev. A* **66**, pp. 052509-1–11.

Kreckel, H., Mikosch, J., Wester, R., et al. 2005a, "Towards state selective measurements of the H$_3^+$ dissociative recombination rate coefficient," *J. Phys.: Conf. Ser.* **4**, pp. 126–133.

Kreckel, H., Motsch, M., Mikosch, J., et al. 2005b, "High resolution dissociative recombination of cold H$_3^+$ and first evidence for nuclear spin effects," *Phys. Rev. Lett.* **95**, pp. 263201-1–4.

Kreckel, H., Tennyson, J., Schwalm, D., Zajfman, D., & Wolf, A. 2004, "Rovibrational relaxation model," *New. J. Phys.* **6**, pp. 151-1–16.

Kroes, G.-J., van Hermert, M. C., Billing, G. D., & Neuhauser, D. 1997, "Photodissociation of CH$_2$: Three-dimensional quantum dynamics of the dissociation through the coupled $2A'$ and $3A'$ states," *J. Chem. Phys.* **107**, pp. 5757–5770.

Krohn, S., Amitay, Z., Baer, A., et al. 2000, "Electron-induced vibrational deexcitation of H$_2^+$," *Phys. Rev. A* **62**, pp. 032713-1–8.

Krohn, S., Kreckel, H., Lammich, L., et al. 2003, "Electron induced vibrational deexcitation of the molecular ions H$_2^+$ and D$_2^+$," in *Dissociative Recombination of Molecular Ions with Electrons*, ed. S. L. Guberman, New York: Kluwer/Plenum Publishers, pp. 127–138.

Krohn, S., Lange, M., Grieser, M., et al. 2001, "Rate coefficients and final states for the dissociative recombination of LiH$^+$," *Phys. Rev. Lett.* **86**, pp. 4005–4008.

Kubach, C., Sidis, V., Fussen, D., & van der Zande, W. J. 1987, "Decay of the $A\,^2\Sigma^+$ and $B\,^2\Pi$ quasibound states of HeH," *Chem. Phys.* **117**, pp. 439–447.

Kulander, K. C., & Guest, M. F. 1979, "Excited electronic states of H$_3$ and their role in the dissociative recombination of H$_3^+$," *J. Phys. B* **12**, pp. L501–L504.

Kulander, K. C., & Heller, E. J. 1978, "Time dependent formulation of polyatomic photofragmentation: Applications to H$_3^+$," *J. Chem. Phys.* **69**, pp. 2439–2449.

Lammer, H., & Bauer, S. J. 1993, "Atmospheric mass loss from Titan by sputtering," *Planet. Space Sci.* **41**, pp. 657–663.

Lammich, L., Altevogt, S., Buhr, H., et al. 2007, "Electron-impact dissociation and transient properties of a stored LiH$_2^-$ beam," *Eur. Phys. J. D* **41**, pp. 103–111.

Lammich, L., Kreckel, H., Krohn, S., et al. 2003a, "Breakup dynamics in the dissociative recombination of H$_3^+$ and its isotopomers," *Rad. Phys. Chem.* **68**, pp. 175–179.

Lammich, L., Strasser, D., Kreckel, H., et al. 2003b, "Evidence for subthermal rotational populations in stored molecular ions through state-dependent dissociative recombination," *Phys. Rev. Lett.* **91**, pp. 143201-1–4.

Lammich, L., Strasser, D., Kreckel, H., et al. 2005, "DR rate coefficient measurements using stored beams of H$_3^+$ and its isotopomers," *J. Phys.: Conf. Ser.* **4**, pp. 98–103.

Lampert, A., Wolf, A., Habs, D., et al. 1996, "High-resolution measurement of the dielectronic recombination of fluorinelike selenium ions," *Phys. Rev. A* **53**, pp. 1413–1423.

Lange, M., Levin, J., Gwinner, G., *et al.* 1999, "Threshold effects and ion-pair production in the dissociative recombination of HD$^+$," *Phys. Rev. Lett.* **83**, pp. 4979–4982.

Laperle, C. M., Mann, J. E., Clements, T. G., & Continetti, R. E. 2004, "Three-body dissociation dynamics of the low-lying Rydberg states of H$_3$ and D$_3$," *Phys. Rev. Lett.* **93**, pp. 153202-1–4.

2005, "Experimentally probing the three-body predissociation dynamics of the low-lying Rydberg states of H$_3$ and D$_3$," *J. Phys.: Conf. Ser.* **4**, pp. 111–117.

Larson, Å., & Orel, A. E. 1999, "Dissociative recombination of HeH$^+$: Product distributions and ion-pair formation," *Phys. Rev. A* **59**, pp. 3601–3608.

2001, "Ion-pair formation and product branching ratios in dissociative recombination of HD$^+$," *Phys. Rev. A* **64**, pp. 062701-1–8.

2005, "Wave packet study of the products formed in dissociative recombination of HeH$^+$," *Phys. Rev. A* **72**, pp. 032701-1–13.

Larson, Å., Roos, J., & Orel, A. E. 2006, "Ion-pair formation in electron recombination with H$_3^+$," *Phil. Trans. R. Soc. A* **364**, pp. 2999–3005.

Larson, Å., Djurić, N., Zong, W., *et al.* 2000, "Resonant ion-pair formation in electron collisions with HD$^+$ and OH$^+$," *Phys. Rev. A* **62**, pp. 042707-1–8.

Larson, Å., Le Padellec, A., Semaniak, J., *et al.* 1998, "Branching fractions in dissociative recombination of CH$_2^+$," *Astrophys. J.* **505**, pp. 459–465.

Larson, Å., Tonzani, S., Santra, R., & Greene, C. H. 2005, "Dissociative recombination of HCO$^+$," *J. Phys.: Conf. Ser.* **4**, pp. 148–154.

Larsson, M. 1995a, "Atomic and molecular physics with ion storage rings," *Rep. Prog. Phys.* **58**, pp. 1267–1319.

1995b, "Dissociative recombination in ion storage rings," *Int. J. Mass Spectrom. Ion Proc.* **149/150**, pp. 403–414.

1997, "Dissociative recombination with ion storage rings," *Annu. Rev. Phys. Chem.* **48**, pp. 151–179.

2000a, "Dissociative electron–ion recombination studies using ion synchrotrons," in *Photoionization and Photodetachment*, Part II, ed. C.-Y. Ng, Advanced Series in Physical Chemistry Vol. 10B, Singapore: World Scientific, pp. 693–747.

2000b, "Experimental studies of the dissociative recombination of H$_3^+$," *Phil. Trans. R. Soc. Lond. A* **358**, pp. 2433–2444.

2001, "Merged-beam studies of electron–molecular ion interactions in ion storage rings," *Adv. Gas Phase Ion Chem.* **4**, pp. 179–211.

2003, "Ion storage rings," in *Encyclopedia of Mass Spectrometry, Volume 1: Theory and Ion Chemistry*, ed. P. Armentrout, Amsterdam: Elsevier, pp. 195–199.

2005, "Molecular ion recombination in merged beams; experimental results on small systems and future perspectives," *J. Phys.: Conf. Ser.* **4**, pp. 50–57.

2006, "Hyzone," *Phil. Trans. R. Soc. A* **364**, pp. 3147–3148.

Larsson, M., & Thomas, R. 2001, "Three-body reaction dynamics in electron–ion dissociative recombination," *Phys. Chem. Chem. Phys.* **3**, pp. 4471–4480.

Larsson, M., Mitchell, J. B. A., & Schneider, I. F. (eds.) 2000, *Dissociative Recombination: Theory, Experiment and Applications IV*, Singapore: World Scientific.

Larsson, M., Carlson, M., Danared, H., Broström, L., Mannervik, S., & Sundström, G. 1994, "Vibrational cooling of D$_2^+$ in an ion storage ring as revealed by dissociative recombination measurements," *J. Phys. B* **27**, pp. 1397–1406.

Larsson, M., Danared, H., Larson, Å., *et al.* 1997, "Isotope and electric field effects in dissociative recombination of D$_3^+$," *Phys. Rev. Lett.* **79**, pp. 395–398.

Larsson, M., Danared, H., Mowat, J. R., et al. 1993a, "Direct high-energy neutral-channel dissociative recombination of cold H_3^+ in an ion storage ring," *Phys. Rev. Lett.* **70**, pp. 430–433.

Larsson, M., Djuric, N., Dunn, G. H., et al. 2003, "Studies of dissociative recombination in CRYRING," in *Dissociative Recombination of Molecular Ions with Electrons*, ed. S. L. Guberman, New York: Kluwer/Plenum Publishers, pp. 87–94.

Larsson, M., Ehlerding, A., Geppert, W. D., et al. 2005, "Rate constants and branching ratios for the dissociative recombination of $C_3D_7^+$ and $C_4D_9^+$," *J. Chem. Phys.* **122**, pp. 156101-1–3.

Larsson, M., Lepp, S., Dalgarno, A., et al. 1996, "Dissociative recombination of H_2D^+ and the cosmic abundance of deuterium," *Astron. Astrophys.* **309**, pp. L1–L3.

Larsson, M., Sundström, G., Carlson, M., et al. 1993b, "Phase-space cooled molecular ions in CRYRING," in *The Physics of Electronic and Atomic Collisions: XVIII International Conference (XVIII ICPEAC, Aarhus, Denmark)*, eds. T. Andersen, B. Fastrup, F. Folkmann, H. Knudsen, & N. Andersen, AIP Conf. Proceedings Vol. 295, New York: American Institute of Physics, pp. 803–810.

Laubé, S., Lehfaoui, L., Rowe, B. R., & Mitchell, J. B. A. 1998a, "The dissociative recombination of CO^+," *J. Phys. B* **31**, pp. 4181–4189.

Laubé, S., Le Padellec, A., Sidko, O., Rebrion-Rowe, C., Mitchell, J. B. A., & Rowe, B. R. 1998b, "New FALP–MS measurements of H_3^+, D_3^+ and HCO^+ dissociative recombination," *J. Phys. B* **31**, pp. 2111–2128.

Le Bourlot, J., Pineau des Forêts, G., Roueff, E., & Schilke, P. 1993, "Bistability in dark cloud chemistry," *Astrophys. J. Lett.* **416**, pp. L87–L90.

Lee, C. M. 1977, "Multichannel dissociative recombination theory," *Phys. Rev. A* **16**, pp. 109–122.

Lee, M.-T., Iga, I., Brescansin, L. M., Machado, L. E., & Machado, F. B. C. 2002, "Theoretical study on electron–free-radical scattering: An application to CF," *Phys. Rev. A* **66**, pp. 012720-1–7.

Lefebvre-Brion, H., & Field, R. W. 1986, *Perturbations in the Spectra of Diatomic Molecules*, Orlando: Academic Press Inc.

Leforestier, C., Bisseling, R. H., Cerjan, C., et al. 1991, "A comparison of different propagation schemes for the time dependent Schrödinger equation," *J. Comp. Phys.* **94**, pp. 59–80.

Lehfaoui, L., Rebrion-Rowe, C., Laubé, S., Mitchell, J. B. A., & Rowe, B. R. 1997, "The dissociative recombination of hydrocarbon ions. I. Light alkanes," *J. Chem. Phys.* **106**, pp. 5406–5412.

Lengsfield III, B. H., & Yarkony, D. R. 1992, "Nonadiabatic interactions between potential energy surfaces: theory and applications," *Adv. Chem. Phys.* **82**, pt. 2, pp. 1–71.

Lennon, J. J., & Sexton, M. C. 1959, "Recombination in xenon and krypton afterglow," *J. Electr. Control* **7**, pp. 123–133.

Leo, W. R. 1987, *Techniques for Nuclear and Particle Physics Experiments*, Berlin: Springer-Verlag.

Le Padellec, A. 2003, "Studies of electron collisions with CN^+, CN^- and HCN^+/HNC^+," in *Dissociative Recombination of Molecular Ions with Electrons*, ed. S. L. Guberman, New York: Kluwer/Plenum Publishers, pp. 109–125.

Le Padellec, A., Sheehan, C., & Mitchell, J. B. A. 1998, "The dissociative recombination of CN^+," *J. Phys. B* **31**, pp. 1725–1728.

Le Padellec, A., Andersson, K., Hanstorp, D., et al. 2001a, "Electron scattering on CN^-," *Phys. Scr.* **64**, pp. 467–473.

Le Padellec, A., Djurić, N., Al-Khalili, A., et al. 2001b, "Resonant ion-pair formation in the recombination of NO$^+$ with electrons: Cross-section determination," *Phys. Rev. A* **64**, pp. 012702-1–7.

Le Padellec, A., Larsson, M., Danared, H., et al. 1998, "A storage ring study of dissociative excitation and recombination of D$_3^+$," *Phys. Scr.* **57**, pp. 215–221.

Le Padellec, A., Laubé, S., Sidko, O., et al. 1997b, "The dissociative recombination of KrH$^+$ and XeH$^+$," *J. Phys. B* **30**, pp. 963–967.

Le Padellec, A., Mitchell, J. B. A., Al-Khalili, A., et al. 1999, "Storage ring measurements of the dissociative recombination and excitation of the cyanogen CN$^+$ ($X\,^1\Sigma^+$ and $a\,^3\Pi$, $v = 0$)," *J. Chem. Phys.* **110**, pp. 890–901.

Le Padellec, A., Rabilloud, F., Pegg, D., et al. 2001c, "Electron-impact detachment and dissociation of C$_4^-$ ions," *J. Chem. Phys.* **115**, pp. 10671–10677.

Le Padellec, A., Sheehan, C., Talbi, D., & Mitchell, J. B. A. 1997a, "A merged-beam study of the dissociative recombination of HCO$^+$," *J. Phys. B* **30**, pp. 319–327.

Le Petit, F., & Roueff, E. 2003, "Dissociative recombination and deuterium fractionation in interstellar clouds," in *Dissociative Recombination of Molecular Ions with Electrons*, ed. S. L. Guberman, New York: Kluwer/Plenum Publishers, pp. 373–383.

Lepp, S. 1993, "Chaos in interstellar clouds," *Nature* **366**, pp. 633–634.

Lepp, S., & Dalgarno, A. 1988, "Heating of interstellar gas by large molecules or small grains," *Astrophys. J.* **335**, pp. 769–773.

Lepp, S., & Shull, J. M. 1984, "Molecules in the early universe," *Astrophys. J.* **280**, pp. 465–469.

Le Teuft, Y. H., Millar, T. J., & Markwick, A. J. 2000, "The UMIST database for astrochemistry 1999," *Astron. Astrophys. Suppl. Ser.* **146**, pp. 157–168.

Leu, M. T., Biondi, M. A., & Johnsen, R. 1973a, "Measurements of the recombination of electrons with H$_3$O$^+\cdot$(H$_2$O)$_n$-series ions," *Phys. Rev. A* **7**, pp. 292–298.

1973b, "Measurements of recombination of electrons with H$_3^+$ and H$_5^+$," *Phys. Rev. A* **8**, pp. 413–419.

1973c, "Measurements of recombination of electrons with HCO$^+$ ions," *Phys. Rev. A* **8**, pp. 420–422.

Leung, K. T. 1998, "Recent developments in electron momentum density measurements of polyatomic molecular (e,2e) spectroscopy," in *Novel Aspects of Electron-Molecule Collisions*, ed. K. H. Becker, Singapore: World Scientific, pp. 199–240.

Light, J. C. 1967, "Statistical theory of bimolecular exchange reactions," *Discussions Faraday Soc.* **44**, pp. 14–29.

Lindsay, C. M., & McCall, B. J. 2001, "Comprehensive evaluation and compilation of H$_3^+$ spectroscopy," *J. Mol. Spectrosc.* **210**, pp. 60–83.

Linsky, J. L. 2003, "Atomic deuterium/hydrogen in the galaxy," *Space Sci. Rev.* **106**, pp. 49–60.

Lis, D. C., Blake, G. A., & Herbst, E. (eds.) 2006, *Astrochemistry: Recent Successes and Current Challenges, IAU Symp. 231*, Cambridge: Cambridge University Press.

Little, D. P., Speir, J. P., Senko, M. W., O'Connor, P. B., & McLafferty, F. W. 1994, "Infrared multiphoton dissociation of large multiply charged ions for biomolecule sequencing," *Anal. Chem.* **66**, pp. 2809–2815.

Liu, D.-J., Ho, W.-C., & Oka, T. 1987, "Rotational spectroscopy of molecular ions using diode lasers," *J. Chem. Phys.* **87**, pp. 2442–2446.

Liu, X., Hwang, D. W., Yang, X. F., Harich, S., Lin, J. J., & Yang, X. 1999, "Photodissociation of hydrogen sulfide at 157.6 nm: Observation of SH bimodal rotational distribution," *J. Chem. Phys.* **111**, pp. 3940–3945.

Lodge, O. 1902, "Mr. Marconi's results in day and night wireless telegraphy," *Nature* **66**, p. 222.
Loeb, L. B. 1939, *Fundamental Processes of Electrical Discharge in Gases*, New York: John Wiley and Sons, Inc.
  1955, *Basic Processes of Gaseous Electronics*, Berkeley: University of California Press.
Löfgren, P., Andler, G., Bagge, L., et al. 2006, "Design of the double electrostatic storage ring DESIREE," in *Proceedings of EPAC 2006, Edinburgh, Scotland*, Edinburgh: European Physical Society Accelerator Group, pp. 252–254.
Lykke, K. R., Murray, K. K., Neumark, D. M., & Lineberger, W. C. 1988, "High-resolution studies of autodetachment in negative ions," *Phil. Trans. R. Soc. Lond. A* **324**, pp. 179–196.
Macdonald, J., Biondi, M. A., & Johnsen, R. 1984, "Recombination of electrons with H_3^+ and H_5^+," *Planet. Space Sci.* **32**, pp. 651–654.
Macko, P., Bánó, G., Hlavenka, P., et al. 2004a, "Afterglow studies of $H_3^+(v=0)$ recombination using time resolved cw-diode laser cavity ring-down spectroscopy," *Int. J. Mass Spectrom.* **233**, pp. 299–304.
Macko, P., Bánó, G., Hlavenka, P., et al. 2004b, "Decay of H_3^+ dominated low-temperature plasma," *Act. Phys. Slov.* **54**, pp. 263–271.
Mahdavi, M. R., Hasted, J. B., & Nakshbandi, M. M. 1971, "Electron–ion recombination measurements in the flowing afterglow," *J. Phys. B* **4**, pp. 1726–1737.
Mannervik, S., Lidberg, J., Norlin, L.-O., et al. 1999, "Lifetime measurement of the metastable 4d $^2D_{3/2}$ level in Sr^+ by optical pumping of a stored ion beam," *Phys. Rev. Lett.* **83**, pp. 698–701.
Marconi, M. G. 1910, "Nobel lecture," in *Les Prix Nobel en 1909*, ed. M. C. G. Santesson, Stockholm: P. A. Nordstedt & Söner, pp. 1–24.
Margenau, H. 1946, "Conduction and dispersion of ionized gases at high frequencies," *Phys. Rev.* **69**, pp. 508–513.
Märk, T. D. 1984, "Ionization of molecules by electron impact," in *Electron–Molecule Interactions and Their Applications* Vol. 1, ed L. G. Christophorou, Orlando: Academic Press, pp. 251–335.
  1992, "Ionization by electron impact," *Plasma Phys. Contr. Fusion* **34**, pp. 2083–2090.
Märk, T. D., & Dunn, G. H. 1985, *Electron Impact Ionization*, Vienna, Springer Verlag.
Martin, D. W., McDaniel, E. W., & Meeks, M. L. 1961, "On the possible occurrence of H_3^+ in interstellar space," *Astrophys. J.* **134**, pp. 1012–1013.
Massey, H. S. W. 1937, "Dissociation, recombination and attachment processes in the upper atmosphere – I," *Proc. Roy. Soc. A* **163**, pp. 542–553.
  1950, *Negative Ions*, second edn, Cambridge: Cambridge University Press.
  1976, *Negative Ions*, third edn, Cambridge: Cambridge University Press.
Matheson, M. S., & Dorfman, L. M. 1969, *Pulse Radiolysis*, Cambridge, MA: MIT Press.
Mathur, D. 2004, "Structure and dynamics of molecules in high charge states," *Phys. Rep.* **391**, pp. 1–118.
Mathur, D., Khan, S. U., & Hasted, J. B. 1978, "Dissociative recombination in low energy e-H_2^+ and e-H_3^+ collisions, *J. Phys. B* **11**, pp. 3615–3619.
Maul, C., & Gericke, K.-H. 1997, "Photo induced three body decay," *Int. Rev. Phys. Chem.* **16**, pp. 1–79.
McCall, B. J. 2001, "Spectroscopy of H_3^+ in laboratory and astrophysical plasmas," Ph.D. Thesis, University of Chicago.

2006, "Dissociative recombination of cold H_3^+ and its interstellar implications," *Phil. Trans. R. Soc. A* **364**, pp. 2953–2963.

McCall, B. J., & Oka, T. 2003, "Enigma of H_3^+ in diffuse interstellar clouds," in *Dissociative Recombination of Molecular Ions with Electrons*, ed. S. L. Guberman, New York: Kluwer/Plenum Publishers, pp. 365–371.

McCall, B. J., Geballe, T. R., Hinkle, K. H., & Oka, T. 1998, "Detection of H_3^+ in the diffuse interstellar medium toward Cygnus OB2 no. 12," *Science* **279**, pp. 1910–1913.

McCall, B. J., Hinkle, K. H., Geballe, T. R., *et al.* 2002, "Observations of H_3^+ in the diffuse interstellar medium," *Astrophys. J.* **567**, pp. 391–406.

McCall, B. J., Huneycutt, A. J., Saykally, R. J., *et al.* 2003, "An enhanced cosmic-ray flux towards ζ Persei inferred from a laboratory study of the $H_3^+ - e^-$ recombination rate," *Nature* **422**, pp. 500–502.

McCall, B. J., Huneycutt, A. J., Saykally, R. J., *et al.* 2004, "Dissociative recombination of rotationally cold H_3^+," *Phys. Rev. A* **70**, pp. 052716-1–13.

McCall, B. J., Huneycutt, A. J., Saykally, R. J., *et al.* 2005, "Storage ring measurements of the dissociative recombination of rotationally cold H_3^+," *J. Phys.: Conf. Ser.* **4**, pp. 92–97.

McCurdy, C. W., & Turner, J. L. 1983, "Wave packet formulation of the boomerang model for resonant electron-molecule scattering," *J. Chem. Phys.* **78**, pp. 6773–6779.

McDaniel, E. W. 1989, *Atomic Collisions: Electron and Photon Projectiles*, New York: Wiley.

McDaniel, E. W., & Manskey, E. J. 1994, "Guide to bibliographies, books, reviews and compendia on data on atomic collisions," *Adv. At. Mol. Opt. Phys.* **33**, pp. 384–463.

McDaniel, E. W., Mitchell, J. B. A., & Rudd, M. E. 1993, *Atomic Collisions: Heavy Particle Projectiles*, New York: Wiley.

McGowan, J. W., & Mitchell, J. B. A. 1984, "Electron–molecular postive-ion recombination," in *Electron–Molecule Interactions and Their Applications* Vol. 2, ed L. G. Christophorou, New York: Academic Press, pp. 65–88.

McGowan, J. W., Caudano, R., & Keyser, J. 1976, "Detailed study of recombination: the importance of the Rydberg state in electron–H_2^+ recombination," *Phys. Rev. Lett.* **36**, pp. 1447–1450.

McGowan, J. W., Mul, P., D'Angelo, V. S., Mitchell, J. B. A., DeFrance, P., & Froelich, H. R. 1979, "Energy dependence of dissociative recombination below 0.08 eV measured with (electron–ion) merged-beam technique," *Phys. Rev. Lett.* **42**, pp. 373–375.

McLain, J. L., Poterya, V., Molek, C. D., & Adams, N. G. 2006, "Determination of neutral product distributions for dissociative electron–ion recombination of hydrocarbon ions using a novel flowing afterglow mass spectrometric technique," *Abstract PHYS 303, 231st American Chemical Society National Meeting, Atlanta, Georgia, March 26–30, 2006.* Available through URL: www.acs.org

McLain, J. L., Poterya, V., Molek, C. D., Babcock, L. M., & Adams, N. G. 2004, "Flowing afterglow studies of the temperature dependencies for dissociative recombination of O_2^+, CH_5^+, $C_2H_5^+$, and $C_6H_7^+$," *J. Phys. Chem. A* **108**, pp. 6704–6708.

McLain, J. L., Poterya, V., Molek, C. D., Jackson, D. M., Babcock, L. M., & Adams, N. G. 2005, "$C_3H_3^+$ isomers: temperature dependencies of production in the H_3^+ reaction with allene and loss by dissociative recombination with electrons," *J. Phys. Chem. A* **109**, pp. 5119–5123.

McLennan, J. C., & Shrum, G. M. 1925, "On the origin of the auororal green line 5577 Å, and other spectra associated with the aurora borealis," *Proc. Roy. Soc. A* **108**, pp. 501–512.

McNab, I. R. 1995, "The spectroscopy of H_3^+," *Adv. Chem. Phys.* **89**, pp. 1–87.

Mehr, F. J., & Biondi, M. A. 1968, "Electron-temperature dependence of electron-ion recombination in argon," *Phys. Rev.* **176**, pp. 322–326.

1969, "Electron temperature dependence of recombination of O_2^+ and N_2^+ ions with electrons," *Phys. Rev.* **181**, pp. 264–271.

Mentzoni, M. H. 1963, "Effective electron recombination in heated nitrogen," *J. Geophys. Res.* **68**, pp. 4181–4186.

1965, "Electron removal during the early oxygen afterglow," *J. Appl. Phys.* **36**, pp. 57–61.

Meulenbroeks, R. F. G., van Beek, A. J., van Helvoort, A. J. G., van de Sanden, M. C. M., & Schram, D. C. 1994, "Argon–hydrogen plasma jet investigated by active and passive spectroscopic means," *Phys. Rev. E* **49**, pp. 4397–4406.

Michels, H. H. 1974, "Theoretical study of dissociative-recombination kinetics," USNTIS AD Rep. No 781195/3GA, AFWL-TR-73–288, pp. 1–127.

1975, "Calculation of energetics of selected atmospheric systems," *Air Force Cambridge Research Laboratory Report*, AFCRL-TR-75–0509, pp. 1–43.

1980, "Theoretical research investigation upon reaction rates to the nitric oxide positive ion," AFGL-TR-80–0072, no. AD-104303, pp. 1–60.

1981, "Electronic structure of excited states of selected atmospheric molecules," *The Excited State in Chemical Physics,* Part 2, Advances in Chemical Physics Vol. 45, ed. J. W. McGowan, New York: Wiley, pp. 225–340.

1984, "Dissociative recombination of $e + H_3^+$. An analysis of reaction product channels," in *International Symposium on the Production and Neutralization of Negative Ions and Beams*, AIP Conf. Proceedings, Vol. 111, New York: American Institute of Physics, pp. 118–124.

1989, "Potential energy curves for dissociative recombination of HeH^+," in *Dissociative Recombination: Theory, Experiment and Applications*, eds. J. B. A. Mitchell & S. L. Guberman, Singapore: World Scientific, pp. 97–108.

Michels, H. H., & Hobbs, R. H. 1984, "Low-temperature dissociative recombination of $e + H_3^+$," *Astrophys. J. Lett.* **286**, pp. L27–L29.

Miescher, E. 1966, "Absorption spectrum of NO molecule. Part VII. Extension of Rydberg series of ns, np, nd, and nf . . . complexes," *J. Mol. Spectrosc.* **20**, pp. 130–140.

Mikhailov, I. A., Kokoouline, V., Larson, Å., Tonzani, S., & Greene, C. H. 2006, "Renner–Teller effects in HCO^+ dissociative recombination," *Phys. Rev. A* **74**, pp. 032707-1–9.

Miku, O., 1978, "Electron temperature dependence of the dissociative recombination coefficient in krypton," *J. Phys. D: Appl. Phys.* **11**, pp. L39–L42.

Millar, T. J. 1983, "Dense cloud chemistry – II. The HCS^+/CS ratio," *Mon. Not. R. Astron. Soc.* **202**, pp. 683–689.

2003, "Deuterium fractionation in interstellar clouds," *Space Sci. Rev.* **106**, pp. 73–86.

2006, "What we know and what we need to know," in *Astrochemistry: Recent Successes and Current Challenges, IAU Symp. 231*, Cambridge: Cambridge University Press, pp. 77–86.

Millar, T. J., Adams, N. G., Smith, D., & Clary, D. C. 1985, "The HCS^+/CS abundance ratio in interstellar clouds," *Mon. Not. R. Astron. Soc.* **216**, pp. 1025–1031.

Millar, T. J., DeFrees, D. J., McLean, A. D., & Herbst, E. 1988, "The sensitivity of gas-phase models of dense interstellar clouds to changes in dissociative recombination branching ratios," *Astron. Astrophys.* **194**, pp. 250–256.

Miller, S., Tennyson, J., Lepp, S., & Dalgarno, A. 1992, "Identification of features due to H_3^+ in the infrared spectrum of supernova SN1987A," *Nature* **355**, pp. 420–421.

Minaev, B. & Larsson, M. 2002, "MCSCF linear response study of the three-body dissociative recombination $CH_2^+ + e \rightarrow C + 2H$," *Chem. Phys.* **280**, pp. 15–30.

Minh, Y. C. & Van Dishoeck, E. F. (eds.) 2000, *Astrochemistry: From Molecular Clouds to Planetary Systems, IAU Symp. 197*, San Francisco: ASP.

Mitchell, J. B. A. 1986, "Dissociative recombination of molecular ions," in *Atomic Processes in Electron–Ion and Ion–Ion Collisions*, ed. F. Brouillard, New York: Plenum Press, pp. 185–222.

  1990a, "The dissociative recombination of molecular ions," *Phys. Rep.* **186**, pp. 215–248.

  1990b, "Dissociative recombination in ion–electron collisions: new directions," in *Physics of Ion Impact Phenomena*, ed. D. Mathur, Berlin: Springer Verlag, pp. 275–286.

  1994, "New results for the dissociative recombination of H_3^+," *Bull. Am. Phys. Soc. Ser. 2* **39**, p. 1456.

  1995, "Electron–molecular ion collisions," in *Atomic and Molecular Processes in Fusion Edge Plasmas*, ed. R. K. Janev, New York: Plenum, pp. 225–262.

  1996, "Dissociative recombination and excitation in ITER divertor plasmas," in *Dissociative Recombination: Theory, Experiment and Applications III*, eds. D. Zajfman, J. B. A. Mitchell, D. Schwalm, & B. Rowe, Singapore: World Scientific, pp. 21–28.

Mitchell, J. B. A., & Guberman, S. L. (eds.) 1989, *Dissociative Recombination: Theory, Experiment, and Applications*, Singapore: World Scientific.

Mitchell, J. B. A., & Hus, H. 1985, "The dissociative recombination and excitation of CO^+," *J. Phys. B* **18**, pp. 547–555.

Mitchell, J. B. A., & McGowan, J. W. 1978, "The dissociative recombination of CH^+ $X\,^1\Sigma^+$ ($v = 0$)," *Astrophys. J. Lett.* **222**, pp. L77–L79.

  1983, "Experimental studies of electron–ion recombination," in *Physics of Ion–Ion and Electron–Ion Collisions*, eds. F. Brouillard & J. W. McGowan, New York: Plenum Press, pp. 279–324.

Mitchell, J. B. A., & Rebrion-Rowe, C. 1997, "The recombination of electrons with complex molecular ions," *Int. Rev. Phys. Chem.* **16**, pp. 201–213.

Mitchell, J. B. A., & Rowe, B. R. 2000, "Electron–molecule collisions: New experiments, new ideas," in *Dissociative Recombination: Theory, Experiment and Applications IV*, eds. M. Larsson, J. B. A. Mitchell, & I. F. Schneider, Singapore: World Scientific, pp. 240–250.

Mitchell, J. B. A., & Yousif, F. B. 1989a, "Merged beam studies of dissociative recombination – recent results," in *Dissociative Recombination: Theory, Experiment and Applications*, eds. J. B. A. Mitchell & S. L. Guberman, Singapore: World Scientific, pp. 109–123.

  1989b, "Molecular ion recombination: branching ratio measurements," in *Microwave and Particle Beam Sources and Directed Energy Concepts, SPIE Vol. 1061*, ed. H. E. Brandt, Bellingham, WA: Optical Society of America, pp. 536–541.

Mitchell, J. B. A., Lipson, R. H., & Sarpal, B. K. 2003, "Dissociative recombination of Xe_2^+ and XeH^+," in *Dissociative Recombination of Molecular Ions with Electrons*, ed. S. L. Guberman, New York: Kluwer/Plenum Publishers, pp. 59–66.

Mitchell, J. B. A., Forand, J. L., Ng, C. T., *et al.* 1983, "Measurement of the branching ratio for the dissociative recombination of $H_3^+ + e$," *Phys. Rev. Lett.* **51**, pp. 885–888.

Mitchell, J. B. A., Ng, C. T., Forand, L., Janssen, R., & McGowan, J. W. 1984, "Total cross sections for the dissociative recombination of H_3^+, HD_2^+ and D_3^+," *J. Phys. B* **17**, pp. L909–L913.

Mitchell, J. B. A., Novotny, O., Angelova, G., *et al.* 2005a, "Dissociative recombination of rare gas hydride ions: II. ArH^+," *J. Phys. B* **38**, pp. L175–L181.

Mitchell, J. B. A., Novotny, O., LeGarrec, J. L., *et al.* 2005b, "Dissociative recombination of rare gas hydride ions: I. NeH^+," *J. Phys. B* **38**, pp. 693–703.

Mitchell, J. B. A., Rebrion-Rowe, C., Le Garrec, J. L., *et al.* 2003, "Branching ratios for the dissociative recombination of hydrocarbon ions. I. The cases of $C_4H_9^+$ and $C_4H_5^+$," *Int. J. Mass Spectrom.* **227**, pp. 273–279.

Mitchell, J. B. A., Van der Donk, P., Yousif, F. B., & Morgan T. J. 1993, "Recent merged beams investigations of hydrogen molecular ion recombination," in *Dissociative Recombination: Theory, Experiment, and Applications*, eds. B. R. Rowe, J. B. A. Mitchell, & A. Canosa, NATO ASI Series B: Physics Vol. 313, New York: Plenum Press, pp. 87–97.

Moiseyev, N., & Corcoran, C. 1979, "Autoionizing states of H_2 and H_2^- using the complex-scaling method," *Phys. Rev. A* **20**, pp. 814–817.

Molek, C. D., Poterya, V., McLain, J. L., & Adams, N. G. 2006, "A novel technique for quantitative identification of neutral product distributions from dissociative electron-ion recombination," *Abstract PHYS 78, 231st American Chemical Society National Meeting, Atlanta, Georgia, March 26–30, 2006.* Available through URL: www.acs.org

Møller, S. 1991, "ASTRID," in *Conference Record of the 1991 IEEE Particle Accelerator Conference (San Francisco)*, ed. K. Berkner, IEEE, New York, pp. 2811–2813.

Møller, S. P. 1997, "ELISA, an electrostatic storage ring for atomic physics," *Nucl. Instr. Meth. Phys. Res. A* **394**, pp. 281–286.

Momozaki, Y., & El-Genk, M. S. 2002, "Dissociative recombination rate coefficient for low temperature equilibrium cesium plasma," *J. Appl. Phys.* **92**, pp. 690–697.

Montaigne, H., Geppert, W. D., Semaniak, J., *et al.* 2005, "Dissociative recombination of the thioformyl (HCS^+) and carbonyl sulfide (OCS^+) cations," *Astrophys. J.* **631**, pp. 653–659.

Mordaunt, D. H., Ashfold, M. N. R., & Dixon, R. N. 1994, "Dissociation dynamics of $H_2O(D_2O)$ following photoexcitation at the Lyman-α wavelength (121.6 nm)," *J. Chem. Phys.* **100**, pp. 7360–7375.

Morgan, L. A., Tennyson, J., & Gillan, C. J. 1998, "The UK molecular R-matrix codes," *Comp. Phys. Comm.* **114**, pp. 120–128.

Mostefaoui, T., Laube, S., Gautier, G., Rebrion- Rowe, C., Rowe, B. R., & Mitchell, J. B. A. 1999, "The dissociative recombination of NO^+: the influence of vibrational excitation state," *J. Phys. B* **32**, pp. 5247–5256.

Motapon, O., Fifirig, M., Florescu, A., *et al.* 2006, "Reactive collisions between electrons and NO^+ ions: rate coefficient computations and relevance for the air plasma kinetics," *Plasma Sources Sci. Technol.* **15**, pp. 23–32.

Mowat, J. R., Danared, H., Sundström, G., *et al.* 1995, "High-resolution, low-energy dissociative recombination spectrum of $^3HeH^+$," *Phys. Rev. Lett.* **74**, pp. 50–53.

Mul, P. M., & McGowan, J. W. 1979a, "Merged electron–ion beam experiments III. Temperature dependence of dissociative recombination for atmospheric ions NO^+, O_2^+ and N_2^+," *J. Phys. B* **12**, pp. 1591–1601.

1979b, "Dissociative recombination of N_2H^+ and N_2D^+," *Astrophys. J.* **227**, pp. L157–L159.

1980, "Dissociative recombination of C_2^+, C_2H^+, $C_2H_2^+$ and $C_2H_3^+$," *Astrophys. J.* **237**, pp. 749–751.

Mul, P. M., McGowan, J. W., Defrance, P., & Mitchell, J. B. A. 1983, "Merged electron–ion beam experiments: V. Dissociative recombination of OH^+, H_2O^+, H_3O^+ and D_3O^+," *J. Phys. B* **16**, pp. 3099–3107.

Mul, P. M., Mitchell, J. B. A., D'Angelo, V. S., *et al.* 1981, "Merged electron–ion beam experiments IV. Dissociative recombination for the methane group $CH^+, \ldots, CH_5^+$," *J. Phys. B* **14**, pp. 1353–1361.

Müller, U., & Cosby, P. C. 1996, Product state distributions in the dissociation of H_3 ($n = 2,3$) Rydberg states," *J. Chem. Phys.* **105**, pp. 3532–3550.

1999, "Three-body decay of the $3s\ ^2A'_1$ ($N = 1, K = 0$) and $3d\ ^2E''$ ($N = 1, G = 0, R = 1$) Rydberg states of the triatomic hydrogen molecule H_3," *Phys. Rev. A* **59**, pp. 3632–3642.

Müller, U., Eckert, Th., Braun, M., & Helm, H. 1999, "Fragment correlation in three-body breakup of triatomic hydrogen," *Phys. Rev. Lett.* **83**, pp. 2718–2721.

Mullikan, R. S. 1964, "Rare-gas and hydrogen molecule electronic states, noncrossing rule, and recombination of electrons with rare-gas and hydrogen ions," *Phys. Rev.* **136**, pp. A962–A965.

Nagaoka, A., Watanabe, N., & Kouchi, A. 2006, "Efficient formation of deuterated methanol by H-D substitution on interstellar grain surfaces," in *Astrochemistry – From Laboratory Studies to Astronomical Observations*, eds. R. I. Kaiser, P. Bernath, Y. Osamura, S. Petrie, & A. M. Mebel, AIP Conf. Proceedings Vol. 855, New York: American Institute of Physics, pp. 69–75.

Någård, M. B., Pettersson, J. B. C., Derkatch, A. M., *et al.* 2002, "Dissociative recombination of D^+ $(D_2O)_2$ water cluster ions with free electrons," *J. Chem. Phys.* **117**, pp. 5264–5270.

Nakamura, H. 1984, "Electronic transitions in atomic and molecular dynamic processes," *J. Phys. Chem.* **88**, pp. 4812–4823.

1991, "What are the basic mechanisms of electronic transitions in molecular dynamic processes?," *Int. Rev. Phys. Chem.* **10**, pp. 123–188.

Nakamura, H., Takagi, H., & Nakashima, K. 1989, "Theoretical study of dissociative recombination of electrons with H_2^+ and CH^+," in *Dissociative Recombination: Theory, Experiment and Applications*, eds. J. B. A. Mitchell, & S. L. Guberman, Singapore: World Scientific, pp. 73–83.

Nakashima, K., Takagi, H., & Nakamura, H. 1987, "Dissociative recombination of H_2^+, HD^+, and H_2^+ by collisions with slow electrons," *J. Chem. Phys.* **86**, pp. 726–737.

Nakashima, K., Nakamura, H., Achiba, Y., & Kimura, K. 1989, "Autoionization mechanism of the NO molecule: Calculation of quantum defect and theoretical analysis of multiphoton ionization experiment," *J. Chem. Phys.* **91**, pp. 1603–1610.

Neale, L., Miller, S., & Tennyson, J. 1996, "Spectroscopic properties of the H_3^+ molecule: A new calculated line list," *Astrophys. J.* **464**, pp. 516–520.

Neau, A., Al-Khalili, A., Rosén, S., *et al.* 2000, "Dissociative recombination of D_3O^+ and H_3O^+: Absolute cross sections and branching ratios," *J. Chem. Phys.* **113**, pp. 1762–1770.

Neau, A., Derkatch, A., Hellberg, F., *et al.* 2002, "Resonant ion pair formation of HD^+: Absolute cross sections for the $H^- + D^+$ channel," *Phys. Rev. A* **65**, pp. 044701-1–3.

Neumark, D. M. 2005, "Probing transition state with negative ion photodetachment: experiment and theory," *Phys. Chem. Chem. Phys.* **7**, pp. 433–442.

Ng, C.-Y. (ed.) 2000a, *Photoionization and Photodetachment, Part I*, Advanced Series in Physical Chemistry Vol. 10A, Singapore: World Scientific.

(ed.) 2000b, *Photoionization and Photodetachment, Part II*, Advanced Series in Physical Chemistry Vol. 10B, Singapore: World Scientific.

2003, "Absolute total state-selected cross sections for ion-molecule reactions of importance in planetary ionospheres. Reactions of O^+ (^4S, ^2D, ^2P)," in *Dissociative Recombination of Molecular Ions with Electrons*, ed. S. L. Guberman, New York: Kluwer/Plenum Publishers, pp. 401–414.

Ngassam, V., & Orel, A. E., 2006, "Dissociative recombination of Ne_2^+ molecular ions," *Phys. Rev. A* **73**, pp. 032720-1–10.

2007, "Resonances in low-energy electron scattering from $HCNH^+$," *Phys. Rev. A* **75**, pp. 062702-1–8.

Ngassam, V., Orel, A. E., & Suzor-Weiner, A. 2005, "*Ab initio* study of the dissociative recombination of $HCNH^+$," *J. Phys.: Conf. Ser.* **4**, pp. 224–228.

Ngassam, V., Florescu, A., Pichl, L., Schneider, I. F., Motapon, O., & Suzor-Weiner, A. 2003, "The short-range reaction matrix in MQDT treatment of dissociative recombination and related processes," *Eur. Phys. D* **26**, pp. 165–171.

Ngassam, V., Motapon, O., Florescu, A., Pichl, L., Schneider, I. F., & Suzor-Weiner, A. 2003, "Vibrational relaxation and dissociative recombination of H_2^+ induced by slow electrons," *Phys. Rev. A* **68**, pp. 032704-1–8.

Nicolet, M. 1954, "Origin of the oxygen green line in the airglow," *Phys. Rev.* **93**, p. 633.

Nielsen, S. B., Lappiere, A, Andersen, J. U., Pedersen, U. V., Tomita, S., & Andersen, L. H. 2001, "Absorption spectrum of the green fluorescent protein chromophore anion *in vacuo*," *Phys. Rev. Lett.* **87**, pp. 228102-1–4.

Nielsen, S. E., & Berry, R. S. 1971, "Dynamic coupling phenomena in molecular excited states. III. Associative ionization and dissociative recombination in H_2," *Phys. Rev. A* **4**, pp. 865–885.

Noren, C., Yousif, F. B., & Mitchell, J. B. A. 1989, "Dissociative recombination and excitation of N_2^+," *J. Chem. Soc. Faraday Trans. 2* **85**, pp. 1697–1703.

Novotný, O., Mitchell, J. B. A., LeGarrec, J. L., et al. 2005a, "The dissociative recombination of fluorocarbon ions: II. CF^+," *J. Phys. B* **38**, pp. 1471–1482.

Novotný, O., Plašil, R., Pysanenko, A., Korolov, I., & Glosík, J. 2006, "The recombination of D_3^+ and D_5^+ with electrons in deuterium containing plasma," *J. Phys. B* **39**, pp. 2561–2569.

Novotný, O., Sivaraman, B., Rebrion-Rowe, C., Travers, D., Mitchell, J. B. A., & Rowe, B. R. 2005b, "Measurement of the recombination of photoproduced PAH ions," *J. Phys.: Conf. Ser.* **4**, pp. 211–215.

Novotný, O., Sivaraman, B., Rebrion-Rowe, C., et al. 2005c, "Recombination of polycyclic aromatic hydrocarbon photoions with electrons in a flowing afterglow plasma," *J. Chem. Phys.* **123**, pp. 104303-1–6.

Oddene, S., Sheldon, J. W., Hardy, K. A., & Peterson, J. R. 1997, "Dissociative recombination of the $A\,^2\Pi_g$ and $X\,^2\Sigma_g$ states of N_2^+ in a glow discharge," *Phys. Rev. A* **56**, pp. 4737–4741.

Ogram, G. L., Chang, J.-S., & Hobson, R. M. 1980, "Dissociative recombination of H_3O^+ and D_3O^+ at elevated electron and gas temperatures," *Phys. Rev. A* **21**, pp. 982–989.

Öjekull, J., Andersson, P. U., Någård, M. B., et al. 2004 "Dissociative recombination of NH_4^+ and ND_4^+ ions: Storage ring experiments and *ab intio* molecular dynamics," *J. Chem. Phys.* **120**, pp. 7391–7399.

Öjekull, J., Andersson, P. U., Någård, M. B., et al. 2006, "Dissociative recombination of ammonia clusters studied by storage ring experiments," *J. Chem. Phys.* **125**, pp. 194306-1–9.

Oka, T. 1980, "Observation of the infrared spectrum of H_3^+," *Phys. Rev. Lett.* **45**, pp. 531–534.
  1983, "The H_3^+ ion," in *Molecular Ions: Spectroscopy, Structure and Chemistry*, eds. T. A. Miller & V. E. Bondybey, Amsterdam: North Holland, pp. 73–90.
  1992, "The infrared spectrum of H_3^+ in laboratory and space plasmas," *Rev. Mod. Phys.* **64**, pp. 1141–1149.
  2000, "H_3^+ in the diffuse interstellar medium: The enigma related to dissociative recombination," in *Dissociative Recombination: Theory, Experiment and Applications IV*, eds. M. Larsson, J. B. A. Mitchell, & I. F. Schneider, Singapore: World Scientific, pp. 13–24.
  2003a, "Help!!! Theory for H_3^+ recombination badly needed," in *Dissociative Recombination of Molecular Ions with Electrons*, ed. S. L. Guberman, New York: Kluwer/Plenum Publishers, pp. 209–220.
  2003b, "Microwave and infrared spectroscopy of molecular ions," in *The Encyclopedia of Mass Spectrometry*, eds. M. L. Gross, & R. Caprioli, Vol. 1, *Theory and Ion Chemistry*, ed. P. B. Armentrout, Amsterdam: Elsevier, pp. 217–226.
  (ed.) 2006a, "Physics, chemistry and astronomy of H_3^+," *Phil. Trans. R. Soc. A* **364**, pp. 2845–3151.
  2006b, "Correction and addendum to Oka 2000: introductory remarks," *Phil. Trans. R. Soc. A* **364**, pp. 3149–3151.
  2006c, "Interstellar H_3^+," *Proc. Natl. Acad. Sci. USA* **103**, pp. 12235–12242.
Okada, T., & Sugawara, M. 1993, "Microwave determination of the coefficient of dissociative recombination of Ar_2^+ in Ar afterglow," *J. Phys. D: Appl. Phys.* **26**, pp. 1680–1686.
O'Malley, T. F. 1966, "Theory of dissociative attachment," *Phys. Rev.* **150**, pp. 14–29.
  1971, "Diabatic states of molecules – quasistationary electronic states," *Adv. At. Mol. Phys.* **7**, pp. 223–249.
  1981, "Rydberg levels and structure in dissociative recombination cross sections," *J. Phys. B* **14**, pp. 1229–1238.
O'Malley, T. F., & Geltman, S. 1965, "Compound-atom states for two-electron systems," *Phys. Rev.* **137**, pp. A1344–A1352.
O'Malley, T. F., & Taylor, H. S. 1968, "Angular dependence of scattering products in electron–molecule resonant excitation and in dissociative attachment," *Phys. Rev.* **176**, pp. 207–221.
O'Malley, T. F., Cunningham, A. J., & Hobson, R. M. 1972, "Dissociative recombination at elevated temperatures II. Comparison between theory and experiment in neon and argon afterglows," *J. Phys. B* **5**, pp. 2126–2133.
O'Neil, R. R., Lee, E. T. P., & Huppi, E. R. 1979, "Auroral O (^1S) productions and loss processes: Ground-based measurements of the artificial auroral experiment Precede," *J. Geophys. Res.* **84**, pp. 823–833.
Oran, E. S., Julienne, P. S., & Strobel, D. F. 1975, "The aeronomy of odd nitrogen in the thermosphere," *J. Geophys. Res.* **80**, pp. 3068–3076.
Orel, A. E. 2000a, "Time dependent wave packet study of the direct low-energy dissociative recombination of HD^+," *Phys. Rev. A* **62**, pp. 020701-1–4.
  2000b, "Wave packet studies of dissociative recombination and dissociative excitation of molecular ions," in *Dissociative Recombination: Theory, Experiment and Applications IV*, eds. M. Larsson, J. B. A. Mitchell, & I. F. Schneider, Singapore: World Scientific, pp. 91–100.
  2005, "Wave packet studies of dissociative recombination," *J. Phys.: Conf. Ser.* **4**, pp. 142–147.

Orel, A. E., and Kulander, K. C. 1983, "Coherence effects in charge transfer collisions," *J. Chem. Phys.* **79**, pp. 1326–1333.
    1988, "Wave packet studies of molecular photofragmentation via strongly coupled electronic surfaces," *Chem. Phys. Lett.* **146**, pp. 428–433.
    1993, "Resonant dissociative recombination of H_3^+," *Phys. Rev. Lett.* **71**, pp. 4315–4318.
Orel, A. E., & Kulander, K. C. 1996, "Resonance-enhanced dissociation of a molecular ion below its electronic excitation level," *Phys. Rev. A* **54**, pp. 4992–4996.
Orel, A. E., Kulander, K. C., & Lengsfield, B. H. 1994, "Triple intersections of H_3 resonance states," *J. Chem. Phys.* **100**, pp. 1756–1758.
Orel, A. E., Kulander, K. C., & Rescigno, T. N. 1995, "Effects of open inelastic channels in the resonant dissociative recombination of HeH^+, *Phys. Rev. Lett.* **74**, pp. 4807–4810.
Orel, A. E., Rescigno, T. N., & Lengsfield, B. H. 1991, "Dissociative excitation of HeH^+ by electron impact," *Phys. Rev. A* **44**, pp. 4328–4335.
Orel, A. E., Schneider, I. F., & Suzor-Weiner, A. 2000, "Dissociative recombination of H_3^+: progress in theory," *Phil. Trans. R. Soc. Lond. A* **358**, pp. 2445–2456.
Orlov, D. A., Sprenger, F., Lestinsky, M., Weigel, U., Terekhov, Schwalm, D., & Wolf, A. 2005, "Photocathodes as electron sources for high resolution merged beam experiments," *J. Phys.: Conf. Ser.* **4**, pp. 290–295.
Orsini, N., Torr, N. G., Brinton, H. C., et al. 1977, "Determination of the N_2^+ recombination rate coefficient in the ionosphere," *Geophys. Res. Lett.* **4**, pp. 431–433.
Oskam, H. J. 1969, "Recombination of rare gas ions with electrons," in *Case Studies in Atomic Collision Physics I*, eds. E. W. McDaniel and M. R. C. McDowell, Amsterdam: North-Holland Publishing Company, pp. 465–523.
Oskam, H. J., & Mittelstadt, V. R. 1963, "Recombination coefficient of molecular rare-gas ions," *Phys. Rev.* **132**, pp. 1445–1454.
Österdahl, F. 2006, "Ionization of molecules at the CRYRING facility," Licentiate Thesis, the Royal Institute of Technology, Stockholm (TRITA-FYS-2006:42).
Österdahl, F., Rosén, S., Bednarska, V., Petrignani, A., Hellberg, F., Larsson, M., & van der Zande, W. J. 2005, "Position and time-sensitive coincident detection of fragments from the dissociative recombination of O_2^+ using a single hexanode delay-line detector," *J. Phys.: Conf. Ser.* **4**, pp. 286–289.
Paal, A., Simonsson, A., Källberg, A., Dietrich, J., & Mohos, I. 2003, "Current measurements of low-intensity beams at CRYRING," in *Proceedings 6th European Workshop on Beam Diagnostics and Instrumentation for Particle Accelerators (DIPAC 2003), Mainz, Germany*, pp. 240–241.
Parkhomchuk, V. V., & Skrinsky, A. N. 1991, "Electron cooling: physics and prospective applications," *Rep. Prog. Phys.* **54**, pp. 919–947.
Peart, B., & Dolder, K. T. 1971, "Collisions between electrons and H_2^+ I. Measurements of cross sections for proton production," *J. Phys. B* **4**, pp. 1496–1505.
    1972a, "Collisions between electrons and H_2^+ ions II. Measurements of cross sections for dissociative excitation," *J. Phys. B* **5**, pp. 860–865.
    1972b, "Collisions between electrons and H_2^+ ions III. Measurements of proton production cross sections at low energies," *J. Phys. B* **5**, pp. 1554–1558.
    1973a, "Measurements of cross sections for the dissociative recombination of D_2^+ ions," *J. Phys. B* **6**, pp. L359–L361.
    1973b, "Collisions between electrons and H_2^+ ions IV. Measurements of cross-sections for dissociative ionization," *J. Phys. B* **6**, pp. 2409–2414.
    1974a, "Collisions between electrons and H_2^+ ions V. Measurements of cross section for dissociative recombination," *J. Phys. B* **7**, pp. 236–243.

1974b, "The production of de-excited H_3^+ ions and measurements of the energies of two electronically-excited states," *J. Phys. B* **7**, pp. 1567–1573.

1974c, "Measurement of the dissociative recombination of H_3^+ ions," *J. Phys. B* **7**, pp. 1948–1952.

1975, "Collisions between electrons and H_2^+ ions VI. Measurements of cross-sections for simultaneous production of H^+ and H^-," *J. Phys. B* **8**, pp. 1570–1574.

Peart, B., Foster, R. A., & Dolder, K. T. 1979, "Measured cross sections for the formation of H^- by collisions between H_3^+ ions and electrons," *J. Phys. B* **12**, pp. 3441–3443.

Pechukas, P., & Light, J. C. 1965, "On detailed balancing and statistical theories of chemical kinetics," *J. Chem. Phys.* **42**, pp. 3281–3291.

Pedersen, H. B., Bilodeau, R., Jensen, M. J., Makassiouk, I. V., Safvan, C. P., & Andersen, L. H. 2001, "Electron collisions with the diatomic fluorine anion," *Phys. Rev. A* **63**, pp. 032718-1–7.

Pedersen, H. B., Buhr, H., Altevogt, S., et al. 2005, "Dissociative recombination and low-energy inelastic electron collisions of the helium dimmer ion," *Phys. Rev. A* **72**, pp. 012712-1–28.

Pedersen, H. B., Djurić, N., Jensen, M. J., et al. 1998, "Doubly charged negative ions of B_2 and C_2," *Phys. Rev. Lett.* **81**, pp. 5302–5305.

Pedersen, H. B., Djurić, N., Jensen, M. J., et al. 1999, "Electron collisions with diatomic anions," *Phys. Rev. A* **60**, pp. 2882–2899.

Peek, J. M. 1967, "Theory of dissociation of H_2^+ by fast electrons," *Phys. Rev.* **154**, pp. 52–56.

1974, "Theory of electron–H_2^+ collisions," *Phys. Rev. A* **10**, pp. 539–549.

Pegg, D. J. 2004, "Structure and dynamics of negative ions," *Rep. Prog. Phys.* **67**, pp. 857–905.

Penetrante, B. M., & Bardsley, J. N. 1986, "Electron heating in microwave-afterglow plasmas," *Phys. Rev. A* **34**, pp. 3253–3261.

Persson, K.-B., & Brown, S. C. 1955, Electron loss process in hydrogen afterglow," *Phys. Rev.* **100**, pp. 729–733.

Peterson, J. R., Devynck, P., Hertzler, C., & Graham, W. G. 1992, "Predissociation of H_3 $n = 2$ Rydberg states: product branching and isotope effectes," *J. Chem. Phys.* **96**, pp. 8128–8135.

Peterson, J. R., Le Padellec, A., Danared, H., et al. 1998, "Dissociative recombination and excitation of N_2^+: cross sections and product branching ratos," *J. Chem. Phys.* **108**, pp. 1978–1988.

Petrie, S. 2001, "Hydrogen isocyanide, HNC: A key species in the chemistry of Titan's atmosphere?," *Icarus* **151**, pp. 196–203.

Petrie, S., & Bohme, D. K. 2003, "Mass spectrometric approaches to interstellar chemistry," *Top. Curr. Chem.* **225**, pp. 37–75.

Petrignani, A., Andersson, P. U., Pettersson, J. B. C., et al. 2005a, "Dissociative recombination of the weakly bound NO-dimer cation: cross sections and three-body dynamics," *J. Chem. Phys.* **123**, pp. 194306-1–11.

Petrignani, A., Hellberg, F., Thomas, R. D., Larsson, M., Cosby, P. C., & van der Zande, W. J. 2005b, "Electron-energy dependent product state distributions in the dissociative recombination of O_2^+," *J. Chem. Phys.* **122**, pp. 234311-1–8.

Petrignani, A., van der Zande, W. J., Cosby, P. C., Hellberg, F., Thomas, R. D., & Larsson, M. 2005c, "Vibrationally resolved rate coefficients and branching fractions in the dissociative recombination of O_2^+," *J. Chem. Phys.* **122**, pp. 014302-1–11.

Peverall, R., Rosén, S., Larsson, M., et al. 2000, "The ionospheric oxygen green airglow: Electron temperature dependence and aeronomical implications," *Geophys. Res. Lett.* **27**, pp. 481–484.

Peverall, R., Rosén, S., Peterson, J. R., et al. 2001, "Dissociative recombination of excitation of O_2^+: cross sections, product yields and implications for studies of ionospheric airglows," *J. Chem. Phys.* **114**, pp. 6679–6689.

Phaneuf, R. A., Crandall, D. H., & Dunn, G. H. 1975, "Production of $D^*(n=4)$ from electron–D_2^+ dissociative recombination," *Phys. Rev. A* **11**, pp. 528–535.

Phaneuf, R. A., Havener, C. C., Dunn, G. H., & Müller, A. 1999, "Merged-beams experiments in atomic and molecular physics," *Rep. Prog. Phys.* **62**, pp. 1143–1180.

Phelps, A. V., & Brown, S. C. 1952, "Positive ions in the afterglow of a low pressure helium discharge," *Phys. Rev.* **86**, pp. 102–105.

Philbrick, J., Mehr, F. J., & Biondi, M. A. 1969, "Electron temperature dependence of recombination of Ne_2^+ ions with electrons," *Phys. Rev.* **181**, pp. 271–274.

Pichl, L., Nakamura, H., & Horacek, J. 2000, "Analytical treatment of singular equations in dissociative recombination," *Comp. Phys. Comm.* **124**, pp. 1–18.

Pineau des Forêts, G., & Roueff, E. 1993, "H_3^+ recombination and bistability in the interstellar medium," *Phil. Trans. R. Soc. Lond. A* **358**, pp. 2549–2559.

Plane, J. M. C. 2003, "Atmospheric chemistry of meteoric metals," *Chem. Rev.* **103**, pp. 4963–4984.

Plašil, R., Glosík, J., Poterya, V., et al. 2002, "Advanced stationary afterglow method for experimental study of recombination of processes of H_3^+ and D_3^+ ions with electrons," *Int. J. Mass Spectr.* **218**, pp. 105–130.

Plašil, R., Glosík, J., Poterya, V., Kudrna, P., Vicher, M., & Pysanenko, A. 2003, "Recombination of H_3^+ and D_3^+ with electrons," in *Dissociative Recombination of Molecular Ions with Electrons*, ed. S. L. Guberman, New York: Kluwer/Plenum Publishers, pp. 249–263.

Plašil, P., Hlavenka, P., Macko, P., Bánó, G., Pysanenko, A., & Glosík, J. 2005, "The recombination of spectroscopically identified H_3^+ ($v=0$) ions with electrons," *J. Phys.: Conf. Ser.* **4**, pp. 118–125.

Platzman, R. L. 1962a, "Superexcited states of molecules and the primary action of ionizing radiation," *Vortex* **23**, pp. 372–385.

1962b, "Superexcited states of molecules," *Radiat. Res.* **17**, pp. 419–425.

Plumb, I. C., Smith, D., & Adams, N. G. 1972, "Formation and loss of O_2^+ and O_4^+ ions in krypton–oxygen afterglow plasmas," *J. Phys. B* **5**, pp. 1762–1772.

Popović, D. B., Djurić, N., Holmberg, K., Neau, A., & Dunn, G. H. 2001, "Absolute cross sections for H^+ formation from electron-impact dissociation of C_2H^+ and $C_2H_2^+$," *Phys. Rev. A* **64**, pp. 052709-1–5.

Porter, R. N. 1982, "H_3 and H_3^+: Correlations of theory and experiment," *Ber. Bunsenges. Phys. Chem.* **86**, pp. 407–413.

Poterya, V., Glosík, J., Plašil, R., Tichý, M., Kudrna, P., & Pysanenko, A. 2002, "Recombination of D_3^+ ions in the afterglow of a He-Ar-D_2 plasma," *Phys. Rev. Lett.* **88**, pp. 044802-1–4.

Poterya, V., McLain, J. L., Adams, N. G., & Babcock, L. M. 2005, "Mechanisms of electron–ion recombination of N_2H^+/N_2D^+ and HCO^+/DCO^+ ions: Temperature dependence and isotopic effect," *J. Phys. Chem. A* **109**, pp. 7181–7186.

Prasad, S. S., & Capone, L. A. 1971, "The Jovian ionosphere: Composition and temperatures," *Icarus* **15**, pp. 45–55.

Pysanenko, A., Plašil, R., & Glosík, J. 2004, "The temperature dependence of electron–ion recombination in hydrogen plasma," *Czech. J. Phys.* **54** (Suppl. C), pp. C1042–C1049.

Quéffelec, J. L., Rowe, B. R., Morlais, M., Gomet, J. C., & Vallée, F. 1985, "The dissociative recombination of N_2^+ ($v=0,1$) as a source of metastable atoms in planetary atmospheres," *Planet. Space Sci.* **33**, pp. 263–270.

Quéffelec, J. L., Rowe, B. R., Vallée, F., Gomet, J. C., & Morlais, M. 1989, "The yield of metastable atoms through dissociative recombination of O_2^+ ions with electrons," *J. Chem. Phys.* **91**, pp. 5335–5342.

Rabadán, I., & Tennyson, J. 1996, "R-matrix calculation of the bound and continuum states of e^-–NO^+ system," *J. Phys. B* **29**, pp. 3747–3761.

1997, "*Ab initio* potential energy curves of Rydberg, valence and continuum states of NO," *J. Phys. B* **30**, pp. 1975–1988; *ibid.* 1998, Corrigendum, **31**, pp. 4485–4487.

Rai Dastidar, K., & Rai Dastidar, T. K. 1979, "Dissociative recombination of H_2^+, HD^+ and D_2^+ molecular ions," *J. Phys. Soc. Japan* **46**, pp. 1288–1294.

Ramos, G. B., Schlamkowitz, M., Sheldon, J., Hardy, K. A., & Peterson, J. R. 1995a, "Observation of dissociative recombination of Ne_2^+ and Ar_2^+ directly to the ground state of the product atoms," *Phys. Rev. A* **51**, pp. 2945–2950.

1995b, "Dissociative recombination studies of Ar_2^+ by time-of-flight spectroscopy," *Phys. Rev. A* **52**, pp. 4556–4566.

Rebrion-Rowe, C., Le Garrec, J. L., Hassouna, M., Travers, D., & Rowe, B. R. 2003, "Experimental evaluation of the recombination rate of cations formed from fluoranthene," *Int. J. Mass Spectrom.* **223–224**, pp. 237–251.

Rebrion-Rowe, C., Lehfoui, L. Rowe, B., & Mitchell, J. B.A 1998, "The dissociative recombination of hydrocarbon ions. II. Alkene and alkyne derived species," *J. Chem. Phys.* **108**, pp. 7185–7189.

Rebrion-Rowe, C., Mostefaoui, T., Laubé, S., Lehfaoui, L., & Mitchell, J. B. A. 2000b, "The recombination of hydrocarbon ions with electrons," in *Dissociative Recombination: Theory, Experiment and Applications IV*, eds. M. Larsson, J. B. A. Mitchell, & I. F. Schneider, Singapore: World Scientific, pp. 36–39.

Rebrion-Rowe, C., Mostefaoui, T., Laubé, S., & Mitchell, J. B. A. 2000a, "The dissociative recombination of hydrocarbon ions. III. Methyl-substituted benzene ring compounds," *J. Chem. Phys.* **113**, pp. 3039–3045.

Redfield, A., & Holt, R. B. 1951 "Electron removal in argon afterglows," *Phys. Rev.* **82**, pp. 874–876.

Rees, M. H. 1984, "Excitation of atomic oxygen (^1S) and emission of 5577 Å," *Planet. Space Sci.* **32**, pp. 373–378.

Rescigno, T. N., & McCurdy, C. W. 1998, "Improvements to the 'standard' complex Kohn variational method: Towards the development of an 'R-matrix theory without a box'," in *Novel Aspects of Electron–Molecule Collisions*, ed. K. H. Becker, Singapore: World Scientific, pp. 325–346.

Rescigno, T. N., Lengsfield, B. H., & McCurdy, C. W. 1995, "The incorporation of modern electronic structure methods in electron–molecule collision problems: variational calculations using the Complex Kohn method," in *Modern Electronic Structure Theory*, Vol. 1, ed. D. R. Yarkony, Singapore: World Scientific, pp. 501–588.

Rescigno, T. N., McCurdy, C. W., Orel, A. E., & Lengsfield, B. H. 1995, "The Complex Kohn variational method," in *Computational Methods for Electron–Molecule Collisions*, eds. W. M. Huo, & F. A. Gianturco, New York: Plenum Press, pp. 1–44.

Roberge, W., & Dalgarno, A. 1982, "The formation and destruction of HeH^+ in astrophysical plasmas," *Astrophys. J.* **255**, pp. 489–496.

Rogelstad, M. L., Yousif, F. B., Morgan, T. J., & Mitchell, J. B. A. 1997, "Stimulated radiative recombination of H^+ and He^+," *J. Phys. B* **30**, pp. 3913–3931.

Rogers, W. A., & Biondi, M. A. 1964, "Studies of the mechanism of electron–ion recombination. II," *Phys. Rev.* **134**, pp. A1215–A1225.

Rosati, R. E., Johnsen, R., & Golde, M. F. 2003, "Absolute yields of CO($a'\ ^3\Sigma^+$, $d\ ^3\Delta_i$, $e\ ^3\Sigma^-$) + O from dissociative recombination of CO_2^+ ions with electrons," *J. Chem. Phys.* **119**, pp. 11630–11635.

2004, "Yield of electronically excited N_2 molecules from the dissociative recombination of N_2H^+ with e^-," *J. Chem. Phys.* **120**, pp. 8025–8030.

Rosati, R. E., Pappas, D., Johnsen, R., & Golde, M. F. 2007, "Yield of electronically excited CN molecules from the dissociative recombination of HNC^+ with electrons," *J. Chem. Phys.* **126**, pp. 154303-1–8.

Rosén, S., Derkatch, A., Semaniak, J., et al. 2000, "Recombination of simple molecular ions studied in storage rings: dissociative recombination of H_2O^+," *Faraday Discuss.* **115**, pp. 295–302.

Rosén, S., Peverall, R., Larsson, M., et al. 1998a, "Absolute cross sections and final state distributions for dissociative recombination and excitation of CO^+ (v=0) using an ion storage ring," *Phys. Rev. A* **57**, pp. 4462–4471.

Rosén, S., Peverall, R., ter Horst, J., et al. 1998b, "A position- and time-sensitive particle detector with subnanosecond time resolution," *Hyperfine Interact.* **115**, pp. 201–208.

Ross, S., & Jungen, Ch. 1987, "Quantum-defect theory of double-minimum states in H_2," *Phys. Rev. Lett.* **59**, pp. 1297–1300.

Roth, M., Maul, C., & Gericke, K.-H. 2004, "Photodissociation dynamics of Cl_2O: Interpretation of electronic transitions," *J. Phys. Chem. A* **108**, pp. 7954–7964.

Roueff, E. 2005, "Microphysics and astrophysical observations: the molecular perspective," *J. Phys.: Conf. Ser.* **4**, pp. 1–9.

Roueff, E., & Gerin, M. 2003, "Deuterium molecules of the interstellar medium," *Space Sci. Rev.* **106**, pp. 61–72.

Roueff, E., & Pineau des Forêts, G. 1993, "Dissociative recombination in interstellar clouds," in *Dissociative Recombination: Theory, Experiment, and Applications*, eds. B. R. Rowe, J. B. A. Mitchell, & A. Canosa, NATO ASI Series B: Physics Vol. 313, New York: Plenum Press, pp. 249–261.

Roueff, E., Le Bourlot, J., & Pineau des Forêts, G. 1996, "Impact of dissociative recombination reactions on dark interstellar cloud models," in *Dissociative Recombination: Theory, Experiment and Applications III*, eds. D. Zajfman, J. B. A. Mitchell, D. Schwalm, & B. Rowe, Singapore: World Scientific, pp. 11–20.

Rowe, B. R., & Rebrion-Rowe, C. 1996, "Measurements of reaction rate constants of aromatic hydrocarbons with electrons," in *Dissociative Recombination: Theory, Experiment and Applications III*, eds. D. Zajfman, J. B. A. Mitchell, D. Schwalm, & B. Rowe, Singapore: World Scientific, pp. 184–194.

Rowe, B. R., Gomet, J. C., Canosa, C., & Mitchell, J. B. A. 1992, "A further study of HCO^+ recombination," *J. Chem. Phys.* **96**, pp. 1105–1110.

Rowe, B. R., Mitchell, J. B. A., & Canosa, A. (eds.) 1993, *Dissociative Recombination: Theory, Experiment, and Applications*, NATO ASI Series B: Physics, Vol. 313, New York: Plenum Press.

Rowe, B. R., Vallée, F., Quéffelec, J. L., Gomet, J. C., & Morlais, M. 1988, "The yield of oxygen and hydrogen atoms through dissociative recombination of H_2O^+ with electrons," *J. Chem. Phys.* **88**, pp. 845–850.

Royal, J., & Orel, A. E. 2005, "Dissociative recombination of He_2^+," *Phys. Rev. A* **72**, pp. 022719-1–8.

2006, "Dissociative recombination of Ar_2^+," *Phys. Rev. A* **73**, pp. 042706-1–12.

Rundel, R. D. 1972, "Proton production in collisions between electrons and H_2^+ ions," *J. Phys. B* **5**, pp. L77–L78.

Safvan, C. P., Jensen, M. J., Pedersen, H. B., & Andersen, L. H. 1999, "Dissociative recombination of the CO^{2+} dication," *Phys. Rev. A* **60**, pp. R3361–R3364.

Saito, M., Haruyama, Y., Tanabe, T., *et al.* 2000, "Vibrational cooling of H_2^+ and D_2^+ in a storage ring studied by means of two-dimensional fragment imaging," *Phys. Rev. A* **61**, pp. 062707-1–7.

Saltpeter, E. E. 1950, "Dissociative cross sections for fast hydrogen molecule ion," *Proc. Phys. Soc. A* **63**, pp. 1295–1297.

Sarpal, B. K., & Tennyson, J. 1993, "Calculated vibrational excitation rates for electron–H_2^+ collisions," *Mon. Not. R. Astron. Soc.* **263**, pp. 909–912.

Sarpal, B. K., Tennyson, J., & Morgan, L. 1994, "Dissociative recombination without a curve crossing: study of HeH^+," *J. Phys. B* **27**, pp. 5943–5953.

Sato, H. 2001, "Photodissociation of simple molecules in the gas phase," *Chem. Rev.* **101**, pp. 2687–2725.

2004, "Photodissociation in the gas phase," *Ann. Rep. Prog. Chem. Sec. C* **100**, pp. 73–98.

Sauer Jr., M. C., & Mulac, W. A. 1971, "Ion–electron and ion–ion recombination coefficients in gases studied by pulse radiolysis," *J. Chem. Phys.* **55**, pp. 1982–1983.

1972, "Studies of light emission in the pulse radiolysis of gases: Electron–ion recombination in gases," *J. Chem. Phys.* **56**, pp. 4995–5004.

1974, "Light emission resulting from ion-recombination in the pulse-radiolysis of argon containing naphthalene or anthracene," *Int. J. Rad. Phys. Chem.* **6**, pp. 55–65.

Sawicka, A., Skursi, P., Hudgins, R. R., & Simons, J. 2003, "Model calculations relevant to disulfide bond cleavage via electron capture influenced by positively charged groups," *J. Phys. Chem. B* **107**, pp. 13505–13511.

Sayers, J., & Kerr, L. W. 1957, "Ionic reactions in gases," in *Proceedings of the Third International Conference on Ionization Phenomena in Gases*, Venice: Italian Society of Physics, pp. 908–911.

Schennach, S., Müller, A., Uwira, O., *et al.* 1994, "Dielectronic recombination of lithium-like Ar^{15+}," *Z. Phys. D* **30**, pp. 291–306.

Schilke, P., Wamsley, C. M., Pineau des Forêts, G., Roueff, E., Flower, D. R., & Guilloteau, S. 1992, "A study of HCN, HNC, and their isotopomers in OMC-1 I. Abundances and chemistry," *Astron. Astrophys.* **256**, 595–612.

Schinke, R. 1993, *Photodissociation Dynamics*, Cambridge: Cambridge University Press.

Schneider, B. I., & Hay, P. J. 1976, "Elastic scattering of electrons for F_2: An R-matrix calculation," *Phys. Rev. A* **13**, pp. 2049–2056.

Schneider, I. F., & Orel, A. E. 1999, "Accurate nonadiabatic couplings for H_3: Application to predissociation," *J. Chem. Phys.* **111**, pp. 5873–5881.

Schneider, I. F., Dulieu, O., & Giusti-Suzor, A. 1991, "The role of Rydberg states in the H_2^+ dissociative recombination with slow electrons," *J. Phys. B* **24**, pp. L289–L297.

1992, "Resonances in the dissociative recombination of H_2^+ with slow electrons," *Phys. Rev. Lett.* **68**, p. 2251.

Schneider, I. F., Orel, A. E., & Suzor-Weiner, A. 2000, "Channel mixing effects in the dissociative recombination of H_3^+ with slow electrons," *Phys. Rev. Lett.* **85**, pp. 3785–3788.

Schneider, I. F., Larsson, M., Orel, A. E., & Suzor-Weiner, A. 2000a, "Dissociative recombination of H_3^+ and predissociation of H_3," in *Dissociative Recombination: Theory, Experiment and Applications IV*, eds. M. Larsson, J. B. A. Mitchell, & I. F. Schneider, Singapore: World Scientific, pp. 131–141.

Schneider, I. F., Rabadan, I., Carata, L., Andersen, L. H., Suzor-Weiner, A., & Tennyson, J. 2000b, "Dissociative recombination of NO^+: calculations and comparison with experiment," *J. Phys. B* **33**, pp. 4849–4861.

Schneider, I. F., Strömholm, C., Carata, L., Urbain, L., Larsson, M., & Suzor-Weiner, A. 1997, "Rotational effects in HD^+ dissociative recombination: theoretical study of resonant mechanisms and comparison with ion storage ring experiments," *J. Phys. B* **30**, pp. 2687–2705.

Schopman, J., Fournier, P. G., & Los, J. 1973, "The dissociation of 10 keV HeH^+ molecular ions. IV. Rotational predissociation of the $X\,^1\Sigma^+$ state," *Physica* **63**, pp. 518–526.

Schramm, A., Weber, J. M., Kreil, J., Klar, D., Ruf, M.-W., & Hotop, H. 1998, "Laser photoelectron attachment to molecules skimmed in a supersonic beam: diagnostics of weak electric fields and attachment cross sections down to 20 μeV," *Phys. Rev. Lett.* **81**, pp. 778–781.

Schulz, G. J. 1973, "Resonances in electron impact on diatomic molecules," *Rev. Mod. Phys.* **45**, pp. 423–486.

Schulz, G. J., & Asundi, R. K. 1967, "^{50}Isotope effect in the dissociative attachment in H_2 at low energy," *Phys. Rev.* **158**, pp. 25–29.

Schulz, P. A., Gregory, D. C., Meyer, F. W., & Phaneuf, R. A. 1986, "Electron-impact dissociation of H_3O^+," *J. Chem. Phys.* **85**, pp. 3386–3394.

Seiersen, K., Al-Khalili, A., Heber, O., et al. 2003c, "Dissociative recombination of the cation and dication of CO_2," *Phys. Rev. A* **68**, pp. 022708-1–6.

Seiersen, K., Bak, J., Bluhme, H., Jensen, M. J., Nielsen, S. B., & Andersen, L. H. 2003b, "Electron-impact detachment of O_3^-, NO_3^- and SO_2^- ions," *Phys. Chem. Chem. Phys.* **5**, pp. 4814–4820.

Seiersen, K., Heber, O., Jensen, M. J., Safvan, C. P., & Andersen, L. H. 2003a, "Dissociative recombination of dications," *J. Chem. Phys.* **119**, pp. 839–843.

Semaniak, J., Larson, Å., Le Padellec, A., et al. 1998, "Dissociative recombination of CH_5^+: absolute cross sections and branching fractions," *Astrophys. J.*, **498**, pp. 886–895.

Semaniak, J., Minaev, B. F., Derkatch, A. M., et al. 2001, "Dissociative recombination of $HCNH^+$: absolute cross-sections and branching ratios," *Astrophys. J. Suppl. Ser.* **135**, pp. 275–283.

Semaniak, J., Rosén, S., Sundström, G., et al. 1996, "Product-state distributions in the dissociative recombination of $^3HeD^+$ and $^4HeH^+$," *Phys. Rev. A* **54**, pp. R4617–R4620.

Sen, A., McGowan, J. W., & Mitchell, J. B. A. 1987, "Production of low-vibrational-state H_2^+ ions for collision studies," *J. Phys. B* **20**, pp. 1509–1515.

Seong, J., & Sun, H. 1996, "Dissociative recombination rates of O_2^+ ions with low energy electrons," *Bull. Korean Chem. Soc.* **17**, pp. 1065–1073.

Sexton, M. C., & Craggs, J. D. 1958, "Recombination in the afterglows of argon and helium using microwave techniques," *Int. J. Electr.* **4**, pp. 493–502.

Sham, T.-K. (ed.) 2002, *Chemical Applications of Synchrotron Radiation, Part I: Dynamics and VUV Spectroscopy*, Advanced Series in Physical Chemistry Vol. 12A, Singapore: World Scientific.

Shauer, S. N., Williams, P., & Compton, R. N. 1990, "Production of small doubly charged negative carbon cluster ions by sputtering," *Phys. Rev. Lett.* **65**, pp. 625–628.

Sheehan, C. H. 2000, Ph.D. thesis, University of Western Ontario.

Sheehan, C. H., & St.-Maurice, J.-P. 2004a, "Dissociative recombination of the methane family ions: rate coefficients and implications," *Adv. Space Res.* **33**, pp. 216–220.
  2004b, "Dissociative recombination of N_2^+, O_2^+, and NO^+: rate coefficients for ground state and vibrationally excited ions," *J. Geophys. Res.* **109**, pp. A03302-1–21.
Sheehan, C., Lennard, W. J., & Mitchell, J. B. A. 2000, "Measurement of the efficiency of a silicon surface barrier detector for medium energy ions using a Rutherford backscattering experiment," *Meas. Sci. Technol.* **11**, pp. L5–L7.
Sheehan, C., Le Padellec, A., Lennard, W. N., Talbi, D., & Mitchell, J. B. A. 1999, "Merged beam measurement of the dissociative recombination of HCN^+ and HNC^+," *J. Phys. B* **32**, pp. 3347–3360.
Shiba, Y., Hirano, T., Nagashima, U., & Ishii, K. 1998, "Potential energy surfaces and branching ratio of the dissociative recombination reaction $HCNH^+ + e^-$: an *ab initio* molecular orbital study," *J. Chem. Phys.* **108**, pp. 698–705.
Shiu, Y.-J., & Biondi, M. A. 1977, "Dissociative recombination in krypton: dependence of the total rate coefficient and excited-state production on electron temperature," *Phys. Rev. A* **16**, pp. 1817–1820.
  1978, "Dissociative recombination in argon: dependence of the total rate coefficient and excited-state production on electron temperature," *Phys. Rev. A* **17**, pp. 868–872.
Shiu, Y.-J., Biondi, M. A., & Sipler, D. P. 1977, "Dissociative recombination in xenon: variation of the total rate coefficient and excited-state production with electron temperature," *Phys. Rev. A* **15**, pp. 494–498.
Shy, J.-T., Farley, J. W., Lamb Jr., W. E., & Wing, W. H. 1980, "Observation of the infrared spectrum of the triatomic molecular deuterium molecular ion D_3^+," *Phys. Rev. Lett.* **45**, pp. 535–537.
Sidis, V. 1971, "Simple expression for the off-diagonal elements of the d/dR operator between exact electronic states of diatomic molecules," *J. Chem. Phys.* **55**, pp. 5838–5839.
Simonsson, A. 1991, *Beam Dynamics and Injection in CRYRING*, Ph.D. thesis, Royal Institute of Technology, Stockholm.
Singh, P. D. (ed.) 1992, *Astrochemistry of Cosmic Phenomena, IAU Symp. 150*, Dordrecht: Kluwer Academic Publishers.
Skrzypkowski, M. P., & Johnsen, R. 1997, "Electron-temperature dependence of the recombination of NH_4^+ (NH_3) ions with electrons," *Chem. Phys. Lett.* **274**, pp. 473–477.
Skrzypkowski, M. P., Gougousi, T., Johnsen, R., & Golde, M. F. 1998, "Measurement of the absolute yield of CO ($a\ ^3\Pi$) + O products in the dissociative recombination of CO_2^+ with electrons," *J. Chem. Phys.* **108**, pp. 8400–8407.
Slanger, T. G., & Black, G. 1982, "Photodissociative channels at 1216 Å for H_2O, NH_3, and CH_4," *J. Chem. Phys.* **77**, pp. 2432–2437.
Smirnov, B. M. 1977, "Cluster ions in gases," *Sov. Phys. Usp.* **20**, pp. 119–133.
Smith, D. 1992, "The ion chemistry of interstellar clouds," *Chem. Rev.* **92.**, pp. 1473–1485
Smith, D., & Adams, N. G. 1979, "Recent advances in flow tubes: Measurement of ion-molecule rate coefficients and product distributions," in *Gas Phase Ion Chemistry*, Vol. 1, ed. M. T. Bowers, New York: Academic Press, pp. 1–44.
  1983, "Studies of ion–ion recombination using flowing afterglow plasmas," in *Physics of Ion–Ion and Electron–Ion Collisions*, eds. F. Brouillard, & J. W. McGowan, New York: Plenum Press, pp. 501–531.
  1984, "Dissociative recombination coefficients for H_3^+, HCO^+, N_2H^+, and CH_5^+ at low temperature: interstellar implications," *Astrophys. J.* **284**, pp. L13–L16.

1987, "Ionic reactions in thermal plasmas," *J. Chem. Soc. Faraday Trans. II* **83**, pp. 149–157.

Smith, D., & Goodall, C. V. 1968, "The dissociative recombination coefficient of O_2^+ ions with electrons in the 180°–630° K," *Planet. Space Sci.* **16**, pp. 1177–1188.

Smith, D., & Španěl, P. 1993a, "Dissociative recombination of H_3^+ and some other interstellar ions: a controversy resolved," *Int. J. Mass Spectrom. Ion Proc.* **129**, pp. 163–182.

1993b, "Dissociative recombination of H_3^+. Experiment and theory reconciled," *Chem. Phys. Lett.* **211**, pp. 454–460.

1994, "Studies of electron attachment at thermal energies using the flowing afterglow-Langmuir probe technique," *Adv. At. Mol. Opt. Phys.* **32**, pp. 307–343.

Smith, D., Adams, N. G., & Ferguson, E. E. 1990, "Interstellar ion chemistry: laboratory studies," in *Molecular astrophysics*, ed. T. W. Hartquist, Cambridge: Cambridge University Press, pp. 181–210.

Smith, D., Adams, N. G., Dean, A. G., & Church, M. J. 1975, "The application of Langmuir probes to the study of flowing afterglow plasma," *J. Phys. D* **8**, pp. 141–152.

Smith, D., Goodall, C. V., Adams, N. G., & Dean, A. G. 1970, "Ion- and electron-density decay rates in afterglow plasmas of argon and argon–oxygen mixtures," *J. Phys. B* **3**, pp. 34–43.

Snyder, L. E., Hollis, J. M., Ulich, B. L., Lovas, F. J., & Buhl, D. 1975, "On the identification of interstellar X-ogen," *Bull. Am. Astron. Soc.* **7**, p. 497.

Solomon, P. M. 1973, "Interstellar molecules," *Physics Today* **32**(3), pp. 32–40.

Solomon, P. M., & Klemperer, W. 1972, "The formation of diatomic molecules in interstellar space," *Astrophys. J.* **178**, pp. 389–421.

Sonnenfroh, D. M., Caledonia, G. E., & Lurie, J. 1993, "Emission from OH(*A*) produced in the dissociative recombination of H_2O^+," *J. Chem. Phys.* **98**, pp. 2872–2881.

Španěl, P., & Smith, D. 1995, "Recent studies of electron attachment and electron–ion recombination at thermal energies," *Plasma Sources Sci. Technol.* **4**, pp. 302–306.

Španěl, P., Dittrichová, L., & Smith, D. 1993, "FALP studies of dissociative recombination coefficients for O_2^+ and N_2^+ within the electron temperature range 300–2000 K," *Int. J. Mass Spectrom. Ion Proc.* **129**, pp. 183–191.

Stamatovic, A., & Schultz, G. J. 1968, "Trochoidal electron monochromator," *Rev. Sci. Instr.* **39**, pp. 1752–1753.

1970, "Characteristics of the trochoidal electron monochromator," *Rev. Sci. Instr.* **41**, pp. 423–427.

Stancil, P. C., Lepp, A., & Dalgarno, A. 1996, "The lithium chemistry in the early universe," *Astrophys. J.* **458**, pp. 401–406.

Stearns, J. W., Berkner, K. H., Pyle, R. V., Briegleb, B. P., & Warren, M. L. 1971, "Dissociation cross sections for 0.5- to 1-MeV HeH^+ ions in H_2, He, N_2, and Ne gases," *Phys. Rev. A* **4**, pp. 1960–1964.

Stein, R. P., Scheibe, M., Syverson, M. W., Shaw, T. M., & Gunton, R. C. 1964, "Recombination coefficient of electrons with NO^+ ions in shock-heated air," *Phys. Fluids* **7**, pp. 1641–1650.

Stephens, J. A., & Greene, C. H. 1995, "Rydberg state dynamics of rotating, vibrating H_3 and the Jahn–Teller effect," *J. Chem. Phys.* **102**, pp. 1579–1591.

Sternberg, A., Dalgarno, A., & Lepp, S. 1987, "Cosmic-ray-induced photodestruction of interstellar molecules in dense clouds," *Astrophys. J.* **320**, pp. 676–682.

Stibbe, D. T., & Tennyson, J. 1996, "Time-delay matrix analysis of resonance in electron scattering: e^-–H_2 and H_2^+," *J. Phys. B* **29**, pp. 4267–4283.

Stolyarov, A. V., Pupyshev, V. I., & Child, M. S. 1997, "Analytical approximations for adiabatic and non-adiabatic matrix elements of homonuclear diatomic Rydberg states: applications to the singlet p-complex of the hydrogen molecule," *J. Phys. B* **30**, pp. 3077–3093.

Strasser, D., Lammich, L., Kreckel, H., et al. 2002a, "Breakup dynamics and the isotope effect in H_3^+ and D_3^+ dissociative recombination," *Phys. Rev. A* **66**, pp. 032719-1–13.

Strasser, D., Lammich, L., Kreckel., H., et al. 2004, "Breakup dynamics and isotope effects in D_2H^+ and H_2D^+ dissociative recombination," *Phys. Rev. A* **69**, pp. 064702-1–4.

Strasser, D., Lammich, L., Krohn, S., et al. 2001, "Two- and three-body kinematical correlation in the dissociative recombination of H_3^+," *Phys. Rev. Lett.* **86**, pp. 779–782.

Strasser, D., Levin, J., Pedersen, H. B., et al. 2002b, "Branching ratios in the dissociative recombination of polyatomic ions: the H_3^+ case," *Phys. Rev. A* **65**, pp. 010702-1–4.

Strasser, D., Levin, J., Pedersen, H. B., et al. 2003, "A model for calculating branching ratios in H_3^+ dissociative recombination," in *Dissociative Recombination of Molecular Ions with Electrons*, ed. S. L. Guberman, New York: Kluwer/Plenum Publishers, pp. 235–242.

Strasser, D., Urbain, X., Pedersen, H. B., et al. 2000, "An innovative approach to multiparticle three-dimensional imaging," *Rev. Sci. Instr.* **71**, pp. 3092–3098.

Strauss, C. E. M., & Houston, P. L. 1990, "Correlations without coincidence measurements: deciding between stepwise and concerted mechanisms for ABC $\to$ A + B + C," *J. Phys. Chem.* **94**, pp. 8751–8762.

Strömholm, C., Danared, H., Larson, Å., et al. 1997, "Imaging spectroscopy of recombination fragments of OH^+," *J. Phys. B* **30**, pp. 4919–4933.

Strömholm, C., Schneider, I. F., Sundström, G., et al. 1995, "Absolute cross sections for dissociative recombination of HD^+: comparison of experiment and theory," *Phys. Rev. A* **52**, pp. R4320–R4323.

Strömholm, C., Semaniak, J., Rosén, S., et al. 1996, "Dissociative recombination and dissociative excitation of $^4HeH^+$: absolute cross sections and mechanisms," *Phys. Rev. A* **54**, pp. 3086–3094.

Suits, A. G., & Continetti, R. E. 2001, *Imaging in Chemical Dynamics*, Washington, DC: American Chemical Society.

Sun, H., & Nakamura, H. 1990, "Theoretical study of the dissociative recombination of NO^+ with slow electrons," *J. Chem. Phys.* **93**, pp. 6491–6501.

Sundström, G., Datz, S., Mowat, J. R., et al. 1994a, "Direct dissociative recombination of ground state HeH^+," *Phys. Rev. A* **50**, pp. R2806–R2809.

Sundström, G., Mowat, J. R., Danared, H., et al. 1994b, "Destruction rate of H_3^+ by low energy electrons measured in a storage-ring experiment," *Science* **263**, pp. 785–787.

Surko, C. M., Gribakin, G. F., & Buckman, S. J. 2005, "Low-energy positron interactions with atoms and molecules," *J. Phys. B* **38**, pp. R57–R126.

Suzor-Weiner, A., & Schneider, I. F. 2001, "Mystery of an interstellar ion," *Nature* **412**, pp. 871–872.

Svendsen, A., El Ghazaly, M. O. A., & Andersen, L. H. 2005a, "Molecular size effects in NCO and NCS dianion resonances," *J. Chem. Phys.* **123**, pp. 114311-1–5.

Svendsen, A., Bluhme, H., El Ghazaly, M. O. A., Seiersen, K., Brønsted Nielsen, S., & Andersen, L. H. 2005b, "Tuning the continuum ground state energy of NO_2^{2-} by water molecules," *Phys. Rev. Lett.* **94**, pp. 223401-1–4.

Svendsen, A., Bluhme, H., Seiersen, K., & Andersen, L. H. 2004, "Electron scattering on OH^- $(H_2O)_n$ clusters ($n = 0$–4)," *J. Chem. Phys.* **121**, 4642–4649.

Syrstad, E. A., & Turuček, F. 2005, "Toward a general mechanism of electron capture dissociation," *J. Am. Chem. Soc.* **16**, pp. 208–224.

Tachikawa, H. 1999, "Reaction mechanism of the astrochemical electron capture reaction $HCN^+ + e^- \to HNC + H$: a direct ab initio dynamics study," *Phys. Chem. Chem. Phys.* **1**, pp. 4925–4930.

  2000, "Full dimensional *ab-initio* dynamics calculations of electron capture processes by the H_3O^+ ion," *Phys. Chem. Chem. Phys.* **2**, pp. 4327–4333.

Takagi, H. 1993, "Rotational effects in the dissociative recombination process of $H_2^+ + e$," *J. Phys. B* **26**, pp. 4815–4832.

  1996, "Theoretical study of dissociative processes in HD^+/H_2^+–e collisions with the energy up to 10 eV," in *Dissociative Recombination: Theory, Experiment and Applications III*, eds. D. Zajfman, J. B. A. Mitchell, D. Schwalm, & B. Rowe, Singapore: World Scientific, pp. 174–183.

  2004, "Theoretical study of dissociative recombination of HeH^+," *Phys. Rev. A* **70**, pp. 022709-1–10.

Takagi, H., & Nakamura, H. 1980, "Elastic scattering of electrons from H_2^+: phaseshifts, quantum defects and two-electron excited states," *J. Phys. B* **13**, pp. 2619–2632.

Takagi, T., Kosugi, N., & Le Dourneuf, M. 1991, "Dissociative recombination of CH^+," *J. Phys. B* **24**, pp. 711–732.

Takahashi, H., Clemesha, B. R., Batista, P. P., Sahai, Y., Abdu, M. A., & Muralikrishna, P. 1990, "Equatorial F-region OI 6300 Å and OI 5577 Å emission profiles observed by rocket-borne airglow photometers," *Planet. Space Sci.* **38**, pp. 547–554.

Taketsugu, T., Tajima, A., Ishii, K., & Hirano, T. 2004, "Ab initio direct trajectory simulation with nonadiabatic transitions of the dissociative recombination reaction $HCNH^+ + e^- \to HNC/HCN + H$," *Astrophys. J.* **608**, pp. 323–329.

Talbi, D. 2003, "Dissociative recombination of c-$C_3H_3^+$," in *Dissociative Recombination of Molecular Ions with Electrons*, ed. S. L. Guberman, New York: Kluwer/Plenum Publishers, pp. 203–208.

Talbi, D., & Ellinger, Y. 1993, "A theoretical study of the HCO^+ and HCS^+ electronic dissociative recombination," in *Dissociative Recombination: Theory, Experiment, and Applications*, eds. B. R. Rowe, J. B. A. Mitchell, & A. Canosa, NATO ASI Series B: Physics Vol. 313, New York: Plenum Press, pp. 59–66.

  1998, "Potential energy surfaces for the electronic dissociative recombination of $HCNH^+$: astrophysical implications on the HCN/HNC abundance ratio," *Chem. Phys. Lett.* **288**, pp. 155–164.

Talbi, D., Le Padellec, A., & Mitchell, J. B. A. 2000, "Quantum chemical calculations for the dissociative recombination of HCN^+ and HNC^+," *J. Phys. B* **33**, pp. 3631–3646.

Talbi, D., Pauzat, F., & Ellinger, Y. 1988, "Potential energy surfaces for dissociative recombination reactions of HCO^+ and HCS^+," *Chem. Phys.* **126**, pp. 291–300.

Talbi, D., Hickman, A. P., Pauzat, F., Ellinger, Y., & Berthier, G. 1989, "A tentative interpretation for the difference in the abundance ratios HCO^+/CO and HCS^+/CS in interstellar space," *Astrophys. J.* **339**, pp. 231–238.

Tanabe, T., Chida, K., Noda, K., & Watanabe, I. 2002, "An electrostatic storage ring for atomic and molecular science," *Nucl. Instr. Meth. Phys. Res. A* **482**, pp. 595–605.

Tanabe, T., Chida, K., Watanabe, T., *et al.* 2000, "Dissociative recombination at the TARN II storage ring," in *Dissociative Recombination: Theory, Experiment and Applications IV*, eds. M. Larsson, J. B. A. Mitchell, & I. F. Schneider, Singapore: World Scientific, pp. 170–179.

Tanabe, T., Katayama, I., Inoue, N., et al. 1993, "Dissociative recombination of HeH$^+$ at large center-of-mass energies," *Phys. Rev. Lett.* **70**, pp. 422–425.

Tanabe, T., Katayama, I., Inoue, N., et al. 1994, "Origin of the low-energy component and isotope effect on dissociative recombination of HeH$^+$ and HeD$^+$," *Phys. Rev. A* **49**, pp. R1531–R1534.

Tanabe, T., Katayama, I., Kamegaya, H., et al. 1995, "Dissociative recombination of HD$^+$ with an ultracold electron beam in a cooler ring," *Phys. Rev. Lett.* **75**, pp. 1066–1069.

Tanabe, T., Katayama, I., Kamegaya, H., et al. 1996, "Dissociative recombination of light molecular ions in the storage ring TARN II," in *Dissociative Recombination: Theory, Experiment and Applications III*, eds. D. Zajfman, J. B. A. Mitchell, D. Schwalm, & B. Rowe, Singapore: World Scientific, pp. 84–93.

Tanabe, T., Katayama, I., Ono, S., et al. 1998, "Dissociative recombination of HeH$^+$ isotopes with an ultra-cold electron beam from a superconducting electron cooler in a storage ring," *J. Phys. B* **31**, pp. L297–L303.

Tanabe, T., Noda, K., Honma, T., et al. 1991, "Electron cooling experiments at INS," *Nucl. Instr. Methods Phys. Res. A*, **307**, pp. 7–25.

Tanabe, T., Noda, K., Saito, M., Lee, S., Ito, Y., & Takagi, H. 2003, "Resonant neutral-particle emission in collisions of electrons and peptide ions in a storage ring," *Phys. Rev. Lett.* **90**, pp. 193201-1–4.

Tanabe, T., Noda, K., Saito, M., Starikov, E. B., & Tateno, M. 2004, "Regular threshold-energy increase with charge for neutral-particle emission in collisions of electrons and oligonucleotide anions," *Phys. Rev. Lett.* **93**, pp. 043201-1–4.

Tanabe, T., Noda, K., Saito, M., Takagi, H., Starikov, E. B., & Tateno, M. 2005, "Neutral-particle emission in collisions of electrons with biomolecular ions in an electrostatic storage ring," *J. Phys.: Conf. Ser.* **4**, pp. 239–244.

Tanabe, T., Takagi, H., Katayama, I., et al. 1999, "Evidence of superelastic electron collisions from H_2^+ studied by dissociative recombination using an ultracold electron beam from a storage ring," *Phys. Rev. Lett.* **83**, pp. 2163–2166.

Tanaka, H., & Sueko, O. 2001, "Mechanisms of electron transport in electrical discharges and electron collision cross sections," *Adv. At. Mol. Opt. Phys.* **44**, pp. 1–32.

Tanaka, K. 2003, "The origin of macromolecule ionization by laser irradiation," in *Le Prix Nobel, The Nobel Prizes 2002*, ed. T. Frängsmyr, Stockholm: Edita Nordstedts Tryckeri AB, pp. 197–217.

Tashiro, M., & Kato, S. 2002a, "Predissociation of H_3 2s Rydberg state: quantum dynamics study," *Chem. Phys. Lett.* **354**, pp. 14–19.

2002b, "Quantum dynamics study on predissociation of H_3 Rydberg states: Importance of indirect mechanism," *J. Chem. Phys.* **117**, pp. 2053–2062.

2003, "Quantum dynamical study of H_3^+," in *Dissociative Recombination of Molecular Ions with Electrons*, ed. S. L. Guberman, New York: Kluwer/Plenum Publishers, pp. 243–248.

Teloy, E., & Gerlich, D. 1974, "Intergral cross sections for ion–molecule reactions. I. The guided beam technique," *Chem. Phys.* **4**, pp. 417–427.

Tennyson, J. 1995, "Spectroscopy of H_3^+: planets, chaos and the Universe," *Rep. Prog. Phys.* **58**, pp. 421–476.

Tennyson, J., & Miller, S. 1994, "H_3^+: from first principles to Jupiter," *Cont. Phys.* **35**, pp. 105–116.

2000, "Spectroscopy of H_3^+ and its impact on astrophysics," *Spectrochim. Acta A* **57**, pp. 661–667.

Tennyson, J., & Sutcliffe, B. T. 1982, "The *ab initio* calculation of the vibrational–rotational spectrum of triatomic systems in the close-coupling approach, with KCN and H_2Ne as examples," *J. Chem. Phys.* **77**, pp. 4061–4072.

Tennyson, J., Kostini, M. A., Mussa, H. Y., Polyansky, O. L., & Prosmiti, R. 2000, "H_3^+ near dissociation: theoretical progress," *Phil. Trans. R. Soc. Lond. A* **358**, pp. 2419–2432.

Thaddeus, P., Guélin, M., & Linke, R. A. 1981, "Three new 'nonterrestrial' molecules," *Astrophys. J. Lett.* **246**, pp. L41–L45.

Thomas, R. 2004, "Branching between different decay channels in the dissociative recombination of poly-atomic molecules," *Phys. Scripta* **T110**, pp. 188–192.

Thomas, R., Ehlerding, A., Hellberg, F., *et al.* 2003, "Hot water from cold: dissociative recombination of $D_5O_2^+$," in *23rd International Conference on Photonic, Electronic and Atomic Collisions (XXIII ICPEAC, Stockholm, Sweden), Abstracts of Contributed Papers*, Vol. II, eds. J. Anton *et al.*, Stockholm: Universitetsservice US AB, p. Mo109.

Thomas, R., Rosén, S., Hellberg, F., *et al.* 2002, "Investigating the three-body fragmentation dynamics of water via dissociative recombination and theoretical modeling calculations," *Phys. Rev. A* **66**, pp. 032715-1–16.

Thomas, R. D., Ehlerding, A., Geppert, W., *et al.* 2005a, "The effect of bonding on the fragmentation of small systems," *J. Phys.: Conf. Ser.* **4**, pp. 187–190.

Thomas, R. D., Hellberg, F., Neau, A., *et al.* 2005b, "Three-body fragmentation dynamics of amidogen and methylene radicals via dissociative recombination," *Phys. Rev. A* **71**, pp. 032711-1–16.

Thomson, J. J. 1911, "Rays of positive electricity," *Phil. Mag.* **21**, pp. 225–249.

  1912, "Further experiments on positive rays," *Phil. Mag.* **24**, pp. 209–253.

  1913, *Rays of Positive Electricity and Their Application to Chemical Analyses*, first edn, London: Longman, Green and Co.

  1921, *Rays of Positive Electricity and Their Application to Chemical Analyses*, second edn, London: Longman, Green and Co.

  1934, "Heavy hydrogen," *Nature* **133**, pp. 280–281.

  1937, *Recollections and Reflections*, New York: The Macmillan Company.

Thomson, J. J., & Rutherford, E. 1896, "On the passage of electricity through gases exposed to Röntgen rays," *Phil. Mag.* **42**, pp. 392–407.

Tolliver, D. E., Kyrala, G. A., & Wing, W. H. 1979, "Observation of the infrared spectrum of the helium-hydride molecular ion $^4HeH^+$," *Phys. Rev. Lett.* **43**, pp. 1719–1722.

Tomashevsky, M., Herbst, E., & Kraemer, W. P. 1998, "Classical and quantum-mechanical calculations of the $HCO^+ + e \rightarrow CO(v) + H$," *Astrophys. J.* **498**, pp. 728–734.

Tomita, S., Andersen, J. U., Gottrup, C., Hvelplund, P., & Pedersen, U. V. 2001, "Dissociation energy for C_2 loss from fullerene cations in a storage ring," *Phys. Rev. Lett.* **87**, pp. 073401-1–4.

Torr, D. G., & Orsini, N. 1978, "The effect of N_2^+ recombination on the aeronomic determination of the charge exchange coefficient of O^+ (^2D) with N_2," *Geophys. Res. Lett.* **5**, pp. 657–659.

Torr, M. R., & Torr, D. G. 1979, "Recombination of NO^+ in the mid-latitude trough and the polar ionization hole," *J. Geophys. Res.* **84** (A8), pp. 4316–4329.

Torr, M. R., St.-Maurice, J. P., & Torr, D. G. 1977, "The rate coefficient for the $O^+ + N_2$ reaction in the ionosphere," *J. Geophys. Res.* **82**, pp. 4829–4833.

Trafton, L., Lester, D. F., & Thompson, K. L. 1989, "Unidentified emission lines in Jupiter's northern and southern 2 micron aurorae," *Astrophys. J.* **343**, pp. L73–L76.

Trajmar, S., & McConkey, J. W. 1994, "Benchmark measurements of cross sections for electronic collisions: analysis of scattered electrons," *Adv. At. Mol. Opt. Phys.* **33**, pp. 63–96.

Trajmar, S., Register, D. F., & Chutjian, A. 1983, "Electron scattering by molecules II. Experimental data and methods," *Phys. Rep.* **97**, pp. 219–356.

Tsuji, M., Nakamura, M., Nishimura, Y., & Obase, H. 1995, "Nascent rovibrational distribution of CO($A\ ^1\Pi$) produced in the recombination of CO_2^+ with electrons," *J. Chem. Phys.* **103**, pp. 1413–1421.

1998, "Nascent rovibrational distributions of CO($d\ ^3\Delta_i, e\ ^3\Sigma^-, a'\ ^3\Sigma^+$) produced in the dissociative recombination of CO_2^+ with electrons," *J. Chem. Phys.* **108**, pp. 8031–8038.

Turner, B. E. 1989, "Dissociative electronic recombination in astrophysics and astrochemistry," in *Dissociative Recombination: Theory, Experiment and Applications*, eds. J. B. A. Mitchell & S. L. Guberman, Singapore: World Scientific, pp. 329–340.

Turuček, F., Polášek, M., Frank, A. J., & Sadílek, M. 2000, "Transient hydrogen atom adducts to disulfides. Formation and energetics," *J. Am. Chem. Soc.* **122**, pp. 2361–2370.

Uggerud, E. 2004, "Electron capture dissociation of the disulfide bond–a quantum chemical model study," *Int. J. Mass Spectrom.* **234**, pp. 45–50.

Ullrich, J., Moshammer, R., Dorn, A., Dörner, R., Schmidt, L. P. H., & Schmidt-Böcking, H. 2003, "Recoil-ion and electron momentum spectroscopy: reaction-microscopes," *Rep. Prog. Phys.* **66**, pp. 1463–1545.

Unser, K. 1981, "A toroidal DC beam current transformer with high resolution," *IEEE Trans. Nucl. Sci.* **28**, pp. 2344–2346.

Urbain, X., Djuric, N., Safvan, C. P., *et al.*, 2005, "Storage ring study of dissociative recombination of He_2^+," *J. Phys. B* **38**, pp. 43–50.

Urbain, X., Safva, C. P., Jensen, M. J., & Andersen, L. H. 2000. "Storage ring studies of the dissociative recombination of He_2^+," in *Dissociative Recombination: Theory, Experiment and Applications IV*, eds. M. Larsson, J. B. A. Mitchell, & I. F. Schneider, Singapore: World Scientific, pp. 261–262.

Vâlcu, B., Schneider, I. F., Raoult, M., Strömholm, C., Larsson, M., & Suzor-Weiner, A., 1998, "Rotational effects in low energy dissociative recombination of diatomic ions," *Eur. Phys. J. D* **1**, pp. 71–78.

Vallée, F., Rowe, B. R., Gomet, J. C., Quéffelec, J. L., & Morlais, M. 1986, "Observation of the fourth positive system of CO in dissociative recombination of vibrationally excited CO_2^+," *Chem. Phys. Lett.* **124**, pp. 317–320.

Van der Donk, P., Yousif, F. B., & Mitchell, J. B. A. 1991, "Dissociative recombination and excitation of D_3^+," *Phys. Rev. A* **43**, pp. 5971–5974.

Van der Donk, P., Yousif, F. B., Mitchell, J. B. A., & Hickman, A. P. 1991, "Dissociative recombination of H_2^+," *Phys. Rev. Lett.* **67**, pp. 42–45.

Van der Zande, W. J. 2000, "Dissociative recombination of diatomics: do we understand product state branching?" in *Dissociative Recombination: Theory, Experiment and Applications IV*, eds. M. Larsson, J. B. A. Mitchell, & I. F. Schneider, Singapore: World Scientific, pp. 251–260.

Van der Zande, W. J., Semaniak, J., Zengin, V., *et al.* 1996, "Dissociative recombination: product information and very large cross sections of vibrationally excited H_2^+," *Phys. Rev. A* **54**, pp. 5010–5018.

Van Dishoeck, E. F. 1990, "Diffuse cloud chemistry," in *Molecular Astrophysics*, ed. T. W. Hartquist, Cambridge: Cambridge University Press, pp. 55–83.

(ed.) 1997, *Molecules in Astrophysics: Probes and Processes, IAU Symp. 178*, Dordrecht: Kluwer Academic Publishers.

Van Dishoeck, E. F., & Black, J. H. 1986, "Comprehensive models of diffuse interstellar clouds: Physical conditions and molecular abundances," *Astrophys. J. Suppl. Ser.* **62**, pp. 109–145.

Van Hemert, M. C., & Peyerimhoff, S. D. 1991, "Resonances and bound rovibrational levels in the interacting X, A, C, and D states of HeH, HeD, ^{3}HeH, and ^{3}HeD," *J. Chem. Phys.* **94**, pp. 4369–4383.

Vanroose, W., McCurdy, C. W., & Rescigno, T. N. 2002, "Interpretation of low-energy electron-CO_2 scattering," *Phys. Rev. A* **66**, pp. 032720-1–10.

Vardya, M. S., & Tarafdar, S. P. (eds.) 1987, *Astrochemistry, IAU Symp. 120*, Dordrecht: Reidel Publishing Company.

Vejby-Christensen, L., Andersen, L. H., Heber, O., et al. 1997, "Complete branching ratios for the dissociative recombination of H_2O^+, H_3O^+, and CH_3^+," *Astrophys. J.* **483**, pp. 531–540.

Vejby-Christensen, L., Kella, D., Pedersen, H. B., & Andersen, L. H. 1998, "Dissociative recombination of NO^+," *Phys. Rev. A* **57**, pp. 3627–3634.

Viggiano, A. A., Ehlerding, A., Arnold, S. T., & Larsson, M. 2005a, "Dissociative recombination of hydrocarbon ions," *J. Phys.: Conf. Ser.* **4**, pp. 191–197.

Viggiano, A. A., Ehlerding, A., Hellberg, F., et al. 2005b, "Rate constants and branching ratios for the dissociative recombination of CO_2^+," *J. Chem. Phys.* **122**, pp. 226101-1–3.

Vikor, L., Al-Khalili, A., Danared, H., et al. 1999, "Branching fractions in the dissociative recombination of NH_4^+ and NH_2^+ molecular ions," *Astron. Astrophys.* **344**, pp. 1027–1033.

Vinci, A., & Tennyson, J. 2004, "Continuum states of CO^+," *J. Phys. B* **37**, pp. 2011–2031.

Vogler, M., & Dunn, G. H. 1975, "Dissociative recombination of electrons and D_2^+ to yield $D(2p)$," *Phys. Rev. A* **11**, pp. 1983–1987.

Vogt, E. & Wannier, G. H. 1954, "Scattering of ions by polarization forces," *Phys. Rev.* **95**, pp. 1190–1198.

von Busch, F., & Dunn, G. H. 1972, "Photodissociation of H_2^+ and D_2^+: experiment," *Phys. Rev. A* **5**, pp. 1726–1743.

Vuitton, V., Yelle, R. V., & Anicich, V. G. 2006, "The nitrogen chemistry of Titan's upper atmosphere revealed," *Astrophys. J. Lett.* **647**, pp. L175–L178.

Wahlgren, U., Liu, B., Pearson, P. K., & Schaefer, H. F. 1973, "Theoretical support for the assignment of X-ogen to the oxomethylium molecular ion," *Nature Phys. Sci.* **246**, pp. 4–6.

Waite Jr., J. H., Niemann, H., Yelle, R. V., et al. 2005, "Ion neutral mass spectrometer results from the first flyby of Titan," *Science* **308**, pp. 982–985.

Wallis, M. K. 1978, "Exospheric density and escape fluxes of atomic isotopes from Venus and Mars," *Planet. Space Sci.* **26**, pp. 949–953.

Walls, F. L., & Dunn, G. H. 1974, "Measurement of total cross sections for electron recombination with NO^+ and O_2^+ using ion storage techniques," *J. Geophys. Res.* **79**, pp. 1911–1915.

Wang, X.-B., Yang, X., Nicholas, J. B., & Wang, L.-S. 2001, "Bulk-like features in the photoemission spectra of hydrated doubly charged anion clusters," *Science* **294**, pp. 1322–1325.

Warke, C. S. 1966, "Nonradiative dissociative electron capture by molecular ions," *Phys. Rev.* **144**, pp. 120–126.

Watson, W. D. 1973, "Rate of formation of interstellar molecules by ion-molecule reactions," *Astrophys. J.* **183**, pp. L17–L20.

   1974, "Ion–molecule reactions, molecule formation, and hydrogen-isotope exchange in dense interstellar clouds," *Astrophys. J.* **188**, pp. 35–42.

Wauchop, T. S., & Broida, H. P. 1972, "Lifetime and quenching of CO ($a\,^3\Pi$) produced by recombination of CO_2 ions in a helium afterglow," *J. Chem. Phys.* **56**, pp. 330–332.

Wayne, R. P. 2000, *Chemistry of Atmospheres*, third edn, Oxford: Oxford University Press.
Weiner, J., Masnou-Seeuws, F., & Giusti-Suzor, A. 1989, "Associative ionization: experiment, potentials and dynamics," *Adv. At. Mol. Opt. Phys.* **26**, pp. 210–296.
Weller, C. S., & Biondi, M. A. 1967, "Measurements of dissociative recombination of CO_2^+ with electrons," *Phys. Rev. Lett.* **19**, pp. 59–61.
  1968, "Recombination, attachment, and ambipolar diffusion of electrons in photo-ionized NO afterglows," *Phys. Rev.* **172**, pp. 198–206.
Wendt, G. L., & Landauer, R. S. 1920, "Triatomic hydrogen," *J. Am. Chem. Soc.* **42**, pp. 930–946.
Wenthold, P. G., & Lineberger, W. C. 1999, "Negative ion photoelectron spectroscopy studies of organic reactive intermediates," *Acc. Chem. Res.* **32**, pp. 597–604.
Wheeler, M. D., Orr-Ewing, A. J., & Ashfold, M. N. R. 1997, "Predissociation dynamics of the $A\,^2\Sigma^+$ state of SH and SD," *J. Chem. Phys.* **107**, pp. 7591–7600.
Whitaker, M., Biondi, M. A., & Johnsen, R. 1981a, "Electron-temperature dependence of dissociative recombination of electrons with $CO^+ \cdot (CO)_n$ series ions," *Phys. Rev. A* **23**, pp. 1481–1485.
  1981b, "Electron-temperature dependence of dissociative recombination of electrons with $N_2^+ \cdot N_2$ ions," *Phys. Rev. A* **24**, pp. 743–745.
Wigner, E. P. 1948, "On the behavior of cross sections near thresholds," *Phys. Rev.* **73**, pp. 1002–1009.
Wilkins, R. L. 1966, "Monte Carlo calculations of cross sections of electron–positive-molecular-ion dissociative recombination," *J. Chem. Phys.* **44**, pp. 1884–1888.
Williams, T. L., Adams, N. G., Babcock, L. M., Herd, C. R., & Geoghegan, M. 1996, "Production and loss of the water-related species H_3O^+, H_2O and OH in dense interstellar clouds," *Mon. Not. R. Astron. Soc.* **282**, pp. 413–420.
Wilson, H. A. 1931, "Electrical conductivity of flames," *Rev. Mod. Phys.* **3**, pp. 156–189.
Wilson, L. N., & Evans, E. W. 1967, "Electron recombination in hydrogen–oxygen reactions behind shock waves," *J. Chem. Phys.* **46**, pp. 859–863.
Winstead, C., & McKoy, V. 1996, "Highly parallel computational techniques for electron-molecule collisions," *Adv. At. Mol. Opt. Phys.* **36**, pp. 183–219.
  2000, "Electron-molecule collisions in low-temperature plasmas: the role of theory," *Adv. At. Mol. Opt. Phys.* **43**, pp. 111–145.
Witase, O., Dutuit, O., Jilensten, J., *et al.* 2002, "Prediction of a CO_2^{2+} layer in the atmopshere of Mars," *Geophys. Res. Lett.* **29**, pp. 104-1–4.
Witase, O., Dutuit, O., Jilensten, J., *et al.* 2003, "Correction to 'Prediction of a CO_2^{2+} layer in the atmosphere of Mars'," *Geophys. Res. Lett.* **30**, p. 12-1.
Wolf, A., Lammich, L., & Schmelcher (eds) 2005, "Sixth International conference on Dissociative Recombination: Theory, Experiments and Applications (DR2004)," *J. Phys.: Conf. Ser.* **4**, pp. 1–299.
Wolf, A., Kreckel, H., Lammich, L., *et al.* 2006, "Effects of molecular rotation in low-energy electron collisions of H_3^+," *Phil. Trans. R. Soc. A* **364**, pp. 2981–2997.
Wolf, A., Lammich, L., Strasser, D., *et al.* 2004, "Storage ring experiments with cold molecular ions: the H_3^+ puzzle," *Phys. Scripta* **T110**, pp. 193–199.
Woods, R. C., Dixon, T. A., Saykally, R. J., & Szanto, P. G. 1975, "Laboratory microwave spectrum of HCO^+," *Phys. Rev. Lett.* **35**, pp. 1269–1271.
Wooten, A., Boulanger, F., Bogey, M., *et al.* 1986, "A search for interstellar H_3O^+," *Astron. Astrophys.* **166**, pp. L15–L18.
Yamasaki, K., Okada, S., Koshi, M., & Matsui, H. 1991, "Selective product channels in the reaction $NH(a\,^1\Delta)$ and $NH(X\,^3\Sigma^-)$ with NO," *J. Chem. Phys.* **95**, pp. 5087–5096.

Yee, J.-H., & Killeen, T. L. 1986, "Thermospheric production of O (^{1}S) by dissociative recombination of vibrationally excited O_2^+," *Planet. Space Sci.* **34**, pp. 1101–1107.

Yee, J.-H., Abeu, V. J., & Colwell, W. B. 1989, "Aeronomical determination of the quantum yields of O (^{1}S) and O (^{1}D) from dissociative recombination of O_2^+," in *Dissociative Recombination: Theory, Experiment and Applications*, eds. J. B. A. Mitchell, & S. L. Guberman, Singapore: World Scientific, pp. 286–302.

Young, R. A., & St. John, G. 1966, "Recombination coefficient of NO^+ with e," *Phys. Rev.* **152**, pp. 25–28.

Yousif, F. B., & Mitchell, J. B. A. 1988, "The dissociative recombination of HeH^+," *Bull. Am. Phys. Soc. Ser. 2* **33**, p. 1010.

1989, "Recombination and excitation of HeH^+," *Phys. Rev. A* **40**, pp. 4318–4321.

1995, "Electron-impact dissociative excitation of H_2^+: low energy studies," *Z. Physik D* **34**, pp. 195–197.

Yousif, F. B., Rogelstadt, M., & Mitchell, J. B. A. 1995, "Rydberg state formation in H_3^+ formation," in *Atomic and Molecular Physics: 4th US/Mexico Symposium*, eds. I. Alvarez, C. Cisneros, & T. J. Morgan, Singapore: World Scientific, pp. 343–351.

Yousif, F. B., Van der Donk, P., & Mitchell, J. B. A. 1993, "Ion-pair formation in the dissociative recombination of H_3^+," *J. Phys. B* **26**, pp. 4249–4255.

Yousif, F. B., Mitchell, J. B. A., Rogelstadt, M., Le Padellec, A., Canosa, A., & Chibisov, M. I. 1994, "Dissociative recombination of HeH^+: A reexamination," *Phys. Rev. A* **49**, pp. 4610–4615.

Yousif, F. B., Van der Donk, P. J. T., Orakzai, M., & Mitchell, J. B. A. 1991, "Dissociative excitation and recombination of H_3^+," *Phys. Rev. A* **44**, pp. 5653–5658.

Zajfman, D., Heber, O., & Strasser, D. 2003, "Time resolved cameras," in *Imaging in Molecular Dynamics*, ed. B. Whitaker, Cambridge: Cambridge University Press, pp. 122–137.

Zajfman, D., Schwalm, D., & Wolf, A. 2003, "Molecular physics in storage rings: From laboratory to space," *Hyperfine Interact.* **146/147**, pp. 265–268.

Zajfman, D., Amitay, Z., Broude, C., et al. 1995, "Measurement of branching ratios for the dissociative recombination of cold HD^+ using fragment imaging," *Phys. Rev. Lett.* **75**, pp. 814–817.

Zajfman, D., Amitay, Z., Lange, M., et al. 1997, "Curve crossing and branching ratios in the dissociative recombination of HD^+," *Phys. Rev. Lett.* **79**, pp. 1829–1832.

Zajfman, D., Graber, T., Kanter, E. P., et al. 1991, "Measurement of the distribution of bond angles in H_2O^+," *J. Chem. Phys.* **94**, pp. 2543–2547.

Zajfman, D., Mitchell, J. B. A., Schwalm, D., & Rowe, B. R. (eds.) 1996, *Dissociative Recombination: Theory, Experiment, and Applications III*, Singapore: World Scientific.

Zajfman, D., Strasser, D., Lammich, L., et al. 2003, "Breakup dynamics in H_3^+ and D_3^+ dissociative recombination," in *Dissociative Recombination of Molecular Ions with Electrons*, ed. S. L. Guberman, New York: Kluwer/Plenum Publishers, pp. 265–274.

Zajfman, D., Wolf, A., Schwalm, D., et al. 2005, "Physics with colder molecular ions: The Heidelberg cryogenic storage ring CSR," *J. Phys.: Conf. Ser.* **4**, pp. 296–299.

Zhang, J. Z. H., & Miller, W. H. 1987, "New method for quantum reactive scattering, with applications to the 3-D $H+H_2$ reaction," *Chem. Phys. Lett.* **140**, pp. 329–337.

Zhaunerchyk, V., Ehlerding, A., Geppert, W. D., et al. 2004, "Dissociative recombination study of Na^+ (D_2O) in a storage ring," *J. Chem. Phys.* **121**, pp. 10483–10488.

Zhaunerchyk, V., Geppert, W. D., Larsson, M., et al. 2007, "Three-body breakup in dissociative recombination of the covalent triatomic molecular ion O_3^+," *Phys. Rev. Lett.* **98**, pp. 223201-1–4.

Zhaunerchyk, V., Hellberg, F., Ehlerding, A., et al. 2005, "Dissociative recombination study of PD_2^+ at CRYRING: absolute cross-section, chemical branching ratios and three-body fragmentation dynamics," *Mol. Phys.* **103**, pp. 2735–2745.

Zhdanov, V. P. 1980, "Dissociative recombination of e–H_2^+ collisions," *J. Phys. B* **13**, pp. L311–L313.

Zhdanov, V. P., & Chibisov, M. I. 1978, "Dissociative recombination of electrons with molecular hydrogen (+1) and molecular deuterium (+1) ions with the formation of highly excited atoms," *Zh. Eksp. Teor. Fiz.* **74**, pp. 75–85.

Zipf, E. C. 1970, "The dissociative recombination of O_2^+ ions into specifically identified final atomic states," *Bull. Amer. Phys. Soc.* **15**, p. 418.

1978, "N (^2P) and N (^2D) atoms: their production by e-impact dissociation of N_2 and destruction by associative ionization," *Eos Trans AGU* **59**, p. 336.

1979, "The OI (^1S) state: Its quenching by O_2 and formation by the dissociative recombination of vibrationally excited O_2^+ ions," *Geophys. Res. Lett.* **6**, pp. 881–884.

1980a, "The dissociative recombination of vibrationally excited N_2^+ ions," *Geophys. Res. Lett.* **7**, pp. 645–648.

1980b, "A laboratory study of the dissociative recombination of vibrationally excited O_2^+ ions," *J. Geophys. Res. A* **85**, pp. 4232–4236.

1984, "Dissociation of molecules by electron impact," in *Electron-Molecule Interactions and Their Applications*, Vol. 1, ed. L. G. Christophorou, New York: Academic Press, pp. 335–401.

1988, "The excitation of the O (^1S) state by dissociative recombination of O_2^+ ions: electron temperature dependence," *Planet. Space Sci.* **36**, pp. 621–628.

Ziurys, L. M., & Turner, B. E. 1986, "$HCNH^+$: A new interstellar molecular ion," *Astrophys. J. Lett.* **302**, pp. L31–L36 (1986).

Zong, W., Dunn, G. H., Djurić, N., et al. 1999, "Resonant ion pair formation in electron collisions with ground state molecular ions," *Phys. Rev. Lett.* **83**, pp. 951–954.

Zubarev, R. A. 2003, "Reactions of polypeptide ions with electrons in the gas phase," *Mass Spectrom. Rev.* **22**, pp. 57–77.

Zubarev, R. A., Kelleher, N. L., & McLafferty, F. W. 1998, "Electron capture dissociation of multiply charged protein cations. A nonergodic process," *J. Am. Chem. Soc.* **120**, pp. 3265–3266.

Zubarev, R. A., Haselman, K. F., Budnik, B., Kjeldsen, F., & Jensen, F. 2002, "Towards an understanding of the mechanism of electron-capture dissociation: a historical perspective and modern ideas," *Eur. J. Mass Spectrom.* **8**, pp. 337–349.

Zubarev, R. A., Kruger, N. A., Fridriksson, E. K., et al. 1999, "Electron capture dissociation of gaseous multiply-charged proteins is favored at disulfide bonds and other sites of high hydrogen atom affinity," *J. Am. Chem. Soc.* **121**, pp. 2857–2862.

Index

absence of curve crossing 124
abundance ratio 247, 250
adiabatic expansion technique 38
air-breathing engine 277, 281
airglow 2–3
   green 157, 159
   red 157
ambipolar diffusion 55
analytical application 283
angle 232–233
angular distribution 178–179
antisymmetric stretch mode 203
antisymmetrized product 83
associative ionization 287
ASTRID 31
astrochemistry 315
   database 318
atmospheric ions 154
atomic collisions 287
attachment 4
auroral green line 2–3, 157
autoionization 223, 311
   width 83
avoided crossing 125

beam techniques 302
benchmark molecules 143
bistable solutions 318
blackbody infrared radiative dissociation 283
Born–Oppenheimer approximation 71
branching ratio 218
breakup dynamics 211
   concerted 233
      asynchronous 233
      synchronous 233, 239, 244
   linear breakup 214
   three-body 227, 231
Breit–Wigner form 78

Car–Parinello simulations 70
Cassini–Huygen spacecraft 319
cavity ring-down spectroscopy 205, 219, 221
chemical double slits 295

chemical reaction dynamics 287
cluster ions 298
collision-induced dissociation 283
collisional radiative recombination 144, 191, 224
Coulomb explosion technique 115, 203
comets 316
complex coordinate method 109
complex Kohn variational method 84, 292
compound state 302
conical intersection 235, 236
Co-ordinated Research Project 277
cosmic-ray ionization 226
Coulomb orbital 88
cross section
   measurement 287
   vibrationally resolved 155
crossed beams 8, 289
CRYRING 30
current transformer 39

Dalitz plot 214, 216
de-ionization 202
detection efficiency 26–27
detuning energy 16
detuning velocity 16
diabatic representation 294
dianion 300
dication 182
dielectronic recombination 14
diffuse interstellar clouds 104
dissociative attachment 300, 304
dissociative detachment 310
dissociative excitation 28, 121, 288, 292
   resonant 292, 297
dissociative ionization 288, 293
dissociative processes 288
dissociative recombination
   calculation of 74
   database 10
   direct 7, 71, 217, 246, 251
   first mention 2
   high-energy 196

dissociative recombination (*cont.*)
  indirect 7, 71, 177, 217
    new type 136–138
  products 66
  product states 152
  resonant 293
  super-dissociative-recombination 154, 272, 273, 274
  tunneling mode 126
disulfide bonds 285
drag force 40
duplasmatron ion source 197

early universe 140
electron affinity 296
electron attachment 287
electron capture dissociation 284, 309
electron cooler 34
electron-impact detachment 296, 299
electron–molecular ion scattering 300
electron–molecule collisions 287
electron scattering 70
electron velocity distribution 18–19
electronic excitation 302
ELISA 49
energetics 227
energy level diagram 205
enthalpy 282
ergodic 284
Euler angles 232

Fabry–Perot spectrometer 158
fast ion beam laser spectroscopy 309
Fermi's Golden Rule 88
Feshbach projection operator method 109
Feschbach resonance 302
flame study 257
flowing afterglow technique 8, 59
form factor 23
formation of O (^{1}S) 157
Fourier transform ion cyclotron resonance 283, 309
frame transformation 94
Franck–Condon region 296
Franck–Hertz experiment 287
free radicals 281, 283

grain surface reaction 319
Green's function 79
grid technique 45–46, 297

half-collision 313
heating mechanism 36
Hellman-Feynman
  couplings 92
  theorem 90
high-energy peak 122
hypersonic speed 277
hyperspherical coordinates 99, 218
hyperspherical radius 218
hyzone 186

imaging detector 46–48
impurity gas 63
inclined-beam technique 187, 290
infrared laser absorption spectroscopy 191
infrared multiphoton dissociation 283
infrared spectrum 185
interaction energy 12
International Conference on Photonic, Electronic and Atomic Collisions 288
International Conference on the Physics of Electronic and Atomic Collisions 288
interstellar molecular ions 244, 247
ion chemistry 65
ion–molecule reaction 316
ion-pair formation 294
  resonant 294
ion sources
  at storage rings 44
ion storage ring 30
ion trap 8
  radio-frequency 207
ionization potential 171
ionosphere 2
  ionospheric models 267
  Martian 263
isomeric forms 282
isosceles triangle 185
isotope effect 127, 200, 202
isotope fractionation 318, 319
isotopolog 104, 200

Jahn–Teller
  coupling 99
  distortion 217
Jost function 75
Jupiter 319

KEK ring 50
kinematic compression 15
kinetic energy 232

Landau–Zener approximation 103
Landau–Zener–Stückelberg model 295
Langmuir probe 59, 192
laser-induced fluorescence 259
Lippman–Schwinger equation 97
low vibrational state model 151

magnetic rigidity 33
Mars 170, 266, 319
mass spectrometer 6, 54
mass spectrometry 184, 283
MEIBE 14
merged beam technique 8, 187
  single-pass 14
metal-containing polyatomic molecular ions 283
microwave techniques 5
molecular astrophysics 315
molecular cloud chemistry 316
molecular spectroscopy 308

momentum conservation 228
multichannel quantum-defect theory 9, 93
multiconfiguration time-dependent Hartree 100

negative ion resonances 303
negative ion source development 302
negative molecular ions 296
nonadiabatic coupling 89
nonergodic 284

optical observation 6
over-the-barrier model 298

penetration terms 84
Penning ionization 187, 222
peptide amide bond 284
photocathode technique 38
photochemistry 308
photodetachment
  processes 311
  spectroscopy 310
photodissociation 287
  dynamics 311
  spectrum 309
photoelectron spectroscopy 311
photofragment kinetic energy spectroscopy 309
photoionization 311
  of ions 310
planetary atmospheres 104, 316
plasma-enhanced processing 302
plasmas in tokamaks 104
polycyclic aromatic hydrocarbon ions 280
positrons 288
potential curves
  adiabatic 124
  diabatic 80
potential scattering 75
predissociation 216
  direct 73
problem in molecular astrophysics 133
product branching ratios 257
proton affinity 316

quantum chemistry packages 70
quantum defect 176
quantum defect theory 91
quasiclassical trajectory calculations 235

R-matrix method 85
radiation chemistry 308
radiative recombination 3
radio-frequency quadrupole 31
ramjet 277
rate coefficient 18
  dependence on electron temperature 56–59, 278
  partial 157
  radiative 190
  temperature independent 248
  weak temperature dependence 267

recombination of atomic ions 287
reduced mass 12
region
  external 94
  short-range 93
relative speed 12
Renner–Teller coupling 99
Renner–Teller effect 246–247
resonance
  electronic parameter 83
  position 83
  width 88
resonant state 71
rotational cooling 36
rotational coupling 99, 111
rotational excitation 204, 287, 302
Rydberg
  atoms 40
  core excited state 72
  molecules 40

S-matrix 75
Saturn 319
Schottky detector 34
scramjet 277
semiclassical theories 196
semiconductor processing industry 283
shape resonance 302
shock tube technique 8, 68, 257
Siegert pseudostates 99
slow rotation limit 82
slow-heating methods 283
sounding rockets 158
spin-orbit interaction 161
SQUID 199
stationary afterglow technique 51
statistical model 211
storage rings 10
superelastic collision 36
superexcited state 311–313
supersonic jet source 205
surface barrier detector 24
swarm techniques 302
symmetric breathing mode 203

TARN II 30
temporary negative ion state 302
thermonuclear fusion reactor 277, 281
three-body recombination 143, 195
threshold phenomena 288
threshold photoelectron spectroscopy for attachment 306
Titan 170, 319
toroidal region 39, 199
trajectory calculations 251, 263
transition state spectroscopy 310
translational beam cooling 34
triatomic hydrogen 185
TSR 30
turbojet 277

valence bonds 259
Venus 266, 319
vibrational angular momentum 261
vibrational cooling 36
vibrational excitation 302
   resonant 304

vibrational relaxation 190
vibrational state distribution 168, 290
von Busch and Dunn distribution 107, 290

wave packet methods 99, 295
Wronskian 98